PORK

PORK
Meat Quality and Processed Meat Products

Edited by
Paulo E. S. Munekata, Mirian Pateiro, Daniel Franco, and José M. Lorenzo

CRC Press
Taylor & Francis Group
Boca Raton London New York

CRC Press is an imprint of the
Taylor & Francis Group, an **informa** business

First edition published 2022
by CRC Press
6000 Broken Sound Parkway NW, Suite 300, Boca Raton, FL 33487-2742

and by CRC Press
2 Park Square, Milton Park, Abingdon, Oxon, OX14 4RN

© 2022 Taylor & Francis Group, LLC

CRC Press is an imprint of Taylor & Francis Group, LLC

Library of Congress Cataloging-in-Publication Data
A catalog record has been requested for this book

ISBN: 978-0-367-34123-7 (hbk)
ISBN: 978-1-032-05979-2 (pbk)
ISBN: 978-0-429-32403-1 (ebk)

Typeset in Times
by MPS Limited, Dehradun

Table of Contents

Preface

The processing of pork is a common technological practice that modifies the taste, flavor, texture, and color of raw pork meat. Due to its accessible price and versatility, the manufacture of products from pork is an important strategy that offers different options to consumers by meat industries in this sector. Those industries also improve their profits and expand their businesses to new markets at the local, regional, and international levels. The current diversity of pork products can be attributed to local cultures and history, and the preservation of traditional processing practices. This combination of factors has led to the production of several pork products that can be classified in one of the following categories: fresh meat, fermented sausages, dry-cured products, blood sausages, and cooked sausages. Each category has specific requirements for raw material pre-processing (entire cut vs. minced pork meat), ingredients (e.g., sodium chloride, starter cultures, blood, seasoning, and spices), processing conditions (such as salting, drying, thermal processing, and fermenting stages), and consumption habits (cooking, slicing, spreading, and use as an ingredient in meals, for instance). Consequently, a wide variety of pork meat products are currently being produced worldwide that includes chorizo, salami, dry-cured ham, morcilla and butifarra sausages, salchichón, Bologna sausages, lacón, and cooked ham.

In Chapter 1, the importance of pork meat in the development of mankind and data about worldwide production and pork meat consumption and prospects are covered. The most important porcine breeds, both industrial and autochthonous, are briefly highlighted, as is the effect of the main productive factors, pre- and post-slaughter, on the pork meat quality. Other crucial aspects of modern pig farming production, such as animal welfare, sustainability, genetic improvement, and precision livestock are highlighted. Finally, the main pork meat products that are elaborated worldwide are reviewed. These products will be more extensively treated in subsequent chapters.

Dry-cured ham is a high-quality product made around the world. Chapter 2 aims to describe the processing stages and the reactions that take place during the elaboration of dry-cured ham. An overview to extend knowledge about the wide range of hams around the world is also shown. Lacón is a traditional meat product made in the North West of Spain following a manufacturing technology very similar to that used in the manufacture of dry-cured ham. Chapter 3 offers some information on the history of this meat product, the description of its manufacturing process, and a review of the existing information in the literature regarding its chemical characteristics.

Chapter 4 is organized into sections including fresh pork sausage ingredients, manufacturing equipment, sausage composition, and determination of product shelf life. Additionally, this chapter will enable the reader to identify ingredients and equipment that are used in fresh pork sausage production to produce a high-quality product with optimum shelf life. Chapter 5 is dedicated to chorizo, which is usually made with minced pork, air-dried for either a short (hours–days; fresh or semi-dried sausages) or long period (weeks–months, dry-cured sausages), and characterized by the use of red paprika.

Traditional pork products are prepared locally by the people following unique, preparation methods of which knowledge has been transmitted from one generation to the next. These products are prepared mostly on a domestic scale by judicious and efficient utilization of locally available raw materials and cater to the sensory and nutritive demands of a particular area. Chapters 6, 8, 9, 10, 11, 12, and 13 give an overview of the chemical composition, aromatic profiles, and sensory characteristics of sausages that include Asian, Serbian, Cuore di Spalla, Bulgarian, Brazilian, Indian, Italian salami, and Salchichón, respectively.

Cooked ham is produced as a whole muscle product made from individual muscles free of visible fat and connective tissues and packed in plastic materials. The quality of cooked ham is influenced by many outstanding factors, which include quality of meat, composition of brine and brine injection level, mechanical treatment, and cooking/cooling treatment. Chapter 14 primarily deals with manufacturing aspects of cook-in ham technology, completing the various stages of production. Bacon is generally described as a salt-cured pork product that is marketed and served in thin slices. Chapter 15 discusses bacon as a processed meat product and highlights the many recent research efforts dedicated to bacon. Bologna sausages are emulsified meat products with great economic importance for the meat industry in several countries. The main ingredients and additives, and the precautions required in the selection of the raw material and the manufacturing process to produce high-quality and safe Bologna sausages are compiled in Chapter 16 Moreover, the chemical composition of Bologna sausages produced around the world and the shelf life–limiting factors are discussed.

Morcilla and butifarra sausages are very well known and produced throughout Spain. They are cooked sausages in which blood, mainly from swine or cattle, is the characteristic ingredient. These sausages are mainly produced by artisans and small producers using traditional recipes for local distribution and consumption. The elaboration process and scientific data are discussed in Chapter 17 Finally, Morcela de Arroz is a Portuguese blood sausage that is unique for having, beyond pork and blood, rice in its composition. It is usually cooked, not smoked, and ready to eat. In Chapter 18 the manufacturing process, chemical composition and shelf life of Morcela de Arroz compared to other blood sausages are discussed.

About the Editors

José M. Lorenzo is Head of Research at the Meat Technology Centre of Galicia (CTC), Ourense, Spain. He received his M.S. in Food Science and Technology (University of Vigo). He obtained his Ph.D. in Food Science and Technology (University of Vigo) in 2006. He started his scientific career in the Department of Food Science and Technology at the University of Vigo, first on a research scholarship, then, since April 2006, as academic researcher. In 2006–2005, from October to March, he completed a stage of his research project at the Stazione Sperimentale per L´Industria delle Conserve Alimentaria, (Parma, Italy). He has developed numerous projects, many related to agro-industry and meat companies, and acquired extensive experience in the field of food technology. During this period, he completed analytic training in LC and GC, developing methods to quantify levels of lipid/protein oxidation, lipid fractionation by SPE, vitamins with HPLC/FD/DAD, and volatiles by GC/MS. These have focused on 1) characterization of products from different species raised under different rearing conditions, such as pigs, poultry and horses; 2) extension of food shelf life using natural extracts with antioxidant and antimicrobial capacities from agro-products; 3) understanding physico-chemical, biochemical, and microbial changes during the technological processes applied to meat products; and 4) development of new, healthier meat foods based on fat and salt reduction or improving lipid profile modification, replacement of fat, or incorporating functional compounds. Currently, he is involved in identifying proteomes and biomarkers associated with pastiness in dry-cured ham and their consequences for meat quality, using proteomic 2-DE techniques for protein separation and subsequent identification and quantification applying HPLC/MS/MS. As a result of this experience in the research, he has published more than 480 papers in well-recognized peer-reviewed international journals (SCI), with 68% of them in the first quartile (number of publications in Q1 is higher than 312), and 305 communications to congresses, mostly international. His h-index is 47 with 8,215 cites in Scopus. He was principal researcher in three European projects, three national projects, 38 in regional projects (Galicia, NW Spain), and he was lead investigator in 25-project development with meat enterprises and industry. In addition, he has participated in more than 65 projects as research collaborator. He has edited 15 international books and one national book, and contributed 64 and 8 chapters to them, respectively. In addition, he has one national patent as an inventor. Finally, he has supervised six doctoral theses.

Daniel Franco's scientific career started in 1997 with a European project (development of new processes for the extraction of oils and active products from nonconventional oilseeds and vegetables for the pharmaceutical and food industries [CEE (1997/PEO16)] at the Chemical Engineering Department of the University of Santiago of Compostela. During the Ph.D. stage, he acquired extensive knowledge of solid–liquid extraction from food matrices rich in oil and compounds with antioxidant activity. He also obtained experience in technical instrumental near-infrared spectroscopy and liquid chromatography (HPLC/DAD), used for determining ethanol–oil micelles and polyphenolic compounds. This research led to his Ph.D. degree in Chemistry in 2002. From 2003 to 2005, he worked as a technologist in the IIM-CSIC. During this time, he acquired skills focused on the management of R and D, and technology transfer for SMEs. From 2005 to 2009, he obtained a postdoctoral position in the Department of Animal Production at the Agricultural Research Center in Mabegondo. His research focused on improvements in animal feed using natural antioxidants and on understanding the biochemical mechanism behind meat ageing. During this period, he gained experience in the field of animal production, mainly focused on cattle, and methodological aspects of meat quality such as determination of texture parameters, and sensorial and nutritional analysis. Since 2009 he has worked at the Meat Technology Centre where he is responsible for leading the "differential and nutritional quality" research line. During this time, he has been PI of several projects of R and D and innovation related to meat science and food technology. He has developed numerous projects, many related to agro-industry and meat companies, and acquired extensive experience in the field of food technology. As a result of this experience, he has published more than 100 research papers in well-recognized peer-reviewed international journals (SCI) and more than 150 communications to congresses, mostly international. In addition, he participated in more than 50 research projects. He has two Spanish patents, has edited three books and written 16 book chapters, and has supervised two doctoral theses.

Paulo E. S. Munekata obtained his Ph.D. in Science of Food Engineering from the University of São Paulo, Brazil, in 2016 with one pre-doctoral research internship at the Meat Technology Center of Galicia (CTC), Ourense, Spain, during 12 months in 2014–2015. He is currently carrying out post-doctoral research activities at the same research center. He has collaborated in several research projects in Food Technology and Meat Science in the last 8 years, with major focus on 1) extraction, identification, and quantification of compounds of technological and health interest from natural sources; 2) evaluation of physico-chemical, biochemical, and microbiological characteristics of meat and meat products during storage; and 3) development of healthier meat products by reducing/replacing target ingredients and adding functional ingredients. He has published 26 scientific articles and five review articles in high-impact peer-reviewed journals and five book chapters in the food science and biochemistry areas. He recently edited a scientific book in the meat science area and has also written 23 scientific communications to national and international meetings, congresses, and symposiums. His h-index is 12 with 309 citations in the Scopus database.

Mirian Pateiro has degrees in Agricultural Technical Engineering (1998–2002) and in Food Science and Technology (2002–2005). She obtained her Ph.D. in Food Science and Technology from the University of Vigo (Spain) in 2010. She has been developing her research activity for 9 years as Superior Technician at the Meat Technology Center of Galicia (CTC), Ourense, Spain. She has collaborated in several research projects in Food Technology and Meat Science in the last 9 years with major focus on 1) reduction of additives in meat products; 2) development of new, healthier meat products; 3) use of agro-food industry by-products as potential sources of valuable and bioactive compounds; 4) use of emerging and green technologies (supercritical fluid extraction and pulse electric fields) for the recovery of biological active compounds from several plant matrices and by-products; 5) incorporation of bioactive compounds and development of functional foods; 6) use of active packaging to protect foods against oxidative degradation; and 7) evaluation of carcass, meat and meat product quality from different species. Her scientific production includes one hundred and seven scientific research papers in well-recognized peer-reviewed international journals (SCI), and 80 communications to congresses, mostly international. Her h-index is 29 with 2,149 cites in Scopus. She has participated in more than 25 projects as a research collaborator. She is editing three international books, and she has contributed 16 chapters to international and national books.

Contributors

A.J. Pham-Mondala
Mississippi State University
Mississippi, USA

Akhilesh K. Verma
Department of Livestock Products Technology, College of Veterinary and Animal Sciences, Sardar Vallabhbhai Patel University of Agriculture and Technology
Uttar Pradesh, India

Alexandra Esteves
Veterinary and Animal Research Centre and Department of Veterinary Sciences, School of Agrarian and Veterinary Sciences, Universidad de Trás-os-Montes e Alto Douro
Vila Real, Portugal

Alexandre José Cichoski
Universidade Federal de Santa Maria, Santa Maria
Rio Grande do Sul, Brazil

Ana María Díez Maté
Universidad de Burgos, Facultad de Ciencias
Burgos, Spain

Antonio Iglesias
Department of Anatomy, Animal Production and Veterinary Clinical Sciences, University of Santiago de Compostela, Santiago de Compostela
Galicia, Spain

Beatriz Melero Gil
University of Burgos, Department of Biotechnology and Food Science
Burgos, Spain

Benjamin M. Bohrer
Department of Food Science, University of Guelph
Guelph, Ontario, Canada

Bettit K. Salvá
Facultad de Industrias Alimentarias, Universidad Nacional Agraria La Molina
Lima, Peru

Bibiana Alves Dos Santos
Universidade Federal de Santa Maria, Santa Maria
Rio Grande do Sul, Brazil

Branislav Šojić
University of Novi Sad, Faculty of Technology Novi Sad
Novi Sad, Serbia

Carolina Pugliese
Department of Agriculture, Food, Environment and Forestry, Section of Animal Sciences, University of Firenze
Firenze, Italy

Chiara Aquilani
Department of Agriculture, Food, Environment and Forestry, Section of Animal Sciences, University of Firenze
Firenze, Italy

Chunbao Li
Key Laboratory of Meat Processing and Quality Control, MOE; Key Laboratory of Meat Processing, MOA; College of Food Science, Nanjing Agricultural University
Nanjing, China

Cristina Pérez-Santaescolástica
Centro Tecnológico de la Carne de
  Galicia
Ourense, Spain

Cui Zhiyong
Key Laboratory of Meat Processing and
  Quality Control, MOE; Key Labo-
  ratory of Meat Processing, MOA;
  College of Food Science, Nanjing
  Agricultural University
Nanjing, China

Daniel Franco
Centro Tecnológico de la Carne de
  Galicia
Ourense, Spain

Daphne D. Ramos-Delgado
Laboratorio de Salud Pública y Salud
  Ambiental, Facultad de Medicina
  Veterinaria, Universidad Nacional
  Mayor de San Marcos
Lima, Peru

Devendra Kumar
Division of Livestock Products Tech-
  nology, Indian Veterinary Research
  Institute
Uttar Pradesh, India

Djekic Ilija
Food Safety and Quality Management
  Department, University of Belgrade
Belgrade, Serbia

Đorđević Vesna
Institute of Meat Hygiene and Technology
Belgrade, Serbia

Emanuela Zanardi
Dipartimento di Scienze degli Alimenti
  e del Farmaco, Università degli Studi
  di Parma
Parma, Italy

Enrico Novelli
Dipartimento di Scienze degli Alimenti
  e del Farmaco, Università degli Studi
  di Parma
Parma, Italy

Enrique A. Cabeza-Herrera
Departamento de Microbiología, Facultad
  de Ciencias Básicas, Universidad de
  Pamplona
Pamplona, Colombia

Eva María Santos López
Universidad Autónoma del Estado de
  Hidalgo
Hidalgo, Mexico

Francesco Sirtori
Department of Agriculture, Food, Environ-
  ment and Forestry, Section of Animal
  Sciences, University of Firenze
Firenze, Italy

Heloísa Valarine Battagin
Universidade de Sao Paulo Faculdade de
  Zootecnia e Engenharia de Alimentos
São Paulo, Brazil

Irma Caro
Departamento de Pediatría, Inmunología,
  Obstetricia-Ginecología, Nutrición-
  Bromatología, Psiquiatría e Historia de
  la Ciencia, Universidad de Valladolid
Valladolid, Spain

Isabel Jaime Moreno
Universidad de Burgos, Facultad de
  Ciencias
Burgos, Spain

Javier Carballo
Área de Tecnología de los Alimentos,
  Facultad de Ciencias de Ourense,
  Universidad de Vigo
Ourense, Spain

Javier Castro Rosas
Instituto de Ciencias Básicas e Ingeniería,
    Universidad Autónoma del Estado de
    Hidalgo
Hidalgo, Mexico

Javier Mateo
Departamento de Higiene y Tecnología
    de los Alimentos, Universidad de
    León
León, Spain

Jesús Cantalapiedra
Farm Counselling Services. Consellería
    do Medio Rural, Xunta de Galicia
Lugo, Spain

Jordana Lima da Rosa
Universidade Federal de Santa Maria,
    Santa Maria
Rio Grande do Sul, Brazil

Jordi Rovira Carballido
Universidad de Burgos, Facultad de
    Ciencias
Burgos, Spain

Jorge A. Pereira
Universidade do Algarve, Instituto
    Superior de Engenharia, Campus da
    Penha, Faro, Portugal, Mediterranean
    Institute for Agriculture, Environment
    and Development, Universidade do
    Algarve
Faro, Portugal

José Ángel Pérez-Álvarez
IPOA Research Group. Agri-Food Tech-
    nology Department, Orihuela Poly-
    technical High School, Miguel
    Hernández University
Alicante, Spain

José M. Lorenzo
Centro Tecnológico de la Carne de
    Galicia, Ourense, Spain, Área de

Tecnología de los Alimentos, Facultad
    de Ciencias de Ourense, Universidad
    de Vigo
Ourense, Spain

Juana Fernández-López
IPOA Research Group. Agri-Food Tech-
    nology Department, Orihuela Pol-
    ytechnical High School, Miguel
    Hernández University
Alicante, Spain

Juliana Cristina Baldin
Universidade de Sao Paulo Faculdade de
    Zootecnia e Engenharia de Alimentos
São Paulo, Brazil

Laura Purriños
Centro Tecnológico de la Carne de
    Galicia
Ourense, Spain

Lídia Dionísio
Mediterranean Institute for Agriculture,
    Environment and Development, Uni-
    versidade do Algarve, Faro, Por-
    tugal, Universidade do Algarve,
    Faculdade de Ciências e Tec-
    nologia, Campus de Gambelas
Faro, Portugal

Luis Patarata
Centro de Ciência Animal e Veterinária,
    Universidade de Trás-os-Montes e
    Alto Douro
Vila, Portugal

Luz H. Villalobos-Delgado
Instituto de Agroindustrias, Universidad
    Tecnológica de la Mixteca, Huajuapan
    de León
Oaxaca, Mexico

M.W. Schilling
Mississippi State University
Mississippi, USA

Magdalena Isabel Cerón Guevara
Universidad Autónoma del Estado de
    Hidalgo
Hidalgo, Mexico

Manuel Viuda-Martos
IPOA Research Group. Agri-Food Tech-
    nology Department, Orihuela Poly-
    technical High School, Miguel
    Hernández University
Alicante, Spain

Marco Antonio Trindade
Universidade de Sao Paulo Faculdade de
    Zootecnia e Engenharia de Alimentos
São Paulo, Brazil

María López-Pedrouso
Department of Zoology, Genetics and
    Physical Anthropology, University of
    Santiago de Compostela, Santiago de
    Compostela
Galicia, Spain

Marija Jokanović
University of Novi Sad, Faculty of
    Technology Novi Sad
Novi Sad, Serbia

Mirian Pateiro
Centro Tecnológico de la Carne de
    Galicia
Ourense, Spain

Nitin Mehta
Department of Livestock Products Tech-
    nology, College of Veterinary Science,
    Guru Angad Dev Veterinary and
    Animal Sciences University
Punjab, India

Paulo Cezar Bastianello Campagnol
Universidade Federal de Santa Maria,
    Santa Maria
Rio Grande do Sul, Brazil

Paulo E. S. Munekata
Centro Tecnológico de la Carne de
    Galicia
Ourense, Spain

Pavan Kumar
Department of Livestock Products Tech-
    nology, College of Veterinary Science,
    Guru Angad Dev Veterinary and
    Animal Sciences University
Punjab, India

Pramila Umaraw
Department of Livestock Products Tech-
    nology, College of Veterinary and
    Animal Sciences, Sardar Vallabhbhai
    Patel University of Agriculture and
    Technology
Uttar Pradesh, India

Roberto Bermúdez
Centro Tecnológico de la Carne de
    Galicia
Ourense, Spain

Roberto González-Tenorio
Instituto de Ciencias Agropecuarias,
    Universidad Autónoma del Estado
    de Hidalgo
Hidalgo, Mexico

Rubén Agregán
Centro Tecnológico de la Carne de
    Galicia
Ourense, Spain

Ruben Dominguez
Centro Tecnológico de la Carne de
    Galicia
Ourense, Spain

Sergio Soto
Instituto de Ciencias Agropecuarias,
    Universidad Autónoma del Estado
    de Hidalgo
Hidalgo, Mexico

Simunović Stefan
Institute of Meat Hygiene and Technology
Belgrade, Serbia

Siyuan Chang
Institute of Food and Nutrition Development, MOA, Haidian
Beijing, China

T.R. Jarvis
Mississippi State University
Mississippi, USA

T.T. Dinh
Mississippi State University
Mississippi, USA

Teodora Popova
Agricultural Academy, Institute of Animal Science
Kostinbrod, Bulgaria

Teresa J.S. Matos
LEAF, Tapada da Ajuda, Lisbon, Portugal, Instituto Superior de Agronomia, University of Lisbon
Lisbon, Portugal

Tomašević Igor
Department of Animal Source Food Technology, University of Belgrade
Belgrade, Serbia

Tomović Vladimir
Faculty of Technology, University of Novi Sad
Novi Sad, Serbia

V. P. Singh
Department of Livestock Products Technology, College of Veterinary and Animal Sciences, Sardar Vallabhbhai Patel University of Agriculture and Technology
Uttar Pradesh, India

Wangang Zhang
College of Food Science and Technology, Nanjing Agricultural University
Nanjing, China

Y.L. Campbell
Mississippi State University
Mississippi, USA

Yana Jorge Polizer Rocha
Universidade de Sao Paulo Faculdade de Zootecnia e Engenharia de Alimentos
São Paulo, Brazil

1 Pigs

Breeds, Production, Meat Quality and Market: An Overview

Rubén Agregán,[1] José M. Lorenzo,[1] María López-Pedrouso,[2] Jesús Cantalapiedra,[3] Antonio Iglesias,[4] Wangang Zhang,[5] and Daniel Franco[1]

[1]Centro Tecnológico de la Carne de Galicia, Avd. Galicia 4, Parque Tecnológico de Galicia, San Cibrao das Viñas, 32900 Ourense, Spain

[2]Department of Zoology, Genetics and Physical Anthropology, University of Santiago de Compostela, Santiago de Compostela, Galicia, Spain

[3]Farm Counselling Services. Consellería do Medio Rural, Xunta de Galicia, Lugo, Spain

[4]Department of Anatomy, Animal Production and Veterinary Clinical Sciences, University of Santiago de Compostela, Santiago de Compostela, Galicia, Spain

[5]College of Food Science and Technology, Nanjing Agricultural University, Nanjing 210095, China

CONTENTS

1.1   INTRODUCTION

1.1.1   PIG ORIGINS AND PERCEPTIONS OF IT AS A FOOD SOURCE

The modern pig belongs to the order *Artiodactyla*, family *Suidae* and gender *Sus*, which includes the domestic pig and the European wild boar. The roots of the common pig come from a class of animals called *Hyotherium*, whose existence dates to ancient times. Historical data indicate that the pig was domesticated in China around 4900 BC and brought to Europe around 1500 BC. It was subsequently introduced to North America by Hernando de Soto in 1539 (Moeller & Crespo, 2009).

Pig has always played a vital role in food for humanity, providing meat for survival, but it is also an animal that has had, and it still has, a fundamental role in socio-cultural and religious traditions. Nowadays, pigs continue to be a fundamental source of food, due in part to their adaptability to multiple environmental conditions, allowing it to thrive in many habitats around the planet. An important factor that makes it so adaptable is that it is omnivorous, able to eat both forage-based and cereal grain-based feed. (Moeller & Crespo, 2009).

There is an extended culture of pork consumption in Western countries, whose origin seems to be in Greek and Roman civilizations, according to archaeology remains and documents in which the names of meat cuts are preserved. It is well known that Roman legions consumed manufactured products from pork, and they even carried out slaughters in some military bases (Swatland, 2010). Today, pork is consumed in many areas, playing an important economic role, especially in China, the UE and the USA (Ngapo et al., 2007). In the EU, pork is the most consumed, having a traditional place in the diet (Verbeke et al., 2011). However, pork does not have the same acceptance in all countries; for instance, in Muslim or Jewish ethnic populations, pork consumption is forbidden for religious and cultural reasons (Rosenblum, 2010). The appreciation of pork also varies even within countries, showing different degrees of consumption by different perceptions according to population groups.

A recent survey has shown different consumption levels and typology of products bought by pork consumers in the EU, taking into account factors such as social position, sex or location, observing that most of the pork consumers are families with high or medium-high incomes (Verbeke et al., 2010). Indeed, meat consumption data are an important indicator of the standard of living, according to sociologists and economists (Soare & Chiurciu, 2017). The survey states that the low or even absence of consumption is seen in women predominantly, noting that sex is an important factor that influences consumption. Moreover, this survey highlights that people living in rural areas of northern Europe are the biggest

Total production (years 1961-2007)   Total production (years 2007-2017)

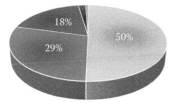
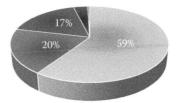

· Asia · Europe · Americas · Africa · Oceania   · Asia · Europe · Americas · Africa · Oceania

FIGURE 1.1   Distribution of swine production worldwide in two periods (Faostat, 2020).

consumers of pork, with a frequency of up to several times a day, without pre-dilection for any product. According to the survey, factors such as living in rural areas seem to increase pork consumption (Verbeke et al., 2010).

1.1.2   Production and Consumption of Pork Worldwide: Prospects

According to FAO, the swine production worldwide has increased steadily from 1961 to 2007 (Figure 1.1; Faostat, 2020); Asia is the continent with the highest production, with around the 50% of the world total. Among countries, China leads production in this continent and worldwide. Europe, with more than 25% of the world production, is the continent with the second-highest production, and Germany, Poland and Spain are countries with a remarkable rate of swine production. North and South America produce close to 20% of the total pig production worldwide, led by the USA and Brazil (Figure 1.1; Faostat, 2020).

Between 2007 and 2017 production increased in Asia, becoming the highest swine producer (almost 60% of the world total) with China assuming most of the production: more than 3,700 million pigs were produced during this time. This proves the strategic importance that the pig sector has for the culture of this area of the world (Szymańska, 2017), enhanced through a system of subsidies that stimulate production (Soare & Chiurciu, 2017). Europe and America have similar production percentages, highlighting the USA as the main producer. The swine production in the United States is concentrated in Minnesota, Iowa, Nebraska, Missouri, Illinois, Indiana and Ohio, due to factors such as large field areas available to cultivate feed, adequate climate and satisfactory economic conditions. Currently the USA is making advances technology, leading to the introduction of innovations in the sector. Following the USA, the largest pig-producing countries are Brazil, Mexico and Canada in the Americas, and Germany, Spain, Russia and France in Europe. EU member states have increased pork production in recent years. Brazil has consolidated itself as the third largest swine producer worldwide, and its position in the world ranking continues to strengthen. The rapid increase in swine production in this country is associated with higher possibilities of exporting pork meat to Russia, as well as to cover the greater demand in the domestic market (Szymańska, 2017). Between 2007 and 2017, a decrease of 13 million pigs was observed (Figure 1.2; Faostat, 2020). However, these data are very strongly influenced by Chinese rate production. The number of pigs in China significantly decreased during

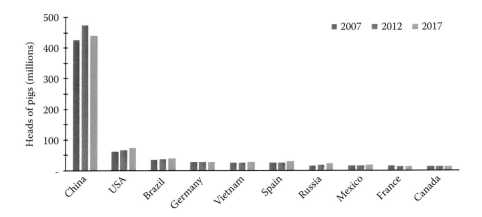

FIGURE 1.2 Evolution of pig production in the top ten producing countries between 2007 and 2017 (Faostat, 2020).

the last 4 years, especially, due to the application of an Environmental Protection Law, which restricts pig production in the southern part of the country. However, production is expected to increase in the following years (OECD/FAO, 2017).

Regarding pork consumption, it increased from 1961 to 2013, with the Asian continent at the top of this growth, from 2.8 million tons in 1961 to 67.4 million tons in 2013, mainly due to consumption in China. Indeed, in 2013, pork demand in that country accounted for 80% of the Asian continent and around 50% worldwide. To a lesser extent, countries such as Japan, Vietnam and the Philippines stand out as large consumers of pork. On the other hand, there iwas also an increase in pork consumption in Europe and America during this period, although more gradually. Countries such as the USA, the former USSR, and now Russia and Germany, were the largest consumers. In 2013, total consumption in the USA accounted for more than 50% of the pork consumed throughout the Americas. Finally, consumption in Africa and Oceania was very low, reaching only 2% worldwide (Figure 1.3; Faostat, 2020).

There is a correlation between production and pork consumption, and the biggest producers are also the biggest consumers. Other countries are also worthy of mention since, although they do not stand out as pork producers, they have a high per-capita consumption, demonstrating a great tradition of consuming pork products, as in Hungary and Austria. Indeed, Hungary had, until the 1990s, the highest pork consumption per capita worldwide, reaching historical highs of almost 80 kg/inhabitant/year in the 1980s. Austria, similar to Germany, is known for the consumption of pork-derived products. Since the 1990s its consumption has been decreasing gradually, although it is still at values higher than 50 kg/inhabitant/year, at the level of the countries with the highest consumption, such as China and Hong Kong (Faostat, 2020).

According to data published by the OECD, there have been few changes in 2019 versus 2013 in the total consumption of pork, and China and the USA remain the main consumers, with 53 and almost 10 million tons, respectively. The EU countries also maintain a pork consumption of around 20 million tons, led by Russia, Germany, Spain and France. (Figure 1.4)

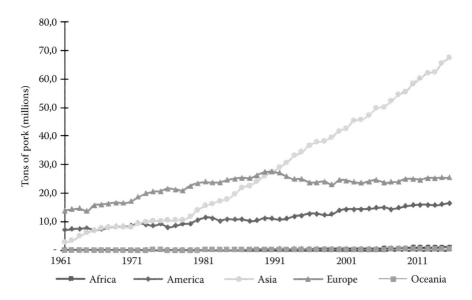

FIGURE 1.3 Evolution of pork consumption (tons) worldwide between 1961 and 2013 (Faostat, 2020).

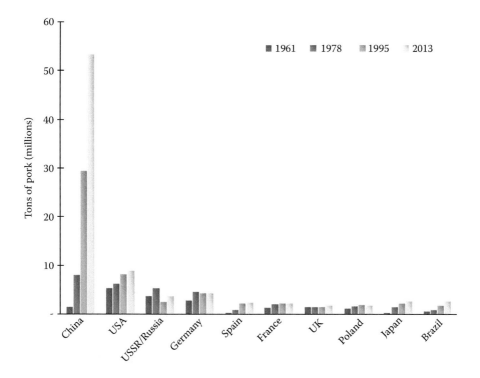

FIGURE 1.4 Evolution of pork consumption worldwide in the top ten consumer countries between 1961 and 2013 (Faostat, 2020).

Swine production is expected to increase in the next three decades due to increased demand. This need is driven by continued population growth, as well as the dietary transition toward more animal protein per capita (Lassaletta et al., 2019). According to the report on agricultural prospects for the 2019–2028 period published by OECD and FAO (OECD/FAO, 2019), Asia will continue to be the main pork producer, and China will provide half of the worldwide production in the coming years. However, production is also expected to decrease in the EU in the next decade due to environmental and public concerns about manure management, in addition to the fact that pork is not considered an essential food in the diet of many Western countries.

Regarding pork consumption for the next decade, OECD predicts that it will increase globally until 2028 by almost 11 million tons. Despite this increase, a slowdown in consumption is observed with respect to the years prior to 2014, as a result of the estimated deceleration in pig production for this period. China will remain as the country where more tons of pork will be consumed, assuming around 45% of the world total. On the other hand, according to OECD prospects for the coming years, the EU will eat less pork, indicating a reduction in consumption of 325,800 tons between 2021 and 2028, which coincides with the decrease in production estimated for this period (OECD/FAO, 2019). (Figure 1.5)

1.2   PORCINE BREEDS

1.2.1   INDUSTRIAL PORCINE BREEDS

Pigs underwent an evolution over time due to their interaction with people, who are always looking for sources of food and income. These needs lead to the raising of

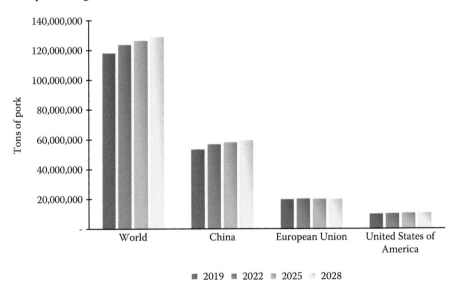

FIGURE 1.5   Estimate of trends in pork consumption in the world between 2019 and 2028 (Faostat, 2020).

animal species, such as the pig, whose meat and by-products are highly appreciated worldwide. The current different pig breeds have their origins in diverse environments, needs and different stocks to initiate genetic enhancement (Buchanan & Stalder, 2011). In some cases, the adjudication of breeds is somewhat problematic due to the different criteria used to establish the differentiation, but for pigs generally, the classification is by taxonomic category (Moeller & Crespo, 2009). Nowadays, the most recognized and commonly used pig breeds for industrial production are Duroc, Landrace, Pietrain, Yorkshire (also known as Large White), Berkshire, Chester White, Hampshire, Spotted and Poland China.

- **Duroc:** The origins of the Duroc breed date to before 1865. It comes from a breed of red pigs raised in the eastern United States (Flanders, 2011). Pigs of this breed have a characteristic reddish color, which can vary from coppery gold to deep burgundy. The animals are large, with drooping ears (McCosker, 2009). Due to their good musculature, meat quality, and hardiness they are often used as terminal crosses in crossbreeding programs (Klober, 2009).
- **Landrace:** The Landrace breed emerged in Denmark as the result of a crossing native pigs with those of the Large White breed. The Danish government maintained strict control of the improvement of this breed for years, until, with the arrival of World War II, it allowed its export (McCosker, 2009). These animals are characterized by their soft, white coat and pink skin. The body is long, with the back and ears dropping. The animals of this breed have lean meat and, from their long bodies, up to 16 or 17 pairs of ribs, more cuts than are obtained from other pig breeds (Cooper, 2011). It is a docile breed, frequently reared indoors, and its litters are extensive (Klober, 2009).
- **Pietrain:** The Pietrain breed of pig comes from the Belgian village of Pietrain, from which it takes its name. It emerged in the early 1950s, and after building a reputation, it was exported one decade later to Germany. Anatomically, the animals of this breed are medium-sized, with short legs and well-developed hams. They have erect ears and are white with black spots. This breed is in high demand for breeding programs owing to its extraordinary high lean-to-fat ratio (Gillespie & Flanders, 2009).
- **Yorkshire (Large White):** Pigs of this breed are more than two centuries old. They were developed in England in the late 1700s and have their origin in the crossing of a small Chinese pig with white pigs from the county of Yorkshire and others nearby (McCosker, 2009). The body of the Yorkshire is white, with very erect ears, and it has black eyes (Klober, 2009). The females of this breed are known to be extraordinary milk producers and have large litters (Cooper, 2011).
- **Berkshire:** The Berkshire breed emerged in England in the counties of Berkshire, from which it takes its name, and Wiltshire (Gillespie & Flanders, 2009). It is believed that it was discovered by Oliver Cromwell's troops when they arrived in Reading (a Berkshire county) during the English Civil War in the 17th century (Cooper, 2011). It is a predominantly black breed. Animals have white spots and splashes, mostly located on the face and lower half of the body, and have pretty short, erect ears (Klober, 2009). They are pigs with

a very fatty meat and abundant infiltration, which is not suitable for butchers, but is accepted in restaurants and sells well at farmers markets (McCosker, 2009). The Berkshire breed has good growth and a reasonable reproduction rate, but its litters are limited in number (Cooper, 2011).

- **Chester White:** This breed of pig comes from the state of Pennsylvania, United States, and its origins date back to the early 19th century, when the English Chester, Lincolnshire, and Yorkshire breeds were crossed. Pigs of this breed are white with a thick coat, and are medium-sized, with a slightly dished face and floppy ears (Cooper, 2011). Their robustness makes them appropriate for simple facilities, where animals need to be outdoors. Moreover, they stand out for their maternal abilities, milking and litter size (Klober, 2009).

- **Hampshire:** The Hampshire breed is native to the county of Hampshire, England. Pigs are distinguished by a characteristic white belt on a black body, which surrounds the forepart, including the forelegs. They have erect ears. These animals stand out for their foraging ability, muscling and carcass leanness. Hampshire is a popular breed and is often used in crossbreeding programs (Flanders, 2011).

- **Spotted:** This breed of pig was developed in Indiana, United States. It is the result of crossing Poland China pigs with spotted, whose crosses were in turn crossed with pigs of English origin called Gloucester Old Spots (Gillespie & Flanders, 2009). Pigs of this breed have a pattern of black and white spots and whites across the body, as the name suggests, and dropped ears. Spotted sows are extremely skilled mothers, and they give birth to exceptionally large litters (Cooper, 2011). These pigs offer long, trim carcasses (Klober, 2009).

- **Poland China:** The Poland China breed comes from the counties of Butler and Warren, in the American state of Ohio, and its development took place between 1800 and 1850. At first the name of the breed was Warren County Hog. The name Poland China was officially adopted by the National Swine Breeders Convention in 1872 (Gillespie & Flanders, 2009). These pigs are black with six white spots, located on the four feet, the face, and the tip of the tail. They have forward-drooping ears and large bodies. It is, in fact, one of the largest swine breeds (Flanders, 2011). The Poland China breed is mostly a muscular breed, producing lean meat (Cooper, 2011).

1.2.2 AUTOCHTHONOUS PORCINE BREEDS

Pig breeds such as those described above are widely used for productive purposes. Based on this economic model, the food industry focused on some pig breeds for meat production and discarded the least profitable, putting them in danger of extinction. However, there are breeds produced in lower numbers that are considered important for historical reasons or as a source of unique genetic material; therefore, a multitude of breeds must be conserved for scientific or cultural reasons (Buchanan & Stalder, 2011). Indigenous breeds suffered a significant decrease in numbers due to the introduction of improved breeds and their crosses. But in the development of native breeds and their typical products, it is advisable to assess the role and values of traditional livestock production systems.

The use of indigenous breeds as an alternative pig production system has important advantages, since these breeds are closely related to the environment and help to maintain biodiversity and sustainable agricultural production, especially in depressed areas. This fact led to efforts to preserve these breeds, although most of them are not currently well protected (Čandek-Potokar et al., 2019). Some of the most recognized autochthonous pig breeds in the world are Alentejano, Bísaro, Ibérico, Celta, Cinta Senese, Nero Siciliano, Mangalitsa, Moura, Criollo, Meishan, and Guinea.

Alentejano: This breed originated in the southwest of the Iberian Peninsula. It belongs to the Iberian type trunk and today adds economic, ecological and social value to the Alentejo region in Portugal. The animals are medium in size and black in color, with scarce black, blond or reddish hair. The head is long and thin, with a pronounced frontonasal angle, and with small, thin and forward-facing ears, triangular and slightly sloping. Limbs are short and thin, with small feet and uniform black pigmented hooves. Even though this breed has low prolifigacy and growth rate, its meat and fat are considered excellent for fresh meat production and also for the manufacture of dry-cured products. Alentejano pig is well adapted to the environment and uses natural resources appropriately (Charneca et al., 2019).

Bísaro: This pig breed belongs to the Celtic line and is native to Portugal, distributed throughout the north of the country. Pigs are big, with convex back, long legs, strong shoulders and a big head. They have very long ears that cover the eyes and a long and concave snout. They present several skin colours depending on the variety. In the Minho region, the white-spotted variety can be found, while in the Minho, Trás os Montes and Beiras regions the grey or black-spotted variety is most common. Bísaro has high prolifigacy and a docile character but presents some drawbacks, such as slow growth and unfavourable carcass conformation. Conversely, its meat has excellent sensory qualities and is suitable for the manufacture of meat products such as traditional smoke-cured products (for example, chouriços, salpicões or presunto) (Santos Silva et al., 2019).

Ibérico: This breed originated in the Iberian Peninsula and it is derived from ancestral domestic pig populations of this area. Nowadays, this type of pig can be found in the Andalusia (west) and Extremadura regions, and in the Salamanca province in Spain. Today, there are different varieties of this breed, but Retinto is the most common. In general, pigs are medium in size. The hair is weak and quite scarce in Entrepelado varieties and absent in hairless or Lampiño varieties. They have thin and resistant legs, with dark hooves, except in the Torbiscal variety, which can present depigmented or whitish-striped legs (Nieto et al., 2019). Ibérico pigs graze in a Spanish type of forest known as *dehesa*, feeding on pastures and acorns. Due to this way of life, they are rustic and well adapted to difficult environmental conditions (Galián, 2007). These pigs tend to accumulate fat, and lean tissue is lower than other, more conventional, breeds. In addition, their muscles are redder, darker and show more fat infiltrated in the longissimus muscle. The production of this pig breed is mainly focused on the production of cured products, such as hams, shoulders and loins. Other charcuterie products of minor relevance are also manufactured, such as chorizo and salchichón (Nieto et al., 2019).

FIGURE 1.6   Celta pig breed.

Celta: This autochthonous breed has its origins in the northwest of Spain and the existing specimens are currently distributed throughout the Galicia region (Franco et al., 2014). Three varieties are distinguished: Santiaguesa, Barcina, and Carballina (Figure 1.6).

These names are related to their places of origin, and their differences are based solely and exclusively on their geographical location and the presence or absence of pigmentation. Morphologically, they are narrow, long and arched pigs. They have a large and elongated head, with large, drooping ears that cover their small eyes. The color of the Santiaguesa variety is pink with total absence of pigmentation. The Barcina variety has little pigmentation, while in the Carballina variety it is extensive, black and shiny, and sometimes cover the entire body (Lorenzo et al., 2012). The muscles of these pigs provide meat with very good characteristics both for fresh consumption and for the manufacture of dry-cured meat products (Lorenzo et al., 2012).

Cinta Senese: This breed originated in the Italian province of Siena, around the localities of Monteriggioni, Sovicille and Poggibonsi. Today, most of the specimens are in the same province and the rest are distributed elsewhere in the Tuscany region. Cinta Senese is a medium-sized pig, with a light but solid skeleton. It has black skin and hair, except for a white band surrounding the trunk at shoulder level, including the forelimbs. The head is medium in size, with ears directed forward and down, and the limbs are thin but robust. Pigs are rustic and perfectly adapted to life outdoors and are fed with natural resources, such as acorns and chestnuts. This type of rearing gives a distinctive flavour and aroma to the meat of this breed, allowing the production of high-quality meat products such as dry-cured ham, Tuscan salami, pancetta, cardo and capocollo (Pugliese et al., 2019).

Nero Siciliano: The Nero Siciliano pig comes from the island of Sicily, in southern Italy, and is raised mainly in the province of Messina, particularly in the Monti

Nebrodi. The origin of this breed dates back to the Carthaginian occupation of this territory, and after a period of stagnation during the Arab domination, its breeding spread throughout the island. The animals of this breed are medium-sized, black in color, robust and with a strong skeleton. They have small ears directed obliquely towards the top, with the tips forward horizontally. Some of these animals can be found with all or part of the face in a white color. Their meat is described as very tasty, ruby red in color and suitable for making traditional products, such as the salami of S. Angelo, Troinese sausage, Nebrodi bacon and Nicosia ham (Bozzi, 2019).

Mangalitsa: The Mangalitsa breed originated from the ancient Serbian pig breed Šumadinka. This breed is present in the countries of Serbia, Austria, the Czech Republic, Germany, Romania, Hungary, Slovakia and Switzerland. It is medium-sized, with a characteristic thick, woolly coat like a sheep. It has a relatively small head, with large ears that hang in front over the eyes and face. The chest and body are wide and deep, extending to just below the elbow. The back and loin are straight and well developed with a long and cylindrical abdomen. The limbs are long, wide and muscular, with black hooves. Skin is black or brown in colour, with bright and curly bristles, which can be from grey-yellow to reddish; the tip is always black. Mangalitsa is a late pig breed, which accumulates high percentages of fat, with slow growth, low fertility and long lactation periods. On the contrary, pigs are very resistant and well-adapted to extensive conditions of rearing, feeding with pastures and acorns. The meat and fat from Mangalitsa, as well as derived meat products, such as Kulen and Sremska, the country's most popular fermented sausages, are very appreciated in Serbia, (Radović et al., 2019).

Moura: This pig breed is native to southern Brazil and originated from the Duroc, Canastra and Canastrao breeds. The main physical characteristic of the breed is the smooth grey or blue coat. The Moura breed is also known as Estrela, Estralense because pigs have a white star on the forehead. The head is medium-sized, with a sub-concave to rectilinear profile, a wide snout of medium length, a small double chin and, often, wattles on the short neck. The ears are droopy and medium in size. They have a wide back of a medium length and the legs are short and strong (Porter, 2016). Pigs from this breed and their crosses show slower growth and higher fat amount in the carcass than commercial pigs (Bertol et al., 2013). However, their feed-to-meat conversion index is relatively efficient and these pigs are considered a primary source of edible fat or lard (Carneiro et al., 2014).

Criollo: The Criollo pig descends from Ibérico pig, introduced in America on the second voyage of Cristobal Colón in 1493. For more than 500 years, this breed has evolved due to geographical adaptation and crosses with other, more commercial breeds. It is widely distributed throughout Latin America and different ethnic groups can be found, such as the Pampa Rocha, Mamellado, and Casco de Mula ecotypes in Uruguay or Entrepeluda-pelú and Lampiña-chinos in Cuba. Criollo pigs are well adapted to living outdoors and show great rusticity. For many years they have survived in poor conditions with little productive control. Therefore, they have been forced to adapt to an unfavourable environment (food shortage or attack by pathogenic micro-organisms). The sows of this breed are precocious mothers, but in contrast they are poorly reproductive. These animals show good meat potential, abundant fat content, and adequate organoleptic and nutritional characteristics (Linares et al., 2011).

Meishan: This pig breed is of Chinese origin and can be divided into three subgroups according to the size of the animal: big Meishan, middle Meishan, and small Meishan. In the late part of the 20th century, the number of pigs of this breed fell considerably due to the introduction of Western commercial breeds to China. As a consequence, the big Meishan pigs are extinct. The difference in body weight between a middle Meishan and a small Meishan pig is approximately 50 kg (Sun et al., 2018). This breed is famous for its wrinkled face and skin. The body is medium-sized, and it has large drooping ears. These pigs are known worldwide for their high reproduction rate and considered one of the most prolific pig breeds, with females giving birth to litters of 15–16 piglets (Nakajima et al., 2006).

Guinea: The provenance of the Guinea pig is uncertain. Its origins are believed to date back to West Africa at the time of the slave trade. These pigs are black, and have erect ears and a very rough coat (Klober, 2009). Other characteristics of this breed are its crooked tail and straight back. They are grazing animals, very small in size, weighing less than 30 kg, which makes them very manageable (Cooper, 2011).

1.3 CURRENT CHALLENGES OF PIG HOLDINGS

Since the middle of the 20th century, pig production has become established as an important driver of rural economies and as one of the most important sources of animal protein, with a trend towards increasingly larger and smaller farms (Park et al., 2017), that are more efficient (Hoste, 2017) and use more technological management (Benjamin & Yik, 2019). In addition, animals with more weight are sought in a context of globalized and highly competitive markets which demand safer and higher quality products (Edwards, 2005; Verbeke, 2009). This intensification of production has generated great ethical concerns in terms of animal treatment, environmental issues (carbon footprint, water consumption, increase of waste and pollution due to the excretion of nitrogen and phosphorus), and health concerns due to the use of growth promoters and preventive drugs that threaten human and animal health because of the increase in microbial resistance (Dumont et al., 2019; Yang, 2007).

1.3.1 ANIMAL WELFARE AND SUSTAINABILITY

Animal welfare (AW) and sustainability are, without question, two of the great challenges facing the pig industry today. On the one hand, improvements to the animals' living conditions have been mandated to increase their comfort, and, on the other hand, producers must adapt to ever-increasing legal and environmental requirements. AW is recognized as a central component of responsible livestock praxis (FAO, 2014). Good animal welfare practices were defined by the UK's Farm Animal Welfare Council (FAWC) in 1993, and include five freedoms: freedom from disease, pain and injury; the prevention and reduction of fear or anguish; the absence of hunger, thirst and malnutrition through the provision of adequate diets; environmental conditions which prevent physical or thermal discomfort; and the freedom of the animals in such a way that they can develop their normal behaviour patterns. This concern is shared by more and more consumers and operators of the food chain (European Commission, 2016; Kjaernes et al., 2007), and different surveys show a growing support for animal production that protects and improves the welfare of farm animals. This reality is not always reflected by consumers in the choice of purchase or in the price they would be willing to pay for it, although they consider that meat produced using animal-friendly practices is healthier, safer, tastier, more hygienic, authentic, ecologically positive and traditional (Alonso et al., 2020). Consumers are also increasingly demanding clear information and labelling of these products. A study conducted in Australia shows that the provision of additional information on the living conditions of animals significantly increases the purchase intention over conventional welfare products in women, young people and families with lower incomes (Cornish et al., 2020).

Compliance with AW standards worldwide and their legal equivalence are frequently being included in transnational trade agreements (European Commission, 2020; FAO, 2014), and consequently, they are affecting commercial transactions between countries. A clear example is the modification of the facilities of sows on factory farms due to the prohibition in Europe of the use of cages for their confinement, which has caused exporting countries to implement plans for their elimination in the coming years.

AW improvements in pig raising have focused, in recent years, on the design of the floors and farrowing pens, environmental enrichment, castration and the elimination of tail docking (Godyń et al., 2019; Mul et al., 2010). Worldwide, all animal production systems, including the pig industry, will face important challenges of economic, environmental and social sustainability (van Holsteijn et al., 2016). With an estimated world population of around 9 billion inhabitants (Alonso et al., 2020) by the middle of the 21st century, the sustainability challenges on farms will increase considerably due to the escalation in demand for animal products caused by the growth of the population, the increase in the standard of living in growing and emerging countries, and changes in socio-cultural and nutritional habits (Herrero et al., 2013; Makkar, 2013). The increase in animal protein intake is also associated with a considerable increase in the area intended to produce animal feed, changes in traditional agricultural systems (Altieri, 2004, 2009), deforestation problems, especially in South America (Barona et al., 2010), and an increase in the volatility of the prices of agricultural products (FAO, 2011).

Regarding environmental sustainability, the livestock sector contributes 14.5% to total greenhouse gas emissions (Gerber et al., 2013), of which 668 million tons of CO_2 correspond to pigs, with 95% of production in Asia, Europe and America. In this regard, it should be noted that food processing, storage and transportation consume a large amount of energy (FAO, 2011). Another major problem is the high environmental load due to the excretion of nitrogen and phosphorus generated in areas of high pig cattle intensity (Garcia-Launay et al., 2014), which has forced a revaluation of nutritional programs to reduce the excretion of nutrients (Pomar & Remus, 2019). Improving sustainability in future pig production raises two possible scenarios, the first is a high-yield system based on sustainable intensification with minimal environmental impact, maximizing animal protein and production efficiency in a limited land area, and a second scenario with a reduced entry and exit system based on the selection of animals, looking for those that are more resistant to climate change and perfectly adapted to transform low-quality feed, local raw materials, food co-products, or food waste into meat (Rauw et al., 2020). It is clear that animal production must contribute positively to the sustainability of food production systems (van Holsteijn et al., 2016), and in the case of the pig industry, a prioritization of objectives based on understanding between consumers, society and producers is required, both from a local and global perspective (Yang, 2007).

1.3.2 GENETIC IMPROVEMENT AND PRECISION LIVESTOCK

For several centuries, the domestic pig preserved the phenotypic characteristics of its ancestors with natural adaptations to the environment. The creation of hybrid swine breeds, which emerged in the 19th century in England, in the context of the profound transformation of agriculture and livestock during the Industrial Revolution, was a fundamental stage of the genetic improvement process since they were much more profitable for the meat industry, replacing traditional pig breeds.

In the mid-1960s, changes in the productive structure of the pig industry began due to the growth of the population in urban areas, which caused an increase in demand to satisfy nutritional needs. Then the selection and crossbreeding organizations appeared, which controlled large groups of animals, becoming the main protagonists of swine genetic improvement.

The pressure of genetic selection carried out on these breeds, based on interbreeding, caused an appreciable evolution in their morphology and ethology. Subsequently, its development has resulted in the creation of new populations, commercially known as lines. The selection programs are based on obtaining data from pig populations which involve productive results of growth, performance and reproductive efficiency, in addition to the characteristics and quality of the carcass. The correct handling of these data then facilitates, with adequate treatment, the guarantee of the expected genetic progress, which should be specified in the achievement of global efficiency in the farms.

The implementation of computer programs, increasingly complete, allowed the analysis of the data record and the control of the yields obtained in the selection centres. The combination of these records facilitated the performance of a common genetic evaluation by applying the Best Linear Unbiased Prediction (BLUP), to predict the genetic value of each individual candidate for selection as a future breeder. In the process, the records of all collaterals and ancestors are used to determine the previously chosen selection objectives. The information generated made it possible to increase the precision of the selection and at that time meant the possibility of approaching the selection systems from an economic perspective.

The important advances in molecular genetics in recent years, since the discovery of the polymerase chain reaction (PCR), facilitated working with sequences such as DNA microsatellites, from which it was possible to search for Quantative trait locis (one or more loci which influence traits of economic importance) throughout the genome. After the studies were carried out, several mutations were identified, some of which use the application of molecular marker-assisted selection (MAS).

The first linkage maps in pigs were published in the first half of the 1990s (Ellegren et al., 1994). The first QTL identified by linkage analysis refers to the amount of back and abdominal fat; the locus involved is located on porcine chromosome 4 (Andersson et al., 1994). At present, there are numerous QTLs, which have been discovered and made available to researchers, listed in the Pig QTL database (Hu et al., 2019). These QTL indicate the location of the genes responsible for the important quantitative traits in pig production and are mainly related to the quality of the meat and the carcass. Of the identified mutations, some of them are used in the application of MAS, which simultaneously uses phenotypic and molecular data. The development of Single Nucleotide Polymorphism (SNP) markers allowed the application of Genomic Wide Selection (GWS), aimed at identifying genetically superior individuals in order to maximize the efficiency of genetic improvement and achieve high selection precision. SNP mutations are by far the most common mutations in the porcine genome.

Recently, the advances produced in genomics, which studies the complete genome of an individual, its functions, and the interactions between genes with environmental factors, have made possible the knowledge of the productive characteristics of pigs from new perspectives based on the understanding of the relationships between genes and the environment.

Epigenetics studies all those transmissible modifications from one generation to the next, in the expression of genes, without altering their DNA sequences. The genetic sequence is constant throughout the life of the individual but there is also an epigenetic code that affects the expression levels of genes.

In pig cattle, it provides new knowledge related to the effects of the environment on the mechanisms of inheritance. It is known that stressors, especially those which occur after the weaning period, would affect their eating behaviour, generating a decrease in food consumption at this stage of production. Heat stress is also known to affect muscle development and meat quality in food animals, causing a negative impact on meat quality. The epigenetics information can be incorporated into the process aimed at obtaining the genetic value of the breeding swine which are destined to be the best candidates for reproduction (González-Recio et al., 2015; Varona, 2017).

Nutritional genomics, also known as nutrigenomics, studies the complex interactions of genes with nutrition, establishing molecular relationships between nutrients and the response of genes. Studies have been carried out on the influence of dietary fat on changes in gene expression and its impact on changes in intramuscular fat deposit and in the thickness of back fat that establish interactions between nutrition and gene expression in pigs (Yin & Li, 2009).

Currently, there are numerous chips which allow simultaneous and massive genotyping; their analysis provides abundant information about economically important characters in pig production. The application of nutritional genomics will be much more efficient as specific gene chips are being developed to analyse different tissues. This research will also help develop appropriate diets to maintain and improve meat quality.

The development of nutrigenomic studies has generated new technologies of massive analysis; currently emerging, they are the so-called "omic" technologies, such as transcriptomics, which studies the set of RNAs present in cells; proteomics, which investigates the structure and function of proteins; and metabolomics, which deals with the structure and function of metabolites. These technologies are being consolidated as important research tools in swine nutrition.

The application of process-engineering principles and techniques to livestock in order to monitor, model and manage animal production using the animals themselves as sensors (Banhazi et al., 2012; Wathes, 2009) is called Precision Livestock Farming (PLF). The data that is normally collected in these systems is obtained directly from the animals (behaviour, physiology, growth, food consumption, reproduction, etc.) and from the environment, with quantitative measurements of parameters such as air temperature, relative humidity, and the flow of water or polluting gases (Jensen, 2016; Piñeiro et al., 2019). The PLF in pig farming is a very useful tool in the traceability and productivity of animals, by reducing costs in food, energy and medicines, while improving sustainability and the economic technical management of farms, allowing control of more individualized animals in increasingly intensified herds (Vranken & Berckmans, 2017).

In pig farms, the most widely used sensors are water, weight, feed supply meters, cameras to measure activity, distribution, identification, respiratory diseases, etc., and the big data obtained on farms can be transformed into valuable information for decision-making by farmers and technicians (Piñeiro et al., 2019).

1.4   FACTORS AFFECTING PORK QUALITY

The organoleptic attributes of pork products, such as colour, flavour and texture, could determine its acceptance or rejection by consumers. Additionally, other pork characteristics that are less visible, such as pH, and fat or protein content, are crucial for product acceptance by retailers and butchers, and finally by the general population. Therefore, some sensory or technological defects might cause an increase in meat discards, causing an irreparable economical loss, since inputs of animal feeding and labour during animal husbandry would be unrecoverable. This problem concerns the entire meat industry, which is aware that many of the intrinsic and extrinsic factors during the rearing, the production line and in the sales points may affect the meat quality. In the following lines, we will discuss the main factors affecting fresh pork quality.

1.4.1   Ante-Mortem Factors

Pig farming comprises a series of activities that are carefully selected with the aim of improving not just the pork meat quality, but also efficiency and yield of animal production. At the same time, these activities try to be respectful of the environment and provide for animal welfare. In addition to these management practices during farming, other factors, such as breed, sex or slaughter age of pigs, may affect their meat quality. Some of the most important ante-mortem factors affecting pork quality are the following:

Breed: A high number of studies have reported the influence of breed in the final quality of pork. Certainly, significant improvements in the body composition of pigs have been made through genetic selection (Latorre et al., 2008), and crossbreeding is employed to improve lean growth without decreasing pork eating quality (Lo et al., 1992). Moreover, some pig breeds are considered superior to others regarding eating qualities (Bonneau & Lebret, 2010). Lee et al. (2012) studied a total of 243 commercial purebred pigs from the Berkshire, Duroc, Landrace and Yorkshire lines. They found significant differences among them in quality traits such as pH, colour, cooking loss, or muscle fibre, as well as in sensory aspects such as texture, flavour and smell, reporting that the Berkshire breed provided tenderer and more flavourful loins than did Landrace and Yorkshire pigs.

The flavours and aromas of pork are usually strongly associated with lipids, more specifically with phospholipids (Ngapo & Gariépy, 2008). In this sense, intramuscular fat plays a key role, affecting flavour and texture features. Indeed, several reports place the Duroc breed as one of the best, mainly because of its higher intramuscular fat content compared to white European breeds (Barton-Gade, 1987; Ngapo & Gariépy, 2008). An optimum intramuscular fat content is necessary for the processing of traditional meat products, especially the production of high-quality dry-cured ham (Oliver et al., 1993); hence, the Duroc breed is usually employed in Spain for Iberian pig production.

Age/weight at slaughter: Age and weight at the slaughter moment are related factors, and farmers play with them according to the meat's demanded characteristics. For instance, to produce pigs with high lean meat content, it is necessary to decrease the fattening period, consequently lowering the slaughter age. For the elaboration of dry-cured meat products, such as ham, lacón, or dry-cured loin, heavier slaughter weights are required (Franco et al., 2016). It has been demonstrated that the increase of age/slaughter weight can enhance some quality traits such as intramuscular fat content. Indeed, it was observed that increasing the slaughter age in purebred Duroc pigs increased this type of fat, also varying the fatty acid profile (significant increase and decrease of 4% in oleic and linoleic acids, respectively) (Bosch et al., 2012). However, the increase of age and weight is not always synonymous with improvement, because undesirable flavours, aromas, or textures in fresh meat have been attributed to pigs of an excessive age at the time of slaughtering (Marin et al., 2012). These authors observed a decrease in the protein content and an increase in dry matter and fat content, affecting pork quality, when the age and weight of animals increased. This lowers meat acceptance by consumers when the products are consumed fresh. Overall, it is generally accepted that high intramuscular fat content enhances the eating quality of pork. Indeed, Fernandez et al. (1999) fixed a maximum of 3.5% of intramuscular fat for fresh meat, because higher levels were associated with a significant risk of rejection by consumers.

Sex: Numerous studies show that sex is a variable that predominantly affects pork chemical composition, nutritional value, and quality (Franco & Lorenzo, 2013; Lorenzo et al., 2012). Indeed, Babicz et al. (2019) observed that males pigs reached 0.3% more in intramuscular sirloin fat than females. In this line, Zemva et al. (2015) indicated higher intramuscular fat content in barrows than in gilts of the indigenous Krškopolje breed. In another study, however, gender effect did not affect intramuscular fat content (Franco & Lorenzo, 2013). Additionally, it has been proved that sex has a strong influence on the fatty acid profile, modifying the percentage of saturated and monounsaturated fatty acids in the intramuscular fat (Zemva et al., 2015). Also, it has been reported that the gender effect has an influence on meat colour and sensorial attributes. Indeed, sex significantly affected luminosity and the yellowness parameter in *longismius dorsi* muscles, showing that meat from females was darker and less yellow than from barrows (Franco & Lorenzo, 2013). Regarding sensorial features, Rodrigues and Teixeira (2014), working with Preto Alentejano and commercial breeds, revealed a higher odour and flavour in males; meanwhile females, particularly commercial females, presented a higher hardness than males. In the study of Franco and Lorenzo (2013) with castrated barrows and entire females of the Celta breed, there were significant differences in odour, taste, fibrousness and juiciness attributes. Odour and taste traits were significantly higher in meat from females, while, steaks from barrows were the most fibrous and had the most hardness, and, consequently, were less juicy. Overall, sex is a variable that may affect pork quality; hence, it should be considered.

Feeding: The type of feed is usually employed to modify muscle composition (protein and fat content as well as fatty acid profile) of monogastric animals. Pigs are animals with a monocavitary stomach; hence, they do not substantially modify the fat consumed during the digestion of the feed and deposits this fat in the tissues with little or no modification. Because of this, the fatty acid profile of the lipid deposits shows a composition like the fat found in the feed. Compounds coming from feeds contribute to the final flavour of fresh meat, especially dry-cured meat products.

It was observed that modifications in the protein-energy ratio could be used to increase the infiltration of fat in muscle. A correct energy supply to pig diets, with an adequate protein amount or deficiency in lysine during the growing or finishing stages, has been successful in increasing intramuscular fat deposition, improving meat tenderness and juiciness (Lebret, 2008). The use of chestnuts is widely used in the feeding of autochthonous breeds from Spain and Italy, such as Celta, Iberian and Cinta Senese, in extensive production systems. These studies have been shown that the use of chestnuts in the pig diet increases the degree of fat unsaturation in the different carcasses' locations and meat (Domínguez et al., 2015; Pugliese et al., 2013). Overall, multiple studies in the literature have constated the influence of diet on pork quality.

Production system: The aim of the production system is to improve the pork quality, maintaining animal welfare and sustainability. Factors affecting the production system have an impact on pork quality. Typically, livestock production systems can be divided into extensive, semi-extensive or intensive systems. In the extensive system, pigs are reared outdoors, having access to pasture, acorns, chestnuts and other natural resources, and animals can be supplemented with commercial feeding in the last months prior to slaughter. Conversely, the intensive systems are based on indoor feedlots, where pigs are fed *ad libitum* using commercial feed. Semi-extensive systems are in an intermediate position. There can be variations within production systems between those using local breeds and extensive breeding conditions and those using conventional breeds under intensive conditions. Another pig production system is the organic production system, which is based on feeding pigs with forages and decreasing their digestible energy intake, together with particular housing and husbandry specifications. As we discussed before, the lack of certain amino acids created by dietary restrictions and other diet imbalances may increase the intramuscular fat deposition, affecting pork quality (Bonneau & Lebret, 2010).

1.4.2 Peri-Mortem Factors

The transport and slaughter of the animal are potentially stressful procedures. Different pig handling operations at this point in the production chain might act as triggers for uncomfortable situations, such as lack of food and water, fatigue, pain due to accidental shocks, or even fear of unfamiliar surroundings (Terlouw, 2005). These drawbacks might cause a decrease in meat quality and, as a result, an increase in the cost/benefit ratio.

Fasting: Fasting aims to prevent meat contamination with the intestinal content during the carcass cutting process, which is crucial for obtaining quality meat. The fasting time usually ranges between 16 and 24 hours from the farm to the slaughter of the pig (Faucitano & Goumon, 2018). It has been reported that longer times increase pig aggressivity, contributing to increase the incidence of dark, firm and dry (DFD) meats (Sionek & Przybylski, 2016).

Loading: Loading is considered one of the most crucial stages. A badly designed loading facility, and problems dividing group of pigs into the finishing pens by inadequate handling or group size may affect pig welfare, causing stress. Handing of a small group of pigs is preferred to diminish animal stress, even though more time is necessary (Faucitano & Goumon, 2018). Other aspects to consider are animal density and mixing with unfamiliar individuals. In the former factor, it has been reported that an excessive filling leads to aggression and fights, while a low animal density causes injuries due to falls. The mixing with animals from different pens or batches can raise fights and aggression, increasing the animals' stress and causing skin damage, other carcass defects, and even pale, soft, exudative (PSE) meat (Sionek & Przybylski, 2016). It was observed that those animals that are used to leaving their pens periodically during fattening do not feel so threatened at the time of loading, improving handling (Geverink, 1998).

Transport: During the journey from the farm to the slaughterhouse, pigs are exposed to stressful situations, such as unfamiliar and loud noises, new smells, vibrations, reduced space, sudden speed changes and frequent contact with handlers. It is necessary to avoid these uncomfortable situations to guarantee carcass and pork quality; hence, transport duration must be as short as possible. It has been indicated that rough and long journeys resulted in lower carcass pH at 45 minutes with a high incidence of PSE meat (Faucitano & Goumon, 2018). The risk of poor animal welfare during the trip can be minimized by a proper driving, allowing the pig to adopt an upright or lying position to cope with the high level of vibrations (Driessen et al., 2013).

Unloading: In general terms, unloading is less problematic than loading. However, there are a number of situations that must be avoided in order to maintain adequate animal welfare. The truck should be unloaded gradually and appropriately, treating the animals with kindness. Poor lighting and improper design and location of the unloading area can lead to refusal of pigs to get off the truck. It is also important to pay attention to the colours used in this area, as well as with the shadows, since they might scare the animals. Pigs must walk directly towards the lairage pen, always avoiding the presence of corners (Driessen et al., 2013).

Lairage: The optimal residence time for pigs in the lairage pens is believed to be between 2 and 3 hours. It is estimated that it is enough time for regeneration, after being subjected to a stressful trip, to a greater or lesser extent. The size of the pens, the number of animals per pen, the type of barriers, the type of floor and the noise level may be decisive factors in achieving optimal carcass quality (Sionek & Przybylski, 2016).

Stunning: Pigs can be stunned through two methods, electrical and carbon dioxide. Electric stunning is considered a stressful method, causing a rapid drop in pH shortly after the animal's death due to accelerated *post-mortem* glycolysis. A variation of this method, based on the same principle, but with electrodes placed from head to back (sternum), prevents pigs from recovering consciousness before bleeding. Depending on the way the electric shock is applied, different effects on the bodies of pigs have been reported. An excess of energy applied can fracture the spine, whereas prolonged application times could favour the appearance of PSE meat. Carbon dioxide is considered less stressful than electrical stunning, resulting in reduced incidence of bruises and the bone fractures associated with convulsive movements, improving carcass and meat quality (Sionek & Przybylski, 2016).

1.4.3 Post-Mortem Factors

Once the pig is slaughtered, its carcass undergoes diverse manipulation processes and, depending on the way they are carried out, may have adverse effects on meat quality. After stunning, bleeding should be carried out as soon as possible in order to prevent faulty meat (Adzitey & Huda, 2012). Temperature and cooling rate are important factors affecting the whole post-mortem process.

Fast cooling of the carcass is an effective way of slowing metabolic processes and decreasing pH, thus preventing the occurrence of PSE meats. In addition, rapid cooling reduces microbial growth in the carcass surface and improves lean colour and water-holding capacity. However, if the cooling process is carried out before rigor mortis status and below 10ºC, it produces cold shortening, a meat defect that causes abnormal hardness during cooking. There are some practices on the carcass that may affect positively meat quality. One of them is electrical stimulation, which causes muscle contraction, decrease in pH and accelerates the offset of rigor mortis, reducing the risk of cold shortening and favouring meat tenderization. Another possibility is hot processing, which consists of removing pieces of carcass while it is still hot. With this procedure, more uniform meat colour and an increase in water-holding capacity are achieved. However, joint deformation is promoted and tenderness could decrease. Finally, the ageing procedure is widely used, which consists of storing the carcass at refrigeration temperature after slaughter, thus improving tenderness and meat flavour (Sionek & Przybylski, 2016). Several studies have reported that storage from 6 to 10 days has a positive effect on pork, especially at

7 days, where a strong increase in flavour occurs. However, a period of ageing greater than 10 days might have the contrary effect, and unusual flavours have been described after 12 days of storage (Ngapo & Gariépy, 2008).

1.5   PORK MEAT AS A COMPONENT OF MULTIPLE PRODUCTS WORLDWIDE

Fresh pork has allowed the proliferation of an endless number of products that have been part of the human diet since ancient times. There is no absolute certainty about when these products began to be consumed, but it is estimated that it was around 3000 BCE, when salt was discovered. Salt gained importance over the years, becoming an essential ingredient for the preparation of meat products, providing them with interesting flavours and increasing their shelf life. Preserving methods based on thermal processes such as sun drying allowed people to keep meat products in good condition for longer periods of time. Later, it was discovered that fire increased the possibilities of conservation through smoking and cooking processes. Today, we know that this improvement in meat conservation is due to changes in protein structure and certain bacteriostatic action (Illescas et al., 2012).

Nowadays, pork is an important part of the diet of many cultures because of its great versatility and the number of foods that can be manufactured from it, from sausage to ham. Preservation techniques, such as drying, smoking and salting, allow the diversification of pork meat, creating delicious products whose consumption is an essential pillar in the subsistence of many peoples. However, this consumption goes beyond the purely nutritional and has already become an element of the culture of many regions, as much as art or literature can be. A great variability of pork products is available in supermarkets and traditional butcheries (Figure 1.7).

Products manufactured from fresh pork: This refers to those products made from fresh pork meat without thermal treatment; by-products can be added, as well as meat extenders and additives according to current legislation in the marketing area, and they can be stuffed, cured or smoked. Within fresh pork, two subgroups can be distinguished, the fresh meat products based on stuffed and non-stuffed minced meat (de Oña et al., 2012).

The stuffed products correspond to a wide group of products denominated commonly as fresh sausages. Their manufacture is relatively easy due to the absence of a cooking process, and consists of the addition of salt and a mixture of herbs and spices to minced meat. Subsequently, the sausages are kept under refrigeration to be fried, barbecued, boiled or steamed (Marianski & Marianski, 2011). Because these products are not cooked before their sale, suitable handling is essential to avoid microbiological spoilage (Feiner, 2006). Different fresh sausages can be found worldwide, containing a wide variety of cuts and flavours. For instance, Nuremberg bratwurst sausages are well-known uncooked sausages from Germany, made exclusively of pork and additives such as phosphates, salt and spices (pepper, cardamom, mace, ginger, marjoram, and lemon). Traditional Nuremberg bratwurst is marketed fresh with a very short shelf life, often 2 days from its manufacture, although this type of sausage is sometimes presented cooked at the point of sale.

FIGURE 1.7   Traditional butchery selling dry-cured hams and other pork products.

Depending on the food culture of each area, certain attributes of sausages, such as length or colour, can vary, as can their names. In countries such as Argentina, Uruguay and Spain, the sausage called *chorizo criollo*, made with a mixture of beef and pork and various seasonings, is widely consumed. In South Africa, a famous fresh sausage known as *boerewors* is marketed. This sausage is composed of muscle and fat from pork and other meat types (beef or mutton), spices and vinegar (Van

Schalkwyk et al., 2013). The English breakfast sausages and Cumberland sausages are traditional sausages much consumed in England. They are composed of lean pork shoulder, jowls, pork fat trimmings, rusk, and additives such as salt, phosphates and spices (Feiner, 2006).

Regarding the group of pork products based on non-stuffed minced meat, burger highlights over the rest. This product is widely consumed all over the world, mainly due to its pleasant sensory characteristics and its practicality to be cooked (Heck et al., 2019). Burgers and patties are marketed in multiple ways using different types of meat. Usually, beef is employed, although there are other variants of this product with chicken or a mixture of beef and pork (de Oña et al., 2012). There are several ways to prepare patties, some of them include meat and salt, sometimes spices are added and, in other cases, the recipe is more complete, including fat, water, phosphates and flavour enhancers (Feiner, 2006).

Fermented sausages: Fermented sausages are an excellent way of preserving meat. They are mainly composed of minced meat, salt and spices, and there are endlesss variations in recipes and manufacturing processes. The basic recipe uses a proportion of two-thirds pork meat and one-third fat, usually pork back fat, mixed with salt, sugar, sodium nitrite and starter culture (for example, lactic acid bacteria, S. xylosus or S. carnosus, among others). These microorganisms are responsible for the final organoleptic characteristics, such as colour, texture and taste of the sausage (Toldrá et al., 2007). The mixture is stuffed into a natural or artificial casing, allowed to ferment, and then it is dried in a smoke chamber and drying room (Holck et al., 2017).

There is a wide tradition in the consumption of dry-fermented sausages in northern Europe, especially in Germany, which produces around 40% of the fermented meat products of the European continent, including Bregenwurst, a semidry and spreadable pork sausage from Lower Saxony, and the Frankfurter Rindswurt, a smoked sausage made exclusively from pork (Vignolo et al., 2010). Fermented sausages from pork are produced in Poland, including *krakowscha sucha*, made of pork and beef, or *kabanosy*, made exclusively with pork (Toldrá et al., 2007).

Spain and Portugal highlight the traditional chorizo and salchichón made of pork and pork back fat, and seasoned with spices of which paprika and black pepper stand out as key ingredients. Other flavourings and ingredients, such as chilli, garlic, salt, nitrite, and ascorbate, are also added. Another relevant fermented pork product from Spain, specifically from Catalonia, is the fuet. This sausage is seasoned with spices, such as paprika and garlic, which provide a sweet and aromatic flavour. Fuet is fermented and dried, but unlike chorizo, an innocuous mould is applied to the surface (Feiner, 2006).

In Italy, the traditional Italian salami is very popular, and it can be found fermented (raw or cooked) or not. Raw, fermented salamis are usually made of pork lean, fatty trimmings and pork fat mixed with salt, spices (pepper and garlic), nitrites, and antioxidants (ascorbic acid, ascorbate, erythorbate, and occasionally tocopherol). During the manufacturing process, selected starter cultures of the genera *lactobacillus*, *staphylococcus*, *pediococcus*, and *micrococcus* are added, primarily to acidify and provide the characteristic colour and flavour. They are introduced into the sausage mass at the beginning of the cutting or mixing process. In addition to bacteria, yeasts and moulds also participate in the salami manufacturing. They are added through a spray or dipping the sausage in a solution of some of these groups of microorganisms, preferably using the mould genus *penicillium*. Then, salamis are left to ferment and dry, and often they are smoked in order to generate typical smoked colour and flavour (Feiner, 2006). Regarding cooked fermented salamis, they can be semi-cooked or fully cooked. The first type is usually produced by reducing the time of fermentation through a fast-fermentation process. In the case of complete cooking, the salami is produced as fast as possible. The destination of this type of salami is as pizza topping or for sandwiches (Feiner, 2006). Italy is the main market of salami, where, depending on the area of the country, different varieties are marketed, such as *salame Brianza, di Varzì, Piacentino*, or *di Calabria* (Aquilanti et al., 2012). However, sausage is also marketed in other countries that have their own recipes, such as in Hungary, where salami called Hungarian salami is produced, or in North America, where pepperoni is made of pork and/or beef and seasoned with red pepper and cayenne pepper, among other ingredients. Dry fermented sausages on the American continent were introduced by European settlers during the colonial era, who brought their own recipes and manufacturing processes (Toldrá et al., 2007).

Pork fermented sausages are also traditional products in other parts of the world such as in Eastern Asia (China, Thailand, Philippines, Sri Lanka, and Nepal). In China, a sausage made of pork and pork back fat, spices, soy sauce, and alcoholic beverages, known as Lap Cheong, is made (Vignolo et al., 2010). In Thailand, sausages (northern Thai sausage and northeastern sour Thai-style sausage) made from pork, Thai curry, cooked rice, and/or chili are also consumed (Toldrá et al., 2007).

Dry-cured pork products: Drying and curing processes have been used in the production of pork products for centuries. It is believed that dry-cured products originated in European regions around the Mediterranean Sea due to its climate, which favours the drying processes naturally, and numerous historical references prove it (Toldrá, 2008). Briefly, the drying and curing process consists of the addition of salt, nitrates/nitrites, and other curing agents such as sugar and ascorbic acid. Firstly, salt penetrates into the meat by solubilization with the water. Then the meat is left to stand to favour the salt diffusion and its redistribution through the entire cut, and the final stage is a drying/ripening, in which water loss and development of multiple biochemical reactions affect organoleptic attributes, such as colour, texture and flavour. Specifically, proteolysis and lipolysis are important reactions responsible for many of these changes. Proteolysis generates small peptides and free amino acids through protein degradation, while lipolysis affects triacyclglycerols and phospholipids, causing free fatty acid formation and the release of aroma compounds (Toldrá, 2004).

There is a wide range of dry-cured pork products on the market. One of the best known is bacon, which is made of pork bellies, salt, sugar, sodium nitrite and sodium erythorbate, and is processed by smoking and/or heat (Flores, 2009). The name *bacon* covers a wide variety of products consumed in different parts of the world, such as belly or streaky bacon in the USA and Wiltshire bacon in the UK (Knipe & Beld, 2014). Apart from bacon, one of the most famous dry-cured pork products in the world is produced in Spain and Italy. Both countries have an extensive tradition in the elaboration of jamón and prosciutto. There are different varieties of ham on the market. Italian prosciuttos such as San Daniele, di Parma or di Modena are well known, while, two of the most famous hams known worldwide are Spanish, Jamón Serrano and Ibérico. The latter is widely recognized worldwide for its exclusive organoleptic characteristics. High infiltration of intramuscular fat, along with a natural diet and exercise as well as careful drying-ripening process, are behind the appreciated flavour and aromas of this product.

Other typical dry-cured products from pork around the world are pancetta from Italy, made of lean, deboned pork belly without skin, using salt, sugar, nitrite, and spices (ground black pepper, chilli, garlic, juniper, and rosemary). Parma Coppa is another product from Italy, made of boneless pork neck without subcutaneous fat, salt, nitrite, ascorbate and sugar. In Germany, a type of dry-cured ham is produced called Black Forest ham. It is made of deboned pork legs with the topside and the hock removed, conserving fat and skin (Feiner, 2006).

Blood sausages: Blood is a very common ingredient in sausage manufacture and is not discarded by many cultures. The first record of these meat products dates to the 9th century BC when it is mentioned in Homer's *Odyssey* (Marianski & Marianski, 2010). Originally, these sausages were created with ingredients of low value, such as pork head, jowls, tongues, pork heart, or its stomach and blood. Generally, the content of these type of sausages are cooked fat pork, ground cooked meat and gelatine, mixed with beef or pork blood and stuffed into a casing. Other possible ingredients are rusk, barley, rice, or potatoes in England and Ireland; groats or bread crumbs in Poland; milk, rice, eggs or cheese in Spain; or wheat gluten called seitan, flour and cornflour in Argentina (Marianski & Marianski, 2011).

In Spain and Latin America, the famous sausage known as morcilla is consumed. This sausage is popular and its consumption is deeply rooted in the culture. It is usually made of rice, onion and spices (for example, black pepper and Spanish paprika). In Northern England, Scotland and Ireland, black pudding is consumed, made of beef suet, pork flare fat or bacon mixed with cooked oatmeal, barley or both, and pork blood (Marianski & Marianski, 2011). Many types of blood sausages can be found in China, according to the region. White blood sausage is famous in the northeast. It mainly consists of pig blood and pork plus other ingredients (for example, ginger and pepper). In northwest and southwest China, sheep's blood can be used as the raw material of blood sausage, with tofu being generally added as supplementary material (Lin & Yi, 2015).

Cooked pork products: The cooking process is a very widespread technique that favours the diversification of numerous products elaborated from pork. Heat generates a series of physicochemical processes that affect parameters such as texture, water-holding capacity, colour and flavour. One example is cooked ham. This pork product is the complete ham, or part of it, cooked at high temperature to coagulate the proteins, conferring specific organoleptic characteristics as well as long shelf life (Arboix, 2004). Another traditional product from pork and submitted to cooking is pork liver pâté. Liver pâtés are widely consumed in many countries, especially in Europe, with Denmark, France, Germany and Spain in the list of main consumers (Agregán et al., 2018). To elaborate pork liver pâté, the ingredients, mainly liver and fat, are finely chopped and heat treated (Estévez et al., 2007). In

France, typical pâtés such as *pâté de campagne* or *de Bretons* are manufactured (Feiner, 2006). The cooking process is also widely used to produce sausages, varying according to the cooking method. These methods can consist of smoking only, smoking and/or poaching in hot water, and cooking the ingredients in water and poaching all the material after the stuffing process. Some of the best-known cooked pork sausages, manufactured totally or partially of pork, are bacon sausage, Chinese sausage, ham sausage, *jagdwurst* from Germany, or *kabanosy* from Poland (Marianski & Marianski, 2010).

REFERENCES

Adzitey, F., & Huda, N. (2012). Effects of post-slaughter carcass handling on meat quality. *Pakistan Veterinary Journal, 32*(2), 161–164.

Agregán, R., Franco, D., Carballo, J., Tomasevic, I., Barba, F. J., Gómez, B., Muchenje, V., & Lorenzo, J. M. (2018). Shelf life study of healthy pork liver pâté with added seaweed extracts from *Ascophyllum nodosum, Fucus vesiculosus* and *Bifurcaria bifurcata. Food Research International, 112*, 400–411. https://doi.org/10.1016/j.foodres.2018.06.063

Alonso, M. E., González-Montaña, J. R., & Lomillos, J. M. (2020). Consumers' concerns and perceptions of farm animal welfare. *Animals, 10*(3), 1–13. https://doi.org/10.3390/ani1 0030385

Altieri, M. A. (2004). Linking ecologists and traditional farmers in the search for sustainable agriculture. *Frontiers in Ecology and the Environment, 2*(1), 35–42. https://doi.org/1 0.1890/1540-9295(2004)002[0035:LEATFI]2.0.CO;2

Altieri, M. A. (2009). Agroecology, small farms, and food sovereignty. *Monthly Review, 61*(3), 102–113. https://doi.org/10.14452/mr-061-03-2009-07_8

Andersson, L., Haley, C. S., Ellegren, H., Knott, S. A., Johansson, M., Andersson, K., Andersson-eklund, L., Edfors-lilja, I., Fredholm, M., Håkansson, J., Lundström, K., Smith, J., Natl, P., Andersson, L., Haley, C. S., Ellegren, H., Knott, S. A., Johansson, M., Andersson, K., … Lundstrom, K. (1994). Genetic mapping of quantitative trait loci for growth and fatness in pigs. *Science, 263*(5154), 1771–1774.

Aquilanti, L., Garofalo, C., Osimani, A., & Clementi, F. (2012). Italian salami: Survey of traditional Italian salami, their manufacturing techniques, and main chemical and microbiological traits. In Y. H. Hui & E. Ö. Evranuz (Eds.), *Handbook of fermented food and beverage technology* (pp. 565–592). CRC Press.

Arboix, J. A. (2004). Ham production. Cooked ham. In C. Devine & M. Dikeman (Eds.), *Encyclopedia of meat sciences* (pp. 562–567). Academic Press.

Babicz, M., Kasprzyk, A., & Kropiwiec-Domańska, K. (2019). Influence of the sex and type of tissue on the basic chemical composition and the content of minerals in the sirloin and offal of fattener pigs. *Canadian Journal of Animal Science, 99*(2), 343–348. https://doi.org/10.1139/cjas-2018-0085

Banhazi, T. M., Babinszky, L., Halas, V., & Tscharke, M. (2012). Precision livestock farming: Precision feeding technologies and sustainable livestock production. *International Journal of Agricultural and Biological Engineering, 5*(4), 54–61. https://doi.org/10.3965/j.ijabe.20120504.006

Barona, E., Ramankutty, N., Hyman, G., & Coomes, O. T. (2010). The role of pasture and soybean in deforestation of the Brazilian Amazon. *Environmental Research Letters, 5*(2), 024002. https://doi.org/10.1088/1748-9326/5/2/024002

Barton-Gade, P. A. (1987). Meat and fat quality in boars, castrates and gilts. *Livestock Production Science, 16*(2), 187–196. https://doi.org/10.1016/0301-6226(87)90019-4

Benjamin, M., & Yik, S. (2019). Precision livestock farming in swinewelfare: A review for swine practitioners. *Animals, 9*(4), 1–21. https://doi.org/10.3390/ani9040133

Bertol, T. M., de Campos, R. M. L. D., Ludke, J. V., Terra, N. N., de Figueiredo, E. A. P., Coldebella, A., dos Santos Filho, J. I., Kawski, V. L., & Lehr, N. M. (2013). Effects of genotype and dietary oil supplementation on performance, carcass traits, pork quality and fatty acid composition of backfat and intramuscular fat. *Meat Science, 93*(3), 507–516. https://doi.org/10.1016/j.meatsci.2012.11.012

Bonneau, M., & Lebret, B. (2010). Production systems and influence on eating quality of pork. *Meat Science, 84*(2), 293–300. https://doi.org/10.1016/j.meatsci.2009.03.013

Bosch, L., Tor, M., Reixach, J., & Estany, J. (2012). Age-related changes in intramuscular and subcutaneous fat content and fatty acid composition in growing pigs using longitudinal data. *Meat Science, 91*(3), 358–363. https://doi.org/10.1016/j.meatsci.2012.02.019

Bozzi, R. (2019). Nero Siciliano pig. In M. Candek-Potokar & R. Nieto (Eds.), *European local pig breeds: Diversity and performance. A study of project TREASURE.* IntechOpen. https://doi.org/10.5772/intechopen.84438

Buchanan, D. S., & Stalder, K. (2011). Breeds of pigs. In M. F. Rothschild & A. Ruvinsky (Eds.), *The genetics of the pig* (2nd ed., pp. 445–472). CABI.

Čandek-Potokar, M., Fontanesi, L., Lebret, B., Gil, J. M., Ovilo, C., Nieto, R., Fernandez, A., Pugliese, C., Oliver, M.-A., & Bozzi, R. (2019). Introductory chapter: Concept and ambition of project TREASURE. In M. Candek-Potokar & R. Nieto (Eds.), *European local pig breeds: Diversity and performance. A study of project TREASURE.* IntechOpen. https://doi.org/http://dx.doi.org/10.5772/intechopen.84246

Carneiro, H., Paiva, S. R., Ledur, M., Figueiredo, E. A. P., Grings, V. H., Silva, F. C. P., & McManus, C. (2014). Pedigree and population viability analyses of a conservation herd of Moura pig. *Animal Genetic Resources/Ressources Génétiques Animales/Recursos Genéticos Animales, 54*, 127–134. https://doi.org/10.1017/s2078633613000362

Charneca, R., Martins, J., Freitas, A., Neves, J., Nunes, J., Paixim, H., Bento, P., & Batorek-Lukač, N. (2019). Alentejano pig. In M. Candek-Potokar & R. Nieto (Eds.), *European local pig breeds: Diversity and performance. A study of project TREASURE.* IntechOpen. https://doi.org/http://dx.doi.org/10.5772/intechopen.83757

Cooper, C. (2011). *The complete guide to raising pigs: Everything you need to know explained simply.* Atlantic Publishing.

Cornish, A. R., Briley, D., Wilson, B. J., Raubenheimer, D., Schlosberg, D., & McGreevy, P. D. (2020). The price of good welfare: Does informing consumers about what on-package labels mean for animal welfare influence their purchase intentions? *Appetite, 148* (January), 104577. https://doi.org/10.1016/j.appet.2019.104577

de Oña, C., Serrano, D., & Orts, M. (2012). *Elaboración de preparados cárnicos frescos: carnicería y elaboración de productos cárnicos (MFO297_2).* IC Editorial.

Domínguez, R., Martínez, S., Gómez, M., Carballo, J., & Franco, I. (2015). Fatty acids, retinol and cholesterol composition in various fatty tissues of Celta pig breed: Effect of the use of chestnuts in the finishing diet. *Journal of Food Composition and Analysis, 37*, 104–111. https://doi.org/10.1016/j.jfca.2014.08.003

Driessen, B., Peeters, E., Thielen, J. Van, & Beirendonck, S. Van. (2013). Practical handling skills during road transport of fattening pigs from farm to slaughterhouse: A brief review. *Agricultural Sciences, 04*(12), 756–761. https://doi.org/10.4236/as.2013.412103

Dumont, B., Ryschawy, J., Duru, M., Benoit, M., Chatellier, V., Delaby, L., Donnars, C., Dupraz, P., Lemauviel-Lavenant, S., Méda, B., Vollet, D., & Sabatier, R. (2019). Review: Associations among goods, impacts and ecosystem services provided by livestock farming. *Animal, 13*(8), 1773–1784. https://doi.org/10.1017/S1751731118002586

Edwards, S. A. (2005). Product quality attributes associated with outdoor pig production. *Livestock Production Science, 94*(1–2), 5–14. https://doi.org/10.1016/j.livprodsci.2004.11.028

Ellegren, H., Chowdhary, B. P., Johansson, M., Marklund, L., Fredholm, M., Gustavsson, I., & Andersson, L. (1994). A primary linkage map of the porcine genome reveals a low rate of genetic recombination. *Genetics, 137*(4), 1089–1100.

Estévez, M., Ramírez, R., Ventanas, S., & Cava, R. (2007). Sage and rosemary essential oils versus BHT for the inhibition of lipid oxidative reactions in liver pâté. *LWT – Food Science and Technology, 40*(1), 58–65. https://doi.org/10.1016/j.lwt.2005.07.010

European Commission. (2016). *Special Eurobarometer Survey 442.* Attitudes of Europeans towards animal welfare.

European Commission. (2020). *Communication from the commission to the European Parliament, the Council, the European Economic and Social Committee and the Committee of the Regions "From farm to fork" strategy for a fair, healthy and environmentally friendly food system COM/2020/381.* Brussels (Belgium).

FAO. (2011). *Price volatility in food and agricultural markets: Policy responses.* Rome, Italy.

FAO. (2014). *Review of animal welfare legislation in the beef, pork, and poultry industries.* Rome, Italy.

Faostat. (2020). Food and agriculture data from Food and Agriculture Organization of the United Nations (FAO). http://www.fao.org/faostat/en/#data

Faucitano, L., & Goumon, S. (2018). Transport of pigs to slaughter and associated handling. In M. Špinka (Ed.), *Advances in pig welfare* (pp. 261–293). Woodhead Publishing Limited. https://doi.org/10.1016/B978-0-08-101012-9.00009-5

Feiner, G. (2006). *Meat products handbook: Practical science and technology.* Woodhead Publishing Limited.

Fernandez, X., Monin, G., Talmant, A., Mourot, J., & Lebret, B. (1999). Influence of intramuscular fat content on the quality of pig meat – 1. Composition of the lipid fraction and sensory characteristics of *m. longissimus lumborum. Meat Science, 53*, 59–65. https://www.sciencedirect.com/science/article/pii/S0309174099000376%0Ahttp://ac.els-cdn.com/S0309174099000388/1-s2.0-S0309174099000388-main.pdf?_tid=645612fc-e448-11e4-b3c1-00000aab0f01&acdnat=1429196210_d031f9676b5a8d9be86e1af942e888d9

Flanders, F. B. (2011). *Exploring animal science.* Delmar.

Flores, M. (2009). Flavor of meat products. In L. M. L. Nollet & F. Toldrá (Eds.), *Handbook of processed meats and poultry analysis* (pp. 385–398). CRC Press.

Franco, D., Carballo, J., Bermñudez, R., & Lorenzo, J. M. (2016). Effect of genotype and slaughter age on carcass traits and meat quality of the Celta pig breed in extensive system. *Annals of Animal Science, 16*(1), 259–273. https://doi.org/10.1515/aoas-2015-0056

Franco, D., & Lorenzo, J. M. (2013). Effect of gender (barrows vs. females) on carcass traits and meat quality of Celta pig reared outdoors. *Journal of the Science of Food and Agriculture, 93*(4), 727–734. https://doi.org/10.1002/jsfa.5966

Franco, D., Vazquez, J. A., & Lorenzo, J. M. (2014). Growth performance, carcass and meat quality of the Celta pig crossbred with Duroc and Landrance genotypes. *Meat Science, 96*(1), 195–202. https://doi.org/10.1016/j.meatsci.2013.06.024

Galián, M. (2007). *Características de la canal y calidad de la carne, composición mineral y lipídica del cerdo Chato Murciano y su cruce con Ibérico. Efecto del sistema de manejo.* University of Murcia.

Garcia-Launay, F., van der Werf, H. M. G., Nguyen, T. T. H., Le Tutour, L., & Dourmad, J. Y. (2014). Evaluation of the environmental implications of the incorporation of feed-use amino acids in pig production using Life Cycle Assessment. *Livestock Science, 161*(1), 158–175. https://doi.org/10.1016/j.livsci.2013.11.027

Gerber, P. J., Steinfeld, H., Henderson, B., Mottet, A., Opio, C., Dijkman, J., Falcucci, A., & Tempio, G. (2013). *Tackling climate change through livestock – A global assessment of emissions and mitigation opportunities.* Food and Agriculture Organization of the United Nations (FAO).

Geverink, N. A. (1998). *Preslaughter treatment of pigs: Consequences for welfare and meat quality*. ID Lelystad, Institute for Animal Science and Health (WIAS).

Gillespie, J. R., & Flanders, F. B. (2009). *Modern livestock and poultry production* (8th ed.). Delmar.

Godyń, D., Nowicki, J., & Herbut, P. (2019). Effects of environmental enrichment on pig welfare—A review. *Animals, 9*, 383. https://doi.org/10.3390/ani9060383

González-Recio, O., Toro, M. A., & Bach, A. (2015). Past, present, and future of epigenetics applied to livestock breeding. *Frontiers in Genetics, 6*(September), 1–5. https://doi.org/10.3389/fgene.2015.00305

Heck, R. T., Saldaña, E., Lorenzo, J. M., Correa, L. P., Fagundes, M. B., Cichoski, A. J., de Menezes, C. R., Wagner, R., & Campagnol, P. C. B. (2019). Hydrogelled emulsion from chia and linseed oils: A promising strategy to produce low-fat burgers with a healthier lipid profile. *Meat Science, 156*(April), 174–182. https://doi.org/10.1016/j.meatsci.2019.05.034

Herrero, M., Havlík, P., Valin, H., Notenbaert, A., Rufino, M. C., Thornton, P. K., Blümmel, M., Weiss, F., Grace, D., & Obersteiner, M. (2013). Biomass use, production, feed efficiencies, and greenhouse gas emissions from global livestock systems. Proceedings of the National Academy of Sciences of the United States of America, *110*(52), 20888–20893. https://doi.org/10.1073/pnas.1308149110

Holck, A., Axelsson, L., McLeod, A., Rode, T. M., & Heir, E. (2017). Health and safety considerations of fermented sausages. *Journal of Food Quality, 2017*. https://doi.org/10.1155/2017/9753894

Hoste, R. (2017). *International comparison of pig production costs 2015; results of InterPIG*. Wageningen Economic Research.

Hu, Z. L., Park, C. A., & Reecy, J. M. (2019). Building a livestock genetic and genomic information knowledgebase through integrative developments of Animal QTLdb and CorrDB. *Nucleic Acids Research, 47*(D1), D701–D710. https://doi.org/10.1093/nar/gky1084

Illescas, J. L., Ferrer, S., & Bacho, O. (2012). *Porcino: guía práctica*. Mercasa.

Jensen, D. B. (2016). *Automatic learning and pattern recognition using sensor data in livestock farming*. University of Copenhagen.

Kjaernes, U., Miele, M., & Roex, J. (Eds.). (2007). *Attitudes of consumers, retailers and producers to farm animal welfare. Welfare quality reports no. 2*. Cardiff University.

Klober, K. (2009). *Storey's guide to raising pigs* In R. Boyd-Owens, S. Guare, & D. Burns (Eds.); 3rd ed.). Storey Publishing.

Knipe, C. L., & Beld, J. (2014). Bacon production. In M. Dikeman & C. Devine (Eds.), *Encyclopedia of meat sciences* (pp. 53–63). Academic Press.

Lassaletta, L., Estellés, F., Beusen, A. H. W., Bouwman, L., Calvet, S., van Grinsven, H. J. M., Doelman, J. C., Stehfest, E., Uwizeye, A., & Westhoek, H. (2019). Future global pig production systems according to the Shared Socioeconomic Pathways. *Science of the Total Environment, 665*, 739–751. https://doi.org/10.1016/j.scitotenv.2019.02.079

Latorre, M. A., Pomar, C., Faucitano, L., Gariépy, C., & Méthot, S. (2008). The relationship within and between production performance and meat quality characteristics in pigs from three different genetic lines. *Livestock Science, 115*(2–3), 258–267. https://doi.org/10.1016/j.livsci.2007.08.013

Lebret, B. (2008). Effects of feeding and rearing systems on growth, carcass composition and meat quality in pigs. *Animal, 2*(10), 1548–1558. https://doi.org/10.1017/S1751731108002796

Lee, S. H., Choe, J. H., Choi, Y. M., Jung, K. C., Rhee, M. S., Hong, K. C., Lee, S. K., Ryu, Y. C., & Kim, B. C. (2012). The influence of pork quality traits and muscle fiber characteristics on the eating quality of pork from various breeds. *Meat Science, 90*(2), 284–291. https://doi.org/10.1016/j.meatsci.2011.07.012

Lin, J. S., & Yi, S. (2015). The study of the Manchu diet culture in Jilin province. *Journal of Beihua University (Social Sciences)*, *16*, 72–75.

Linares, V., Linares, L., & Mendoza, G. (2011). Ethnic-Zootechnic characterization and meat potential of *Sus scrofa* "creole Pig" in Latin America. *Scientia Agropecuaria*, *2*, 97–110. https://doi.org/10.17268/sci.agropecu.2011.02.05

Lo, L. L., McLaren, D. G., McKeith, F. K., Fernando, R. L., & Novakofski, J. (1992). Genetic analyses of growth, real-time ultrasound, carcass, and pork quality traits in Duroc and Landrace pigs: I. Breed effects. *Journal of Animal Science*, *70*(8), 2373–2386. https://doi.org/10.2527/1992.7082373x

Lorenzo, J. M., Fernández, M., Franco, D., García-Calvo, L., Purriños, L., Gómez, M., Domínguez, R., Bermúdez, R., Bragado, C., García, Á., González-Barros, J. A. C., Pérez, C., Rodríguez, I., Carballo, F. J., Martínez, S., Franco, I., Martínez, C. J. R., Feijóo, J., Domínguez, B., & Iglesias, A. (2012). *Manual del Cerdo Celta*. CETECA (Fundación Centro Tecnolóxico da Carne).

Lorenzo, J. M., Montes, R., Purriños, L., Cobas, N., & Franco, D. (2012). Fatty acid composition of Celta pig breed as influenced by sex and location of fat in the carcass. *Journal of the Science of Food and Agriculture*, *92*(6), 1311–1317. https://doi.org/10.1002/jsfa.4702

Makkar, H. P. S. (2013). Towards sustainable animal diets. In H. P. Makkar & D. Beever (Eds.), Proceedings of the FAO symposium optimization of feed use efficiency in ruminant production systems (pp. 67–74). FAO.

Marianski, S., & Marianski, A. (2010). *Home production of quality meats and sausages*. Bookmagic LLC.

Marianski, S., & Marianski, A. (2011). *Making healthy sausages*. Bookmagic, LLC.

Marin, D., Păcală, N., Petroman, I., Petroman, C., Untaru, R., Heber, L., & Ciolac, R. (2012). Influence of age and weight at slaughter over meat quality in conditions of optimum ambient temperature. *Lucrări Ştiinţifice*, *14*(2), 453–458.

McCosker, L. (2009). *Free range pig farming: Starting out*. Lee McCosker.

Moeller, S., & Crespo, F. L. (2009). Overview of world swine and pork production. In R. Lal (Ed.), *Agricultural sciences* (Vol. 1, pp. 195–208). EOLSS Publishers Co.

Mul, M. F., Vermeij, I., Hindle, V. A., & Spoolder, H. A. M. (2010). *EU-welfare legislation on pigs* (Report/Wageningen UR Livestock Research: 273). Wageningen UR Livestock Research. https://edepot.wur.nl/136142

Nakajima, I., Oe, M., Muroya, S., Kanematsu, N., & Chikuni, K. (2006). Establishment of subcutaneous preadipocyte clonal line from Chinese Meishan pig. In D. Troy, R. Pearce, B. Byrne, & J. Kerry (Eds.), *52nd International Congress of Meat Science and Technology (ICoMST)* (pp. 231–232). Wageningen Academic Publishers.

Ngapo, T. M., Martin, J. F., & Dransfield, E. (2007). International preferences for pork appearance: I. Consumer choices. *Food Quality and Preference*, *18*(1), 26–36. https://doi.org/10.1016/j.foodqual.2005.07.001

Ngapo, T. M., & Gariépy, C. (2008). Factors affecting the eating quality of pork. *Critical Reviews in Food Science and Nutrition*, *48*(7), 599–633. https://doi.org/10.1080/10408390701558126

Nieto, R., García-Casco, J., Lara, L., Palma-Granados, P., Izquierdo, M., Hernandez, F., Dieguez, E., Duarte, J. L., & Batorek-Lukač, N. (2019). Ibérico (Iberian) pig. In M. Candek-Potokar & R. Nieto (Eds.), *European local pig breeds: Diversity and performance. A study of project TREASURE*. IntechOpen. https://doi.org/http://dx.doi.org/10.5772/intechopen.83765

OECD/FAO. (2017). *OECD-FAO Agricultural Outlook 2017-2026*. https://doi.org/https://doi.org/https://doi.org/10.1787/agr_outlook-2017-en

OECD/FAO. (2019). *OECD-FAO Agricultural Outlook 2019-2028*. http://www.embase.com/search/results?subaction=viewrecord&from=export&id=L20044855

Oliver, M. A., Gispert, M., & Diestre, A. (1993). The effects of breed and halothane sensitivity on pig meat quality. *Meat Science*, *35*(1), 105–118. https://doi.org/10.1016/03 09-1740(93)90073-Q

Park, H. S., Min, B., & Oh, S. H. (2017). Research trends in outdoor pig production: A review. *Asian-Australasian Journal of Animal Sciences*, *30*(9), 1207–1214. https:// doi.org/10.5713/ajas.17.0330

Piñeiro, C., Morales, J., Rodríguez, M., Aparicio, M., Manzanilla, E. G., & Koketsu, Y. (2019). Big (pig) data and the internet of the swine things: A new paradigm in the industry. *Animal Frontiers*, *9*(2), 6–15. https://doi.org/10.1093/af/vfz002

Pomar, C., & Remus, A. (2019). Precision pig feeding: A breakthrough toward sustainability. *Animal Frontiers*, *9*(2), 52–59. https://doi.org/10.1093/af/vfz006

Porter, V. (2016). Pigs. In V. Porter, L. Alderson, S. J. G. Hall, & D. P. Sponenberg (Eds.), *Mason's world encyclopedia of livestock breeds and breeding*. CABI.

Pugliese, C., Sirtori, F., Acciaioli, A., Bozzi, R., Campodoni, G., & Franci, O. (2013). Quality of fresh and seasoned fat of Cinta Senese pigs as affected by fattening with chestnut. *Meat Science*, *93*(1), 92–97. https://doi.org/10.1016/j.meatsci.2012.08.006

Pugliese, C., Bozzi, R., Gallo, M., Geraci, C., Fontanesi, L., & Batorek-Lukač, N. (2019). Cinta Senese pig. In M. Candek-Potokar & R. Nieto (Eds.), *European local pig breeds: Diversity and performance. A study of project TREASURE*. IntechOpen. https://doi.org/ http://dx.doi.org/10.5772/intechopen.83762

Radović, Č., Savić, R., Petrović, M., Gogić, M., Lukić, M., Radojković, D., & Batorek-Lukač, N. (2019). Mangalitsa (Swallow-Belly Mangalitsa) pig. In M. Candek-Potokar & R. Nieto (Eds.), *European local pig breeds: Diversity and performance. A study of project TREASURE*. IntechOpen. https://doi.org/http://dx.doi.org/10.5772/intechopen.83773

Rauw, W. M., Rydhmer, L., Kyriazakis, I., Øverland, M., Gilbert, H., Dekkers, J. C. M., Hermesch, S., Bouquet, A., Gómez Izquierdo, E., Louveau, I., & Gomez-Raya, L. (2020). Prospects for sustainability of pig production in relation to climate change and novel feed resources. *Journal of the Science of Food and Agriculture*, *100*(9), 3575–3586. https://doi.org/10.1002/jsfa.10338

Rodrigues, S., & Teixeira, A. (2014). Effect of breed and sex on pork meat sensory evaluation. *Food and Nutrition Sciences*, *05*(07), 599–605. https://doi.org/10.4236/fns.2 014.57070

Rosenblum, J. D. (2010). "Why do you refuse to eat pork?": Jews, food, and identity in Roman Palestine. *Jewish Quarterly Review*, *100*(1), 95–110. https://doi.org/10.1353/ jqr.0.0076

Santos Silva, J., Araújo, J. P., Cerqueira, J. O., Pires, P., Alves, C., & Batorek-Lukač, N. (2019). Bísaro pig. In M. Candek-Potokar & R. Nieto (Eds.), *European local pig breeds: Diversity and performance. A study of project TREASURE*. IntechOpen. https://doi.org/http://dx.doi.org/10.5772/intechopen.83759

Sionek, B., & Przybylski, W. (2016). The impact of ante- and post-mortem factors on the incidence of pork defective meat: A review. *Annals of Animal Science*, *16*(2), 333–345. https://doi.org/10.1515/aoas-2015-0086

Soare, E., & Chiurciu, I.-A. (2017). Study on the pork market worldwide. *Scientific Papers: Management, Economic Engineering in Agriculture & Rural Development*, *17*(4), 321–326. https://ezproxy.lib.uconn.edu/login?url=https://search.ebscohost.com/login.aspx? direct=true&db=aph&AN=127579911&site=ehost-live

Sun, H., Wang, Z., Zhang, Z., Xiao, Q., Mawed, S., Xu, Z., Zhang, X., Yang, H., Zhu, M., Xue, M., Liu, X., Zhang, W., Zhen, Y., Wang, Q., & Pan, Y. (2018). Genomic signatures reveal selection of characteristics within and between Meishan pig populations. *Animal Genetics*, *49*(2), 119–126. https://doi.org/10.1111/age.12642

Swatland, H. J. (2010). Meat products and consumption culture in the West. *Meat Science*, *86*(1), 80–85. https://doi.org/10.1016/j.meatsci.2010.04.024

Szymańska, E. J. (2017). The development of the pork market in the world in terms of globalization. *Journal of Agribusiness and Rural Development*, *16*(4), 843–850. https://doi.org/10.17306/j.jard.2017.00362

Terlouw, C. (2005). Stress reactions at slaughter and meat quality in pigs: Genetic background and prior experience: A brief review of recent findings. *Livestock Production Science*, *94*(1–2), 125–135. https://doi.org/10.1016/j.livprodsci.2004.11.032

Toldrá, F. (2004). Curing. Dry. In C. Devine & M. Dikeman (Eds.), *Encyclopedia of meat sciences* (pp. 360–365). Academic Press.

Toldrá, F. (2008). *Dry-cured meat products*. Wiley-Blackwell.

Toldrá, F., Nip, W.-K., & Hui, Y. H. (2007). Dry-fermented sausages: An overview. In F. Todrá (Ed.), *Handbook of fermented meat and poultry* (pp. 321–358). Blackwell Publishing.

van Holsteijn, F. H., de Vries, M., & Makkar, H. P. S. (2016). *Identification of indicators for evaluating sustainability of animal diets* (No. 15).

Van Schalkwyk, C. P. B., Hugo, A., Hugo, C. J., & Bothma, C. (2013). Evaluation of a natural preservative in a boerewors model system. *Journal of Food Processing and Preservation*, *37*(5), 824–834. https://doi.org/10.1111/j.1745-4549.2012.00706.x

Varona, L. (2017). Mejora genética porcina. Más allá de la genética. *Suis*, *137*, 12–17.

Verbeke, W. (2009). Stakeholder, citizen and consumer interests in farm animal welfare. *Animal Welfare*, *18*(4), 325–333.

Verbeke, W., Pérez-Cueto, F. J. A., Barcellos, M. D. de, Krystallis, A., & Grunert, K. G. (2010). European citizen and consumer attitudes and preferences regarding beef and pork. *Meat Science*, *84*(2), 284–292. https://doi.org/10.1016/j.meatsci.2009.05.001

Verbeke, W., Pérez-Cueto, F. J. A., & Grunert, K. G. (2011). To eat or not to eat pork, how frequently and how varied? Insights from the quantitative Q-PorkChains consumer survey in four European countries. *Meat Science*, *88*(4), 619–626. https://doi.org/10.1016/j.meatsci.2011.02.016

Vignolo, G., Fontana, C., & Fadda, S. (2010). Semidry and dry fermented sausages. In F. Toldrá (Ed.), *Handbook of meat processing* (pp. 379–378). Blackwell Publishing.

Vranken, E., & Berckmans, D. (2017). Precision livestock farming for pigs. *Animal Frontiers*, *7*(1), 32–37. https://doi.org/10.2527/af.2017.0106

Wathes, C. M. (2009). Precision livestock farming for animal health, welfare and production. In A. Aland & F. Madec (Eds.), *Sustainable animal production: The challenges and potential developments for professional farming* (pp. 411–420). Wageningen Academic Publishers. https://doi.org/10.3920/978-90-8686-685-4

Yang, T. S. (2007). Environmental sustainability and social desirability issues in pig feeding. *Asian-Australasian Journal of Animal Sciences*, *20*(4), 605–614. https://doi.org/10.5713/ajas.2007.605

Yin, J., & Li, D. (2009). Nutrigenomics approach-a strategy for identification of nutrition responsive genes influencing meat edible quality traits In swine. *Asian-Australasian Journal of Animal Sciences*, *22*(4), 605–610. https://doi.org/10.5713/ajas.2009.r.05

Zemva, M., Ngapo, T. M., Malovrh, S., Levart, A., & Kovac, M. (2015). Effect of sex and slaughter weight on meat and fat quality of the Krškopolje pig reared in an enriched environment. *Animal Production Science*, *55*(9), 1200–1206. https://doi.org/10.1071/AN14059

2 Dry-Cured Ham

Cristina Pérez-Santaescolástica,[1]
Paulo E. S. Munekata,[1] Mirian Pateiro,[1]
Paulo Cezar Bastianello Campagnol,[2]
Daniel Franco,[1] Ruben Dominguez,[1] and
José M. Lorenzo[1,3]

[1]Centro Tecnológico de la Carne de Galicia, Avd. Galicia
nº 4, Parque Tecnológico de Galicia, San Cibrao das Viñas,
32900 Ourense, Spain
[2]Universidade Federal de Santa Maria, CEP 97105-900, Santa
Maria, Rio Grande do Sul, Brazil
[3]Área de Tecnología de los Alimentos, Facultad de Ciencias
de Ourense, Universidad de Vigo, 32004 Ourense, Spain

CONTENTS

33

2.1   INTRODUCTION

In countries around the world, dry-cured ham is considered a high-quality product. It refers to the ham made with the hind limb, cut at the level of the ischiopubic symphysis, with leg and bone, which includes the whole musculoskeletal piece, from adult pigs, subjected to the process of salting and drying-ripening (Order 2484/1967). The dry-curing process has been used since antiquity as a conservation method; the addition of salt reduces the availability of water for a microbial population. It allows stable products, extending shelf-life at the same time that the specific organoleptic characteristics are developed as a consequence of the drying process, which entails physico-chemical and biochemical reactions.

In the beginning, the production of ham was limited to Mediterranean areas (Blasco 1998), where it took advantage of the climate to satisfy the required conditions for an appropriate process. Nowadays, technological advances make it possible to expand the area of ham manufacture and to eliminate climatic restrictions. Thus, outside Europe, it can be found in countries that, little by little, have begun to produce their own type of ham. Such is the case in China and the United States of America (USA). In spite of the small geographical areas of production, consumption is extremely widespread; more than one hundred and fifty countries consume dry-cured ham. Even so, Europe is still the largest producer of ham, with Spain in first place followed by Italy, France and Portugal.

Due to the appreciable rise in ham quality derived from years of studies focused on improving and optimizing the process, ham manufacture has become an important economic resource. Spain went from exporting about 22 tonnes in 2007 to exporting nearly 44 tonnes in 2017, with its products going to 134 countries in total. Today, close to 20% of the total production of dry-cured ham is exported, mainly to countries of the Europe Union (82–84%) due to localization, proximity and similarity of culture between countries. In this regard, Germany is in first place, importing almost 12 million kg, followed by France (around 11 million kg). Among non-European countries, Mexico imports around 1300 tonnes of Spanish dry-cured ham per year, and the USA and Australia both import close to 700 tonnes per year (Rodríguez Marín 2018).

On the other hand, with lifestyle changes and the necessity to supply consumer demands, dry-cured ham can be found in a great range of formats in markets (whole

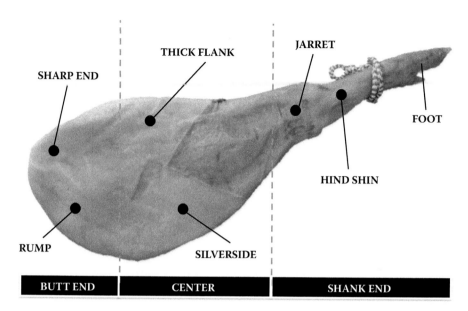

FIGURE 2.1 Main areas of dry-cured ham.

pieces, sliced, small pieces, portions, etc.) and using different packaging techniques, like vacuum or modified atmosphere. Those differences can be classified into different parts with different characteristics (Figure 2.1).

- The knuckle, which includes the hind shin and the jarret areas, is located at the end of the leg and is composed of tibia and fibula as bone supports and biceps as ligamentous and tendinous intersections. The texture in these areas is harder and the taste is quite different compared with other zones.
- The centre zone is considered the most important part. It is positioned over the femur, and its muscular components are the end of the *gluteus medius* (GM), *tensor fasciae latae* (TFL) muscle, *biceps femoris* (BF), *semitendinosus* (ST) and *semimembranosus* (SM). Silverside and thick flank are included in it. The silverside is more valued by consumers because the salt content is lower due to its IMFty contents. On the other hand, thick flank has less lean content and is less juicy, the texture is more fibrous and the taste saltier being curing perception more pronounced.
- The butt end, including the sharp end and rump areas, is made up of coxal bone and the top of the GM and TFL muscles.

Therefore, the piece of ham is heterogeneous due to the different structure and properties of the muscles that form it (Figure 2.2). In weight, the BF, SM and ST muscles are the most important muscles, being representative of an internal, external and intermedium muscle respectively. Some organoleptic differences are observed from slice to slice as well as in the same slice, including colour variations and salt concentration based on the level of drying. Higher lean redness and fat

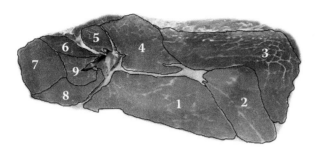

1: *M. biceps femoris*
2: *M. semitendinosus*
3: *M. semimembranosus*
4: *M. adductor*
5: *M. pectineous*
6: *M. vastus medialis*
7: *M. rectus femoris*
8: *M. vastus lateralis*
9: *M. vastus intermedious*

FIGURE 2.2   Dry-cured ham muscles.

yellowness values were shown in SM muscle, whereas BF muscle presented more marbling and greater values of lightness (Bermúdez et al. 2014). The localization of each muscle will determine how the physio-chemical and biochemical reactions would unfold. For instance, higher and faster desiccation is observed in muscles that are not covered by fat or skin (SM) and tend to be more adhesive due to the higher proteolytic activity (Monin et al. 1997). Deep muscles present higher water content (48–19%) than do external muscles (41–45%).

2.2   DRY-CURED HAM MANUFACTURE

The objective of dry-cured ham manufacture is to obtain value-added products, without defects, sanitary, secure and with genuine and appreciated organoleptic characteristics. To satisfy the purpose, it is necessary to exhaustively control the quality of the raw material and use a suitable selection of technological conditions that are applied to each stage of the curing process. These decisions will decide the differences between the different kinds of ham. But, independently of little differences in manufacture from producer to producer, the global process is similar (Figure 2.3). The general process consists of five principal stages, which are explained below, including selection and classification of the raw material, salting and post-salting, drying-ripening and, in most of the cases, an ageing period to obtain extra-quality products. Sometimes producers opted to expose hams to a smoking process, as happens in the case of Country Ham.

2.2.1   PREPARATION, SELECTION AND CLASSIFICATION OF THE RAW MATERIAL

Slaughtering, quartering, outlining and bleeding are included in this phase as a first step to prepare legs for the rest of the process. Previously, pH was measured to check the viability of the meat, allowing early detection of pale, soft and exudative (PSE) meats and dark, firm and dry (DFD) meats. To prevent future problems, pH values 24 hours after slaughter must be between 5.6 and 6.2. Besides that, pieces that present *petechiae* or/and hematomas are removed that could have been caused by anomalous stunning or bone fractures. Both can cause spoiled problems due to microbial growth in addition to the fact that external marks are negatively evaluated.

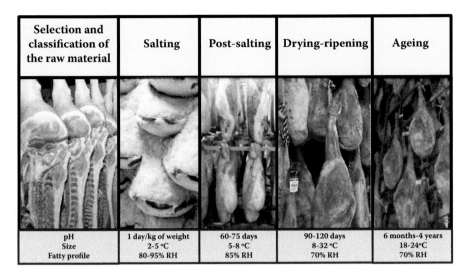

Selection and classification of the raw material	Salting	Post-salting	Drying-ripening	Ageing
pH Size Fatty profile	1 day/kg of weight 2-5 °C 80-95% RH	60-75 days 5-8 °C 85% RH	90-120 days 8-32 °C 70% RH	6 months-4 years 18-24°C 70% RH

FIGURE 2.3 Manufacturing process of dry-cured ham.

After selection, the piece is outlined, removing part of the muscle, fat and skin to get the desired form. Then bleeding is performed by either manual or mechanical pressure, removing as much blood possible, because any remaining blood would damage the ham in the following stage if it decomposed. Finally, legs are kept in refrigerated at 0–3°C for 1 or 2 days to reduce internal temperature. Pieces are classified by weight and fat level to determine the appropriate salting time.

2.2.2 SALTING

The objective is to incorporate in the piece enough salt, curing salts and sometimes additives such as ascorbate and sugar, to inhibit microbial population at the same time that the organoleptic characteristics are developed. Sodium chloride gives the product the salty taste, and nitrite helps red colour development since it reacts with myoglobin and creates the typical pigment of raw meat called nitrosomyoglobin.

There are two ways to aport the curing salts: by an unspecified amount or by the exact amount of curing salts. The first option, an unspecified amount of curing salts, consists of piling hams previously rubbed with a first layer of curing salt mix into a chamber. Hams are positioned in the chamber alternately with curing salts and are periodically turned to get a homogeneous distribution. The required time varies as a function of the weight, fat content and temperature of the chamber, usually around 1 day per kilogram of weight with a maximum between 4 and 7 days.

The second alternative, using a determined amount of curing salts, consists of the addition of a calculated amount of salts based on the hams' weight, using between 35 and 90 g per kg. This process is longer than the first one due to the process conclude when hams obtain the desired amount, usually in 2–4 weeks (Arnau 1993). But this method is less labour intensive and more standardized. This phase occurs at 2–5°C and 80–95% of relative humidity to inhibit undesirable microorganisms and to prevent the external desiccation of hams. During this phase, osmosis takes place and internal water leaves, causing a weakening of 3–7% and leading the salt diffusion to the internal muscles. In the end, the excess salt is removed and pieces are brushed and cleaned. After that, legs are kept for a maximum time of 24 hours before they begin the post-salting stage.

2.2.3 POST-SALTING

At this time, salt has spread uniformly into the muscles causing partial dehydration, that is, a significant water activity reduction which inhibits undesirable microbial growth. Additionally, both salt content and dehydration regulate proteolysis and lipolysis reactions, which will lead to the characteristic flavour development. The temperature must be lower than 5°C until the water activity reaches 0.96 throughout the piece for obtaining a microbiological stabilization; after that it can be established around 5–8°C. The relative humidity needs to be high enough to prevent crusting and enough low to remove the superficial humidity in a short time and thus prevent the *remelo* formation; usually the fixed value is 85%. Equally important is to maintain adequate air circulation to guarantee a good renewal of air. Legs are kept in these conditions for 60–75 days approximately.

2.2.4 DRYING-RIPENING

In this point, hams are transferred to a drying chamber, a room in which the hams are hung in racks, usually by the feet. At the beginning of the stage, the temperature starts low (below 15°C) and the humidity is moderately high (70% approximately). Over time the temperature is increased to 32°C, occasionally reaching 35°C. The objective is to continue dehydration and to intensify the biochemical reactions corresponding to proteolysis and lipolysis, which influence to a great extent the characteristics of texture, odour and taste. Another important point at this time is the fat fusion, which impregnates muscular fibres causing retention of a large part of odour compounds (Arnau 1998).

2.2.5 AGEING

The ageing happens in warehouses where the biochemical and enzymatic reactions already underway continue for a minimum period of 10 months. At this time, the action of microorganisms, basically yeasts and moulds, has an important role in final flavour development. For that, temperatures around 15–20°C together with about 65–75% relative humidity facilitate microbial action. Hams keep here until their commercialization or their preparation for packing, up to 4 years.

2.3 NUTRITIONAL SIGNIFICANCE

From a nutritional point of view, dry-cured ham is a complete product that has a great variety of essential nutrients. In spite of popular thinking about its salt and fat contents, it can be considered a healthy food as detailed below (Jiménez-Colmenero et al. 2009). Its water content represents values up to 45% and the caloric value is about 240–250 kcal/100g, but it varies as a function of the fat content present in the piece.

As a meat product, dry-cured ham has a very good proportion of proteins (30%) with high biological value since most of the essential amino acids are present. The high level of free amino acids is due to the intense proteolysis during the manufacturing process (Toldrá et al. 2000) and their presence helps to improve mineral absorption. At the end of the process, the most abundant free amino acids are alanine, arginine, glutamic acid, leucine and lysine (Martín et al. 2001). Free amino acid increments enhance the digestibility degree of proteins, especially due to the lysine content ($\approx$700 mg/100 g) (Toldrá and Aristoy 1993). The benefits of essential amino acids to proper body functioning are well known, such as the action of tryptophan against mental fatigue and its sedative effect. But not only do essential amino acids play an important role in the body, but some non-essential amino acids take part in important functions, as well. For instance, glutamine has a part in metabolic processes and collaborates in recovery from mental fatigue; taurine is important for nerve function; tyrosine plays an important role in depressive status and regulates blood pressure; and cysteine has power over oxidative stress (Ventanas 2006).

On the other hand, lipid fraction, which represents around 10–20% (Jiménez-Colmenero et al. 2010), is not too high compared with other meat products and, furthermore, it is underlined its fatty acid profile. From the total lipid amount, saturated fatty acids (SFA) comprise 30–40%, monounsaturated fatty acids (MUFA) represent about 45–55% and polyunsaturated fatty acids (PUFA) are 10–15%. SFA consumption increases the content of low-density lipoproteins (LDL) in blood plasma, which in high amounts are related to cardiovascular risk. Since not all SFA have the same impact, the effect of meristic acid seems to be notable; however, in dry-cured hams, predominant palmitic and stearic acid present low and non-existent effects respectively. Conversely, the ingestion of PUFAs reduces LDL levels but also the content of high-density lipoproteins (HLDL), which have an effect against cardiovascular diseases. Among MUFAs, oleic acid is the main one present in hams (around 50% of total fatty acids) and this also presents the capacity of reducing LDL levels but, in this case, HLDL levels are kept intact (Rebollo et al. 1999). Hams also show relatively low values of cholesterol (around 100 g of cholesterol per 55 g) compared to other common foods, such as eggs or liver, which have about 600 mg and 360 mg of cholesterol per 100 g, respectively. And, of course, fat levels vary depending on the genetic traits of the pigs, feeding characteristics and even production methods.

Regarding vitamins, dry-cured ham is an important source of B-group vitamins. The table below shows the content of every vitamin present in dry-cured ham and the recommended daily allowances (RDA) of each one, and describes its function in the human metabolism together with the consequences for its deficit in the diet. The thiamine content is very important, since it shows that pig meat is the best source of this vitamin (Jiménez-Colmenero et al. 2009). Likewise, dry-cured ham has great content in riboflavin, niacin, pyridoxine and cobalamin, all of them belong to the B vitamins, which are strongly involved in enzymatic reactions related to the proper functioning of the central nervous system. On the other hand, fat-soluble vitamins are presented in much smaller amounts. Vitamins A, D, K and E are only present in trace amounts, as is Vitamin C. However, Vitamin E content is important from a sensorial point of view, as much as a nutritional one, due to its antioxidant capacity. (Table 2.1)

Dry-cured ham not only constitutes a good source of heme iron but also enhance the bioavailability of non-heme iron from vegetable products when they are eaten together. There is a high concentration of zinc, which similar to iron, and its bioavailability seems to benefit when animal proteins are ingested in conjunction with vegetables. It is interesting to note the good amount of other minerals, which can be seen in Table 2.2. For instance, note the amount of magnesium, which is necessary for most metabolic reactions as a cofactor, as well as the presence of selenium, which has a notable effect as antioxidant collaborating against cancer (Fleet and Cashman 2003; Higgs 2000). Nevertheless, consumption of dry-cured ham has been adversely affected by concerns regarding the negative consequences of excessive sodium consumption and the resulting changes in lifestyle and eating habits. This evidence has caused producers to look for new strategies to reduce the use of salts, being careful not to provoke unintended, potentially dangerous, consequences or damage quality (Armenteros et al. 2009). Nowadays, the common amount of sodium present in dry-cured hams is around 11–18 mg/g of product, nearly half of the recommendation. Also keep in mind that sodium is fundamental to maintaining a correct electrolytic balance and has many beneficial compounds.

Finally, the content of carbohydrates are not too relevant: only a small portion of free sugars are present such as glucose (0.3%), fructose, maltose and ribose. Glycogen is the only interesting compound, but its content falls after slaughtering due to its conversion into lactic acid.

2.4   PHYSICO-CHEMICAL AND BIOCHEMICAL REACTIONS

During the curing process, several reactions take place. The way that reactions progress will define the organoleptic attributes at the end of the product manufacture. The main reactions consist of desiccation (and the related salt diffusion), lipolysis and lipid oxidations, proteolysis, and Maillard and Strecker reactions.

2.4.1   Desiccation

During ham manufacture, the water content decreases from initial values of 70–75% to 50–60% (Armenteros et al. 2012). Water losses come with increments in the percentage of fat and protein, creating an add-value due to more protein content. The desiccation starts during salting due to osmotic dehydration and it continues during the following stages of the process. The intensity of the desiccation varies, depending on the protection level of the piece (fat or skin area) and the environmental relative humidity. The process takes place only through muscles that are exposed to air, hence a small area, which depends on the cut selected in the outlined time.

In the post-salting stage, the dry mainly occurs by dragging of vapour and for salt diffusion into the piece, as well as, for water diffusion outside the leg. At this

TABLE 2.1

Vitamins Present in Dry-Cured Ham, Recommended Daily Allowances, Function on Body Metabolism and Consequences of Its Deficit

Vitamin	[a]Amount per 100 g	[b]RDA	[a]Function	[a]Deficit
Water soluble vitamins				
Thiamine (B1) (mg)	0.57–0.84	1.4	Coenzyme in enzymatic reactions	Beriberi disease
Riboflavin (B2) (mg)	0.20–0.25	1.6	Oxidizing agent in biological system	Optical diseases
Niacin (B3) (mg)	4.5–11.8	18	Part of metabolic reactions: anaerobic glycolytic pathway, oxidative phosphorylation and fatty-acid biosynthesis and oxidation	Weakness, loss of appetite and, in extreme cases, pellagra
Pyridoxine (B6) (mg)	0.22–0.42	2	Part of more than hundred reactions from amino acids metabolism	Central nervous system changes
Cobalamin (B12) (µg)	Tr-15.68	1	Coenzyme in metabolic reactions with an important role in normal function of all cells	Anaemia and physical and neurological problems
Folic acid (µg)	Tr-13.49	200	Part of the polymeric structure of glutamic acid	Weakness, sleeplessness, etc.
Vitamin C (mg)	Tr	60	Cellular antioxidant	Hair loss, poor cicatrisation and haemorrhage
Fat-soluble vitamins				
Vitamin A (µg)	Tr-5	800	Important for teeth, tissues, mucous membranes and skin development	Night blindness, conjunctival xerosis of the eyes, diarrhoea and respiratory diseases
Vitamin D (µg)	Tr-0.3	5	Promotes calcium and phosphorous absorption	Skeleton deformation
Vitamin K (µg/ kg body weight)	Tr-10	1	Collaborates in blood clotting	Rarely be in human
Vitamin E (mg)	0.08–1.5	10	Prevents lipid oxidation	Neurologic, sensorial nervous and retina problems

Notes

[a] Jiménez-Colmenero et al. (2010); Toldrá (2002).

[b] Council Directive 90/496/CEE (1990)

TABLE 2.2
Minerals Present in Dry-Cured ham, Recommended Daily Allowances,
Function on Body Metabolism and Consequences of Its Deficit

Mineral	[a]Amount per 100 g	[b]RDA	[a]Function	[a]Deficit
Calcium (Ca) (mg)	12–35	800	To maintain bones, blood clotting and to transmit nervous impulse	Spasm, cramp, crawling sensation and osteoporosis
Iron (Fe) (mg)	1.8–3.3	14	Take part in the process of cellular breathing	Anaemia and fatigue
Zinc (Zn) (mg)	2.2–3.0	15	To maintain cell membrane, immune system and brain function	Prasad-Halsted syndrome and hypogonadism
Magnesium (Mg) (mg)	17–18	300	Cofactor in enzymatic and metabolic pathways	Cardiovascular deseases, osteoporosis, diabetes, etc.
Potassium (K) (mg)	153–160	2000	Necessary for osmotic balance	Spasm, cramp, fatigue and constipation
Phosphor (P) (mg)	157–180	800	To maintain cellular and tissues	Rickets, osteoporosis, osteomalacia and muscular weakness
Selenium (Se) (µg)	29	55	To protect against free-radical injury	Cardiomyopathy, cardiovascular risk and liver cancer
Sodium (Na) (mg)	1100–1800	2000	Necessary for osmotic balance	Nausea, vomiting, headache and memory loss

Notes
[a] Jiménez-Colmenero et al. (2010); Toldrá (2002)
[b] Council Directive 90/496/CEE (1990); World Health Organization (2003)

point, if the relative humidity is too high or there is not enough ventilation, the microbial population can grow as a consequence of a slower drying. In contrast, if the water evaporation is higher than the arrival of the inner water (which happens with a very low relative humidity or when the air velocity is not well adjusted) external muscles undergo a deep desiccation that can create crusted zones while internal muscles seem like raw meat, and this is combined with a higher salt taste in the final product (Ventanas and Cava 2001). The drying-ripening stage is the stage with the strongest dehydration due to both the length of the stage and the environmental conditions, which correspond to greater temperature and lesser relative humidity than in the previous stage. Increments of temperature cause increments in the the dehydration of the piece, because the internal water is thrown out at different rates and the evaporation is affected. Salt absorption and water losses lead to sodium chloride increments along the curing process, with the final amount being from 5% to 12% approximately (Virgili et al. 2007) The desiccation also can imply texture changes, favouring a muscular tissue hardening.

2.4.2 CURING SALT EFFECTS

During the process of making ham, curing agents, mainly nitrites and nitrates, are added. Nitrites and nitrates are responsible for the colour development because the nitric oxide originating from their decomposition gives rise to the subsequent reaction with the myoglobin present in pig muscles (Figure 2.4). First, nitrates are ionized in a watery environment to form nitrate ions, which is reduced to nitrite ions by the action of nitrate-reductase enzymes from bacterias like *lactobacillus, micrococcus, vibrium*, etc. After that, the action of bacterias together with a moderate acid medium (5.6–6.0), degrade nitrite ions to nitric oxides which have the capacity to react with myoglobin. In the presence of nitrites, myoglobin is reduced to metmyoglobin which is characterized by a brown colour. Later the metmyoglobin form reacts with nitric oxide, giving rise to nitroso-metmyoglobin; however, this reaction does not occur completely (from 35% to 75% of metmyoglobin turns into nitroso-metmyoglobin) (Ranken 2000). To make this mechanism possible, it is necessary to create a reduced network, which is made possible by the action of enzymes that are naturally on meat. Even so, occasionally producers promote the

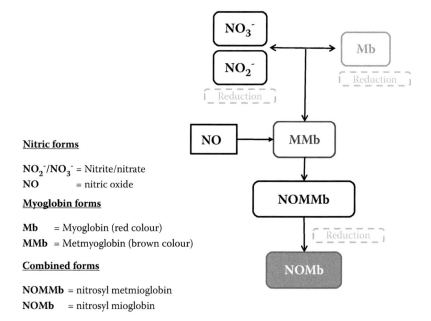

Nitric forms

NO_2^-/NO_3^- = Nitrite/nitrate
NO   = nitric oxide

Myoglobin forms

Mb   = Myoglobin (red colour)
MMb = Metmyoglobin (brown colour)

Combined forms

NOMMb = nitrosyl metmioglobin
NOMb   = nitrosyl mioglobin

FIGURE 2.4   Nitrosyl myoglobin formation.

action of enzymes by a chemical reducing agent; the most commonly used is ascorbate since it increases colour formation and improves intensity and uniformity.

On the other hand, in the absence of nitrification substances, myoglobin is affected mainly by the oxidation action of salt and it can be modified, leading to changes in colour. In this sense, muscles show a light colour (pinky-red) due to the complex formed by Zn protoporphyrin (Moller et al. 2007). Recently, it has been observed that this complex is correlated with the content of intramuscular fat (IMF). The better the marbling is, the better Zn protoporphyrin is formed, which may be due to fatty acids and phospholipids from IMF interacting with the enzyme responsible for the complex formation (Bou et al. 2018).

2.4.3 LIPOLYSIS AND LIPID OXIDATION

Lipolysis is an enzymatic process in which the ester links between lipids and glycerol molecules are hydrolysed causing free fatty acids (Figure 2.5). These free fatty acids, in turn, will be substrates for other oxidative reactions that will origin aromatic volatile compounds on dry-cured products.

The fat present in hams could be divided into two types: IMF and adipose tissue. IMF includes mainly triglycerides (around 90%) and phospholipids (Antequera and Martín 2001), while, adipose tissue is mainly triglycerides (99%). The principal endogenous enzymes related to lipolysis reactions in IMF are lipases, including lysosomal acid lipase, neutral lipase, and phospholipases A1, A2, C and D (Table 2.3). While lipases work on triglycerides at a pH between 5.5 and 6.2 (Motilva et al. 1992), phospholipases are responsible for hydrolysing phospholipids,

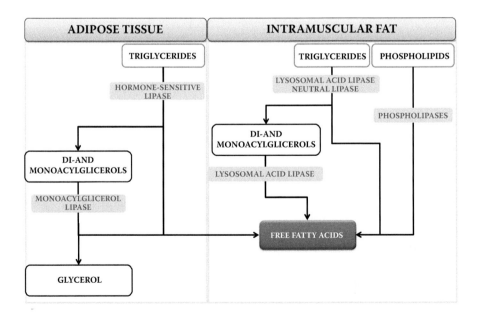

FIGURE 2.5 Principal steps of dry-cured ham lipolysis.

TABLE 2.3

Main Enzymes Involved in Lipolysis and Proteolysis Processes

	Enzyme	pH	Active Against	Observation
Lysosomal proteinases	Cathepsin B	Acid	Myosin and actin	Stable and active throughout the process (5–10% of recoveries after 15 months of processing)
	Cathepsin L	Acid	Myosin heavy chain, actin, tropomyosin, α-actinin and troponins T and I	
	Cathepsin H	Acid	-	
	Cathepsin D	Acid	Myosin heavy chains, titin M and C proteins, tropomyosin, troponins T and I	Disappears between 6 and 10 months of process
Neutral proteinases	Calpain	Neutral	Troponin T and I tropomyosin, C-protein, filamin, desmin and vinculin as well as titin and nebulin	Requiring calcium ions for activity. Its activity is lost after the salting
Muscle aminopeptidases	Aminopeptidase	Neutral	Aromatic, aliphatic and basic aminoacyl-bonds	Activity still recovered after more than 8 or even 12 months
	BAlanyl Pyroglutamyl Aminopeptidases	Neutral		
		Neutral		
		Neutral		
	Leucyl aminopeptidase	Basic	-	
Adipose tissue lipases	Lipoprotein lipase	Basic	Primary ester	Active during salting and post-salting
	Monoacylglycerol	Basic	1 or 2 monoacylglycerols with no positional specificity	
	Hormone sensitive lipase	Neutral	Ester-bond in triacylglycerols and the resulting diacylglycerols	Active during the ripening-drying
Muscle lipases	Lysosomal acid lipase	Acid	Hydrolyses tri-, di- and monoacylglycerols, although it has a marked preference for primary ester bonds of triacylglycerol	Stable throughout the process

Toldrá et al. (1997)

giving off fatty acids saturated as much as mono- and polyunsaturated (Ripollés et al. 2011). On the other hand, in adipose tissue, the main enzyme corresponds to hormone-sensitive lipase working over triglycerides and generating glycerol, and mono- and diglycerides at the same time that a large number of fatty acids are given off. Also, triacylglycerides can be hydrolysed by lysosomal acid esterase and cytosol neutral esterase, but their activity is limited due to the lack of available substrate (Belfrage et al. 1984).

The greater part of lipolysis takes place during the drying-ripening stage and, to a lesser extent, in the following ageing period. This fact could be because the temperature is so high at the end of the drying, therefore lipolysis reactions are reduced and oxidative reactions have more and more importance by that time. The evolution of lipolysis can be followed by the evolution of free fatty acids along the process. Saturated fatty acids are the most stable fraction, which shows a progressive increment during the process; meanwhile, monounsaturated and polyunsaturated fatty acid contents drop along the curing stage; polyunsaturated is the one the most affected, with a reduction from up to 88% of the initial content. Both reductions, monounsaturated and polyunsaturated fatty acids, are attributed to oxidative reactions. However, an increment of unsaturated fatty acids, mainly oleic and linoleic, is seen at the end of the curing process, probably due to their antioxidant characteristics (Antequera et al. 1993).

Lipid oxidation occurs as a consequence of lipolysis and it involves a group of chemical reactions called autoxidation, including primary reactions in which lipid peroxides are formed, as much as a secondary reaction in which volatile compounds are originated. The molecular oxygen reacts with the free fatty acids created by enzymatic hydrolysis. The mechanism happens in three steps as can be observed in Figure 2.6: initiation, propagation and termination.

- During the initiation, free radicals are created by the homolytic breakdown of the C-H links of fatty acids or abstraction of a hydrogen atom by a catalyst (light, metallic ions, heat) or by enzymatic (cyclooxygenase or lipoxygenase) action.
- Propagation is the phase in which free radicals from initiation can react with oxygen to form peroxide radicals ($ROO^\bullet$) and these, in turn, give rise to hydroperoxide ($ROOH$) by the hydrogen capture from a fatty acid. That is how the chain reaction is expanded. Hydroperoxides can be broken by heat or by catalysis induced by transition metals, causing new radicals which start the oxidation process again. These hydroperoxides are tasteless and odourless,

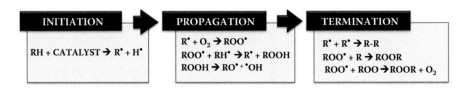

FIGURE 2.6 Lipid oxidation process.

and are characterized by their instability because they quickly decompose in volatile compounds such as aldehydes, ketones or alcohols, all of which have high impact on sensory characteristics.

- Termination consists of the reaction between hydroperoxides and proteins, peptides or amino acids, and their subsequent polymerization (Buscailhon et al. 1993) or their breakdown into volatile compounds with a low molecular weight such as aldehydes, ketones, alcohols, hydrocarbons and acids.

The triglyceride composition consists of 49–51% of monounsaturated fatty acids, oleic acid being the main one, with around 45% of the total, 7–15% of poly-unsaturated and 36–41% of saturated fatty acids; palmitic acid is the primary one. Of the phospholipids, 40–48% are polyunsaturated, 35% are saturated, and 18–25% are unsaturated fatty acids: primarily arachidonic acid, palmitic acid and oleic acid. Because unsaturated fatty acids are oxidized faster than saturated fatty acids, and phospholipids contain more unsaturated fatty acid than triglycerides, oxidative re-actions are initiated in phospholipidic fraction (Gray and Pearson 1987). The number, position and geometry of double links will also influence the velocity of the oxidative process. In this sense, cis connections are oxidized more than the trans, and con-jugated double links are more reactive than non-conjugated (Flores 1997). Independently of unsaturations, fatty acids are oxidized faster when they are in free form than when are connected to glycerol. In addition, oxidation is favoured by the presence of some pro-oxidant compounds such metallic ions, salt, globular protein and oxygen, and at the same time, other compounds reduce it, like tocopherols, ni-trites, ascorbic acid, free amino acids and peptides. Pro-oxidant and anti-oxidant compound balance varies as a function of the moment along the curing process.

The evolution of the oxidation process can be followed by the evolution of per-oxides and aldehydes, since they are the primary and secondary products respectively of lipid oxidation. According to Martín (1996), the highest increment in peroxides happens during the drying stage, but at the beginning of the ripening, peroxides as much as aldehydes show a reduction. In the last month of the ripening/ageing, the increase in the temperature leads again to a large increment of peroxides.

Lipid oxidation is a dynamic process in which peroxides are formed, and at the same time, they are decomposed, originating secondary products.

2.4.4 PROTEOLYSIS

Proteolysis refers to enzymatic reactions in which proteins are broken into small peptides, amino acids and other compounds that have a direct influence on texture and an indirect effect on odour development (Toldrá 2006). The primary enzymes implicated in the process are described in Table 2.3 and consist of endopeptidases, which are related to protein hydrolysis, and exopeptidases, which are associated with small peptides and free amino acid generation (Toldrá et al. 2009).

The process consists of two steps. The first one is the hydrolysis of myofibrillar and sarcoplasmic proteins by muscular endopeptidases (cathepsins and calpains) that act on myosin and troponin, enhancing tenderness and originating big peptides (Nagaraj and Santhanam 2006). At the second step, the big peptides are broken

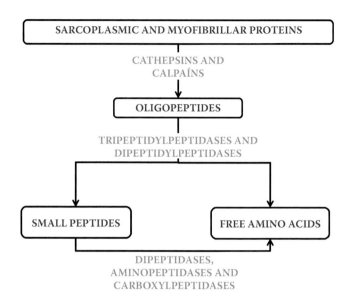

FIGURE 2.7   Principal steps of dry-cured ham proteolysis.

down to smaller peptides and free amino acids by exopeptidases (Figure 2.7). Exopeptidases can mainly be included in two groups:

- Aminopeptidases, dipeptidases and tripeptidases, which break down the peptidic chain from the amino extreme, giving off an amino acid, a dipeptide or a tripeptide, respectively.
- Other common enzymes that are less known in meat products. These enzymes, called carboxypeptidases, act on the final carboxy group originating free amino acids, peptidyldipeptidases that give off dipeptides from the final carboxy of polypeptides, and dipeptidases and tripeptidases that hydrolyze dipeptides and tripeptides respectively (Toldrá et al. 1997).

Proteolysis reactions start before the curing process; endopeptidases begin protein degradations at a post-mortem period favouring the tenderness of the meat. At the beginning of the curing process, salting and the post-salting stage predominate the proteolytic reaction related to peptide formation that turns into a progressive increment of peptide hydrolysis over time; that is, the amino acid generation gets longer. The increase in temperature and progressive desiccation help the hydrolysis of peptides, perhaps due to the tissue aminopeptidases' activity, since its optimal temperature is 37°C (Toldrá et al. 1997). For that, in the drying stage, are the most amino acid increments (Martuscelli et al. 2009), which slows down in the next ageing stage. The peptide generation decreases in the ageing stage, probably for the peptidases' inactivation due to salt concentration. Some extrinsic factor, including the drying level, pH, temperature and salt content, will affect enzymatic activity. In this sense, cathepsins B, D and L are active at slightly acid pH, while calpains I and

II and cathepsin H are active at neutral pH. On the other hand, the optimal temperature is 30°C for cathepsin B and 3°C for cathepsins H and L (Sánchez 2003). During the drying, cathepsins activity is reduced due to the decrease in water activity; the higher salt content, the lower the proteolytic activity (Antequera and Martín 2001). However, residual activities of 5–10% have been shown for cathepsins B, H and L after 15 months, but not for cathepsin D whose activity seems to disappear after 6 months of the process (Toldrá and Aristoy, 1993). Solubility losses of proteins can influence the proteolytic process, stimulating or inhibiting it, and modifying some functional properties such as water-holding capacity (WHC) that enhance the desiccation. Solubility changes are attributed essentially to the salt effect (Larrea et al. 2006) although heavy metals from contaminants in curing salts can catalyze oxidative reactions, originating molecular links that give room to resistant structures.

The level of amino acids originated at the end of the process is a good indication of the extension and depth of the proteolysis that takes place during the process. The highest amounts in the final product correspond to glutamic acid, alanine and lysine (Jurado et al. 2007), amino acids that are described as precursors of pleasant odours. Nevertheless, basic amino acids, like histidine or arginine, show a minor increment because they are involved in other followed degradative reactions, such as volatile compounds and amines generation. In addition to that, free amino acids are the substrate of other reactions, such as Maillard and/or Strecker degradation, having an indirect influence on the final odour. On the other hand, as a result of extreme proteolysis, total volatile basic nitrogen can appear, which is integrated by amines. This fact is not desirable for dry-cured ham elaborations being considered as an alteration index. However, ammoniacal nitrogen fraction showing values below 1.1 mg/g of dry matter are not considered a disturbing value (Martín et al. 1997). Since low-molecular-weight nitrogenous substances are generated, pH increments have been observed in different dry-cured products (Lorenzo et al. 2008; Lorenzo 2014).

2.4.5 Secondary Reactions

Maillard reactions are complex non-enzymatic glucosylation reactions of proteins, in which an amine group from proteins, free amino acids and/or amines reacts with a reducing sugar, giving room to compounds that, in turn, can react with each other or with other carbonyl groups from lipid-causing volatile compounds. In general, these volatile compounds can be involved in three groups: sugar fragmentation products like furans, pyrazines, cyclopentane, carbonyls and acids; amino acid degradation products such as aldehydes and sulphur compounds; and products from secondary reactions, such as as pyrroles, pyridines, imidazoles, thiazoles and alcoholic condensation compounds. Processes in which the drying period is long and water activity presents low values could favour these reactions (Flores 2018). In the mechanism of the Maillard reaction is found the Strecker pathway, which also takes part in the development of volatile compounds (Cremer and Eichner 2000). Aminoacids experience oxidative deamination and decarboxylation in the presence of α—dicarbonyls to produce aminoketones, aldehydes and carbon dioxide

(Hidalgo and Zamora 2016). The volatile compounds that are formed can contribute to odour development, such as 2-methyl propanal, 2-methyl butanal and 3-methyl butanal, which come from the amino acids valine, isoleucine and leucine respectively, as well as other sulphur compounds from sulphur amino acids like methionine and cysteine.

The achievement of these reactions depends on proteolysis, lipolysis and lipid oxidation, since compounds involved in all of them are widely interconnected, as well as on processing conditions, such as temperature, time and pH, which exercise a greater influence over the course of these reactions (Zamora et al. 2015).

2.5 FACTORS AFFECTING PRODUCT QUALITY AND PROCESS

Organoleptic characteristics decide the acceptability level of products by consumers and are the consequence of a complex process in which numerous variables can exercise influence and can modify it. For instance, volatile compounds, as well as free amino acids, will determine the odour and taste of the product, and they come from previously cited chemical and biochemical reactions that will be influenced by raw material (breed, feed, environmental parameters, etc.) and by technologic conditions (salt and additives, temperature, relative humidity, process time, etc.) (González and Ockerman 2000).

Although many producers have attempted to maintain traditional production methods, ham manufacturing must constantly change to satisfy the new requirements of consumers. Currently the market trend is to produce healthier products while trying to maintain the quality. For instance, how to improve the profile of fatty acids by breed or feed modifications has been widely studied, as has salt reduction or its partial replacement, due to the large content of salt in hams and its relationship with cardiovascular diseases (Armenteros et al. 2012). But every change in raw materials, just as in the production chain, is accompanied by an important impact on sensory traits, so it must reach a balance. To express the complexity of the impact, Figure 2.8 shows how all the parameters are interconnected.

2.5.1 FACTORS RELATED TO RAW MATERIAL

2.5.1.1 Ante-Mortem

Ante-mortem factors such as breed, feed, sex and age have a direct influence on carcass composition and, indirectly, on chemical and biochemical transformations during the curing process. Breed and carcass characteristics have a powerful impact on the final product, and their properties as raw material are considered the main reasons for the heterogeneity of the processed products. Over the years, the relationship between pig breeds and final meat quality has been the object of study for a lot of researchers, since good breed selection can prevent several technical problems and allow producers to assure some specific characteristics in the final product. Thus, it is good to know the peculiarities that each breed can aport to the product.

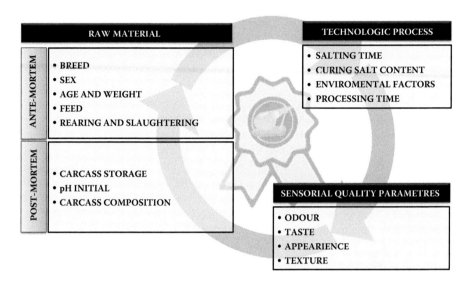

FIGURE 2.8   Relationship between raw materials, technologic factors and the final quality of dry-cured ham.

For instance, a study focused on the influence of genetic parameters on pig meat quality indicates that the muscle tissues of Berkshires are darker, and Durocs show high values in the final pH and present a greater proportion of fat, making the meat more tender. In contrast, Hampshire and Landrace muscles are pale pink but have a smaller proportion of fat than others, and both show lower values of final pH (Goodwin 1994). It is observed that Duroc animals present a quick growth and a good conversion ratio (Blasco et al. 1994), which converts into a more-than-suitable option for dry-cured products, while Pietrain and Belgian Landrace have shown a noticeable inclination to suffer stress, and that fact can increase the risk of exudative meat. Pig meat manufacturers cross different breeds to balance out the possible negative traits of one genotype, improving product quality as much as technological processes do. Their tendency to do this also depends to a large extent on genotype, since breed establishes the predisposition of the meat to undergo lipolytic and proteolytic reactions. Like this, it is checked more tenderness meat in females together with higher moisture values than males (Tabilo et al. 1999). Castrates present higher values of IMF, as well as subcutaneous fat, than boars, which show minor losses during the process (Gou et al. 1995). However, more fat content means more difficult oxygen diffusion and slower salt diffusion. Also, meat colour is perceived as less red in castrates due to the amount of intermuscular fat present, and boars can exude sexual hormone odours derived from a large escaterol or androsterone content. On the other hand, hams from females and castrates are leaner and perceived to be lower in salt, a fact that suggests the product is of higher quality (Armero et al. 1999; Bañón et al. 2003).

The feed in most cases is closely related to the rearing. Pigs from different rearing systems (extensive, semi-extensive or intensive) show changes in some parameters such as sarcoplasmic myoglobin, level of protein denaturation, pH, marbling, etc. Rearing influences the animals' physical activity, which will also affect the final composition of the carcass. Extensive systems result in higher IMF content and also a greater variability due to the wide range of forage available; this affects the fatty acid profile of the animal fat, leading to specific volatile compounds modifying the odour positively or negatively. Because unsaturated fatty acids tend to oxidize and can result in off flavours, it is desirable to obtain meat with more saturated fatty acid. However, from a consumer point of view, the high content of PUFA is considered healthier due to its effect against cardiovascular disease. In this regard, lipid oxidation can be controlled by antioxidant contribution in the diet (Ventanas et al. 2007), but sensorial attributes can be changed (Santos et al. 2008). The level of protein in the diet influences the IMF, so a diet rich in protein results in a lesser IMF (Tang et al. 2010). Conversely, diets based on medium or low amounts of protein have resulted in hams with up to 14–17% less subcutaneous fat compared with hams that were fed a conventional diet (Gallo et al. 2016).

An important purpose of ham manufacture is to slaughter animals with adequate weight and age to allow for optimal development of carcass composition, achieving a convenient content of IMF. It is well known that the older the animal, the higher the IMF it contains, and meat colour turns darker and darker. Also, older animals give rise to meat with more flavour due to the high content of volatile compounds. In addition, differences are observed in enzyme activities as a function of the age, among which a significant predisposition for undergoing proteolysis has been observed in young pigs, due to the higher activity of cathepsin B. Since moisture is reduced over time, meat from old pigs trends to express a lower activity from hydrolytic enzymes (Tibau et al. 2002). On the other hand, stress is as importa ant factor in rearing as slaughtering. During housing, conflicts between animals can cause bruises, wounds and even bone fractures. Therefore, meat from these animals can present visual damage, reducing global quality.

The same problem is seen during transport to slaughter, and is worse when the stress occurs at the moment before killing; a long exposure to stress depletes the glycogen reserves. So the abnormal pH declination can provoke values of pH beyond 6, resulting in meats that are DFD. Again, if the stress is just before the slaughter, or there is not enough glycogen, the pH can decrease very quickly, resulting in meat that iss PSE. Among the reasons for stress are long fasts, animal conflicts, inappropriate temperatures, bad handling and also unsuitable building design. The method that is selected to stun can also induce stress. The common method used to stun pigs is electricity, but the use of CO_2 leads to stress reduction, resulting in higher-quality meat (Barton-Gade 1993). The effectiveness of the exsanguination is also influenced by the stun operation because the volume of removed blood is affected by a suitable suspension of the carcass as well as by the time that elapses between the stun and exsanguination operations. If the exsanguination is not well done, the content of myoglobin in muscle would be too high and sensorial changes of the meat are happened, muscle turns deep red and the taste becomes more metallic.

2.5.1.2 Post-Mortem

The raw material is classified to optimize the process and to adapt technological parameters for the raw meat characteristics. The technological properties of raw meat used to be explained by some factors such as pH value, WHC and enzymatic activity. As previously discussed, the pH of the carcass at the end of rigor mortis defines the meat's inclination to be defective, either DFD or PSE. It has been shown that there is a direct relationship between pH and moisture: high pH leads to high moisture and vice versa (García-Rey et al. 2004; Guerrero et al. 1999).

Similarly, the final composition of the carcass also varies according to pre-mortem factors. Fat, protein and water, among others, will be determined by factors such as diet, animal exercise and stress. The proportion of the different fractions determines how desiccation, lipolysis and proteolysis reactions take place to a greater or lesser degree in the same environmental conditions. Salt diffusion and water loss dynamics are affected by the adipose tissue, since subcutaneous fat protects against fast desiccation and faulty salt penetration. The higher the content in subcutaneous fat, the less salt is able to penetrate inside the muscles during the salting stage, making it necessary to extend the processing time. At the same time, surface evaporation becomes difficult, resulting in hams with higher water content, turning its storage more challenging. In addition to that, the intramuscular fraction is extremely important, especially for taste and juiciness attributes. Enough IMF prevents excessive salty taste and provides more juicy hams with a suitably softer texture. Conversely, an excess of IMF can lead to hams with extremely soft and pasty textures. It has been shown that below 3% of fat, there is a large decrease in palatability, while a proportion between 3% and 6% improves the palatability.

The content of IMF can vary between muscles, and is higher in muscles closer to the subcutaneous fat, such as BF muscle. Regarding the lean fraction, muscle protein is responsible for holding water, so juiciness and tenderness can be changed when this fraction is modified by other factors such as pH, since it affects the isoelectric point of proteins (Kerry et al. 2002). Muscles are formed by different muscle fibres with different metabolic characteristics that can be classified as fibres type I, IIA and IIB (Brooke and Kaiser 1970). Type I fibres have a high oxidative capacity and scarce or absence glycolytic capacity, type IIB fibres show a glycolytic or anaerobic metabolism, and type IIA fibres present an intermediate metabolism. According to that, colour depends on the fibrillar composition being oxidative fibres, redder since oxidative fibres have higher myoglobin content than glycolytic fibres. Oxidative fibres also present a greater amount of phospholipids and polyunsaturated fatty acids, which contribute to the development of odour (Cava et al. 2000).

The conditioned temperature of the carcass will influence how the post-mortem changes happen in muscular tissue, and these, in turn, affect the WHC of the muscular fibres. Elevated temperatures reduce the protein capacity for connecting molecules of water, because metabolism of muscular cells is accelerated, causing increments to the anaerobic glycolysis velocity; it also leads to the fast development of rigor mortis. This fact provokes myofibrillar shortening and hardening, and this is the reason the immediate cooling of meat is needed after slaughtering and must be maintained until the end of the salting and post-salting stages. Increasingly, the legs are frozen after slaughtering to supply the process necessity of the manufacturer. In the thawing, an excessive exudation in the surface of hams appears, solubilizing the salt and increasing its penetration. In this case, it should be necessary to recalculate the salting time to prevent the risk of an overly salty taste.

2.5.2 TECHNOLOGICAL PARAMETERS

The type and amount of curing salt required is influenced by certain enzymatic activities that either activate or inhibit it. Common curing salts used in the dry-cured process are sodium chloride and nitrates and nitrites, but other additives, like ascorbic acid, may be used to a lesser extent. Most of the aminopeptidases, cathepsins, neutral lipase and neutral esterase are inhibited in the presence of great concentrations of sodium chloride while this substance activates the acid lipase. As a result, salt addition could be a good way to reduce problems derived from excessive proteolysis, but a high content of sodium chloride is undesirable due to its role in cardiovascular risk. Salt absorption can be made easier when the skin is partially removed because the brine is then in contact with the lean surface for more time (Arnau 2007). However, this fact increments water losses (García-Gil et al. 2012). Also, salting time affects the final salt intake; a longer time means the final salt content is higher (Martín et al. 1997).

On the other hand, to control the temperature throughout the process is essential, not only for monitoring the enzyme activities but also to prevent microbial growth in the early stages. In the curing process, temperature together with humidity will mark the velocity of the dehydration and the diffusion of curing salts in each stage. In this sense, at the salting period, the absorption and diffusion of salt are better at a high temperature and high relative humidity. However, after salting, the humidity needs to be slightly reduced to prevent the appearance of mites and moulds, and the fomration of slime or phosphate crystals. Additionally, several studies have demonstrated that the longer the time of ripening, the more aromas are developed (Careri et al. 1993; Buscailhon et al. 1994; Ruiz et al. 1998). It has also been shown that up to 420 days salt taste increases while there is a reduction in sweetness. Also, time can affect fat, and muscle colour and texture, based on the locality on the overall piece. The ripening time also changes lightness, redness and yellowness, and may make the yellowness in SM more noticeable, whereas shear force and firmness are enhanced in both muscles, SM and BF, according to the time (Bermúdez et al. 2014).

2.6 SENSORIAL ATTRIBUTES AND THEIR DEVELOPMENT

Sensorial characteristics of dry-cured hams are mainly derived from the reactions occurring during the manufacturing process. The principal reactions that have the greatest impact on sensorial quality are lipolysis and proteolysis and their secondary reactions. Variations in the process conditions, as much as factors affecting the raw material, will influence these reactions, leading to changes in volatile and non-volatile compounds that will define the specific characteristics from ham to ham.

Protein hydrolisis derived from proteolysis has an important effect on texture at the same time that the originated peptides and free amino acids affect taste and participate in odour development. Similarly, lipolysis produces free fatty acids from triacylglycerols and phospholipids, which will be the precursors of several structures related to odour.

2.6.1 COLOUR DEVELOPMENT

The colour of lean is due to the reaction that occurs between myoglobin pigment and the nitric oxide from nitrites. The common desirable colour in dry-cured hams is a pinkish-red tone, and the intensity depends on the pigment amount present in muscular tissue. A uniformity of colour is high valued because it means that the technological process has been applied properly. In the case of fat, it can be deep yellow due to the oxidative process that happens during curing. However, in fat areas that are less exposed to oxidation factors, the colour is essentially white with some yellow notes. The intensity of yellow colour seems to be unrelated to product quality.

2.6.2 TASTE DEVELOPMENT

There are five basic tastes: acid, sweet, salty, bitter and umami, and their interactions will determine the overall taste of the foods (Figure 2.9). Recently, a sixth taste, called oleogustus, has been proposed to define the fat sensation that is noticed when animal products are ingested (Running et al. 2015). Sodium chloride contributes to a salty taste when its concentration is up to 0.1 M. However, this taste perception can vary according to the specific characteristics of the product, such as fat content or the formation of a complex between ions and proteins, which captures part of the salt content, reducing the level of the salty taste. The presence of amino acids from proteolysis can be an enhancer or suppressor in addition to the fact that amino acids can add taste. For instance, the perceived saltiness of the salts added during the manufacturing process could be influenced by glutamic and aspartic acid formation. In that case, the taste is mainly due to the presence of non-volatile peptides and amino acids coming from biochemical processes that take place during

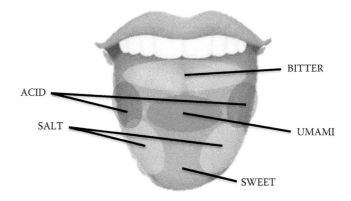

FIGURE 2.9 Main tastes and the location of receptors.

the curing and that can vary according to the variable parameters of the raw materials, such as pH, water activity, enzymatic activity, captured salt amount and process time. As a result, the length of the drying time will influence the amount and type of the amino acids at the end of the process. It has been shown that lysine and tyrosine are related to the characteristic aged taste of Parma hams (Careri et al. 1993), while in Serrano hams, cured and pig tastes are related to the presence of aspartic acid, arginine, glutamic acid, histidine, leucine, isoleucine, valine, lysine, tryptophan, methionine and phenylalanine (Flores 1997). Even then, it is also shown that a high level of proteolysis (more than 30%) can lead to off flavours, like increases in metallic or bitter tastes (Pérez-Santaescolástica et al. 2018a).

Since each amino acid is related to a specific taste, the overall taste will be a function of the proteolysis intensity, that is, the amount of each amino acid and its threshold. A summary of the relationships between specific tastes and the responsible amino acid is shown in Table 2.4. A sweet taste is created by the presence of alanine, serine, proline, glycine and hydroxylproline; bitterness is derived from phenylalanine, tryptophan, arginine, methionine, valine, leucine and isoleucine. The configuration of the amino acid structures is also an influence: series D is mainly sweet and series L mainly bitter. We know that L-tryptophan and L-tyrosine are the

TABLE 2.4

Taste, Threshold and Content of Amino Acids Presented in Dry-cured Hams

Amino acid	[a]Taste	[a]Threshold (mM)	[b,c,d]Amount (mg/ 100 DM)
Phenylalanine	Bitter	58	375–400
Leucine	Bitter	12	560–620
Isoleucine	Bitter	11	340–370
Methionine	Bitter	5	190–220
Valine	Bitter	21	380–400
Tyrosine	Bitter	5	180–200
Proline	Sweet	26	275–290
Arginine	Bitter	75	365–400
Alanine	Sweet	8	400–420
Threonine	Sweet	40	210–225
Histidine	Bitter	48	95–105
Glycine	Acid-Sweet	30	180–200
Serine	Sweet	30	175–205
Glutamic acid	Umami	3	430–460
Aspartic acid	Umami	4	165–185
Taurine	-	>100	80–100
Cysteine	-	2	270–440
Lysine	Bitter	85	250–270

[a] Schlichtherle-Cerny and Grosch (1998)
[b-d] Pérez-Santaescolástica et al. (2018a; 2018b; 2019a)

bitterest amino acids, while D-tryptophan is the sweetest (Fennema 1993). Glutamic acid and aspartic acid also collaborate with histidine and asparagine to create perceptions of sourness, and glutamic acid influences the umami taste and can cover up the perception of bitterness.

On the other hand, bitter taste can be created by the presence of peptides that contain one or two hydrophobic amino acids (phenylalanine, leucine, isoleucine, valine or tryptophan), while the complex glutamic-tyrosine is related to cured taste and valine-glutamic and glycine-glutamic with sour taste (Sentandreu et al. 2003).

2.6.3 Odour Development

Much research has focused on the relationship between volatile compounds and the odour of dry-cured ham, and has identifyied more than 200 compounds, but not all have the same impact. Each one is perceptible in a different amount depending on its threshold (Pérez-Santaescolástica et al. 2019a; Narváez-Rivas et al. 2012; García-González et al. 2008). Some factors derived from compounds belong to the raw material and others formed during the curing process affect odour development. Many compounds ingested in the diet can be incorporated into animal bodies and can remain in the raw material, including toluene, xylene isomers, terpenes and some ramified hydrocarbons. However, one-fifth of odour compounds come from amino acids and free fatty acids. During salting and post-salting, volatile compounds derive from lipid oxidation, but when curing advances, the degradation of amino acids increases. Long curing processes lead to hams with high odour intensity as a consequence of a large number of volatile compounds formed by lipid oxidation as well as by amino acid degradation (Buscailhon et al. 1993; 1994; Coutron-Gambotti et al. 1999).

In general, aldehydes are one of the most important families contribuiting to odour, due to their low threshold. Linear aldehydes derive from unsaturated fatty acid oxidation and contribute either strong and irritating odours (short-chain linear aldehydes) or citric notes (long-chain linear aldehydes) (Théron et al. 2010). Among linear aldehydes, the most important compound in hams is hexanal; its presence in very small concentrations has a high effect on aroma, and it provides the ham with its characteristic grassy odour. The contribution of octanal is important for aroma development due to the pleasant notes that it presents and for its low threshold. In contrast, saturated aldehydes such as nonanal, 2,4-decadienal and 2-pentilfuran contribute unpleasant notes. Pasture-fed pigs are characterized by high contents of oleic acid, whose oxidation creates high amounts of octanal and nonanal, contributing greasy, oily, woody and nutty notes, whereas hams from pigs that have been given feed have large amounts of hexanal and unsaturated aldehydes from linoleic acid oxidation (2,4-nonadienal and 2,4-decadienal), which contribute unpleasant notes described as rancid or fry-like. It can include other aldehydes such as cyclic, ramified and sulphurous, and sulphur compounds, which are formed by oxidative deamination-decarboxylation of amino acids by Strecker degradation (López et al. 1992).

Some characteristic odours of ham are due to compounds like 1-octen-ol and 1-penten-3-ol, but come also from lipid oxidation; especially important is the presence of 1-octen-3-ol due to its characteristic mushroom-like odour. Alcohols contribute greatly with peculiar odours and are mainly derived by lipid oxidation (Rivas-Cañedo et al. 2011), as do ketones, which come from lipid autoxidation and fermentation by microorganisms (Belitz et al. 1997). For instance, 2-heptanone, 2-decanoate and 2-undecanone have fruity notes, while 2-dodecanone contributes an oil-like odour. On the other hand, methyl-ketones are formed by decarboxylation of the secondary products of hydroperoxide decomposition. Also, during lipid oxidation, enzymatic and non-enzymatic esterification of free fatty acids and alcohols lead to esters compounds, which contribute to appreciate notes. Esters appear in high amounts in hams which do not incorporated nitrites and nitrates, like Parma hams, because these compounds inhibit lipid oxidation.

Other compounds that influence the odour of dry-cured ham are furans, responsible for cooked meat odour (Shahidi et al. 1986); nitrogenous compounds, especially pyrazines formed by Maillard and Strecker pathways, due to its aromatic potential; and terpenes, which are obtained from feed and deposited in fatty tissues (Sabio et al. 1998). Lactones with a threshold between 0.007 and 0.1 ppm, created by hydroxy acids, contribute a typical meat odour, as well as with fruity, fatty, oily and butter notes. In contrast, there are families of volatile compounds that have not much influence on the final odour due to their high threshold, such as hydrocarbons. In spite of the high threshold that is observed in saturated linear hydrocarbons, the threshold is reduced in the presence of unsaturation. Their odour contribution varies greatly, including pine odour (pinene), woodiness, lemon (limonene), etc.

As a summary, in Table 2.5 are listed some of the most common compounds that have been shown to have an important influence on the final odour and could influence consumer acceptance.

2.6.4 FLAVOUR

Flavour refers to the overall perceived sensation when the product is ingested. This sensorial reaction is due to the union of odour, taste and texture, which are developed by the combination of volatile and non-volatile compounds and their interactions. Structural components such as fat and proteins are implicated and determine the consistency of the product and its behaviour during mastication, containing sapid and aromatic compounds. Since proteolysis and lipolysis are the main origins of flavour compounds, and they are reactions that are widely connected with the total process time plus the type, number and amount of each flavour compound, they will be affected by the processing time of each type of ham. In this regard, we see that the longer the ripening, the higher the flavour intensity that is developed (Buscailhon et al. 1994).

Taste contributes sensorial characteristics and, after mastication, this sensorial impression is refined thanks to flavour perceptions created by the retro oral cavity and nasal fossa. Mastication, together with a slight increase of temperature in the mouth, favours the liberation of aromatic compounds in specific amounts and adequate velocity, promoting the production of salivary secretions and contributing to the odour and taste arrival at specific receptors. IMF also contributes to juiciness, due to its action as a lubricant when it comes in contact with saliva.

TABLE 2.5

Volatile Compounds with Demonstrated Relevance on Dry-cured Ham Odour and their Biochemical Origin Sorted by Main Family Categories

Compound	[a],[b],[c]Sensory description	[c],[d]Most probable origin
Aliphatic hydrocarbons		
Hexane	Green-mould, spicy	Lipolysis and lipid oxidation
Heptane	Alkane, sweet	Lipolysis and lipid oxidation
Octane	Alkane, sweet	Lipolysis and lipid oxidation
Branched alkanes	Alkane	Lipolysis and lipid oxidation
Aromatic hydrocarbons		
Styrene	Medicinal	Raw material
o-Xylene	Sweet-fruit candy	Raw material
(*p*- or *m*-)Xylene	Smoked-phenolic	Raw material
α-Pinene	Sharp, pine	Raw material
Limonene	Citric, fresh	Raw material
Methyl benzene	Plastic, glue, strong	Raw material
Ethyl benzene	Dry, glue, unpleasant	Raw material
Furans		
2-ethylfuran	Sweet	Lipolysis and lipid oxidation
2-Pentylfuran	Green fruity, ham-like	Lipolysis and lipid oxidation
Aldehydes		
2-Methylbutanal	Green	Proteolysis and secondary reactions
3-Methylbutanal	Cheesy-green, acorn, fruity, cheesy, salty	Proteolysis and secondary reactions
2-Methylpropanal	Green	Proteolysis and secondary reactions
Pentanal	Pungent	Lipolysis and lipid oxidation
Hexanal	Green-grassy, fatty	Lipolysis and lipid oxidation
Heptanal	Fatty, greasy, ham-like	Lipolysis and lipid oxidation
2-Heptenal	Green, fatty, fruity, almond	Lipolysis and lipid oxidation
Octanal	Meat-like Green-fresh	Lipolysis and lipid oxidation
2-Octenal	Leaves, pungent, fatty, fruity	Lipolysis and lipid oxidation
Nonanal	Rancid, fatty, green	Lipolysis and lipid oxidation
2-Nonenal	Fatty, waxy	Lipolysis and lipid oxidation
Decanal	Citrus, waxy, soapy	Lipolysis and lipid oxidation
2,4-Decadienal	Fatty, rancid	Lipolysis and lipid oxidation
Benzaldehyde	Bitter almond, penetrating	Proteolysis and secondary reactions
Alcohols		
3-Methyl-1-butanol	Woody, acorn, penetrating green	Proteolysis and secondary reactions
2-Methyl-3-buten-2-ol	Earthy	Lipolysis and lipid oxidation

(Continued)

TABLE 2.5 (Continued)

Compound	[a],[b],[c]Sensory description	[c],[d]Most probable origin
2-Methyl propanol	Wine, penetrating	Proteolysis and secondary reactions
Ethanol	Lipid oxidation	Lipolysis and lipid oxidation
2-propanol	Alcoholic, dry, buttery taste	Lipolysis and lipid oxidation
1-butanol	Fruity, medicinal	Lipolysis and lipid oxidation
2-butanol	Winey	Lipolysis and lipid oxidation
1-pentanol	Pungent, strong, balsamic	Lipolysis and lipid oxidation
1-hexanol	Fruity, green, resinous	Lipolysis and lipid oxidation
2-Heptanol	Oily, sweet	Lipolysis and lipid oxidation
1-octanol	Fatty, sharp, mushroom	Lipolysis and lipid oxidation
1-Octen-3-ol	Mushroom-like, earthy, dusty	Lipolysis and lipid oxidation
Ketones		
2-Propanone	Fruity, apple, pear	Lipolysis and lipid oxidation
2-butanone	Buttery, ethereal	Lipolysis and lipid oxidation
2,3-Butanedione	Buttery, vanilla/caramel-like	Proteolysis and secondary reactions
2-heptanone	Spicy, acorn, blue cheese, soapy, fruity	Lipolysis and lipid oxidation
2-octanone	Spicy, acorn, blue cheese, soapy, fruity	Lipolysis and lipid oxidation
Octen-3-one	Spicy, mushroom, dirty	-
2-nonanone	Floral, fruity, blue cheese, hot milk	Lipolysis and lipid oxidation
Esters		
Butyl acetate	Pear, fruity	Microbial
Methyl-2-methyl-butanoate	Floral	Microbial
Ethyl butanoate	Apple	Microbial
Ethyl 2-methylbutanoate	Fruity-strawberry	Microbial
Carboxylic Acids		
Acetic Acid	Vinegar	Microbial
Propanoic acid	Sweat, acid, foot-like	Lipolysis and lipid oxidation
Butanoic acid	Cheesy, rancid	Lipolysis and lipid oxidation
Pentanoic acid	Meaty, roasted, spoiled ham	Lipolysis and lipid oxidation
Hexanoic acid	Fatty, cheese, sweaty	Lipolysis and lipid oxidation
2-Methyl propanoic	Iron, fishy	Proteolysis and secondary reactions
Nitrogen compounds		
Methylpyrazine	Nutty	Proteolysis and secondary reactions
2,6-Dimethylpyrazine	Toasted nuts	Proteolysis and secondary reactions
Sulphur compounds		
Dimethyl trisulphide	Cabbage, sulphur	Proteolysis and secondary reactions

TABLE 2.5 (Continued)

Compound	[a,b,c]Sensory description	[c,d]Most probable origin
Dimethyl disulphide	Dirty socks, cauliflower	Microbial

Notes
[a] Flores et al. (1997)
[b] Toldrá (2002)
[c] Pérez-Santaescolástica et al. (2019a)
[d] Flores (2018)

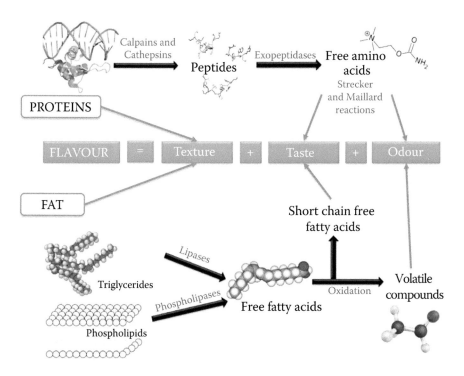

FIGURE 2.10 Flavour development in dry-cured hams.

In conclusion, flavour is the result of the contribution of some chemical compounds: volatile compounds from the degradation of components of raw meat and non-volatile substances such as amino acids, peptides or nucleotides, and other compounds that enhance tastes from other molecules (Figure 2.10). The origin of all these compounds is a global mechanism that includes lipolysis, proteolysis, and Maillard and Strecker reactions that differ among products according to several factors like genetic characteristics, animal age, level of muscular proteolytic and lipolytic enzymes present in raw meat, the free fatty acid profile, and antioxidants from feed contained in muscular tissues.

2.7 POSSIBLE DEFECTS TO COMPROMISE FINAL QUALITY

The concept of quality is subjective and has changed through the ages due to varying consumer preferences and social trends. Likewise, consumer requirements are different from country to country. For instance, Spanish people focus their attention on fat content and visual aspects such as colour; by contrast, in Norway consumers pay more attention to salt levels and curing time (Hersleth et al. 2011). Even with the market differences, several defects can arise that will result in consumers everywhere rejecting the product.

The treatment of the living animals and/or the raw materials is key to preventing many problems that are a detriment to sensorial quality. Sometimes notes from the raw material can be detected, especially sexual or faecal odours in hams produced from non-castrated males. Boar taint is due to the presence of the androsterone hormone and can be increased by the presence of indole and skatole from tryptophan. Similarly, the method and timing of the processing can transfer different odours, such as spiciness due to long curing, or a humid odour due to moulds or bacterial population. It is also possible for final products to lack the appropriate odour when the overall process is not long enough.

During rearing, transportation to slaughtering and slaughtering, the animal can receive blows that result in bruises or surface veining. Poor treatment immediately before slaughter can result in the expulsion of synovial liquid and, during ripening, it can create a bad odour similar to cement. Fatty and connective tissues can present blood spots caused by electric stunning. Also, rested blood in tissues could oxidize and result in green, ochre or brown spots. The excessive oxidation of fat or muscles also generates darker colouring, which diminishes consumer acceptance; it has been shown that the appearance, colour in particular, influences consumer acceptance to a great extent, not only in dry-cured ham but also in a wide range of foods (Delwiche 2004).

Additionally, an incorrect outliner (deficient skinning) can result in problems like higher microbial growth and/or a reduction in the velocity of salt diffusion. Most off odours are generated by microorganisms that contaminate the product, usually between slaughter and the post-salting stage, and provoke alteration. Putrefaction odour could appear due to the bacteria growth, which generates compounds such as ammonia, methanethiol and hydrogen sulphide, accompanied by degradation of the muscle pigments and, consequently, the appearance of green or grey spots. Usually this problem is a consequence of bad hygiene and contamination, such as bad refrigeration or non-homogeneous diffusion of salt. Other unpleasant odours that occur in the presence of bacteria or moulds include potato or earth odour due to pseudomonas, or a phenol-like aroma due to the action of penicilliums. The penetration of microorganisms can create gas in the veins, provoking their inflammation. The presence of air inside the veins, in turn, can trigger mould growth with a resulting anomalous appearance and wrong flavour.

Girón et al. (2015) proposed the use of potentiometric equipment based on a multi-electrode formed by three different metals, Ni, Ag and Cu, as an effective method to detect altered hams instead of the traditional method based on tasting panels. This avoids mistakes due to the fatigue of tasters. Moreover, the use of feeds with a large amount of unsaturated fatty acids, which are susceptible to oxidization, can provoke excessive oxidation and the appearance of off flavours, commonly a rancid odour. Excessive lipid oxidation is connected with preservation problems in food and, in the extreme, can provoke neuromyopathic diseases. To prevent this, it is common to use antioxidants. The most used antioxidant in hams is tocopherol, but it has been shown that it is highly sensitive to light and oxygen, among other external factors. This, together with its insolubility in water, limits the efficacy and efficiency of its application (Aresta et al. 2013). For that reason, and due to increased awareness about environmental problems and food safety, manufacturers had to find alternatives, and a recent study has demonstrated that the application of α-tocopherol-chitosan nanoparticles with chitosan/montmorillonite film enhances product stability and extends the oxidative protection (Yan et al. 2019). Chitosan presents an antibacterial effect and good film-forming properties but its mechanical properties are poor; as a result, the addition of montmorillonite (a natural mineral) could improve the mechanical limits of chitosan.

On the other hand, some practices came from burns, thus, it can be cited the addition of nitrite in the presence of oxygen, salt crystallization or freezing without protection. This burn has a white appearance and makes the absorption of salt difficult because the area is dryer. If the freezing and thawing happen slowly, the resulting ice crystals can break the muscle fibres. Furthermore, if the drying is very slow, a slime can form on the surface. In addition to that, an acid taste is possible when lactic bacteria are present in the product, and a metallic taste could develop in hams with a short curing period or due to an intense abrasion by a metallic material. The enzymatic proteolysis, as well as some microorganism activities, release tyrosine, which precipitates, producing crystals during the curing time as the result of water reduction combined with its high insolubility. Sometimes, in spite of crystalization, tyrosine forms a white film, just as creatinine forms in vacuum packed hams or those from a modified atmosphere.

An important matter to consider is texture. It is essential to prevent textures that are too soft, pasty, superficially crusted and too fibrous, since consumers reject these products (Morales et al. 2008; 2013). It is possible the appearance of pastiness or textures unusually soft that entails technological problems to butchers in hams for slice commercialisation (Arnau 1991). This defect is associated with excessive proteolysis, which causes the breakdown of tissue structures (Wu et al. 2014) probably due to the salt reduction, which tends to enhance enzymatic activity (Armenteros et al. 2012). Some issues are caused by intense proteolysis, for instance, processing conditions in which increases of temperature, as well as low levels of salt or short salting time, could favour the enzyme activity that causes protein breakdown (Martín et al., 1997). Also, it can be caused by greater proteolytic potentials of the raw meat, a pH that is too high or too low, leading to a sticky sensation of touch and elevated viscosity in the mouth when the product contacts the saliva.

In addition to defective texture, excessive production of peptides and amino acids can be associated with the development of bad tastes in the final product. Salty and bitter tastes are derived from anomalous proteolysis when salt is not well penetrated or when the protein breakdown generates peptides with bitter properties. To assure an adequate salt diffusion and appropriate water content, analytical technologies such as irradiation, magnetic resonance imaging, infrared spectroscopy and ultrasounds allow the monitoring salt penetration and predict water activity, allowing a better development of the process (Pérez-Santaescolástica et al. 2019b). However, from a quality point of view, some of these treatments, as well as others like thermal treatments, high-pressure applications or vacuum packing, can enhance the salty sensation or cause organoleptic modifications. On the other hand, a hard texture can result if the drying takes place at an excessive temperature while, at the same time, relative humidity is low and air velocity is insufficient. These conditions are associated, also, with too much fat exudation and, in extreme cases, mites can penetrate and muscles fissures can appear.

In too much relative humidity, moulds also can grow resulting in the development of spots of different colours (black, white, green, red or yellow). Other microorganisms that can generate spots are *Pseudomona libanensis*, which causes the development of blue colouring in the leg, and *Pseudomonas fluorescens* and *Clostridium herbarum* provoke black spots. Mycotoxins could be developed by the actions of fungi, especially from the genus *Aspergillus* (Udomkun et al. 2017). Aflatoxin B1 is considered the most common, with a significant toxic effect. Similarly, products with a large content of salt, like dry-cured ham, can contain high amounts of ochratoxin A, the production of which is due to *Aspergillus waterdijkiae* and linked to high sanitary risk (Vipotnik et al. 2017). There are some strategies to combat toxigenic fungi; chemical procedures, such as the use of NaCl to inhibit growth (Schmidt-Heydt et al. 2011), and biological techniques such as the use of moulds (Andrade et al. 2014).

There is also a traditional defect called coxofemoral coquera, which is characterized by the formation of cracks inside the muscular tissues during drying, resulting in an unstructured tissue, and a brown and malodorous slime along the cracks. The appearance is very dry due to intense drying, and in some cases, there is a development of mites. In addition to mites, insects can infest the ham, and the larvae can cause intestinal problems if they are ingested, due to their resistance to acidity. This is the case with Acari, which can provoke allergic symptoms. Ageing time has a significant effect on these infestations; it has been demonstrated that in hams with fewer than three months of ripening has no or very low contamination, while ageing up to five months entails higher risk. Traditionally, methyl bromide was the main compound used to control pests, due to its wide spectrum and fast action, but 1992 it was included in the list of substances that contribute to ozone depletion and significant reduction in its use by 2005 was proposed (Osteen 2003). Since then, some researchers have focused on alternatives. In this context, phosphine, sulfuryl fluoride, carbon dioxide and ozone were the objects of studies, which concluded that the use of the first three substances was not viable (Sekhon et al. 2010a; 2010b; Schilling et al. 2010). Although phosphine was proposed as the most suitable alternative due to its high effectivity, the lack of residues, the low cost and the ease of application (Fields and White 2002). As well, some physical methods have been proposed to control microbial growth. Among these methods are: Gamma irradiation (Jin et al. 2012), electron beam radiation (Kong et al. 2017), high pressure (Rubio-Celorio et al. 2016), ultraviolet C irradiation (Lah et al. 2012), controlled atmosphere (Sánchez-Molinero et al. 2010), etc. Even though some of them are not suitable due to excessive cost.

2.8 TYPES OF DRY-CURED HAM

Nowadays dry-cured ham is elaborated in several countries around the world, but each has unique characteristics due to the influence of factors associated with the raw meat as well as the conditions of the technical process, which varies among producers.

As noted at the beginning, Europe is the main producer of dry-cured ham. For that reason, we will separate the principal types of European high-quality hams from other elaborations whose social impact is lower.

2.8.1 PRINCIPAL EUROPEAN HAMS

In Europe, there are some quality standards for protecting and creating ad-value at the same time that promoting these high-quality products. There are three standards: Protective Denomination Origin (PDO), Protective Geographical Indication (PGI) and Traditional Speciality Guaranteed (TSG). The purpose of PDO is to protect a product characterizing for a differentiating quality linked to a specific geographical localization. Product quality must be a consequence of its production localization in where are done the raw material production, its transformation and the entire process of elaboration. In contrast, the PGI standard establishes that it is enough one of the technological stages is realized into the designated area meanwhile TSG involves others product that not keep to all the requirements for PDO and PGI but economic reasons and the process localization is worthy of legal protection.

2.8.1.1 Spanish Quality Hams

Three types of ham are produced in Spain: Curado, Serrano and Iberian. Curado and Serrano are hams from a commercial breed reared in an intensive system. While the first one is cured for a minimum of 150 days, the second keeps curing for almost 210 days. The most common presentation in the market used to be with the bone, but now there are also boneless hams and vacuum packaged slices. Serrano ham includes the first ham protected by a Quality Standard, the *Teruel* ham. This ham is produced from a white breed in the northeast region of Spain of the same name. Pigs are fed until 120–130 kg of live weight, obtaining hams with weights below 11.5 kg which are cured for 12–14 months at 800 m of altitude. *Trevelez* hams are also included into Serrano type and constitute one of the first hams to which historical references can be found, specifically in 1862 when Queen Isabel II gave it the honour of carrying the royal crown stamp. These hams, of 8 kg approximately, are protected by a PGI category and are produced in Granada at more than 1200 m of altitude.

On the other hand, the Iberian ham is the most representative ham in Spain comprising 8% of total hams. The pigs come from the Iberian breed or are an Iberian cross with Duroc, a breed characterized by its high lipogenic potential that leads to pieces with great content of not only subcutaneous fat but also IMF. These significant fat contents complicate the salt diffusion, so the post-salting stage needs to be extended to at least 80–100 days. Animals are reared in an extensive free-range of tree-covered meadows where they can move and feed; they are slaughtered at 160 kg and cured more than 600 days (BOE 2014). The slaughter weight and type of rearing result in muscles that are intensely coloured due to the high content of myoglobin. Moreover, the high content of antioxidant prevents excessive oxidation of fat, and the fatty acid profile presents a large amount content of oleic acid, giving slices a bright surface and making the fat very soft. Finally, texture, odour and taste differentiate this type of ham from others. In this regard, there are four PDO that require a minimum of 75% of Iberian breed purity; they are Ham of Guijuelo, Ham of Huelva or Jabugo Ham, Ham of the Extremadura pasture and Ham of Pedroches. Also, one PGI is produced, called Serón.

2.8.1.2   Italian Quality Hams

Italian ham manufacture is concentrated in the north of the country. There are two PDO (Parma and San Daniellehams), production of which has real relevance due to the other six protected varieties (Prosciutto di Modena, Prosciutto di Véneto Berico-Euganeo, Prosciutto di Carpegna, Prosciutto Toscano and Prosciutto di Valle de Aosta or Jambon de Bosses), which are mainly consumed in Italy. The processing of Parma ham is the most studied among the European dry-cured hams, and the ham is characterized by its less consistent texture compared with other hams due to the high humidity. Hams come from Landrace, Large White or Duroc pigs, slaughtered at 160 kg at a minimum of 9 months of age and, under intense monitoring, cured for 10–12 months based on the leg weight. Parma hams have two peculiarities. The first is the lack of any additives in the salting stage; the only compound that is added is salt. The second is the inclusion in the process of an extra step, the so-called greasing stage, which consists of covering the pieces with a mix of pork fat, pepper and, sometimes, rice flour. This layer prevents fast drying in external parts and causes extra water loss. The final product is characterized by its taste, which is lightly salty in comparison with other hams. A similar production process is used for the San Daniel variety, even though the curing time is shorter at almost 9 months. Salt is applied one time per week during one month, accompanied by a manual massage. Like Parma ham, San Daniel is subjected to a greasing stage and, after that, pieces are cured for 12 months, which is in some cases extended up to 2 years. Compared to Parma ham, San Daniel hams are saltier and drier.

On the other hand, the processing of Prosciutto Toscano is unique because the salting mixture is made with a combination of sea salt, pepper, garlic and aromatic herbs. Moreover, after a week of salting, the surface of the hams is spread with a mix of flour, salt, pepper, butter and natural odorants from vegetables. The result is a ham with a long curing time and a characteristic lightly spicy taste and special herbal odour, together with a high intensity of white and red colours. Also, the texture of Toscano ham is considered more hard and fibrous due to the higher content of salt than the Parma variety. Similarly, Prosciutto di Valle d´Aosta or Jambon de Bosses use a salting a mix of sea salt and spices like garlic, salvia, rosemary and other local aromatic herbs, which provide the product with an aromatic and rustic flavour. Finally, Prosciutto di Norcia, the only PGI from Italy, has a production process similar to the *Toscano* and Valle d´Aosta varieties. For salting, it uses a mix of some spices, especially pepper, and hams are subjected to drying ripening after being spread with butter and pepper.

In general, Italian hams, in contrast with hams from other countries, present lower rancid odours, probably due to the fact that ripening time is appreciably shorter (Laureati et al. 2014).

2.8.1.3   French Quality Hams

Dry-cured ham production in France is located in the south of the country, while in the north, producers are focused on cooked or smoked hams. The most popular ham among all produced in France is Jambon de Bayonne, a unique type that has quality recognition. It is a ham with PGI a designation produced from pig breeds that include Large White, Pietrain or the recovered breed Basque, slaughtered at 120 kg. This kind of ham follows a typical production process including the addition of nitrites in the salting stage, but it differs from others in its shorter curing time. Nevertheless, today it is required to cure the hams for almost 9 months at a minimum. In consequence of that, the final product presents higher moisture levels and relatively lower pigmentation content.

In the southwest of France can be found Jambon Noir de Bigorre, included in a DOP in 2017 (CE/2017/1554). Rearing is carried out in a semi-extensive system where natural food sources are at the disposal of animals, which are fed for a minimum of 12 months and slaughtered at weights around 160 kg. Unlike Bayonne style, Noir de Bigorre hams come from legs with up to 5 cm of subcutaneous fat and a good amount of IMF and antioxidant, which necessitates longer drying times, usually above 18 months. Large differences exist between both French DOP hams, derived principally from the breed and ripening time. The high IMF content results in hams with high juiciness and an intense aromatic profile. In this regard, it could be said that Jambon Noir de Bigorre production methods are closer to the Spanish type than to the traditional French products.

2.8.1.4   Portuguese Quality Hams

In Portugal, hams are one of the principal traditional meat products. The technological process is characterized by shorter curing times and higher salt levels. Under the PDO standard, Presunto de Barrancos and Presunto do Alentejo are found, while PGI hams include Presuntos de Campo Maior y Elvas, Barroso, Santana de Serra and Vinhais. With the exception of Barroso and Vinhais, the hams come from the breed called Alentejana. Barroso and Vinhais use an antique breed called Bísara, a local protected breed that was on the verge of extinction.

In the case of Vinhais type, pepper and local oil are spread on the external parts that are not covered with skin. After that, hams are lightly smoked before ripening for a maximum of 12 months.

2.8.2   OTHERS TYPES OF HAMS

Today ham manufacture extends throughout the world, and its production highlights differences among countries that lead to several styles of hams. Besides the hams already discussed, other European counties manufacture less-popular dry-cured hams. For instance, Norway produces a wide range of dry-cured meats from different animals. From pigs come Spekeskine ham, including bulk dry-cured ham production and the good elaboration. Hams from bulk production use weights of 9–10 kg with a subcutaneous fat layer not more than 10 mm wide. These characteristics, together with a significant enzymatic activity, allow for very short production times. The process consists of a massive salting for 2–4 weeks, followed by a short resting time. In some cases, legs are subjected to a smoking process, and after that are dried for around 3 months. On the other hand, for high quality hams, legs from old and heavy pigs are cured for less time than bulk hams but the resting stage is extended up to 4 months. To protect hams from excessive drying, hams are covered with a layer formed by a mixture of rice flour and fat. The entire process entails between 7 and 24 months, developing an intense flavour and odour. Among the German productions, the most popular dry-cured ham is Schwarzwälder Schinken, which is produced in the Black Forest in the southwest and is governed by a quality standard recognition. In this procedure, a mix of spices, together with salt, are used for the salting stage, which takes 2 weeks; after that hams rest for 2–7 weeks and then pieces are cold smoked for 2–3 weeks and finalized with a ripening step for several weeks.

The most popular ham producers outside Europe are likely the USA and China. Ham produced in the USA is called Country Ham. Influenced by European methods, it has been produced since colonial times. At present, the greater part of American production is located in the Carolinas, Kentucky, Tennessee and Virginia. The curing salts use a mixture of salt, sugar (either white or brown) and paprika or pepper (red and/or black), in three additions 1 week apart; pieces are rotated at each application (Rentfrow et al. 2012). At the end of the process, legs are subjected to a smoking process at mild temperatures to preserve endogenous enzyme activity. The ripening time varies from producer to producer, but in any case, is less than 6 months, a very short time compared with the European hams. Less important is Mexican production because the native pork (Hairless Mexican) contains too much fat, and by the time it undergoes a devaluation (Cenobio 1993). This is the reason its production is minimal.

China produces two typical hams: Jinhua and Xuanwei. The first has a characteristic flavour due to the high concentration of carboxylic acids obtained from ripening at 37°C for 40 days (Zhang et al. 2006). This ham is produced in the same-name district in the province of Zhejiang and dates back to the Tang Dynasty. The entire curing process takes 8–15 months (Wu et al. 1959). Xuanwei hams have a strong flavour derived from intense lipolysis. Less popular are hams produced in

Japan, such as Japanese Lachschinken, which is quite sweet and has a lot of umami taste, and in Croatia, where it can be found in three PGI (Drniški, Dalmatinski and Krčki pršut) and one PDO (Istarski pršut). Istarski and Krčki pršut incorporate spices into the salting stage, whereas the other two types are smoked before drying.

2.9 CONCLUSION

This chapter reviewed the main technological aspects of the dry-cured ham production process and how they are associated with the meat's sensorial characteristics. It explained the importance of the main reactions that take place during the process and noted the possible consequences of unsuitable procedures. The wrong lipolysis and proteolysis development could result in a defective product with off flavours, bad taste or anomalous textures that can appear, as well as microbial contaminations. Considering that ham is an extremely important product becuase of its global popularity, unfavourable results entails economic losses due to consumer rejection. Even though ham technology has been studied for a long time, today the manufacture is still experimenting with and applying new technologies to improve the process.

REFERENCES

Andrade, M. J., Thorsen, L., Rodríguez, A., Córdoba, J. J., and Jespersen, L. 2014. Inhibition of ochratoxigenic moulds by Debaryomyces hansenii strains for biopreservation of dry-cured meat products. *International journal of food microbiology* 170: 70–77.

Antequera, M. T., and Martín, L. 2001. Reacciones químicas y bioquímicas que se desarrollan durante la maduración del jamón Ibérico. In *Tecnología del jamón Ibérico*, 293–322. Grupo Mundi-Prensa.

Antequera, T., Córdoba, J. J., Ruiz, J., Martín, L., García, C., Bermúdez, M. E., and Ventanas, J. 1993. Liberación de ácidos grasos durante la maduración del jamón ibérico. *Revista Española de Ciencia y Tecnología de Alimentos* 33: 197–208.

Aresta, A., Calvano, C. D., Trapani, A., Cellamare, S., Zambonin, C. G., and De Giglio, E. 2013. Development and analytical characterization of vitamin (s)-loaded chitosan nanoparticles for potential food packaging applications. *Journal of nanoparticle research* 15: 1592.

Armenteros, M., Aristoy, M. C., Barat, J. M., and Toldrá, F. 2012. Biochemical and sensory changes in dry-cured ham salted with partial replacements of NaCl by other chloride salts. *Meat science* 90: 361–367.

Armenteros, M., Aristoy, M. C., and Toldrá, F. 2009. Effect of sodium, potassium, calcium and magnesium chloride salts on porcine muscle proteases. *European food research and technology* 229: 93–98.

Armero, E., Flores, M., Toldrá, F. *et al.* 1999. Effects of pig sire type and sex on carcass traits, meat quality and sensory quality of dry-cured ham. *Journal of the science of food and agriculture* 79: 1147–1154.

Arnau, J. 1991. Aportaciones a la calidad tecnológica del jamón curado elaborado por procesos acelerados. PhD diss., Universitat Autonoma de Barcelona.

Arnau, J. 1993. Tecnología de elaboración del jamón curado. *Microbiologia Sem* 9: 3–9.

Arnau, J. 1998. Principales problemas tecnológicos en la elaboración del jamón curado. En: *El Jamón curado: Tecnología y análisis de consumo. Simposio Especial - 44th ICoMST.* ed I.R.T.A and Eurocarne: 71–86.

Arnau, J. 2007. Factores que afectan a la salazón del jamón curado. *Eurocarne: La revista internacional del sector cárnico* 160: 59–76.

Bañón, S., Gil, M. D., and Garrido, M. D. 2003. The effects of castration on the eating quality of dry-cured ham. *Meat science* 65: 1031–1037.

Barton-Gade, P. A. 1993. Effect of stunning on pork quality and welfare: Danish experience. In *Allen D. Leman Swine Conference*, 173–178.

Belfrage, P., Frederikson, G., Stralfors, P., and Thornqvist, H. 1984. *Adipose tissue lipases.* Elsevier Science Publisher.

Belitz, H. D., Grosch, W., and Schieberle, P. 1997. *Química de los alimentos.* Acribia.

Bermúdez, R., Franco, D., Carballo, J., and Lorenzo, J. M. 2014. Physicochemical changes during manufacture and final sensory characteristics of dry-cured Celta ham. Effect of muscle type. *Food control* 43: 263–269.

Blasco, A., Gou, P., Gispert, M. *et al.* 1994. Comparison of five types of pig crosses. I. Growth and carcass traits. *Livestock productionscience* 40: 171–178.

Blasco, J. G. 1998. Importancia del jamón a lo largo de la historia. In *El jamón curado: tecnología y análisis de consumo: Simposio especial.* 44th ICOMST, 112–124. Estrategias Alimentarias.

BOE. 1967. Order 2484/1967 which approves the Spanish Food Code.

BOE. 2014. *Order by which approves the Spanish quality standard for Iberian meat, dry-cured ham, dry-cured shoulder and dry-cured loin.*

Bou, R., Llauger, M., Arnau, J., and Fulladosa, E. 2018. Zinc-protoporphyrin content in commercial Parma hams is affected by proteolysis index and marbling. *Meat science* 139: 192–200.

Brooke, M. H., and Kaiser, K. K. 1970. Muscle fiber types: how many and what kind? *Archives of neurology* 23: 369–379.

Buscailhon, S., Berdagué, J. L., Bousset, J., Cornet, M., Gandemer, G., Touraille, C., and Monin, G. 1994. Relations between compositional traits and sensory qualities of French dry-cured ham. *Meat science* 37: 229–243.

Buscailhon, S., Berdagué, J. L., and Monin, G. 1993. Time-related changes in volatile compounds of lean tissue during processing of French dry-cured ham. *Journal of the science of food and agriculture* 63: 69–75.

Careri, M., Mangia, A., Barbieri, G., Bouoni, L., Virgili, R., and Parolari, G. 1993. Sensory property relationships to chemical data of Italian-type dry-cured ham. *Journal of food science* 58: 968–972.

Cava, R., Ventanas, J., Tejeda, J. F., Ruiz, J., and Antequera, T. 2000. Effect of free-range rearing and α-tocopherol and copper supplementation on fatty acid profiles and susceptibility to lipid oxidation of fresh meat from Iberian pigs. *Food chemistry* 68: 51–59.

Cenobio, S. L. 1993. Evaluación del comportamiento reproductivo de un lote de cerdas Pelón Mexicano en la etapa de lactancia en el altiplano. PhD diss., UNAM. México, DF.

Council Directive 90/496/CEE. 1990. Relating to labeling on properties nutritives of food products. *DOCE* 276: 40.

Coutron-Gambotti, C., Gandemer, G., Rousset, S., Maestrini, O., and Casabianca, F. 1999. Reducing salt content of dry-cured ham: effect on lipid composition and sensory attributes. *Food chemistry* 64: 13–19.

Cremer, D. R., and Eichner, K. 2000. The reaction kinetics for the formation of Strecker aldehydes in low moisture model systems and in plant powders. *Food chemistry* 71: 37–43.

Delwiche, J. 2004. The impact of perceptual interactions on perceived flavor. *Food quality and preference* 15: 137–146.

Executive Regulation (UE) 2017/1554 by which a name is registered in the Register of Protected Designations of Origin and Protected Geographical Indications [Jambon noir de Bigorre (DOP)].

Fennema, O. R. 1993. *Quimica de los alimentos*. Acribia.

Fields, P. G., and White, N. D. 2002. Alternatives to methyl bromide treatments for stored-product and quarantine insects. *Annual review of entomology* 47: 331–359.

Fleet, J. C., and Cashman, K. D. 2003. Magnesio. In Conocimientos actuales sobre nutrición. *Publicación Cientifico y Técnica* 592, 318–329.

Flores, J. 1997. Mediterranean vs northern European meat products. Processing technologies and main differences. *Food chemistry* 59: 505–510.

Flores, M. 2018. Understanding the implications of current health trends on the aroma of wet and dry cured meat products. *Meat science* 144: 53–61.

Gallo, L., Dalla Bona, M., Carraro, L., Cecchinato, A., Carnier, P., and Schiavon, S. 2016. Effect of progressive reduction in crude protein and lysine of heavy pigs diets on some technological properties of green hams destined for PDO dry-cured ham production. *Meat science* 121: 135–140.

García-Gil, N., Santos-Garcés, E., Muñoz, I., Fulladosa, E., Arnau, J., and Gou, P. 2012. Salting, drying and sensory quality of dry-cured hams subjected to different pre-salting treatments: skin trimming and pressing. *Meat science* 90: 386–392.

García-González, D. L., Tena, N., Aparicio-Ruiz, R., and Morales, M. T. 2008. Relationship between sensory attributes and volatile compounds qualifying dry-cured hams. *Meat science* 80: 315–325.

García-Rey, R. M., Garc a-Garrido, J. A., Quiles-Zafra, R., Tapiador, J., and De Castro, M. L. 2004. Relationship between pH before salting and dry-cured ham quality. *Meat science* 67: 625–632.

Girón, J., Gil-Sánchez, L., García-Breijo, E., Pagán, M. J., Barat, J. M., and Grau, R. 2015. Development of potentiometric equipment for the identification of altered dry-cured hams: a preliminary study. *Meat science* 106: 1–5.

González, C. B., and Ockerman, H. W. 2000. Dry-cured mediterranean hams: long process, slow changes and high quality: a review 1. *Journal of muscle foods* 11: 1–17.

Goodwin, R. N. 1994. Genetic parameters of pork quality traits. PhD diss., Iowa State University, Ames.

Gou, P., Guerrero, L., and Arnau, J. 1995. Sex and crossbreed effects on the characteristics of dry-cured ham. *Meat science* 40: 21–31.

Gray, J. I., and Pearson, A. M. 1987. Rancidity and warmed-over flavour. In *Advances in meat research*: USA.

Guerrero, L., Gou, P., & Arnau, J. 1999. The influence of meat pH on mechanical and sensory textural properties of dry-cured ham. *Meat science* 52: 267–273.

Hersleth, M., Lengard, V., Verbeke, W., Guerrero, L., and Næs, T. 2011. Consumers' ac-ceptance of innovations in dry-cured ham: impact of reduced salt content, prolonged aging time and new origin. *Food quality and preference* 22: 31–41.

Hidalgo, F. J., and Zamora, R. 2016. Amino acid degradations produced by lipid oxidation products. *Critical reviews in food science and nutrition* 56: 1242–1252.

Higgs, J. D. 2000. The changing nature of red meat: 20 years of improving nutritional quality. *Trends in food science and technology* 11: 85–95.

Jiménez-Colmenero, F., Ventanas, J., and Toldrá, F. 2009. El jamón curado en una nutrición saludable. In Proceedings of *V International Congress of Dry-cured Ham*, Aracena, Spain.

Jiménez-Colmenero, F., Ventanas, J., and Toldrá, F. 2010. Nutritional composition of dry-cured ham and its role in a healthy diet. *Meat science* 84: 585–593.

Jin, S. K., Kim, C. W., Chung, K. H., *et al.* 2012. Physicochemical and sensory properties of irradiated dry-cured ham. *Radiation physics and chemistry* 81: 208–215.

Jurado, Á., García, C., Timón, M. L., and Carrapiso, A. I. 2007. Effect of ripening time and rearing system on amino acid-related flavour compounds of Iberian ham. *Meat science* 75: 585–594.

Kerry, J. P., Kerry, J. F., and Ledward, D. (Eds.). 2002. *Meat processing: improving quality*. CRC Press.

Kong, Q., Yan, W., Yue, L., *et al.* 2017. Volatile compounds and odor traits of dry-cured ham (*Prosciutto crudo*) irradiated by electron beam and gamma rays. *Radiation pysics and chemistry* 130: 265–272.

Lah, E. F. C., Musa, R. N. A. R., and Ming, H. T. 2012. Effect of germicidal UV-C light (254 nm) on eggs and adult of house dustmites, *Dermatophagoides pteronyssinus* and *Dermatophagoides farinae* (Astigmata: Pyroglyhidae). *Asian Pacific journal of tropical biomedicine* 2: 679–683.

Larrea, V., Hernando, I., Quiles, A., Lluch, M. A., and Pérez-Munuera, I. 2006. Changes in proteins during Teruel dry-cured ham processing. *Meat science* 74: 586–593.

Laureati, M., Buratti, S., Giovanelli, G., Corazzin, M., Fiego, D. P. L., and Pagliarini, E. 2014. Characterization and differentiation of Italian Parma, San Daniele and Toscano dry-cured hams: a multi-disciplinary approach. *Meat science* 96: 288–294.

López, M. O., de la Hoz, L., Cambero, M. I., Gallardo, E., Reglero, G., and Ordóñez, J. A. 1992. Volatile compounds of dry hams from Iberian pigs. *Meat science* 31: 267–277.

Lorenzo, J. M. 2014. Changes on physico-chemical, textural, lipolysis and volatile compounds during the manufacture of dry-cured foal "cecina". *Meat science* 96: 256–263.

Lorenzo, J. M., Fontán, M. C. G., Franco, I., and Carballo, J. 2008. Biochemical characteristics of dry-cured lacon (a Spanish traditional meat product) throughout the manufacture, and sensorial properties of the final product. Effect of some additives. *Food control* 19: 1148–1158.

Martín, L. 1996. Influencia de las condiciones del procesado sobre los cambios madurativos en el jamón Ibérico. *PhD diss.*, Universidad de Extremadura.

Martín, L., Antequera, T., Ventanas, J., Ben tez-Donoso, R., and Córdoba, J. J. 2001. Free amino acids and other non-volatile compounds formed during processing of Iberian ham. *Meat science* 59: 363–368.

Martín, L., Córdoba, J. J., Antequera, T., Timón, M. L., and Ventanas, J. 1997. Effects of salt and temperature on proteolysis during ripening of Iberian ham. *Meat science* 49: 145–153.

Martuscelli, M., Pittia, P., Casamassima, L. M., Manetta, A. C., Lupieri, L., and Neri, L. 2009. Effect of intensity of smoking treatment on the free amino acids and biogenic amines occurrence in dry cured ham. *Food chemistry* 116: 955–962.

Moller, J. K., Adamsen, C. E., Catharino, R. R., Skibsted, L. H., and Eberlin, M. N. 2007. Mass spectrometric evidence for a zinc–porphyrin complex as the red pigment in dry-cured Iberian and Parma ham. *Meat science* 75: 203–210.

Monin, G., Marinova, P., Talmant, A., *et al.* 1997. Chemical and structural changes in dry-cured hams (Bayonne hams) during processing and effects of the dehairing technique. *Meat science* 47: 29–47.

Morales, R., Guerrero, L., Aguiar, A. P. S., Guàrdia, M. D., and Gou, P. 2013. Factors affecting dry-cured ham consumer acceptability. *Meat science* 95: 652–657.

Morales, R., Guerrero, L., Claret, A., Guàrdia, M. D., and Gou, P. 2008. Beliefs and attitudes of butchers and consumers towards dry-cured ham. *Meat science* 80: 1005–1012.

Motilva, M. J., Toldrá, F., and Flores, J. 1992. Assay of lipase and esterase activities in fresh pork meat and dry-cured ham. *Zeitschrift für Lebensmittel-Untersuchung und Forschung* 195: 446–450.

Nagaraj, N. S., and Santhanam, K. 2006. Effects of muscle proteases, endogenous protease inhibitors and myofibril fragmentation on postmortem aging of goat meat. *Journal of food biochemistry* 30: 269–291.

Narváez-Rivas, M., Gallardo, E., and León-Camacho, M. 2012. Analysis of volatile compounds from Iberian hams: a review.*Grasas y aceites* 63: 432–454.

Osteen, C. D. 2003. *Methyl bromide phaseout proceeds: users request exemptions* (No. 1490-2016 127176).

Pérez-Santaescolástica, C., Carballo, J., Fulladosa, E., Garcia-Perez, J. V., Benedito, J., and Lorenzo, J. M. 2018a. Effect of proteolysis index level on instrumental adhesiveness, free amino acids content and volatile compounds profile of dry-cured ham. *Food research international* 107: 559–566.

Pérez-Santaescolástica, C., Carballo, J., Fulladosa, E., José, V. G. P., Benedito, J., and Lorenzo, J. M. 2018b. Application of temperature and ultrasound as corrective measures to decrease the adhesiveness in dry-cured ham. Influence on free amino acid and volatile compound profile. *Food research international* 114: 140–150.

Pérez-Santaescolástica, C., Carballo, J., Fulladosa, E., et al.. 2019a. Influence of high-pressure processing at different temperatures on free amino acid and volatile compound profiles of dry-cured ham. *Food research international* 116: 49–56.

Pérez-Santaescolástica, C., Fraeye, I., Barba, F., et al. 2019b. Application of non-invasive technologies in dry-cured ham: an overview. *Trends in food science and technology* 86: 360–374.

Ranken, M. D. 2000. *Handbook of meat product technology*. Blackwell Science.

Rebollo, A. G., Cansado, A. O., Botejara, E. A. M., and Blanco, P. M. 1999. Influencia del consumo de jamón ibérico de bellota sobre el perfil lipídico aterogénico. *Solo Cerdo Ibérico* 2: 107–112.

Rentfrow, G., Chaplin, R., and Suman, S. P. 2012. Technology of dry-cured ham production: science enhancing art. *Animal frontiers* 2: 26–31.

Ripollés, S., Campagnol, P. C. B., Armenteros, M., Aristoy, M. C., and Toldrá, F. 2011. Influence of partial replacement of NaCl with KCl, CaCl$_2$ and MgCl$_2$ on lipolysis and lipid oxidation in dry-cured ham. *Meat science* 89: 58–64.

Rivas-Cañedo, A., Juez-Ojeda, C., Núñez, M., and Fernández-García, E. 2011. Effects of high pressure processing on the volatile compounds of sliced cooked pork shoulder during refrigerated storage. *Food chemistry* 124: 749–758.

Rodríguez Marín, P. 2018. La exportación de jamón curado en España. *Distribución y consumo* 2: 18–24.

Rubio-Celorio, M., Fulladosa, E., Garcia-Gil, N., and Bertram, H. C. 2016. Multiple spectroscopic approach to elucidate water distribution and water–protein interactions in dry-cured ham after high pressure processing. *Journal of food engineering* 169: 291–297.

Ruiz, J., Ventanas, J., Cava, R., Timón, M. L., and Garc a, C. 1998. Sensory characteristics of Iberian ham: influence of processing time and slice location. *Food research international* 31: 53–58.

Running, C. A., Craig, B. A., and Mattes, R. D. 2015. Oleogustus: the unique taste of fat. *Chemical senses* 40: 507–516.

Sabio, E., Vidal-Aragon, M. C., Bernalte, M. J., and Gata, J. L. 1998. Volatile compounds present in six types of dry-cured ham from south European countries. *Food chemistry* 61: 493–503.

Sánchez, F. 2003. Modificaciones tecnológicas para mejorar la seguridad y calidad del jamón curado. PhD diss., Universitat de Girona.

Sánchez-Molinero, F., García-Regueiro, J. A., and Arnau, J. 2010. Processing of dry-cured ham in a reduced-oxygen atmosphere: effects on physicochemical and microbiological parameters and mite growth. *Meat science* 84: 400–408.

Santos, C., Hoz, L., Cambero, M. I., Cabeza, M. C., and Ordóñez, J. A. 2008. Enrichment of dry-cured ham with α-linolenic acid and α-tocopherol by the use of linseed oil and α-tocopheryl acetate in pig diets. *Meat science* 80: 668–674.

Schilling, M. W., Phillips, T. W., Aikins, J. M., Hasan, M. M., Sekhon, R. K., and Mikel, W. B. 2010. Research update: evaluating potential methyl bromide alternatives for their efficacy against pest infestations common to dry cured ham. In *National Country Ham Association Annual Meeting*. Asheville, NC: A.J. Edwards & Associates INC.

Schlichtherle-Cerny, H., and Grosch, W. 1998. Evaluation of taste compounds of stewed beef juice. *Zeitschrift für Lebensmitteluntersuchung und-Forschung A* 207: 369–376.

Schmidt-Heydt, M., Graf, E., Batzler, J., and Geisen, R. 2011. The application of transcriptomics to understand the ecological reasons of ochratoxin a biosynthesis by Penicillium nordicum on sodium chloride rich dry cured foods. *Trends in food science and technology* 22: S39–S48.

Sekhon, R. K., Schilling, M. W., Phillips, T. W., Aikins, M. J., Hasan, M. M., and Mikel, W. B. 2010a. Sulfuryl fluoride fumigation effects on the safety, volatile composition, and sensory quality of dry cured ham. *Meat science* 84: 505–511.

Sekhon, R. K., Schilling, M. W., Phillips, T. W., *et al.* 2010b. Effects of carbon dioxide and ozone treatments on the volatile composition and sensory quality of dry-cured ham. *Journal of food science* 75: C452–C458.

Sentandreu, M. A., Stoeva, S., Aristoy, M. C., Laib, K., Voelter, W., and Toldrá, F. 2003. Identification of small peptides generated in Spanish dry-cured ham. *Journal of food science* 68: 64–69.

Shahidi, F., Rubin, L. J., D'Souza, L. A., Teranishi, R., and Buttery, R. G. 1986. Meat flavor volatiles: a review of the composition, techniques of analysis, and sensory evaluation.*Critical reviews in food science and nutrition* 24: 141–243.

Tabilo, G., Flores, M., Fiszman, S. M., and Toldra, F. 1999. Postmortem meat quality and sex affect textural properties and protein breakdown of dry-cured ham. *Meat science* 51: 255–260.

Tang, R., Yu, B., Zhang, K., *et al.* 2010. Effects of nutritional level on pork quality and gene expression of µ-calpain and calpastatin in muscle of finishing pigs. *Meat science* 85: 768–771.

Théron, L., Tournayre, P., Kondjoyan, N., Abouelkaram, S., Santé-Lhoutellier, V., and Berdagué, J. L. 2010. Analysis of the volatile profile and identification of odor-active compounds in Bayonne ham. *Meat science* 85: 453–460.

Tibau, J., Gonzalez, J., Soler, J., Gispert, M., Lizardo, R., and Mourot, J. 2002. Influence du poids a l'abattage du porc entre 25 et 140 kg de poids vif sur la composition chimique de la carcasse: effets du genotype et du sexe. *Journees de la Recherche Porcine en France* 34: 121–127.

Toldrá, F. 2002. *Dry-cured meat products*. Food & Nutrition Press, INC.

Toldrá, F. 2006. The role of muscle enzymes in dry-cured meat products with different drying conditions. *Trends in food science and technology* 17: 164–168.

Toldrá, F., and Aristoy, M. C. 1993. Availability of essential amino acids in dry-cured ham. *International journal of food sciences and nutrition* 44: 215–219.

Toldrá, F., Aristoy, M. C., and Flores, M. 2000. Contribution of muscle aminopeptidases to flavor development in dry-cured ham. *Food research international* 33: 181–185.

Toldrá, F., Aristoy, M. C., and Flores, M. 2009. Relevance of nitrate and nitrite in dry-cured ham and their effects on aroma development. *Grasas y aceites* 60: 291–296.

Toldrá, F., Flores, M., and Sanz, Y. 1997. Dry-cured ham flavour: enzymatic generation and process influence. *Food chemistry* 59: 523–530.

Udomkun, P., Wiredu, A. N., Nagle, M., Müller, J., Vanlauwe, B., and Bandyopadhyay, R. 2017. Innovative technologies to manage aflatoxins in foods and feeds and the profitability of application: a review. *Food control* 76: 127–138.

Van Boekel, M. A. J. S. 2006. Formation of flavour compounds in the Maillard reaction. *Biotechnology advances* 24: 230–233.

Ventanas, J. 2006. *El jamón ibérico*. Mundi-Prensa Libros.

Ventanas, J., and Cava, R. 2001. Dinámica y control del proceso de secado del jamón Ibérico en secaderos y bodegas naturales y en cámaras climatizadas. In *Tecnología del jamón Ibérico,* 255–292. Mundi-Prensa.

Ventanas, S., Ventanas, J., Tovar, J., García, C., and Estévez, M. 2007. Extensive feeding versus oleic acid and tocopherol enriched mixed diets for the production of Iberian dry-cured hams: effect on chemical composition, oxidative status and sensory traits. *Meat science* 77: 246–256.

Vipotnik, Z., Rodríguez, A., and Rodrigues, P. 2017. Aspergillus westerdijkiae as a major ochratoxin A risk in dry-cured ham based-media. *International journal of food microbiology* 241: 244–251.

Virgili, R., Saccani, G., Gabba, L., Tanzi, E., and Bordini, C. S. 2007. Changes of free amino acids and biogenic amines during extended ageing of Italian dry-cured ham. *LWT-food science and technology* 40: 871–878.

World Health Organization. 2003. *Diet, nutrition, and the prevention of chronic diseases: report of a joint WHO/FAO expert consultation.* World Health Organization.

Wu, A. F., Sun, C. Y., and Sun, G. Q. 1959. *Jinhua ham.* Light Industry Press.

Wu, H., Zhang, Y., Long, M., Tang, J., Yu, X., Wang, J., and Zhang, J. 2014. Proteolysis and sensory properties of dry-cured bacon as affected by the partial substitution of sodium chloride with potassium chloride. *Meat science* 96: 1325–1331.

Yan, W., Chen, W., Muhammad, U., Zhang, J., Zhuang, H., and Zhou, G. 2019. Preparation of α-tocopherol-chitosan nanoparticles/chitosan/montmorillonite film and the anti-oxidant efficiency on sliced dry-cured ham. *Food control* 104: 132–138.

Zamora, R., Navarro, J. L., Aguilar, I., and Hidalgo, F. J. 2015. Lipid-derived aldehyde degradation under thermal conditions. *Food chemistry* 174: 89–96.

Zhang, J., Wang, L., Liu, Y., Zhu, J., and Zhou, G. 2006. Changes in the volatile flavour components of Jinhua ham during the traditional ageing process. *International journal of food science and technology* 41: 1033–1039.

3 Dry-Cured Shoulder *Lacón*

Manufacturing Process, Chemical Composition and Shelf Life

Laura Purriños,[1] Roberto Bermúdez,[1] Daniel Franco,[1] José M. Lorenzo,[1,2] and Javier Carballo[2]

[1]Centro Tecnológico de la Carne de Galicia, Rúa Galicia n° 4, Parque Tecnológico de Galicia, San Cibrao das Viñas, 32900 Ourense, Spain
[2]Área de Tecnología de Alimentos, Facultad de Ciencias, Universidad de Vigo. Campus Universitario, s/n, 32004 Ourense, Spain

CONTENTS

3.1   INTRODUCTION

Dry-cured lacón is a traditional meat product made in the North West of Spain from the foreleg of the pig cut at the humerus–blade joint, following a manufacturing technology very similar to that used in the manufacture of dry-cured ham. After a dry salting using a large amount of coarse salt, pieces are dried and ripened for varying times.

The term *lacón* is a Galician (North West Spain) word derived from the Latin word *lacca*, which refers to the foodstuff made from the foreleg of the pig by drying and ripening processes. References to this meat product are abundant in books devoted to the Galician traditional gastronomy (Cunqueiro 1973; Puga y Parga 1905). However, scientific interest in this meat product is very recent. Traditionally, this product was manufactured following guidelines based on empiric knowledge, with limited control on the duration of the different steps and on the environmental conditions in the manufacturing rooms. As consequence, the chemical and organoleptic characteristics, and therefore the quality, of the final products were very variable, a fact that negatively affected the commercial image of this product and demand for it by consumers.

In the last 20 years, extensive research was performed on the biochemical, microbiological, and sensory characteristics of this meat product, on the biochemical and microbiological changes that take place during its manufacture, and on the effect of some features of the manufacturing process on these changes. This allowed us to know, in depth, the chemical characteristics of this meat product and the effect on these characteristics of some modifications made in the traditional manufacturing method. Overall, this knowledge allowed the production of more homogeneous and high-quality products, which notably improved the image of this product in markets. In the Galicia region, this product has been awarded a Geographically Protected Identity (Official Journal of the European Communities 2001).

In this chapter, after giving some information about the history and the peculiarities of the process to manufacture dry-cured lacón, the chemical characteristics of this meat product are reviewed and some observations on its shelf-life are made.

3.2 MANUFACTURING PROCESS

It begins by cutting the fore extremity of the pig at the shoulder blade–humerus joint (Figure 3.1) and the stages of the process closely resemble those followed in the manufacture of dry-cured hams (Figure 3.2). The raw pieces, once selected and classified by weight (raw pieces usually weigh between 3.5 and 6.5 kg), are dry salted with coarse salt, forming mounds, alternating pieces with salt (Figure 3.3); the period of time the pieces spend in the mound is about 1 day per kg of weight, the temperature of the salting room being between 2 and 5°C and the relative humidity between 80 and 90%. After the salting stage, the pieces are taken from the mound, brushed (sometimes washed) to eliminate the salt from the surface and transferred to a post-salting room where they stay for about two weeks (minimum 7 days) at a temperature of 2–5°C and a relative humidity of around 85%. Once the post-salting stage has finished, the pieces undergo a drying–ripening process for which they are transferred to a room at 12°C and 70% relative humidity. They remain in this room for about a month and half (the length of the drying–ripening process is very variable depending on the needs of the market, with a minimum of 15 days). The

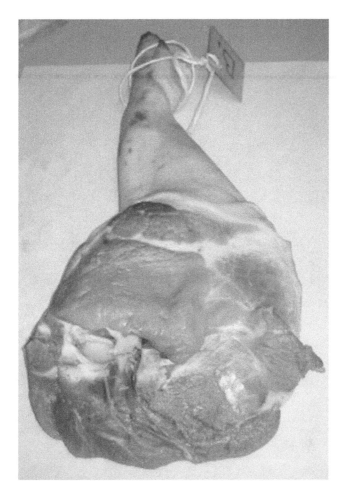

FIGURE 3.1   Raw piece used in the manufacture of dry-cured lacón.

final product (Figure 3.4), depending on the degree of ripening, can be eaten raw or cooked; the most ripened pieces can be consumed raw.

3.3   CHEMICAL COMPOSITION OF DRY-CURED LACÓN AND OF THE RAW PIECES USED IN ITS MANUFACTURE

3.3.1   PROXIMATE COMPOSITION

Information reported in literature on the proximate composition and on the values of the main physico-chemical parameters of the raw pieces used in the manufacture of dry-cured lacón are shown in Table 3.1. Data reported by the different authors in their works reasonably agree for each parameter. Differences in moisture, protein and fat contents reported are probably due to differences in the sampling procedure; Veiga et al. (2003) analyzed only the meat (muscular portion) of the pieces, whereas

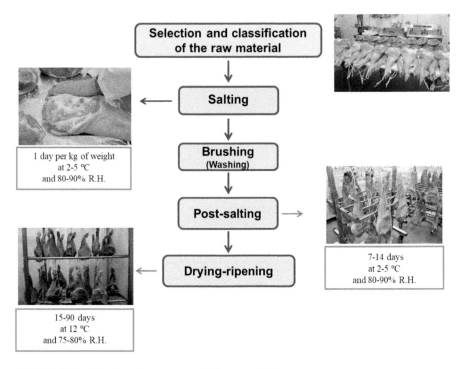

FIGURE 3.2 Manufacturing process of dry-cured lacón.

FIGURE 3.3 Dry-salting process during the manufacture of dry-cured lacón.

in the case of Lorenzo et al. (2003; 2008a) pieces were deboned and skinned, and all the edible portion (both muscles and subcutaneous and intermuscular fats) was minced for further analysis. The inclusion of the subcutaneous and intermuscular fats in the samples for analysis is probably the cause of the lower protein and moisture contents and the higher fat contents reported by Lorenzo et al. (2003; 2008a) when compared to Veiga et al. (2003).

The proximate composition and the main physico-chemical parameters of the dry-cured lacón (end product ready for consumption) are summarized in Table 3.2. During the manufacture of the dry-cured lacón, besides the breakdown processes in the protein and lipid fraction, which we will discuss later, a considerable drop in the moisture content takes place, due to the drying process in the last stages of

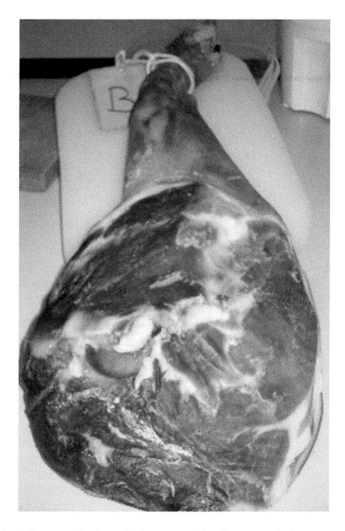

FIGURE 3.4 Dry-cured lacón ready for consumption (raw or cooked).

manufacture. This drying, together with the incorporation of NaCl during the salting stage, and the generation of low molecular weight nitrogen compounds as a consequence of the protein breakdown processes (proteolysis), is responsible for the sharp drop in the values of water activity (a_w) that guarantees the preservation and the microbiological stability of this meat product.

A decrease in the protein and fat contents (expressed as % of the total solids) is observed during the manufacturing process. This decrease appears to be fundamentally due to the sharp increase in the NaCl content during the manufacture (in the salting and post-salting stages), which would decrease the contribution of the protein and fat to the total solid content.

TABLE 3.1

Chemical Composition and Physico-Chemical Parameters of the Raw Pieces Used in the Manufacture of Dry-Cured Lacón

	Veiga et al. (2003)	Lorenzo et al. (2003)	Lorenzo et al. (2008a)
Moisture (g/100g)	76.52 ± 0.82	70.51 ± 1.31	70.30 ± 6.23
Protein ($N \times 6.25$)[a]	76.87 ± 5.44	63.92 ± 1.37	71.50 ± 2.55
Fat[a]	17.64 ± 5.77	30.07 ± 2.47	25.4 ± 2.15
Ash[a]	5.07 ± 0.52	2.72 ± 0.51	3.48 ± 0.73
NaCl[a]		1.58 ± 0.49	0.47 ± 0.02
Total carbohydrates[b]		0.35 ± 0.03	0.11 ± 0.01
Hydroxyproline[a]		0.75 ± 0.09	0.72 ± 0.17
Nitrate[c]		10.14 ± 2.77	37.5 ± 7.19
pH	6.29 ± 0.22	6.02 ± 0.19	6.36 ± 0.27
Titratable acidity[d]		0.27 ± 0.06	0.29 ± 0.11
a_w	0.97 ± 0.01	0.992 ± 0.001	0.997 ± 0.003

Notes
[a] Expressed as g/100 g of total solids
[b] Expressed as g of glucose/100 g of total solids
[c] Expressed as ppm
[d] Expressed as g lactic acid/100 g

Values of pH and titratable acidity show small changes during the manufacture. The slight variation of the pH values during the manufacture appears to indicate that the lacón, as occurs with other raw-cured meat products made from whole cuts such as ham (Bermúdez et al. 2014), Spanish cecina (García et al. 1995) and Italian bresaola (Cantoni and Calcinardi 1967; Cantoni et al. 1968) do not undergo a true lactic fermentation during the manufacturing process.

Differences in moisture contents and in a_w values in the final products, as reported by different authors, reflect differences in the duration of the manufacturing process, above all in the duration of the drying–ripening stage. Pieces analyzed by Veiga et al. (2003) were submitted to a drying–ripening period of 15 days, whereas pieces analyzed by other authors were dried and ripened for longer times (from 45 to 90 days) (Lorenzo et al. 2003; 2008a; Marra et al. 1999). The duration of the manufacturing process is crucial in the achievement of the typical organoleptic characteristics of the final product; in fact, a minimum duration of 15 days for the drying–ripening stage was established for the product protected by the Geographical Indication (I.G.P.) Lacón Gallego (Official Journal of the European Communities 2001). The very short manufacturing processes with very short drying–ripening periods gives rise to products hardly ripened and having severe deficiencies in flavor, color and texture (Lorenzo et al. 2002).

Dry-cured lacón was traditionally manufactured using only coarse salt, without any other additives. Recently, the industry started to use some common additives

TABLE 3.2

Chemical Composition and Physico-Chemical Parameters of Dry-Cured Lacón

	Marra et al. (1999)	Lorenzo et al. (2003)	Veiga et al. (2003)	Lorenzo et al. (2008a)
Moisture (g/100 g)	50.44 ± 4.14	39.62 ± 7.40	60.97 ± 3.24	42.02 ± 2.81
Protein ($N \times 6.25$)[a]	56.40 ± 5.53	52.80 ± 4.21	63.35 ± 3.77	58.0 ± 5.09
Fat[a]	19.73 ± 7.53	22.31 ± 8.04	14.76 ± 4.14	20.7 ± 1.15
Ash[a]	20.66 ± 4.31	23.31 ± 3.80	21.64 ± 2.91	15.7 ± 5.50
NaCl[a]	16.2 ± 4.18	19.50 ± 1.72	18.67 ± 6.52	13.1 ± 4.90
Total carbohydrates[b]	0.58 ± 0.23	0.16 ± 0.04		0.03 ± 0.00
Hydroxyproline[a]	0.76 ± 0.23	0.58 ± 0.07		0.70 ± 0.17
Nitrate[c]	41.1 ± 58.1	58.37 ± 3.58	66.87 ± 26.82	39.0 ± 6.04
pH	6.14 ± 0.19	6.16 ± 0.14	5.91 ± 0.22	6.40 ± 0.22
Titratable acidity[d]	0.08 ± 0.03	0.72 ± 0.07		0.15 ± 0.08
a_w	0.872 ± 0.030	0.767 ± 0.052	0.91 ± 0.01	0.876 ± 0.075
Nitrosyl–heme pigments[c]	34.2 ± 24.5			
Total heme pigments[c]	279 ± 183			
P.C.[e]	13.5 ± 9.36	16.44 ± 2.09		17.50 ± 1.27

Notes

[a] Expressed as g/100 g of total solids

[b] Expressed as g of glucose/100 g of total solids

[c] Expressed as ppm

[d] Expressed as g lactic acid/100 g

[e] P.C. = Percent conversion to cured meat pigments

with the aim of improving the appearance and the quality of the final product (i.e., development of the typical red color of the cured meats and inhibition of the growth of molds on the surface). The use of additives (glucose –2 g/kg-, sodium nitrite –125 mg/kg-, sodium nitrate –175 mg/kg-, sodium ascorbate –500 mg/kg-, and sodium citrate –100 mg/kg-), which are rubbed on the surface of the muscular side of the pieces before the salting, only affect the carbohydrate and nitrate contents, and the percentage of transformation of the pigments (haem to nytrosil-haem), which are significantly higher in the products manufactured with additives (Lorenzo et al. 2008a). The use of such additives, however, seems to improve the color and odor of the final product (Lorenzo et al. 2008a).

Table 3.3 shows the mineral (macro- and microelements) composition of the raw pieces and of the final product. Data reported in the different works agree that Na, K and P are the main macroelements in the raw pieces, whereas Zn and Fe are by far the main microelements. In the dry-cured lacón there is a notable increase in Na content in relation to the fresh piece as a result of the salting process; the use of

TABLE 3.3

Mineral Contents of Dry-Cured Lacón and of the Raw Pieces Used in Its Manufacture

	Raw Pieces		Dry-Cured Lacón	
	Lorenzo et al. (2003)	Lorenzo (2006)	Lorenzo et al. (2003)	Lorenzo (2006)
Ca[a]	0.28 ± 0.16	0.19 ± 0.05	0.38 ± 0.14	0.29 ± 0.08
P[a]	10.37 ± 1.89	4.72 ± 1.12	4.64 ± 1.54	2.55 ± 1.05
Na[a]	15.23 ± 2.56	6.74 ± 1.33	88.49 ± 14.75	41.18 ± 15.46
Mg[a]	0.69 ± 0.12	1.20 ± 0.48	0.65 ± 0.29	0.52 ± 0.16
K[a]	13.63 ± 0.88	13.32 ± 5.66	12.08 ± 1.43	8.84 ± 1.94
Zn[b]	80.76 ± 23.27	98.75 ± 24.67	68.92 ± 8.01	62.64 ± 4.01
Cu[b]	7.03 ± 3.58	5.42 ± 2.46	9.57 ± 4.71	2.45 ± 0.16
Mn[b]	0.79 ± 0.25	0.49 ± 0.15	1.97 ± 1.09	0.36 ± 0.05
Fe[b]	53.36 ± 19.05	43.88 ± 10.11	83.15 ± 44.80	27.00 ± 4.77

Notes
[a] Expressed as g kg^{-1} total solids
[b] Expressed as mg kg^{-1} total solids

additives did not influence in a significant way the mineral content of the dry-cured lacón (Lorenzo 2006).

3.3.2 CHARACTERISTICS OF THE LIPID FRACTION

Table 3.4 shows the composition and characteristics (parameters indicative of the degrading processes) of the fat in the raw pieces and in the dry-cured lacón. The percentage of glycerides remains practically unaltered during the manufacturing process, whereas the phospholipid fraction decreases and the free fatty acid fraction increases. There is a close relationship between the decrease in the phospholipid content and the increase in the free fatty acid content, which seems to indicate that the phospholipids are more intensely degraded than the glycerides during the manufacturing process, and that most of the fatty acids freed during manufacture come from the phospholipids.

Data reported for the parameters that indicate the degree of lipolysis (free fatty acids -FFA-) and the autooxidation of the fat (peroxide value -PV-, and thiobarbituric acid value -TBA-) reasonably agree among the works in literature and reveal that appreciable lipolytic and oxidative changes take place during the manufacture of dry-cured lacón (Lorenzo et al. 2003; 2008a; Marra et al. 1999; Rodríguez et al. 2001; Veiga et al. 2003). When comparing these values with those reported for other dry-cured meat products such as hams after a similar curing period (Antequera et al. 1992; Astiasarán et al. 1988; Flores et al. 1985), data indicated a moderate lipolysis but a strong lipid oxidation during the manufacturing process of dry-cured lacón. This

TABLE 3.4

Lipid Fractions and Fat Parameters of Dry-Cured Lacón and of the Raw Pieces Used in Its Manufacture

	Raw Pieces					Dry-Cured Lacón		
	Veiga et al. (2003)	Lorenzo et al. (2003)	Lorenzo et al. (2008a)	Marra et al. (1999)	Rodríguez et al. (2001)	Veiga et al. (2003)	Lorenzo et al. (2003)	Lorenzo et al. (2008a)
Glycerides[a]	71.61 ± 5.70					72.24 ± 3.78		
Free fatty acids[a]	5.44 ± 2.64					10.55 ± 2.13		
Phospholipids[a]	22.95 ± 4.25					17.09 ± 2.89		
TBA[b]	1.66 ± 1.12		0.95 ± 0.26		6.99 ± 1.95	4.01 ± 1.50		10.39 ± 3.18
F.F.A[c]		0.99 ± 0.30	1.49 ± 0.62	2.33 ± 0.17	3.16 ± 1.60		4.99 ± 0.67	2.77 ± 1.15
P.V.[d]				37.7 ± 10.1	30.50 ± 26.54			

Notes

[a] Expressed as % of total fat

[b] Thiobarbituric acid value, expressed as mg malonaldehyde/kg meat

[c] Free fatty acids, expressed as % oleic acid

[d] Peroxide value, expressed as meq O_2/kg of fat

circumstance could be related with the lower NaCl contents in the hams and with the lower size of the dry-cured lacón pieces that causes an increase of the surface per unit of weight and therefore the surface of contact with the air, and with the oxygen consequently, during the manufacture process.

In the manufacture of the dry-cured lacón, usually, and following the guidelines established for the manufacturing process of dry-cured ham, a day of salting per kg of weight of the raw piece is applied. The increase of the duration of the salting time significantly increases the peroxide and the TBA values in the final product, which seems to indicate a clear prooxidant effect of the NaCl (Garrido et al. 2009). The increase in the salting time seems, in general, to have an inhibitory effect on the lipolytic phenomena; however, this effect was not consistent in all the sampling times in the batches studied (Garrido et al. 2009).

The fatty acid composition of the lipid fractions in the raw pieces and in the dry-cured lacón is shown in Table 3.5 (Veiga et al. 2003). In the glyceride fraction of the raw pieces, the most abundant fatty acids were the monounsaturated (41.8%), followed by the saturated (37.64%) and by the polyunsaturated (20.59%) fatty acids; the main individual fatty acids in the glyceride fraction of the raw pieces were oleic (34.20%), followed by palmitic (24.9%), linoleic (15.5%) and stearic (11.2%). Glycerides of the dry-cured lacón (end product) showed a fatty acid profile close to that of the raw piece, but having higher saturated (39.03%) and monounsaturated (44.08%) fatty acid, and lower polyunsaturated fatty acids (16.90%).

In the phospholipid fraction, the fatty acid profile was very similar in the raw piece and the final product. The polyunsaturated fatty acids were the main group, followed by the saturated and the monounsaturated; the main fatty acid in this fraction was linoleic acid, followed by palmitic, oleic, arachidonic and steraic acids. The phospholipids of the dry-cured lacón showed lower contents of polyunsaturated fatty acids than the phospholipids of the raw pieces.

Regarding the free fatty acid fraction, the saturated fatty acids were dominant in the raw pieces (38.10%), followed by the monounsaturated (31.02%), and the polyunsaturated (30.87%) fatty acids; palmitic (27.80%), linoleic (23.98%), oleic (23.48%) and stearic (8.39%) are the main free fatty acids. The free fatty acid profile in the dry-cured lacón basically coincided with that of the raw piece, but showed higher contents of polyunsaturated fatty acids, above all arachidonic and eicosapentaenoic acids.

Table 3.6 shows the free fatty acid contents in raw pieces and in dry-cured lacón according to Lorenzo et al. (2003; 2008b). Free fatty-acid content in the raw pieces (around 3–4% of the fat) is slightly lower but reasonably agrees with data reported by Veiga et al. (2003) (5%). The initial degree of lipolysis, and the free fatty acid content, therefore, is strongly dependent on the freshness of the raw pieces and on the history of preservation (above all, temperature of storage) from the quartering to the start of the manufacture. Differences in sampling procedure, as previously indicated, among authors could also be in part responsible for this slight difference.

Regarding the free fatty acid content of the dry-cured lacón, data reported by Lorenzo et al. (2003) (18% of the fat) and by Lorenzo et al. (2008b) (15.7% of the fat) are notably higher than those reported by Veiga et al. (2003) (10.55% of the fat). The lower free fatty acid contents reported by Veiga et al. (2003) seems to

TABLE 3.5

Fatty Acid Profiles (% of Total Fatty Acids) of Glyceride, Phospholipid and Free Fatty Acid Fractions of Dry-Cured Lacón and of the Raw Pieces Used in its Manufacture (Veiga et al. 2003)

	Raw Pieces			Dry-Cured Lacón		
	Glycerides	Phospholipid	Free fatty acids	Glycerides	Phospholipid	Free fatty acids
C:14:0	1.27 ± 0.43	0.35 ± 0.46	1.54 ± 0.36	1.34 ± 0.20	0.32 ± 0.16	1.38 ± 0.24
C16:0	24.94 ± 2.03	25.06 ± 2.68	27.80 ± 3.59	23.83 ± 1.54	28.23 ± 3.09	27.29 ± 1.80
C16:1n-9	0.42 ± 0.10	0.25 ± 0.12	1.48 ± 0.37	0.42 ± 0.06	0.28 ± 0.09	1.16 ± 0.27
C16:1n-7	2.76 ± 0.45	0.70 ± 0.24	2.14 ± 0.95	2.74 ± 0.29	0.83 ± 0.23	2.19 ± 0.37
C17:0	0.24 ± 0.07	0.43 ± 0.49	0.37 ± 0.09	0.33 ± 0.08	0.41 ± 0.08	0.41 ± 0.09
C17:1n-9	0.23 ± 0.06	0.17 ± 0.10	0.32 ± 0.10	0.32 ± 0.07	0.22 ± 0.08	0.42 ± 0.11
C18:0	11.19 ± 2.14	9.15 ± 0.85	8.39 ± 1.41	13.54 ± 2.27	9.35 ± 0.69	7.14 ± 0.79
C18:1n-9	34.20 ± 2.91	12.45 ± 2.46	23.48 ± 3.56	36.41 ± 2.74	13.68 ± 2.17	22.10 ± 2.19
C18:1n-7	3.57 ± 0.38	2.91 ± 0.42	3.26 ± 6.08	3.47 ± 0.40	3.66 ± 0.43	3.27 ± 0.38
C18:2n-6	15.46 ± 2.29	33.45 ± 3.24	23.98 ± 3.55	12.91 ± 2.24	31.96 ± 2.62	25.42 ± 2.41
C18:3n-3	0.56 ± 0.15	0.61 ± 0.08	1.05 ± 0.39	0.60 ± 0.10	0.63 ± 0.20	1.15 ± 0.35
C20:1n-9	0.59 ± 0.21	0.17 ± 0.07	0.34 ± 0.18	0.73 ± 0.11	0.22 ± 0.07	0.34 ± 0.11
C20:2n-6	0.38 ± 0.10	0.32 ± 0.16	0.43 ± 0.25	0.49 ± 0.10	0.33 ± 0.08	0.33 ± 0.15
C20:3n-6	0.35 ± 0.14	0.97 ± 0.21	0.54 ± 0.24	0.27 ± 0.05	0.81 ± 0.16	0.62 ± 0.16
C20:4n-6	2.78 ± 1.01	9.48 ± 1.91	3.77 ± 1.36	1.89 ± 0.60	7.21 ± 1.41	5.20 ± 1.22
C20:5n-3	0.14 ± 0.10	0.52 ± 0.21	0.23 ± 0.17	0.09 ± 0.03	0.38 ± 0.12	0.43 ± 0.25
C22:4n-6	0.40 ± 0.18	1.16 ± 0.39	0.29 ± 0.22	0.30 ± 0.08	0.66 ± 0.27	0.37 ± 0.13
C22:5n-3	0.39 ± 0.22	1.15 ± 0.53	0.38 ± 0.34	0.28 ± 0.08	0.67 ± 0.27	0.57 ± 0.30
C22:6n-3	0.12 ± 0.08	0.35 ± 0.23	0.20 ± 0.24	0.09 ± 0.05	0.18 ± 0.10	0.28 ± 0.15
SFA	37.64 ± 1.75	35.00 ± 3.09	38.10 ± 3.60	39.03 ± 2.08	38.32 ± 3.08	36.21 ± 2.03
MUFA	41.77 ± 2.94	16.63 ± 2.87	31.02 ± 4.72	44.08 ± 3.10	18.87 ± 2.52	29.46 ± 2.87
PUFA	20.59 ± 3.74	48.37 ± 4.09	30.87 ± 4.56	16.90 ± 3.03	42.80 ± 4.21	34.33 ± 3.78

be related to the shorter drying–ripening period compared to the duration of this stage in the dry-cured lacón studied by Lorenzo et al. (2003; 2008b). Lipolysis takes place fundamentally in the drying–ripening stage, not only due to the greater duration of this stage, but also due to the temperature (12°C) that is more favorable to the action of the tissue enzymes than the temperature during the salting and post-salting stages (2–5°C).

Regarding the free fatty acid profile reported by Lorenzo et al. (2003; 2008b), oleic, stearic, palmitic and linoleic were the main free fatty acids in the raw piece; this free fatty acid profile basically coincided with that of dry-ripened lacón, and in both cases reasonably agreed with that reported by Veiga el al. (2003).

TABLE 3.6

Free Fatty Acids Content (mg/100 g of fat) of Dry-Cured Lacón and of the Raw Pieces Used in its Manufacture

	Raw Pieces		Dry-Cured Lacón	
	Lorenzo et al. (2003)	Lorenzo et al. (2008b)	Lorenzo et al. (2003)	Lorenzo et al. (2008b)
C10:0	31.5 ± 23.8	ND	32.2 ± 6.4	5 ± 4
C12:0	25.1 ±10.5	ND	38.9 ± 11.0	85 ± 40
C14:0	65.8 ± 33.4	101 ± 78	226.1 ± 104.1	284 ± 47
C15:0		69 ± 5		131 ± 23
C16:0	684.4 ± 271.2	231 ± 97	3710.1 ± 1608.8	880 ± 98
C16:1n-9	131.2 ± 97.5	26 ± 16	479.9 ± 157.7	412 ± 28
C17:0		2 ± 2		83 ± 79
C17:1n-9		ND		16 ± 3
C18:0	756.2 ± 334.0	548 ± 178	2401.3 ± 1397.4	1780 ± 397
C18:1n-9	832.1 ± 240.2	1090 ± 698	8582.7 ± 2030.8	7040 ± 2274
C18:2n-6	365.4 ± 147.2	860 ± 191	2236.0 ± 741.9	2660 ± 218
C18:3n-3	50.2 ± 28.9	2 ± 2	33.8 ± 8.2	106 ± 45
C20:0	12.6 ± 17.3	ND	234.6 ± 150.0	ND
C20:1n-9		127 ± 111		297 ± 72
C20:2n-6		73 ± 63		270 ± 97
C20:4n-6	20.6 ± 22.3	194 ± 169	43.8 ± 28.3	403 ± 92
C22:0		ND		413 ± 358
C22:1		87 ± 15		161 ± 76
C24:0		528 ± 81		725 ± 119
Total	2975.1 ± 700.8	3940 ± 760	18020.2 ± 4912.4	15700 ± 3180
SFA		1270 ± 177		3910 ± 864
MUFA		1530 ± 768		8390 ± 2470
PUFA		1130 ± 208		3440 ± 221

ND = Not detected.

The use of additives does not affect in a significant way the quantities and the profiles of the free fatty acids (Lorenzo et al. 2008b). The increase of the duration of the salting stage significantly affects the fatty acid content, particularly that of the polyunsaturated fatty acids that decreased as the salting time increases, with the linoleic acid being the most affected fatty acid (Garrido et al. 2014).

3.3.3 CHARACTERISTICS OF THE PROTEIN FRACTION

Table 3.7 shows the characteristics of the protein fraction in the fresh pieces and in the final product. The extractability of the sarcoplasmic and myofibrillar proteins

TABLE 3.7

Protein Extractability and Nitrogen Fractions (g/100 g of Total Solids) of Dry-cured Lacón and of the Raw Pieces Used in its Manufacture

	Raw Pieces		Dry-Cured Lacón	
	Lorenzo et al. (2003)	Lorenzo et al. (2008b)	Lorenzo et al. (2003)	Lorenzo et al. (2008b)
Sarcoplasmic proteins	25.32 ± 3.28	27.61 ± 5.43	6.06 ± 1.88	3.50 ± 0.37
Myofibrillar proteins	35.30 ± 6.87	40.40 ± 7.43	6.01 ± 1.32	3.55 ± 0.53
Total nitrogen (TN)	10.23 ± 0.22	12.02 ± 1.15	8.45 ± 0.67	8.19 ± 0.81
Non-protein nitrogen (NPN)	0.33 ± 0.03	0.26 ± 0.02	0.77 ± 0.17	0.74 ± 0.06
α-Amino acidic nitrogen (NH_2-N)	0.25 ± 0.02	0.10 ± 0.00	0.39 ± 0.07	0.43 ± 0.01
Volatile basic nitrogen (VBN)	0.04 ± 0.00	0.03 ± 0.01	0.05 ± 0.01	0.11 ± 0.04

decreased from 25–27 g/100 g and 35–40 g/100 g, respectively, in the fresh pieces to 3.5–6.0 g/100 g and 3.55–6.0 /100 g, respectively, in the dry-cured lacón; this indicates that a high proportion of the sarcoplasmic and myofibrillar proteins became insoluble during the manufacturing process. The myofibrillar proteins undergo a greater degree of insolubilization than the sarcoplasmic proteins. As proteolysis is not very intense in this meat product, judging from the non-protein nitrogen values obtained, it seems that the protein insolubilization is mainly due to denaturation processes probably caused by the high NaCl concentrations reached after the post-salting stage.

From the values of the quantities of the different nitrogen fractions (non-protein nitrogen, α-aminoacidic nitrogen and total basic volatile nitrogen), and from the changes in these nitrogen fractions from the raw pieces to the final ripened product, we can conclude that the protein degradation (proteolysis) processes during the manufacture of this meat product are only moderate. The high NaCl contents, the low environmental temperatures during the manufacture (never higher than 12 °C), and the intense dehydration undergone by the pieces during the manufacture (above all during the drying–ripening stage), appear to be the cause of the low protein degradation observed in dry-cured lacón when compared with other dry-cured meat products made from whole cuts, such as hams (Buscailhon et al. 1994; Martín et al. 1998; Ventanas et al. 1992) or Spanish cecina (García et al. 1997; 1998).

The use of additives in the manufacture (Lorenzo et al. 2008b) does not significantly affect the degree of proteolysis in this meat product. However, the intensity of the proteolysis decreases as the length of the salting time increases (Garrido et al. 2012). From the existing information in the literature, it seems that the proteolytic phenomena that take place during the manufacture of the dry-ripened

TABLE 3.8

Free Amino Acids (Expressed as mg/100 g of Total Solids) of Dry-Cured Lacón and of the Raw Pieces Used in Its Manufacture

	Raw Pieces		Dry-Cured Lacón	
	Lorenzo et al. (2003)	Lorenzo et al. (2008b)	Lorenzo et al. (2003)	Lorenzo et al. (2008b)
Asp	9.08 ± 0.94	7.05 ± 0.08	33.19 ± 11.21	53.1 ± 9.36
Glu	55.37 ± 10.94	15.2 ± 3.15	157.03 ± 39.04	282 ± 12.5
Asn	8.01 ± 0.28	8.96 ± 6.13	14.47 ± 2.94	153 ± 18.9
Ser	22.36 ± 3.55	24.8 ± 2.28	61.88 ± 13.01	193 ± 8.89
Gln	21.69 ± 4.30	44.4 ± 14.0	11.75 ± 36.57	201 ± 17.9
Gly	22.55 ± 4.84	23.8 ± 4.05	43.90 ± 10.52	125 ± 14.1
His	12.41 ± 3.05	16.0 ± 3.23	36.47 ± 2.22	75.6 ± 3.39
Tau	49.74 ± 11.29	38.2 ± 4.13	97.92 ± 29.68	118 ± 2.16
Gaba	30.18 ± 7.34	12.7 ± 2.64	60.74 ± 13.75	224 ± 9.07
Arg	34.75 ± 16.21	25.8 ± 0.94	61.29 ± 12.32	99.0 ± 3.64
Thr	14.89 ± 2.53	17.5 ± 4.69	40.38 ± 8.82	57.2 ± 34.8
Ala	31.49 ± 16.57	36.8 ± 1.45	112.21 ± 35.57	165 ± 15.8
Pro	25.36 ± 17.06	25.8 ± 11.4	72.46 ± 21.81	177 ± 17.6
Tyr	10.60 ± 2.28	13.1 ± 3.62	33.37 ± 8.50	82.0 ± 8.32
Val	13.73 ± 3.00	15.6 ± 4.28	55.43 ± 17.09	87.1 ± 7.39
Met	1.18 ± 1.29	4.79 ± 0.80	4.01 ± 2.18	30.2 ± 3.93
Cys	6.16 ± .073		2.66 ± 0.33	
Ile	10.94 ± 1.69	14.9 ± 1.54	43.23 ± 11.49	106 ± 5.46
Leu	16.03 ± 2.59	25.3 ± 3.87	59.66 ± 13.70	110 ± 16.8
Phe	11.01 ± 1.26	12.9 ± 1.34	32.26 ± 7.11	89.6 ± 8.05
Trp	3.08 ± 0.55	3.31 ± 1.44	6.34 ± 0.47	22.5 ± 6.64
Lys	26.52 ± 7.63	26.4 ± 13.0	155.73 ± 55.83	430 ± 49.0
Total	437.14 ± 35.23	413 ± 8.52	1302.07 ± 170.28	2881 ± 118

meat products made from whole cuts are mainly due to the action of the muscle proteases (Molina and Toldrá 1992; Toldrá and Flores 1998) and that the high NaCl concentrations are able to limit the activity of these enzymes (Sárraga et al. 1989; Rico et al. 1990).

Table 3.8 shows the data on the content in free amino acids in the fresh piece and in the dry-cured lacón. Data on the total amount of free amino acids in the fresh pieces agree among the works reported in literature and are around 420 mg/100 g of total solids. This content increased during the manufacture as the result of the proteolytic processes resulting in a final value (in the dry-cured product) ranging from 1,300 to 2,900 mg/100 g of total solids. Differences in the values reported for the final product seem to be related with differences in the intensity of the proteolytic processes, which are related to the distinct duration of the drying–ripening period, and also with small differences in the environmental conditions that determine differences in moisture content. The higher moisture content and lower NaCl content in the pieces analyzed by Lorenzo et al. (2008a; 2008b) (see also Table 3.2) could be responsible for the higher intensity of the proteolytic processes, resulting in higher concentrations of free amino acids.

Regarding the content of the individual free amino acids, the amino acid profiles reasonably agree in the works published with this subject. With some minor variations, Glu, Tau, Arg, Ala, Lys and Pro are the main free amino acids in the raw pieces, whereas Glu, Lys, Gln, and Pro are the most abundant free amino acids in the dry-cured lacón. This free amino acid profile is consistent with those reported by different authors for dry-cured ham (Antequera et al. 1994; Buscailhon et al. 1994; Careri et al. 1993; Flores et al., 1985;Martín et al., 2001; Virgili et al. 1999).

All the individual free amino acids increased along the manufacturing process, but with different rates of increase among the different amino acids. According to the data reported by Lorenzo et al. (2008b), approximately 10% of the total increase of the free amino acid content occurs during the salting stage, 25% during the post-salting stage, and 63% during the drying-ripening period. It is well established that the nature and quantity of the amino acids freed during the ripening of the meat products depends on the activity of the muscle aminopeptidases, cathepsins and peptidases, and on the NaCl content and water activity values that affect these enzymatic activities (Flores et al., 1985; Toldrá and Flores 1998).

The concentration of some amino acids (glutamic acid being the most representative within them) in the dry-cured lacón exceeds the established values of threshold of perception; these amino acids are therefore expected to contribute significantly to the flavor of the dry-cured lacón.

The individual and total free amino acid content and the free amino acid profile (comparative quantities of the individual free amino acids) are not significantly affected by the use of additives (Lorenzo et al. 2008b).

The increase of the length of the salting time causes a decrease in the free amino acid content in agreement with the inhibition of the proteolysis caused by the high NaCl concentrations (Garrido et al. 2012). The salting time, however, does not affect the free fatty acid profile (Garrido et al. 2012), which appears to indicate that the same enzymatic activities act during the manufacture of lacón with independence of the salting time; the salting time, and the resulting NaCl concentrations, modulate these activities; these enzymes are inhibited as the NaCl increases.

The biogenic amines are formed as products of amino acid decarboxylation reactions during the manufacture of the fermented/ripened foods; this fact has importance not only because of the unfavorable effect of these compounds on the flavor, but also from a health point of view (Silla Santos 1996). Table 3.9 shows the content in biogenic amines in the fresh pieces and in the dry-cured final product (Lorenzo et al. 2007). During the manufacture of the dry-cured lacón, the total biogenic amine content increased significantly. Tryptamine and spermine are reported as the main biogenic amines in fresh pieces, while tryptamine and cadaverine are the most abundant at the end of the manufacture (Lorenzo et al. 2007). As expected in a meat product in which there is little active microbial metabolism during the manufacturing process, the total biogenic amine content at the end of the

TABLE 3.9

Biogenic Amine Contents (mg/kg) of Dry-Cured Lacón and of the Raw Pieces Used in Its Manufacture

	Raw Pieces (Lorenzo et al. 2007)	Dry-Cured Lacón (Lorenzo et al. 2007)
Agmatine	ND	7.70±6.91
Tryptamine	25.03 ± 11.57	36.92 ± 9.80
2-Phenylethylamine	3.64 ± 1.05	6.59 ± 3.81
Putrescine	3.88 ± 2.00	6.34 ± 1.98
Cadaverine	6.90 ± 3.21	39.15 ± 3.44
Histamine	4.44 ± 0.66	4.01 ± 0.06
Tyramine	3.87 ± 1.21	2.14 ± 1.69
Spermidine	7.14 ± 1.21	7.20 ± 4.04
Spermine	32.30 ± 3.02	18.71 ± 11.30
Total	87.22 ± 10.92	129.03 ± 12.06

ND = Not detected

manufacture was low and never exceeded 200 mg/kg. The use of additives in the manufacture significantly increases the total biogenic amine content and the individual content of tryptamine, tyramine and histamine (Lorenzo et al. 2007). The presence of glucose within the additives added, a fact that favors the growth of the lactic acid bacteria and therefore the decarboxylation activity of this microbial group, could explain the different behavior in the generation of biogenic amines when additives were used in the manufacture of the dry-cured lacón.

3.3.4 DEVELOPMENT OF VOLATILE COMPUNDS DURING THE MANUFACTURE

The volatile compounds and their generation during the manufacture have also been the subject of study (Purriños et al. 2011). In the raw pieces, only 34 volatile compounds are present and at very low levels. The number of volatile compounds and their abundance significantly increases during the manufacture process; at the end of the drying–ripening, 102 volatile compounds are identified: aldehydes (24 compounds), alcohols (9), ketones (15), hydrocarbons (37), esters (4), acids (3), furans (4), sulphur compounds (1), chloride compounds (1) and other compounds (4). Results of this work indicate that the main chemical family of volatile compounds responsible for the flavor at the end of the manufacturing process of the dry-cured lacón are aldehydes, followed by hydrocarbons and ketones; also, according to these results, lipids seem to be the main precursor of the flavor compounds in this meat product.

The increase of the duration of the salting stage causes the increase of the total area of the volatile compounds detected, and a significant increase of the aldehydes (among which hexanal is the main compound), ketones (above all 3-octen-2-one), esters and alcohols; however, no significant effect of the salting duration was detected on total hydrocarbons and total furans (Purriños et al. 2012).

The use of chestnuts in the finishing diet of the pigs significantly affects the volatile compounds in the dry-cured lacón at the end of the manufacture (Lorenzo et al. 2014). Using discriminant analysis techniques, six variables (dodecane, butanediol, pentenol, 2-pentenal, decen-3-one, and pyridine-2-methul) were selected and two discriminant functions were developed that allow us to detect and to verify the presence of chestnuts in the pig diet on the basis of the volatile compound profile of the dry-cured lacón.

3.4 SHELF LIFE

As previously indicated, a minimum ripening time is necessary to obtain the desirable organoleptic quality and very short ripening times give rise to inadequate attributes. In fact, in previous studies (Lorenzo et al. 2002), the samples having the less intense proteolysis and lipolysis showed the lowest scores in the attributes *intensity of the odor* and *cured odor*. In these same studies, a positive and significant correlation was observed between the cured odor and the values of the nitrogen fractions (non-protein nitrogen, α-aminoacidic nitrogen and total basic volatile nitrogen) and between the cured odor and the acidity of the fat. According to Garrido et al. (2012), the product salted for a day/kg of raw piece, and next maintained for 14 days in the post-salting rooms and dry-ripened for 84 days (102 days total duration of the manufacture process), when consumed after desalting by immersion in water for 1.5 days at room temperature and cooking for 2 hours at atmospheric pressure, received the best global valuation based on the attributes of odor (intensity, pleasant, cured, rancid), taste (salty, intensity in the meat, intensity in the fat, pleasant in the meat, pleasant in the fat, cured in the meat, cured in the fat, rancid in the meat, rancid in the fat), and global quality.

No chemical or sensory studies were carried out in dry-cured lacón after extended maturation processes. In the case of ham, Cilla et al. (2005) reported that extended ripening under specific environmental conditions (18°C and 75% relative humidity) did not influence the consumer acceptance from 12 to 22 months, while it decreased significantly until 26 months of ripening. These authors reported that this decrease in acceptability is attributable to the high pastiness and adhesiveness that are the result of an excessive proteolysis. As indicated, due to the particular shape of the piece and the higher salt content, the dry-cured lacón shows a more intense lipid oxidation than does ham. According to these observations, an extended ripening process would foreseeably cause excessive rancidity of the fat, which would motivate rejection by the consumer.

In dry-cured ham, modified atmosphere packaging and vacuum packaging for long-period chilled storage (until 120 days) were essayed, and no significant modifications were observed in chemical, microbiological and sensory characters (García-Esteban et al. 2004; Parra et al. 2010). These packaging procedures could also be useful in the increase of the shelf life of the dry-cured lacón.

3.5 CONCLUSIONS

Dry-cured lacón is characterized by high protein and low fat content. During its manufacturing process, proteins and lipids undergo a moderate hydrolysis; however, according to their concentrations and their perception thresholds, some free amino acids seem to participate actively in the flavor of the final product. Fat auto-oxidation is very intense during the manufacture, and lipid degradation seems to be the origin of most of the volatile compounds present in the ripened product. The use of additives considerably improves the color and overall acceptance of this product, but the degradative changes during the manufacture are not significantly affected. The increase of the duration of salting time significantly limits the proteolytic and lipolytic processes, but it enhances the oxidation of the fat, therefore increasing the amounts of most of the volatile compounds present at the end of the manufacture.

REFERENCES

Antequera, T., C. García, C. López, J. Ventanas, M.A. Asensio and J.J. Córdoba. 1994. Evolution of different physico-chemical parameters during ripening of Iberian ham from Iberian (100%) and Iberian x Duroc pigs (50%). *Revista Española de Ciencia y Tecnología de Alimentos* 34: 178–190.
Antequera, T., C.J. López-Bote, J.J. Córdoba, C. García, M.A. Asensio and J. Ventanas. 1992. Lipid oxidative changes in the processing of Iberian pig hams. *Food Chemistry* 45: 105–110.
Astiasarán, I., M.J. Beriaín, J. Melgar, J.M. Sánchez-Monje, R. Villanueva and J. Bello. 1988. Estudio comparativo de las características de jamones curados de cerdo blanco elaborados con distinta tecnología. *Revista de Agroquímica y Tecnología de Alimentos* 28: 519–528.
Bermúdez, R., D. Franco, J. Carballo and J.M. Lorenzo. 2014. Physicochemical changes during manufacture and final sensory characteristics of dry-cured Celta ham. Effect of muscle type. *Food Control* 43: 263–269.
Buscailhon, S., G. Monin, M. Cornet and J. Bousset. 1994. Time-related changes in nitrogen fractions and free amino acids of lean tissue of French dry-cured ham. *Meat Science* 37: 449–456.
Cantoni, C., M.A. Bianchi, P. Renon and C. Calcinardi, 1968. Studi su alcuni aspetti della maturazione delle bresaole. *Archivio Veterinario Italiano* 19: 269–277.
Cantoni, C. and C. Calcinardi. 1967. Studi sul processo di maturazione delle bresaole. *Archivio Veterinario Italiano* 18: 49–59.
Careri, M., A. Mangia, G. Barbieri, L. Bolzoni, R. Virgili and G. Parolari. 1993. Sensory property relationships to chemical data of Italian type dry-cured ham. *Journal of Food Science* 58: 968–972.
Cilla, I., L. Martínez, J.A. Beltrán and P. Roncalés. 2005. Factors affecting acceptability of dry-cured ham throughout extended maturation under "bodega" conditions. *Meat Science* 69: 789–795.
Cunqueiro, A. 1973. *A cociña galega [The Galician cuisine]*. Vigo, Spain: Ed. Galaxia.
Flores, J., P. Nieto, S. Bermell and M.C. Miralles. 1985. Cambios en los lípidos del jamón durante el proceso de curado, lento y rápido, y su relación con la calidad. *Revista de Agroquímica y Tecnología de Alimentos* 25: 117–123.
García, I., V. Díez and J.M. Zumalacárregui. 1997. Changes in proteins during the ripening of Spanish dried beef "Cecina". *Meat Science* 46: 379–385.

García, I., V. Díez and J.M. Zumalacárregui. 1998. Changes in nitrogen fractions and free amino acids during ripening of Spanish dried beef cecina. *Journal of Muscle Foods* 9: 257–266.

García, I., J.M. Zumalacárregui and V. Díez. 1995. Microbial sucession and identification of *Micrococcaceae* in dried beef cecina, an intermediate moisture meat product. *Food Microbiology* 12: 309–315.

García-Esteban, M., D. Ansorena and I. Astiasarán. 2004. Comparison of modified atmosphere packaging and vacuum packaging for long period storage of dry-cured ham: effects on colour, texture and microbiological quality. *Meat Science* 67: 57–63.

Garrido, R., R. Domínguez, J.M. Lorenzo, I. Franco and J. Carballo. 2012. Effect of the length of salting time on the proteolytic changes in dry-cured lacón during ripening and on the sensory characteristics of the final product. *Food Control* 25: 789–796.

Garrido, R., M. Gómez, I. Franco and J. Carballo. 2009. Lipolytic and oxidative changes during the manufacture of dry-cured lacón. Effect of the time of salting. *Grasas y Aceites* 60: 255–261.

Garrido, R., J.M. Lorenzo, I. Franco and J. Carballo. 2014. Effect of the salting duration on lipid oxidation and the fatty acid content of dry-cured lacón. *Journal of Food Research* 3: 46–60.

Lorenzo, J.M. 2006. *Study of the biochemical and microbiological changes which occur during the manufacture process of dry-cured lacón. Effect of the use of some additives.* Doctoral Thesis, University of Vigo, Spain.

Lorenzo, J.M., D. Franco and J. Carballo. 2014. Effect of the inclusion of chestnuts in the finishing diet on volatile compounds during the manufacture of dry-cured "Lacón" from Celta pig breed. *Meat Science* 96: 211–223.

Lorenzo, J.M., M.C. García Fontán, I. Franco and J. Carballo. 2008a. Biochemical characteristics of dry-cured lacón (a Spanish traditional meat product) throughout the manufacture, and sensorial properties of the final product. Effect of some additives. *Food Control* 19: 1148–1158.

Lorenzo, J.M., M.C. García Fontán, I. Franco and J. Carballo. 2008b. Proteolytic and lipolytic modifications during the manufacture of dry-cured lacón, a Spanish traditional meat product. Effect of some additives. *Food Chemistry* 110: 137–149.

Lorenzo, J.M., M.C. García Fontán, I. Franco, B. Prieto and J. Carballo. 2002. Relationship between biochemical characteristics and sensory quality in dry-cured lacón (a traditional meat product) marketed in Galicia (NW of Spain). Suggestions for improvement. *Alimentaria* 338: 31–37.

Lorenzo, J.M., S. Martínez, I. Franco and J. Carballo. 2007. Biogenic amine content during the manufacture of dry-cured lacón, a Spanish traditional meat product. Effect of some additives. *Meat Science* 77: 287–293.

Lorenzo, J.M., B. Prieto, J. Carballo and I. Franco. 2003. Compositional and degradative changes during the manufacture of dry-cured "lacón". *Journal of the Science of Food and Agriculture* 83: 593–601.

Marra, A.I., A. Salgado, B. Prieto and J. Carballo. 1999. Biochemical characteristics of dry-cured lacón. *Food Chemistry* 67: 33–37.

Martín, L., T. Antequera, J. Ventanas, R. Benítez-Donoso and J.J. Córdoba. 2001. Free amino acids and other non-volatile compounds formed during processing of Iberian ham. *Meat Science* 59: 363–368.

Martín, L., J.J. Córdoba, T. Antequera, M.L. Timón and J. Ventanas. 1998. Effects of salt and temperature on proteolysis during ripening of Iberian ham. *Meat Science* 49: 145–153.

Molina, I. and F. Toldrá. 1992. Detection of proteolytic activity in microorganisms isolated from dry-cured ham. *Journal of Food Science* 57: 1308–1310.

Official Journal of the European Communities. 2001. COMMISSION REGULATION (EC) No 898/2001 of 7 May 2001 L 126, Vol. 44, 8 May 2001 (p 18). Supplementing the Annex to Regulation (EC) No 2400/96 on the entry of certain names in the "Register of protected designations of origin and protected geographical indications" provided for in Council Regulation (EEC) No 2081/92 on the protection of geographical indications and designations of origin for agricultural products and foodstuffs.

Parra, V., J. Viguera, J. Sánchez, J. Peinado, F. Espárrago, J.I. Gutiérrez and A.I. Andrés. 2010. Modified atmosphere packaging and vacuum packaging for long period chilled storage of dry-cured Iberian ham. *Meat Science* 84: 760–768.

Puga y Parga, M.M. 1905. *La cocina práctica [The practical cuisine]*. La Coruña, Spain: Editorial Roel.

Purriños, L., R. Bermúdez, D. Franco, J. Carballo and J.M. Lorenzo. 2011. Development of volatile compounds during the manufacture of dry-cured "Lacón", a Spanish traditional meat product. *Journal of Food Science* 76: C89–C97.

Purriños, L., D. Franco, J. Carballo and J.M. Lorenzo. 2012. Influence of salting time on volatile compounds during the manufacture of dry-cured pork shoulder "lacón". *Meat Science* 92: 627–634.

Rico, E., F. Toldrá and J. Flores. 1990. Activity of cathepsin D as affected by chemical and physical dry-curing parameters. *Zeitschrift für Lebensmittel-Untersuchung und-Forschung* 191: 20–23.

Rodríguez, M.P., J. Carballo and M. López. 2001. Characterization of the lipid fraction of some Galician (NW of Spain) traditional meat products. *Grasas y Aceites* 52: 291–296.

Sárraga, C., M. Gil, J. Arnau, J.M. Monfort and R. Cussó. 1989. Effect of curing salt and phosphate on the activity of porcine muscle proteases. *Meat Science* 25: 241–249.

Silla Santos, H. 1996. Biogenic amines: their importance in foods. *International Journal of Food Microbiology* 29: 213–231.

Toldrá, F. and J. Flores. 1998. The role of muscle proteases and lipases in flavor development during the processing of dry-cured ham. *Critical Reviews in Food Science and Nutrition* 38: 331–352.

Veiga, A., A. Cobos, C. Ros and O. Díaz. 2003. Chemical and fatty acid composition of "Lacón gallego" (dry-cured pork foreleg): differences between external and internal muscles. *Journal of Food Composition and Analysis* 16: 121–132.

Ventanas, J., J.J. Córdoba, T. Antequera, C. García, C. López-Bote and M.A. Asensio. 1992. Hydrolysis and Maillard reactions during ripening of Iberian ham. *Journal of Food Science* 57: 813–815.

Virgili, R., G. Parolari, C. Soresi Bordini, C. Schivazappa, M. Cornet and G. Monin. 1999. Free amino acids and dipeptides in dry-cured ham. *Journal of Muscle Foods* 10: 119–130.

4 Fresh Pork Sausage

Manufacturing Process, Chemical Composition and Shelf Life

*M.W. Schilling, T.T. Dinh, A.J. Pham-Mondala,
T.R. Jarvis, and Y.L. Campbell*

CONTENTS

4.1   INTRODUCTION

Fresh pork sausages are coarsely ground, comminuted, uncured meat products that are usually seasoned with herbs and spices and stuffed into casings that are cylindrical in shape. By definition, fresh pork sausage in the United States must not contain more than 50% fat and is limited to 3% added moisture (9 CFR § 319.141; Code of Federal Regulations CFR 2015). However, most fresh pork sausage in the United States contains between 20 and 30% fat to optimize palatability. Prasky, pork sausage, Polish sausage, Lyons sausage, kielbasa, Italian sausage, and bratwurst can all fall under the umbrella of pork sausage if they are made wholly of pork and meet the definition of fresh pork sausage for the countries in which they are sold. The fresh sausage market is affected by seasonal consumer purchasing habits, which peak during summer. This is a challenge for the industry in the United States because demand exceeds processing capability in the summer but is less for the majority of the year. This leads to the importance of making a product that can be frozen and thawed when consumer demand is greater.

The shelf life of commercial pork sausage is also an important factor in the refrigerated meat case. Raw material selection, ingredient technology, and sanitation are all crucial factors in the manufacture of pork sausage so that it can have an acceptable refrigerated shelf life under retail display lights. The use of pre-rigor pork and the inclusion of antioxidants and antimicrobials are important to meeting shelf-life requirements. Temperature control and sanitation are also essential to produce products with low-spoilage bacterial counts of *Pseudomonas* and lactic acid bacteria. Seasonal consumer purchasing habits, fluctuating costs of raw materials, and the relatively short shelf life of fresh pork sausages rank as some of the greatest challenges in this market. Categories of processed meat products with cleaner labels are widely available and continue to gain prominence in the retail market. However, the production of fresh sausage with clean labels poses a great challenge to the optimization of shelf life. There is a continued research effort to identify cost-effective ingredients that maintain desirable palatability without compromising the quality of fresh pork sausage. Among them, additives such as synthetic antioxidants and natural plant extracts are commonly used to enhance shelf life.

4.2 FRESH PORK SAUSAGE REGULATIONS AND STANDARDS BY COUNTRY

Fresh pork sausage is defined as a fresh, uncooked sausage made entirely of pork and seasoned with salt, pepper, and sage. The specifications and standards of identity of sausage vary by country, some of which are more specific than others. The United States of America (USA) Code of Federal Regulations (CFR) and Canada's Safe Foods for Canadians Regulations (SFCR) consist of very specific regulations for meat and meat products, whereas the regulations in Brazil, China, and Europe regarding sausage are not very specific. The USA CFR states that fresh pork sausage is prepared with fresh and/or frozen pork and does not include pork by-products, but may include mechanically separated meat in accordance with section 319.6 of the CFR (9 CFR § 319.141 & 9 CFR § 319.6; Code of Federal Regulations CFR 2015), which states that mechanically separated pork that has a protein content of not less than 14% and a fat content of not more than 30% may constitute up to 20% of the pork sausage. Substances that may be added to fresh pork sausage as seasonings are regulated in section 318 of the CFR (9 CFR § 318; Code of Federal Regulations CFR 2015). This section states that all ingredients and other articles used in sausage shall be clean, sound, healthful, wholesome, and otherwise, such as not resulting in adulteration. This CFR also states that sheep, goat, or swine intestines are the only acceptable sources of animal casings. In addition, by law, the finished product of fresh pork sausage in the US must not contain more than 50% fat, and water or ice used in mixing or chopping may not exceed 3% of the total ingredients used (9 CFR § 319.141; Code of Federal Regulations 2019).

Canada's SFCR regulates uncooked sausages and sausage meats' mandatory ingredients, optional ingredients, mandatory treatments and processes, and minimum and maximum amounts of specific ingredients. Canada's SFCR requires fresh pork sausage to contain boneless meat, fresh meat by-product, fresh mechanically separated meat, or any combination of those (Canadian Food Inspection Agency 2018). The regulations allow for optional ingredients such as fillers, seasoning, spices, sweetening agents, flavor enhancers, salt, and water and specify that comminuted and mechanically separated meat must be frozen prior to use (Canadian Food Inspection Agency 2018). The SFCR also requires a minimum of 7.5% meat product protein and a minimum of 9% total protein when sold as a fresh meat product (Canadian Food Inspection Agency 2018). If extenders are added to a meat product, a minimum of 9% protein and a maximum of 40% fat are required, and if fillers are added, a minimum of 9.5% meat protein and 11% total protein content are required.

The minimum meat content for a sausage to be labeled pork sausage according to European regulations is 42% (30% for other types of meat sausages), although to be classed as meat, the pork meat cannot contain more than 30% fat and 25%

connective tissue (European Union 2011). A *quantitative ingredient declaration* (QUID) is required, stating the percentage of each ingredient. For sausage, this QUID does not need to state the specific parts of the hog but, instead, *pork (42%)* is acceptable labeling. The Brazilian Government Regulatory Instruction for sausage is regulatory instruction number 4 of March 31, 2000 (Legislação: Instrução Normativa SDA – 4 2000). This regulatory body defines sausage as an industrialized meat product obtained from butcher's meat, with or without added adipose tissues and ingredients, being embedded in a natural or artificial wrap, and subjected to the appropriate technological process. In addition, fresh sausages must have a maximum of 70% moisture and 30% fat, a minimum of 12% protein, and a maximum of 0.1% calcium (Legislação: Instrução Normativa SDA – 4 2000).

4.3   OVERVIEW OF INGREDIENTS

It is essential to start with high-quality raw materials as close to the harvest date as possible and to monitor the color, pH, and bacterial counts of those raw materials, because the quality of the raw materials will dictate the quality of the finished product (Sindelar 2014). Permissible ingredients differ by product category and most product categories have what is known as the standard of identity (SOI). For fresh pork sausages in the United States, the SOIs include a maximum of 50% fat and 3% of added water based on the finished product formulation. The addition of pork by-products, paprika, binders, or extenders is not allowed. A typical sausage formulation is based on a 100 lb or 100 kg meat block (Pearson and Gillett 1996) with 20–30% fat. Regular pork trimmings are obtained from primal cuts or removed when shaping hams, shoulders, loins, butts, and bellies. Although fat content differs by primal cuts, trimmings can be blended to achieve the desired fat content. The basic non-meat ingredients for sausage include salt, water, sweeteners, spices, spice extracts, natural flavors, and antioxidants. These ingredients stabilize the sausage mixture and add specific characteristics and flavors to the final product. Most fresh pork sausages will contain approximately 50–65% moisture, 20–30% fat, and 12–18% protein.

4.3.1   RAW MEAT: MUSCLE PROTEINS AND LIPIDS

Over 70% of the fresh pork sausage manufactured in the United States is made from hot-boned culled sows. Large processing facilities hot-bone carcasses and make sausage in the same facility within a few hours. Hot-boning is the pre-rigor excision of muscle from carcasses immediately after evisceration, prior to the development of rigor mortis, followed by pre-blending with salt and rapid chilling. The interest in hot-boning is a result of its potential advantages over cold-boning: higher meat yields (1.4%), less labor (20% faster, 4 minutes/carcass), less weight loss during chilling (1.5% less), less drip loss in vacuum packages (0.1–0.6%), more uniform products, a darker color, reduced refrigerator space (50–55%), lower refrigeration costs (40–50%), shorter processing time (40–50%), lower transport costs (primal vs. carcasses), and improvements in the functional properties of the meat (Ockerman and Basu 2014a). Pre-rigor meat provides a higher water-holding capacity (WHC) and better fat binding properties. Thus, it is better suited to the production of comminuted meat products such as sausages (Allais 2010; Bentley et al. 1988). Improvement in functional properties of pre-rigor meat is a result of higher pH and more ATP (adenosine tri-phosphate) in pre-rigor muscle, which inhibits myosin-actin binding, allowing more inter-filament space for greater fiber hydration (Pisula and Tyburcy 1996). Adding salt prior to the decrease in ATP concentration further arrests rigor mortis by imparting ionic strength, increasing electrostatic repulsion, and depolarizing myosin filaments. The use of pre-rigor meat in the manufacture of sausage also maintains myoglobin in the reduced state by preserving metmyoglobin reducing enzyme systems (Savic 1985). Increased sausage microbiological shelf life has also been attributed to hot-boning and rapid salting since the surface bacteria do not have time to proliferate with the salt addition (Ockerman and Basu 2014b).

4.3.1.1 Muscle Proteins

Muscle proteins are classified based on their solubility in aqueous solutions: myofibrillar proteins (salt-soluble), sarcoplasmic proteins (water-soluble), and stromal proteins (insoluble) (Xiong 1997). Myofibrillar proteins play the most essential role in stabilizing the sausage mixture during meat processing as they interact with both water and fat in fresh pork sausage. The chemical interaction between the polar groups from myofibrillar proteins and water molecules is the most important water–protein interaction. Myoglobin is a water-soluble protein responsible for meat color (Suman and Joseph 2014). Myoglobin possesses a globin moiety with a polypeptide chain that forms a coil around the heme prosthetic group. This coil makes this hydrophobic heme group soluble and protects the heme iron from oxidation. The heme group has an iron atom at the center of the porphyrin ring that can exist either in reduced (ferrous/Fe^{2+}) or oxidized (ferric/Fe^{3+}) states (Suman and Joseph 2014). Bonding of the ferrous iron to the tetrapyrollic ring structure fulfills four of the six coordination positions of the iron atom. The fifth position is attached to an imidazole ring of a globin histidine residue (proximal histidine) within myoglobin and connects heme to the globin chain. The iron atom's sixth binding site on the porphyrin ring determines meat color and is accessible to oxygen or other small ligands such as carbon monoxide or nitric oxide. In contrast, water has limited access to the heme group due to the hydrophobic nature of the imidazole ring. In packaged fresh meats, the chemistry of the three forms of myoglobin, oxymyoglobin (Fe^{2+}), metmyoglobin (Fe^{3+}), and deoxymyoglobin (Fe^{2+}), are alike except at the sixth coordination position of heme iron. Oxymyoglobin is in the reduced state where the sixth coordinate of heme iron is occupied by oxygen, providing a bright cherry-red color in pork sausage. Myoglobin bonding is more stable toward oxidation (metmyoglobin) than reduction (deoxymyoglobin) because oxygen needs electrons to maintain hydrogen bonding with the distal histidine residue (Renerre 2000). Metmyoglobin is the oxidized form of myoglobin with a ferric iron atom. The formation of metmyoglobin is associated with meat discoloration Metmyoglobin is incapable of binding oxygen due to the water molecule at the sixth coordinate of the ferric heme. Deoxymyoglobin exhibits a purplish-red color and is also in the reduced state although it either does not have any ligand bound or has water bound with the heme iron. This chemistry is very important because ingredient technology is needed in fresh pork sausage to max-imize color and shelf life by maintaining myoglobin in the oxymyoglobin state as long as possible.

4.3.1.2 Muscle Lipids

Muscle lipids contribute to the taste sensation by the release of desirable flavors in the mouth and have a direct impact on flavor by serving as a solvent and precursor of aroma compounds (Kinsella 1989). In sausage, intermuscular and adipose tissue lipids (visible fat) constitute the major fat fraction and are primarily composed of triglycerides. Phospholipids constitute a small amount of the muscle weight (0.5–1%), yet they play a critical role in maintaining quality due to their suscept-ibility to oxidation, which contributes to the development of off flavors (Undeland et al. 2002). Oleic (18:1 *cis*-9) (40–49%), palmitic (16:0) (25–30%), and stearic (18:0) (7–27%) are the most predominant fatty acids in pork fat. Linoleic (18:2 *cis*-9, 12) (2–20%), palmitoleic (16:1 *cis*-7) (1–7%), linolenic (18:3 *cis*-9, 12, 15) (0.2–0.6%), and arachidonic (20:4 *cis*-5, 8, 11, 14) (0.2–2%) are the major un-saturated fatty acids. Most of the major fatty acids, such as oleic, palmitic, and stearic acids, are derived from *de novo* fatty acid biosynthesis, whereas linoleic and linolenic acids are essential fatty acids derived from plant materials included in the diet. The endogenous elongation and Δ^9 desaturation of palmitic acid contribute to the predominance of oleic acid. The chemical structure of a lipid molecule, parti-cularly the number and location of the double bonds, determines its susceptibility to oxidation. For instance, the carbon–hydrogen bond dissociation energies of fatty acids are lowest at the positions between two adjacent double bonds. Accordingly, these positions are the thermodynamically favored sites for attack by lipid peroxyl radicals. This results in the formation of various hydroperoxides, which decompose to form many compounds that are undesirable for meat flavor, especially short-chain acids, aldehydes, and ketones. Therefore, the fatty acid composition of meat plays a key role in shelf life and quality.

4.3.2 WATER

Water is added to facilitate chopping and mixing with non-meat ingredients, de-crease formulation costs, and reduce mechanical heating of the sausage batter. Water helps extract myofibrillar proteins, which will in turn increase binding, WHC, and production yield. Water also acts as a carrier of many important func-tional ingredients and spices. Water addition is regulated in the US and many other countries, as this can lower the per-unit-weight cost. In the US, water is limited to 3% of the formulation of fresh pork sausage.

4.3.3 SALT

Salt is the most commonly used functional ingredient in the meat industry and is considered a crucial ingredient in sausage manufacture. Salt (NaCl; 1.5–2.5%) is added to fresh pork sausage formulations to (1) solubilize and extract myofibrillar proteins, (2) increase moisture retention, (3) impart flavor, (4) inhibit microbial growth, and (5) extend product shelf life (Romans et al. 2001). Salt facilitates the solubilization and extraction of myofibrillar proteins, myosin and actin, by increasing the electrostatic repulsion between the myofilaments. This results in the swelling of myofibrils and the unfolding of proteins (Rust 1987). These electrostatic repulsions arising from an increase in the negative protein charge as Cl^- ions are more strongly bound to the proteins than Na^+ ions (Allais 2010). This selective binding of the Cl^- ions to myofibrillar proteins loosens the myofibrillar lattice, due to repulsion between myosin filaments (Offer and Knight 1988). Furthermore, Na^+ ions are pulled close to the filament surfaces by the protein's electrical forces. This increases osmotic pressure within the myofibrils, causing the filament lattice to swell. Salt also imparts a salty taste. Moreover, early exposure of undenatured proteins in muscles to salt improves red color stability. As a preservative in the sausage, salt acts as a bacteriostatic agent, inhibiting bacterial growth and subsequent spoilage through reduced water activity (a_w; Pearson and Gillett 1996). Nearly all bacterial cell walls will allow water to pass through the cell wall barrier from the lower salt concentration (in the bacterial cell) to the greater salt concentration (in the muscle tissue) as a result of osmotic pressure (Ockerman and Basu 2014b). This will eventually result in bacterial cell death and lysis due to the lack of moisture.

4.3.4 SWEETENER

A wide variety of sugar-based products such as sucrose, dextrose, corn syrup, and corn syrup solids are commonly included in fresh pork sausage. Sugar provides a sweetening and flavor-enhancing property, which improves the overall appeal of sausage. Sugar counteracts the harshness of salt and caramelizes during cooking to yield a more flavorful end-product. Non-reducing sugars, such as corn syrup and dextrose, promote browning through Maillard reactions with amino acids. Sweeteners also bind water and decrease a_w, thereby enhancing shelf life.

4.3.5 SPICES

A multitude of natural, whole, ground, cracked, and rubbed spices, along with sterilized spices, oleoresins, and some dehydrated vegetables may be included in fresh pork sausage. In fresh pork sausage, spices and flavorings come in many forms and are usually added in their natural form, which may be whole, cracked, flaked, or powdered. Larger spice pieces create bold, if variable, flavors. In contrast, smaller spices (cracked, flaked or powdered) allow uniform flavor distribution. The spice formulations in sausage vary by region. For example, the type and amount of pepper vary by sausage type and intended usage. The characteristic spice profiles of Wisconsin-style fresh pork sausage include spices such as nutmeg, white pepper, mace, ginger, black pepper (42 mesh), and spice extractives of nutmeg and mace (Brown 2009). A portion of the natural ground spices could be replaced with spice extractives to maximize the shelf life. In addition to imparting unique flavors and aromas to sausages, some spices have antioxidative and antimicrobial activities. Spices and flavorings are added to sausages at 250–500 g per 100 kg of meat without regulatory limit in the US, with the exceptions of paprika, turmeric, and saffron (coloring materials), mustard, and hydrolyzed protein (protein fillers). Several issues encountered with the use of spices include discoloration of fresh products, short shelf life of spices (6–8 months with proper storage), high bacterial loads (mainly spores), and flavor variability among the batch. Two advantages of using spice extractives in fresh sausages are that they provide flavor uniformity and that they can be labeled as natural flavorings although having antioxidant and antimicrobial effects.

4.3.6 ANTIOXIDANTS

Antioxidants maintain quality and extend shelf life by enhancing oxidative stability, especially for products that are intended for long-term frozen storage with subsequent retail display. Antioxidants are substances that have the ability to delay the onset or reduce the rate of oxidation in food products. Despite the name implication, antioxidants do not stop oxidation from happening. They only slow the rate of oxidation by inhibiting the cascade of free radicals in the oxidative process, thus extending the shelf life of a food product. The USDA Code of Federal Regulations CFR (2015), 21 CFR 170.3 defines antioxidants more succinctly as "substances used to preserve food by retarding deterioration, rancidity, or discoloration due to oxidation." Antioxidants delay oxidative reactions in foods by various modes of action, including free-radical termination, chelation of metal ions that catalyze lipid oxidation, quenching singlet oxygen, destroying peroxides to prevent radical formation, removing reactive oxygen species such as oxygen radicals, and controlling various oxidation intermediates (Eskin and Przybylski 2000; Shahidi et al. 1992). Based on these mechanisms, antioxidants can be classified as primary or *chain-breaking* antioxidants, secondary or *preventive* antioxidants, or synergists. Some substances have more than one mechanism of antioxidant activity and are referred to as multiple-function antioxidants.

Additives used as antioxidants are classified according to their sources, either chemically synthesized or naturally extracted. Among the most popular commercially available synthetic antioxidants approved by the US Department of Agriculture (USDA) and other regulatory agencies for food applications are butylated hydroxyanisole (BHA), butylated hydroxytoluene (BHT), tertiary-butylhydroxyquinone (TBHQ), and propyl gallate (PG) (Hazen 2005). These compounds are all hydroxyl substituted phenols and act as radical scavengers by donating hydrogen atoms from their hydroxyl moiety to various lipid radicals, converting them into stable nonreactive species and leading to the cessation of the free radical chain reaction (Berdahl et al. 2010).

Both BHA and BHT are oil-soluble, whereas PG is very water-soluble and only slightly soluble in animal and vegetable oils. Additionally, BHT is not as effective as BHA due to the presence of the two t-butyl groups, creating steric hindrance. The Code of Federal Regulations CFR (2015) permits the use of BHA or BHT in meat products at 0.01% on a fat basis or at 0.02% in combination with other antioxidants (9 CFR § 424.2). Their use in the food industry has been recently declining due to consumer demand for natural products (McClements and Decker 2000; Shahidi and

Ambigaipalan 2015). Despite this, these compounds are still being used because of their reliability, ease of use, and cost-effectiveness. As a result, sausage manufacturers are introducing chilled and frozen product options with fewer artificial ingredients as "healthier" alternatives. The interest by the meat industry in the application of natural plant-derived ingredients with both antioxidant and antimicrobial characteristics has increased steadily in recent years.

Plants have been targeted as potential sources of natural antioxidants due to their ability to produce compounds that can be used to counteract the effects of reactive oxygen species on food (Grün 2009). Extracts from spices, rosemary, thyme, and sage have been reported to possess antioxidant properties comparable with or greater than BHA and BHT (Shahidi and Ambigaipalan 2015; Shahidi et al. 1992). The antioxidant properties of extracts from these plant materials can be attributed mostly to phenolic compounds, which are secondary metabolites that are naturally produced and ubiquitously present in plants. In fact, some spice and herbs, such as rosemary, sage, thyme, oregano, turmeric, ginger, pepper, and cloves have been shown to possess phenolic compounds that exhibit similar antioxidative effects to that of BHA, BHT, or α-tocopherol (Martínez-Tomé et al. 2001; Shan et al. 2005). Plant phenolics have been shown to be effective at reducing lipid oxidation and rancidity in a large variety of muscle foods (Decker et al. 2000; Grün 2009).

According to the Code of Federal Regulations CFR (2015), natural flavors include the natural essence, oleoresin, or extractive that contains the flavoring constituents derived from herb, leaf, bark, root, or similar plant material listed in 21 CFR § 182.10, 182.20, 182.40, 182.50, and 184 and the substances listed in 21 CFR § 172.510 for FDA regulated foods (21 CFR § 101.22). The USDA regulations require food manufacturers to disclose in writing on the product packaging the addition of natural flavors to the product, especially if they are present in amounts greater than 0.1%, in which they are considered to have a technical effect in the finished meat and poultry product (9 CFR § 424.21, Code of Federal Regulations CFR (2015). Thus, regulations permit the labeling of plant-derived extracts as natural flavors.

4.3.7 ANTIMICROBIALS

Antimicrobials are used in fresh, refrigerated sausage-type products to extend shelf life by delaying the log phase of bacterial growth. A combination of sodium lactate and sodium acetate, or acetic acid (vinegar) can extend the shelf life of refrigerated sausage. In addition, sodium or potassium lactate and sodium diacetate are bacteriostatic to *Listeria monocytogenes* and spoilage microorganisms in cooked products. Propionic acid is also gaining attention as an antimicrobial since it is more hydrophobic than lactic and acetic acids, therefore more interactive with the hydrophobic nature of bacterial cell membranes. In addition, propionic acid is more disassociated than other organic acids at the pH of processed meats since it has a greater pKa (4.9) than acetic (4.8) and lactic (3.8) acids. For the cleanest possible label, vinegar can be used at up to 2.5% (liquid vinegar) without negatively impacting sensory properties. Naturally occurring lactic acid, citric acid, and acetic acid can be included through using lactic acid starter cultures, lemon juice solids, and cultured corn sugar, respectively. Natural antimicrobials that can be used in processed meat products without negatively impacting the sensory properties of the product include buffered vinegar, cultured corn sugar, cultured dextrose, lemon juice solids, lemon powder, lime powder, and cherry powder. Buffered vinegar is predominantly acetic acid that has been buffered with potassium hydroxide so that it is at a pH that is suitable for the product. Cultured corn sugar predominantly contains lactic acid in the form of sodium and potassium lactate, and some propionic and acetic acids in the form of sodium propionate and sodium acetate. Cultured dextrose is predominantly propionic acid and acetic acid with some lactic and butyric acids. Lemon juice solids, lemon powder, and lime powder contain citric acid and ascorbic acid. Cherry powder is made from acerola cherries and has significant amounts of malic acid and ascorbic acid. All of these natural ingredients function as antimicrobials due to the natural presence of organic acids. Ascorbic acid also acts as an antioxidant in fresh sausage to preserve the red color or serves as a cure accelerator in cured, cooked products, but also provides antimicrobial properties. High hydrostatic pressure and post-packaging pasteurization also provide clean-label antimicrobial applications.

4.3.8 PROCESSING AIDS

A processing aid is a substance used in the production of a food product that may be in the finished product but does not have to be disclosed to the consumer as an ingredient. One example of a processing aid that can be used in the production of fresh pork sausage is potassium carbonate. Potassium carbonate is considered a processing aid because it can be added to the 3% water in the formulation as a pH control agent to raise pH and WHC.

4.3.9 CASINGS

Casings serve as containers and molds for most sausages. Casings are selected to form specific physical characteristics of sausages such as size, shape, and appearance of the color. Historically, animal intestines were the primary material to produce sausage casing. However, the development of technology has enabled the production of synthetic cellulose and collagen casings in the desired size, appearance, and mechanical strength for various stuffing and linking purposes. Natural casings from hogs, sheep, and cattle are commonly used for fresh sausage. They are commercially designated as rounds or small casings (small intestines from sheep, goats, and hogs), runners (small intestines from cattle), middles (large intestines from cattle and hogs), and bungs (cattle caecum or the end of the intestinal tract from hogs). Other anatomical locations of the intestinal tract, such as the esophagus, bladders, and stomachs, are also used for special types of sausage. Cattle intestines are not as popular as those from hogs and sheep. In the United States, natural casings have to be from either goats, sheep, or swine to be used in the production of fresh pork sausage.

The intestines are removed from animals during slaughter and cleaned through extensive washing, including having the mucosa layer removed. Natural casings are also sorted and graded by size and conditions, and packaged with salt in shipping containers. They have a long shelf life under refrigeration, at least several months. Salt is removed and natural casings are soaked in cold water for re-hydration before use. It is recommended that water be applied regularly to maintain the elasticity of natural casings. Natural casings provide old-fashioned appearance and texture. However, they are expensive and often non-uniform, which increases breakage during stuffing and requires more care. Natural casings are difficult to spool onto stuffing tubes, are permeable to water and smoke, and shrink during cooking. Natural casings are often treated with an antimicrobial treatment prior to stuffing to lower microbial counts on the casings so that shelf-life of the fresh sausage can be maximized.

Synthetic casings are produced from either cellulose (inedible) or regenerated collagen (edible). Co-extruded collagen casing has become more and more popular in automatic processing systems. Cellulose casing is usually used for skinless sausage such as frankfurters. Regenerated collagen casing is manufactured from collagen extracted from animal hides or other animal-sourced collagenous materials. Collagen is hydrated and solubilized in acidic solution and then precipitated in concentrated ammonium sulfate and formed into a continuous tube with a specific size and thickness. Regenerated collagen tubes are dried and can be spooled without re-hydration. Co-extruded collagen casing is produced at the same time as sausage production. As the sausage batter is extruded through the filler tube, the gelatinized collagen solution is coated onto sausage links, which are subsequently immersed in a salt bath to set the casing firmly around the sausage links. Cellulose and other additives can be added to allow moisture and smoke permeability as well as fat leakage during cooking. In the United States, natural casings are most popular for fresh pork sausage, but both regenerated collagen and co-extruded collagen are increasing in popularity.

4.4 TRADITIONAL SAUSAGE FORMULATIONS

4.4.1 BRATWURST

Fresh bratwurst is prevalent in the upper Midwest of the USA and is described as German-style, Wisconsin-style, and Swiss-style brats (Romans et al. 2001). Fresh brats are coarsely ground with various seasoning combinations and stuffed into hog casings. One common formulation for bratwurst includes 93 kg of pork trimmings (70–80% lean), 1.8 kg salt, 1.8 kg non-fat dry milk, 2.97 kg water, 0.19 kg black pepper, and 0.09 kg each of mace, coriander, and hot mustard (Romans et al. 2001).

4.4.2 BREAKFAST SAUSAGE

Breakfast sausages are produced from fresh or frozen pork but can be produced with beef or pork and beef. Breakfast sausages are finely ground, and seasoned with sage, salt, and pepper. They are stuffed into a variety of casings and linked to produce various sizes. Some typical casings used for breakfast sausages include narrow or medium sheep casings, narrow hog casings, or regenerated collagen casings (Romans et al. 2001). One possible formulation that would meet the definition of fresh pork sausage in the US consists of 47.7 kg pork jowls, 44.6 kg pork trimmings (80% lean), 3 kg ice, 2.6 kg corn flour, 1.8 kg salt, 0.22 kg white pepper, 0.05 kg nutmeg or mace, and 0.03 kg each of thyme and sage (Romans et al. 2001).

4.4.3 TRADITIONAL FRESH PORK SAUSAGE

Traditional fresh pork sausage is fresh or uncooked sausages produced from pork and seasoned with salt, pepper, and sage at minimum. This type of sausage is usually sold in 1- or 2-lb (0.45 or 0.90 kg) chubs, links, or in patties. One possible formulation for traditional fresh pork sausage that has been adapted from Pearson and Gillett (1996) includes either 94.4 kg of pork trimmings (70% lean) or 94.4 kg whole hog pre-rigor pork, 2.98 kg ice water, 1.8 kg salt, 0.31 kg white pepper, 0.24 kg sucrose, 0.12 kg sage, and 0.05 kg mace.

4.4.4 ITALIAN PORK SAUSAGE

Although other spices may be utilized, Italian sausage is characterized by the spices of pepper, fennel, and/or anise. Italian sausage is typically made from pork but can also be formulated to include beef and/or veal. For fresh Italian pork sausage, a formulation meeting the definition of fresh pork sausage, similar to the one used by Pearson and Gillett (1996), includes 94.4 kg (70% lean) of pork trimmings or whole hog pre-rigor pork, 3 kg ice water, 1.76 kg salt, 0.24 kg each of white pepper, fennel, and crushed red pepper, and 0.12 kg of coriander.

4.4.5 POLISH AND KIELBASA

Polish sausage is a coarse ground sausage. Kielbasa is a generic term for sausage in Poland. Kielbasa may be spicier. Both sausages are usually produced with pork and beef, with pork being the primary meat ingredient. The most distinctive flavor of Polish sausage is garlic. Polish sausages are generally 10–12 cm long and 3–4 cm in diameter and are smoked thoroughly (Romans et al. 2001). Polish sausage and/or kielbasa can fall under the umbrella of pork sausage in the US if the added water is 3% or less, pork is used in the formulation, and the fat content is 50% or less. A formulation adapted from Romans et al. (2001) consists of 92.1 kg of pork trimmings (70–80% lean), 3 kg of ice water, 1.8 kg of salt, 1.0 kg of sugar, 1.8 kg of corn syrup, 0.16 kg of black pepper, 0.05 kg of coriander, 0.05 kg of monosodium glutamate, 0.03 kg of ginger, and 0.01 kg of garlic.

4.4.6 WHOLE HOG SAUSAGE

Whole hog sausage is usually produced from the entire hog, usually heavier hogs. Whole hog sausage is produced from fresh or frozen pork in proportions similar to that of the animal. It can also be produced using pre-rigor meat and is often produced the same day as slaughter. A formulation for whole hog sausage adapted by Pearson and Gillett (1996) contains 93.9 kg of 70–75% lean pre-rigor sow, 3 kg of water, 1.8 kg of salt, 1 kg of sugar, 0.25 kg of white pepper, and 0.05 kg of ginger.

4.4.7 Commercial Ingredients

It is common to include more than just the basic ingredients listed above when making commercial products in order to increase yields, lower costs, extend shelf life, and preserve color. Each of these ingredients is generally recognized as safe (GRAS) according to 21CFR182. Some of these ingredients include butylated hydroxyanisol (BHA), butylated hydroxytoluene (BHT), propyl gallate (PG), monosodium glutamate (MSG), high fructose corn syrup (or dextrose or corn syrup solids), sodium lactate, and buffered vinegar.

Oxidation is one of the most common causes of meat spoilage. To combat this issue, BHA and BHT are added to sausage formulation as antioxidants. According to 21 CFR § 182.3169 and §182.3173, BHA and BHT are GRAS when the total antioxidant is 0.02% or less of the fat or oil content of the sausage. In addition to BHA and BHT, PG can also be included in sausage formulations to retard oxidation and extend shelf life (21 CFR § 170.3). According to 21CFR §184.1660, PG is also GRAS when the total antioxidant is 0.02% or less of the fat or oil content of the sausage. Monosodium glutamate (MSG) is used to improve the savory and umami flavor of the sausage. According to 21CFR, MSG is also GRAS, and according to 21CFR §101.22, it must be declared as MSG on the ingredient label. High fructose corn syrup (dextrose or corn syrup solids) is used as a less expensive sugar source than sucrose or dextrose and is metabolized in a similar mechanism to that of sugar. High fructose corn syrup is also used because it is more stable than sucrose. Sodium lactate can be used up to 4.8% in a meat product formulation as an antimicrobial (Sebranek 2016). Sodium lactate at 2–3% extends product shelf life by slowing the growth of spoilage bacteria during retail display in refrigerated meat cases.

4.4.8 Clean Label Commercial Ingredients

There is no definition or standard of identity for a "clean" product. Organic is defined by the USDA; however, "natural" is loosely defined and the USDA has not defined a clean label. Some restaurants and companies have defined a clean label as not including certain ingredients. Some ingredients that have been removed from meat products for the consideration of clean label have important functional purposes. For example, BHA, BHT, propyl gallate, MSG, high fructose corn syrup (or dextrose or corn syrup solids), and sodium lactate all have specific functions. To replace these ingredients in fresh pork sausage, one or more ingredients must be substituted or added and the formula must be adjusted accordingly. BHA/BHT and propyl gallate can be replaced with the extracts of rosemary, green tea, oregano, and cherry powder or some combination of these extracts. Rosemary extract is commonly used to replace BHA/BHT due to naturally occurring carnosic acid and carnosol. These compounds inhibit lipid oxidation and myoglobin oxidation in sausage, similar to the functions of BHA, BHT, and PG. MSG can be replaced with savory herbs that are related to a specific product, for example, garlic, rosemary, or pepper. MSG can also be replaced with autolyzed yeast extract. High fructose corn syrup (dextrose or corn syrup solids) can be replaced with sugar (turbinado or cane sugar) at a 1:1 ratio since high fructose corn syrup is composed of fructose and glucose. This substitution may not be favored by producers due to an increase in ingredient costs. Lastly, sodium lactate can be replaced with vinegar. Usually buffered liquid or dry vinegar is used. Both ingredients inhibit microbial growth in meat products. Although having similar effectiveness, buffered dry vinegar may be preferred because of the lower shipping cost and ease of use.

4.5 PROCESSING AND EQUIPMENT

Fresh pork sausage is often produced with pre-rigor pork lean and fat trimmings. The advantages of utilizing pre-rigor meat include faster production (slaughter to packaging in less than one hour), improved color and juiciness of the finished product (from less protein denaturation), and slower color and oxidative changes during refrigerated or frozen storage (Sindelar 2014). Pre-rigor meat is salted and chilled almost immediately postmortem to preserve pre-rigor characteristics (high pH and water-holding capacity) of meat. To produce sausage batter, lean trimming meat is mixed with additional salt (if needed) and additives (antimicrobials and antioxidants) in a mixer prior to grinding. Cold water or ice (3%) is added during pre-blending to minimize mechanical heating and maintain meat temperature. The proper mixing of these ingredients is extremely important to obtain a uniform blend. Some ingredients have antioxidants effect (for example BHA, BHT, propyl gallate, natural plant extracts, spices), antimicrobials (for example spices, natural plant extracts), or prooxidants (for example, salt). Utilizing high-quality salt (low in heavy metals) and an effective antioxidant and antimicrobial system are essential for adding days of shelf life. Accordingly, ensuring widespread and uniform dispersion is important in achieving maximum antioxidant and antimicrobial benefits.

4.5.1 Primary Grind

The first step in fresh pork processing is to grind the pork trimmings or pre-rigor meat through an approximately 1.3 cm diameter plate (Figure 4.1). After grinding, the pork trimming needs to be salted with approximately 1.0–1.5% salt if the sausage is not going to be manufactured immediately. Pre-rigor meat is already salted and may not need additional salt after the primary grind.

4.5.2 Mixing or Blending

After the primary grind and prior to stuffing, the meat will be mixed with ice water, spices, sweeteners, antioxidants, and antimicrobials for approximately 2–3 minutes in either a ribbon or paddle mixer. In optimum conditions, a ribbon blender that has both forward and backward mixing options with some paddle action is the most efficient mixer. The ribbon mixer will operate based on pneumatic air pressure and can often mix ingredients for 1 minute forward and 1 minute backward. Ribbon

(a)   (b)

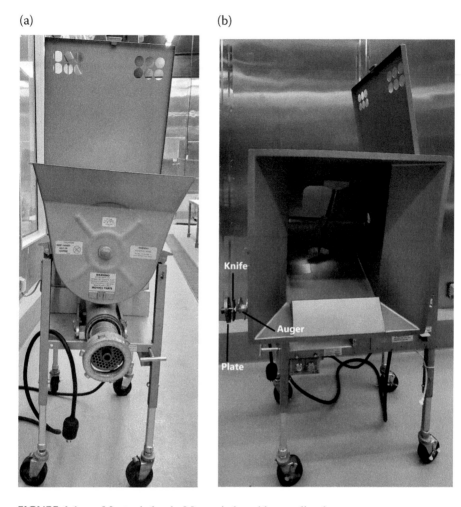

FIGURE 4.1   a: Meat grinder, b: Meat grinder with parts listed.

mixers cause diffusion, convection, and shear of the sausage formulation. The meat and ingredients should be mixed to the point that the meat is slightly tacky, with minimal protein extraction. A paddle mixer works well for smaller batches of sausage and does not mix as aggressively as a ribbon mixer. It may take closer to 3 minutes to mix the ingredients. If the sausage batter is too sticky, there was overmixing and too much protein extraction. If fat smearing occurs, this could be due to either overmixing or reaching a temperature that is too high. A mixer–grinder can also be used that contains both paddle mixing and grinding capabilities. It is important to keep the meat chilled during mixing and grinding, which can be done through CO_2 injection through a mixer hood. This keeps the temperature close to 0°C and prevents fat from smearing.

4.5.3 FINAL GRIND

The final grind is conducted through a 4 mm grinder plate after mixing. It is generally recommended that meat is chilled (0°C) prior to grinding to attain endpoint grinding temperatures below 12°C to minimize fat smearing of the product (Romans et al. 2001). Excessive grinding should be avoided as this can produce a tough product due to a high quantity of extracted salt-soluble proteins. Controlling fat smearing, thereby excessive protein extraction, achieving desirable particle size, and ensuring uniform ingredient distribution are common objectives. Mixing oxygen should also be minimized because it causes oxidation of sausage color and creates off flavors. Fat trimming is to be ground separately and mixed with the lean trimming to disperse fat into protein matrices. Spices can be added during this stage or blended with lean trimming. In some cases such as with whole-hog sausage, there is only one source of meat, and spices are added during blending with additives.

4.5.4 VACUUM STUFFING

After the initial grind, mixing, and the final grind, the sausage batter is stuffed into narrow pork casings or chubs, or put through a patty maker, depending on the desired final form of the fresh pork sausage. Vacuum stuffers are most commonly used to remove gas from the meat, especially oxygen, in order to inhibi oxidation and enhances shelf life (Figure 4.2). Natural casings are often cleaned and treated with an antimicrobial solution prior to stuffing. It is imperative that sausage batter temperatures be at or below 0°C during stuffing to reduce fat smearing. It is also important that casings be filled to maximum capacity to eliminate air pockets and to produce a plump product with the best possible appearance (Rust 1987).

After the meat batter is stuffed into casings, the encased mass is twisted or drawn together to produce links, either by hand or with mechanical devices. The fresh pork sausages are then packaged in expanded polystyrene trays, overwrapped with oxygen-permeable film which allows the product to turn red or bloom, heat-sealed, labeled, and frozen for storage, distribution, and retail sale. It is important to freeze rapidly below the glass transition temperature of pork to obtain small ice crystals and maintain product quality. At retail, fresh sausages are displayed in chilled cabinets and the fridges of grocery retailers, either in the fresh or frozen state, with the fresh form generally considered to be the higher value of the two forms (Sindelar 2014). Controlling fluctuations in temperature (during frozen storage, during retail display, during transport, or any other point in the cold change management) is also crucial. A brief environmental temperature spike has the potential to create a small but detrimental impact on shelf life. Lastly, shelf life and other quality characteristics of the finished product are also affected by hygienic conditions present during processing, such as hygiene of personnel, methods of processing applied, which delays microbiological activity, and the conditions at which the sausages are subsequently held during storage, transport, and sale.

(a)   (b)

FIGURE 4.2   a: Vacuum stuffer, b: Vacuum stuffer with parts listed.

4.5.5 EQUIPMENT SELECTION

Equipment to manufacture sausage needs to be selected based on production speed and volume as well as the layout of the facilities to allow the most efficient flow of raw materials, minimum expenses, and prevention of cross-contamination. Fresh sausage is often produced with pre-rigor meat that is usually not frozen. Therefore, a meat grinder is selected that will provide adequate pressure to extrude meat pieces through a 25 mm grinder plate and cut them with a rotating four-bladed knife.

The degree of communition is determined by the hole diameter. A grinder is usually the first piece of equipment used to reduce the size of meat pieces. A meat grinder may or may not have a bone collection system. It should not operate empty because the friction between the plate and the knife will wear them out quickly. Extruding auger, plate, and knife can be removed for cleaning and lubricated with food-grade lubricants.

In fresh sausage manufacturing, meat pieces are usually not ground below 3–4 mm. A mixer is used to pre-blend meat pieces with non-meat ingredients (additives and spices) and mix pre-blended lean and fat trimmings. Mixers are constructed with a round-bottom vessel with parallel shafts that have agitating paddles. Commercial mixers usually have two shafts to provide adequate back-and-forth agitation to blend meat and other ingredients. Regardless, ingredients must be fed into the mixers slowly and evenly, and mixers can also operate under vacuum to prevent oxidation. To ensure adequate mixing and avoid overheating, grinders and mixers should not operate at maximum capacity. If frozen meats are used for sausage manufacturing, flaking or dicing devices with rotating blades can be employed before grinding or mixing. Sausage batter is stuffed into casings by stuffers, which are either a piston- or pump-filling system. Piston stuffers have a cylindrical vessel equipped with a movable piston to push sausage batter into casings. The piston is moved by compressed air, hydraulic force, or manually. Piston stuffers are suitable for small-scale facilities or home use. Pump-operated stuffers are continuously feeding, allowing refilling during stuffing. Weight and size of sausage links can be controlled electronically through stuffing speed and time. Stuffing speed can be adjusted to reduce breakage and maximize productivity. Stuffers can also be fitted with a loading device and a linker to twist and create sausage links. Linkers are commonly used in high-capacity facilities. Linkers can be a separate piece of equipment, especially in the case of co-extruded collagen casings, because the casing needs to be set firm after stuffing before it can be twisted. In addition to grinding, chopping, mixing, and stuffing equipment, adequate chilling and freezing capacity is important to preserve the quality of meat before processing and quality of sausage before distribution.

4.6   SHELF LIFE

4.6.1   Overview

Shelf life is defined as the duration of a product's life, after which it is no longer considered acceptable (by the manufacturer, consumer, or another party) according to pre-determined thresholds developed for that specific product (Sindelar 2014). The shelf life of sausage products is defined as the period of frozen storage as well as the period after freezing during which the product retains its characteristic properties and remains suitable for consumption or the intended purpose. The deterioration of sausage quality and a decrease in shelf life are caused by lipid and protein oxidation and microbial spoilage. Three main determinants of fresh sausage shelf life in a refrigerated meat case are appearance (red color of oxymyoglobin), flavor (off-flavor notes, mostly aromas), and bacterial growth (total bacterial count). The discoloration, the development of off-flavors, and an increase in the microbial count are simultaneous causes of the end of shelf life. However, when placed under lights, discoloration of redness (brown to tan color) due to myoglobin oxidation is the most common cause of the end of shelf life.

4.6.2   Shelf Life Testing

An example of an industrial method to conduct fresh sausage shelf life includes microbial testing, instrumental color analysis, and sensory analysis of raw appearance, raw aroma, and cooked flavor. For example, a target shelf life of up to 3 months frozen followed by 10 days refrigerated storage at 800 lux would require the following methodology:

1. Store frozen below −20 °C for 3 months.
2. Slack out and store at 4–7 °C for 10 days under 800 lux light intensity.
3. Evaluate total plate counts after 1, 3, 5, 7, 8, 9, and 10 daysof storage. When bacterial counts are greater than 7 log, the product has either reached the end of shelf life if the spoilage bacteria are predominantly *Pseudomonas,* or is approaching the end of shelf life if the bacteria are predominantly lactic acid bacteria.
4. Evaluate the sensory color of blindly labeled golden standard and sausage samples after 1, 3, 5, 7, 8, 9, and 10 days of storage. Sausage has exceeded shelf life when an undesirable brown color is present that consumers find unacceptable.
5. Evaluate sausage for visible microbial growth (mold or visible bacteria) that indicates that the product has exceeded shelf life.
6. Evaluate sausage surface color using either a handheld chromameter or spectrophotometer to determine if the CIE a* value has too little redness.
7. Cook the product per packaging instructions and evaluate the product for off flavors and off textures that indicate shelf life.

4.6.3   METHODS TO DETERMINE SPOILAGE/SHELF LIFE

Microbiological (total viable count) and chemical/instrumental (for example TBARS, gas chromatography) methods have been used to evaluate the shelf life of various sausage products. However, sensory evaluation is a crucial part of all shelf life programs because packaging innovations, ingredient technology, sanitation, and automation mean that sensory defects, including oxidation, often lead to the end of shelf life for many products.

4.6.3.1   Sensory Analyses

A trained or expert panel (6–8 panelists) can be used to determine if differences exist between a control (gold standard or freshly produced product) and an older product near the end of its shelf life. In shelf-life studies, descriptive testing can be used to determine the intensity of sensory descriptors of a product and can be used to identify the end of shelf life (EOSL) for product sensory descriptors and off flavors. This is often the optimum method for the determination of shelf life, but it is very important that this data correlates with consumer perception of the product. Otherwise, the product could be deemed past shelf life but still be acceptable to consumers or vice versa. Other sensory tests commonly used include difference from control and triangle tests, with the difference-from-control test being preferred because it indicates how much the test sample is different from the control.

4.6.3.2   Fresh Sausage Flavor and Off-Flavor Development

The development of flavor and aroma in pork sausage is a complex process. Most of the volatile compounds generated during processing and subsequent storage at refrigerated temperatures are the result of chemical or enzymatic oxidation of unsaturated fatty acids and further interactions with proteins, peptides, and free amino acids. More than 150 volatile compounds have been reported in pork sausage. Thus, the typical aromas of sausage products are mainly associated with the presence of a large number of volatile compounds that are generated by the following reactions that occur during processing and storage: lipid oxidation (aldehydes, ketones, alcohols, esters, acids, and aliphatic hydrocarbons), lipolysis (short-chain free fatty acids), protein oxidation (methyl branched aldehydes), and proteolysis (nitrogen-derived volatile compounds). Many of the volatile compounds originate from spices (terpenes), whereas others were the result of the enzymatic or bacterial activities occurring in the meat during refrigerated storage or retail display. The characteristic aroma and odor threshold of each compound contributes to the final flavor and aroma of the product.

The freeze–thaw process is known to result in protein denaturation, tissue disruption, and damage to muscle fibers, which greatly increases the development of oxidative reactions (Xia et al. 2009). These undesirable processes contribute to significant negative quality changes in sausages during manufacture and storage. The production of off odors and off flavors during the retail display of comminuted meat products has also been attributed to microbial activity. Decomposition products arising via the action of bacterial enzymes during refrigerated storage including free fatty acids, peptides, and amino acids can be further oxidized to form bacterial volatiles, including diacetyl and other methylketones, carboxylic acid esters, fusel alcohols, and 1-octen-3-ol.

Typical spices added to sausages such as pepper, nutmeg, mace, ginger, and rosemary have a high impact on the aroma as the volatiles (terpenes and phenylpropanoids) derived from these ingredients are abundant (Schilling 2019). In black pepper, volatiles comprised mostly monoterpene hydrocarbons, such as sabinene (19% of the total volatile composition), limonene (17%), and β-pinene (10%), and also a large proportion (14%) of β-caryophyllene (sesquiterpene hydrocarbon) and 4% of 1,8-cineole (terpene alcohol). In nutmeg oil, a large proportion of monoterpene hydrocarbons (for example, sabinene, α-pinene, and β-pinene) have been found. In ginger, large quantities of the sesquiterpenes α-zingeribene and *ar*-curcumene, which comprise approximately 38% and 17% of the total ginger oil composition, respectively, have been extracted (Richard 1991). Rosemary essential oil contributed large quantities of aromatic hydrocarbons and alcohols such as *p*-cymene, β-terpineol, and 2-methylanisole in a porcine frankfurter-type product (Estévez, Ventanas, Ramírez, and Cava 2005).

4.6.3.3   Color Changes in Fresh Pork Sausage

In the US, most fresh pork sausage products are packaged in expanded polystyrene trays covered with oxygen-permeable stretch films to permit the rapid oxygenation of the surface myoglobin to develop the desirable red color or bloom. However, this red color only lasts for approximately 7 days when displayed in a retail environment before browning occurs due to metmyoglobin formation. Meat discoloration during frozen storage and thawing compromises its appearance as consumers have a definite aversion to brown meat color. The accumulation of metmyoglobin in meat is influenced by the rate of oxidation and enzymic or nonenzymic reducing systems. The rate of autoxidation of myoglobin is greatly dependent on storage temperature, light intensity on the display area, and method of packaging (James and James 2012).

Andersen and Skibsted (1991) reported the progress of discoloration in light (700 lux on the surface of the product) and in the dark for frozen and salted (1%) pork patties. At the end of the experiment (−18 °C, 31 days), a severe discoloration was observed, as expected for salted pork patties under illumination compared with those in the dark. During extended frozen storage, a dark brown layer of met-myoglobin can form 1–2 mm beneath the surface, which, upon thawing, causes the rapid deterioration of the surface color. Accordingly, when the freeze–thaw cycle reached five times in porcine *longissimus dorsi*, the thawing losses, cooking loss, and b^*-value (yellowness) increased with a concomitant decrease in a^*-value (redness) (Xia et al. 2009). Oxygen concentration is another crucial parameter for color stability. If oxygen permeability of the film is low but sufficient, oxygen utilization, once exceeding oxygen penetration through the film, creates a low partial pressure of oxygen, which promotes oxidation of oxymyoglobin to brown metmyoglobin. The oxidized or ferric form of the myoglobin cannot bind oxygen.

Reduction is another chemical reaction influencing red meat color. After being oxidized to metmyoglobin, myoglobin is converted to deoxymyoglobin in the presence of muscle reducing enzyme systems, the muscle's oxygen scavenging enzymes, and the NADH [Nicotinamide Adenine Dinucleotide (NAD)$^+$ Hydrogen (H)]. The reduced form of myoglobin, deoxymyoglobin, will then be able to bind oxygen again. However, during extended storage, the activities of those three factors continuously decline.

4.6.3.4   Sensory Changes in Fresh Pork Sausage

The flavor of pork sausage can be evaluated based on standard pork flavor and texture descriptors along with any other additional spices. Determining the appropriate storage time for fresh pork sausage also depends on the shelf life, flavors, acceptability, and structural stability of the product. Long-term frozen storage can lead to profound effects on the structural and chemical properties of lipids and proteins, which subsequently influence the quality attributes of the product. While freezing sausages is intended to prolong product shelf life by inhibiting chemical, enzymic, and microbial reactions, undesirable sensory changes in sausages, especially the development of rancid notes, discoloration, and unacceptable texture may still occur during extended frozen storage periods. Depending on the ingredients that are used, somewhere around 36 months is the typical shelf life of frozen fresh sausage. Additionally, the longer the product is stored in the refrigerated retail display, the more off flavors are produced from oxidative deterioration and microbial proliferation.

A progressive loss of sausage sensory quality occurs throughout frozen storage and refrigerated display (Bradley et al. 2011; Crist et al. 2014). A general pattern in off-flavor development in stored meats has been described as the disappearance of the fresh flavors, the appearance and subsequent disappearance of a cardboard flavor, and the final dominance of other flavors representing oxidized or rancid notes (Johnson and Civille 1986). In their study of fresh Italian pork sausage links containing sodium lactate and vinegar derivatives, Crist et al. (2014) showed that spice complex and cooked complex flavors and aromas decreased along with an increase in rancid notes, off flavors, and off odors following 18 days of retail display (1–2°C, 880 lux). Based on their findings, the flavor intensities of the sausages followed a similar pattern of decreasing over time in aroma intensities. Additionally, the findings showed that a decline in umami flavor and an increase in bitterness occurred as retail display proceeded, which correlated with microbial growth in the samples. The development of rancid sensory notes is highly correlated with lipid oxidation in several types of meat products and is a result of the formation of volatile compounds during the advanced stage of the lipid oxidation (Ross and Smith 2006). Some unpleasant tastes such as a bitter or metal aftertaste may appear as a consequence of proteolysis, which generates peptides and free amino acids during the storage of the product. Free amino acids and short-chain peptides are taste-active and may exert a strong influence on the final flavor of the product. This pronounced sensory deterioration could also be linked to bacterial enzymatic activities such as hydrolases from psychrotrophic bacteria (proteinases). Upon thawing, loss of water and flavorful constituents such as some amino acids, which are known to contribute to the sweet (alanine, serine, proline, glycine, and hydroxyproline), bitter (phenylalanine, tryptophan, arginine, methionine, valine, leucine and isoleucine), sour (glutamic and aspartic acids, histidine and arginine), salty (sodium salts of both glutamic and aspartic acids), and umami (glutamic acid/salt of glutamic acid) tastes can lead to diminished product acceptability.

4.6.3.5   Microbiological Changes in Fresh Pork Sausage

Pork sausages are highly susceptible to deterioration and spoilage due to excessive growth of bacteria or molds on the outer surface or within (bacteria only) (Savic 1985). Fresh pork sausage with a high water activity (A_w) and pH, and abundantly available nutrients, create a favorable environment for microbial growth even during extended storage at very low temperatures. Several factors affecting the level of microbial growth in pork sausages during their storage are initial microbial count, the level of contamination, storage period, storage temperature, and the packaging method. In addition to the meat components, sausages have additional sources of contaminating microorganisms through the seasoning and formulation ingredients that are usually added in their production. Many spices have high bacterial counts. Because natural casings are of animal origin, they are inherently contaminated with bacteria, which might include pathogens such as *Streptococci, Enterobacteriaceae*, viruses, coliforms, and *Clostridia*, and are likely to harbor *Listeria*, as well, from the processing environment (Jay et al. 2005; Sebranek 2010). However, research has shown that the salting process is still very effective for controlling the levels of these pathogens as well as most spoilage bacteria (Bakker et al. 1999).

The spoilage of meat is reduced when storage temperatures are reduced to below freezing, as bacteria and mold stop developing below approximately −10 to −12 °C. According to Moorhead (2006), intracellular ice formation and corresponding liquid water removal result in a high concentration of solutes, which has been shown to damage microbial cells via osmotic stress. Furthermore, the decline of both enzymatic and non-enzymatic reactions at frozen storage temperatures contributes to the retardation of the growth of most spoilage and pathogenic microorganisms that are already present on either the meat surface or within the meat product. At −18 °C, frozen meat containing ice in equilibrium will have an a_w of approximately 0.84, which inhibits the growth of most spoilage bacteria (Moorhead 2006). The gradual increase in temperature during thawing favors the growth of psychrophilic organisms, most of which are spoilage bacteria that can cause shelf-life problems. Microbiological and chemical reactions accelerate as temperatures increase (Sindelar 2014). Propagating microorganisms produce proteolytic and hydrolytic enzymes that rapidly degrade the muscle quality. Once residual glucose levels in the meat matrix are depleted, the microorganisms proceed to degrade proteins, which results in the formation of sulfides, amines, acetic acid, lactic acid, isovaleric acid, isobutyric acid, esters, and nitriles (Moorhead 2006; Srinivasan et al. 1997). These by-products produce spoilage odors and flavors. The gradual increase in temperature during the thawing also leads to exudate formation from the meat matrix, which makes moisture and nutrients available to spoilage microbes (Leygonie et al. 2012). Changes in pH of the product, which have also been observed during refrigerated storage, indicate increased levels of microbial growth. Increasing pH levels are typically indicative of an increase in gram-negative bacteria populations, such as *Enterobacteriaceae* and pseudomonads, along with yeasts and molds, which degrade protein and amino acids to ammonia, which consequently increases pH (Nychas et al. 1998). Conversely, a decrease in pH or souring is associated mostly with an increase in gram-positive bacteria populations, such as lactic acid bacteria (LAB), enterococci, and related organisms, or through lipolysis, which liberates free fatty acids in the meat system. Souring results from the utilization of sugars by the organisms during storage, which results in the production of acids.

Depending on the air permeability and characteristics of packaging, either aerobic or anaerobic, different species of bacteria proliferate. The processing environment can also cause contamination. Examples of spoilage bacteria include *Pseudomonas* spp., *Acinetobacter* or *Moraxella* spp., *Aeromonas* spp., *Alteromonas putrafaciens*, *Lactobacillus* spp., and *Brochothrix thermosphacta*. In anaerobically stored meat, the microflora is dominated by slow-growing and fermentative *Lactobacilli,* which have little effect on the organoleptic quality of meat. However, the accumulation of volatile organic acids from fermentation by *Lactobacilli* can lead to the development of spoilage flavors. During aerobic conditions, psychrotrophic bacteria begin to dominate the microflora at low storage temperatures. The predominant species identified are typically *Pseudomonas* spp. (Moorhead 2006) or *Brochothrix thermosphacta* (Jay et al. 2005). Most meat-spoilage molds have a minimum growth temperature near −5 °C, and can lead to conditions known as *black spot, white spot, blue-green mold,* and *whiskers.* Since few meat products are currently sold at these temperatures, mold spoilage is largely a historical problem and thus any reports of mold growth on frozen meats is an indicator of poor storage temperature control (James and James 2012). According to Savic (1985), mold formation on sausage products during refrigerated storage is generally the result of dampness, poor ventilation, or improper packaging.

4.6.3.6   Oxidation

Lipid oxidation. Lipid oxidation in coarse-ground sausages occurs at the fat globule and protein film interface (McClements and Decker 2000; Panya et al. 2012). The primary mechanism of oxidation involves unsaturated fatty acids, which are susceptible to oxygen attack. This leads to unstable intermediates that eventually decompose to form unpleasant flavor and aroma compounds in refrigerated coarse-ground sausage, especially when stored under lights in retail display. The most common and important process by which unsaturated fatty acids and oxygen interact is a free radical mechanism characterized by three main phases: initiation, propagation, and termination (Erickson 2008). During initiation, hydrogen is abstracted from an allylic methylene group of an unsaturated fatty acid to form a free radical and the double bond shift, which in turn reacts with molecular oxygen to form a lipid peroxyl radical (). The free radicals are highly unstable due to their unpaired electrons, which are short-lived.

The direct reaction of a lipid molecule with a molecule of oxygen is thermodynamically unfavorable because the lipid molecule is in a singlet electronic state and the oxygen molecule has a triplet ground state (Gray et al. 1996). However, oxygen can be activated by either a formation of singlet oxygen, formation of partially reduced or reactive oxygen species (ROS) such as hydrogen peroxide (H_2O_2)), superoxide anion ($O_2^-\cdot$), or hydroxyl radical (OH), photooxidation, and formation of active oxygen-iron complexes (ferryl iron or ferric-oxygen-ferrous complex likely present in the food even before it is harvested, and during processing and storage. These radicals react with unsaturated lipids (LH) within the fat globules or at the interface, which leads to the formation of lipid radicals. The lipid oxidation chain reaction propagates as these lipid radicals react with other lipids in their immediate vicinity (McClements and Decker 2000). These hydroxyl radicals are capable of initiating lipid peroxidation cascades in membranes by extracting hydrogen atoms from polyunsaturated fatty acid chains of membrane phospholipids, creating carbon-centered fatty acyl radicals that further interact with oxygen to form peroxyl radicals. Cleavage (β-scission) of the fatty acid chain adjacent to the alkoxyl radicals can produce low-molecular-weight volatile compounds that contribute to rancidity and flavor deterioration.

The second step in decomposition of hydroperoxides involves the carbon-Enterobacteriaceae-carbon bond on the side of the oxygen containing the carbon atom, resulting in formation of an aldehyde and an alkyl radical, while scission of the carbon–carbon between the double bond and the carbon atom produces a hydrocarbon radical and an alkyl oxo-compound (Min and Lee 1999).

According to Gray and Pearson (1994), pure lipid hydroperoxides are stable at physiological temperatures, but in the presence of transition metal complexes like iron salts, their decomposition is accelerated. Since iron and hydroperoxides are ubiquitous in lipid-containing food systems, heme proteins (for example myoglobin, hemoglobin) and non-heme iron are the most important catalysts of lipid oxidation in retail pork sausage and model meat systems (Decker and McClements 2001). Metmyoglobin is oxidized to a porphyrin radical, which can react with hydroperoxide to produce ferryl oxene (Fe^{4+}), which subsequently initiates lipid oxidation. The ability of heme compounds in metmyoglobin, deoxymyoglobin, and oxymyoglobin to activate lipid oxidation through an intermediate species has also been demonstrated (Harel and Kanner 1985). In its non-heme form, iron contributes to the decomposition of hydroperoxides and the production of reactive oxygen species and hydroxyl radicals, via the chemical Fenton reaction. Additionally, this reaction is effective when Fe^{3+} can be recycled to Fe^{2+} by various reducing agents such as oxidoreductase enzymes (for example NADPH oxidase, the cytochrome P450 enzymes, myeloperoxidase, glucose oxidase, xanthine oxidase) and nonenzymic systems (for example Fe^{3+}/O_2/ascorbate or Fe^{3+}/O_2/RSH). Non-heme iron levels are initially low; approximately only 2.4–3.9% of total muscle iron in pork (Hazell 1982). However, increased levels of catalytic low-molecular-weight iron fraction have been found in muscles that have been processed and stored. Rhee et al. (1996) reported that the stability of frozen raw pork was more correlated to the heme pigments and levels of catalase than the amount of polyunsaturated fat, and that heme iron concentration was more important than non-heme iron levels in predicting lipid oxidation in meat. Ground meat products undergo oxidative changes and develop rancidity more quickly than intact muscle because comminution increases exposure of unsaturated fats and proteins to molecular oxygen and allows the interactions among heme compounds, metal pro-oxidants, and enzymes responsible for oxidation. The comminution or any other mechanical disruption method also alters the physical location of oxidative metal catalysts in the muscle tissue through the release of protein-bound iron and potentially dilutes endogenous antioxidants. Likewise, in sausage processing, comminution accelerates oxidative reactions and causes a rapid increase in lipid oxidation from the start of production (Wanous et al. 1989a). This tendency toward rapid oxidation results in major quality problems during subsequent processing and storage.

Fresh pork sausages are generally stored in trays over-wrapped with oxygen-permeable polyvinyl chloride (PVC) films under halogen or fluorescent lamps to facilitate red color blooming. Therefore, fresh pork sausage oxidative stability can be influenced substantially by temperature during storage and retail display. Minimizing temperature variation, especially at the retail level (4–7 °C) can add a few days or up to a week to sausage shelf life (Sindelar 2014). An increase in storage temperature accelerates chemical, enzymic, and microbiological activities. Refrigerated storage (~4 °C) delays or suppresses the oxidation. Frozen storage (−18 to −20 °C) immediately after the production delays these reactions but does not stop them (Erickson 1997; Wanous et al. 1989b). Xia et al. (2009) showed that multiple freeze–thaw cycles had a detrimental effect on the quality of pork as it accelerated lipid and protein oxidation, hastened discoloration, and changed the structure of myofibrillar proteins, leading to the loss of myofibrillar protein functionality in the porcine *longissimus* dorsi muscle. Moreover, Witte et al. (1970) reported less malondialdehyde production at freezing temperature (−20 °C) than at refrigerated temperature (4 °C), and that TBARS values in pork did not significantly change during frozen storage.

Lipid oxidation-induced oxymyoglobin oxidation. Lipid oxidation results in a wide range of secondary oxidation products, which are primarily aldehydes such as n-alkanals, trans-2-alkenals, 4-hydroxy-trans-2-alkenals and malondialdehyde (Lynch and Faustman 2000). These aldehydes covalently bind to myoglobin, accelerating heme oxidation and metmyoglobin formation, and thus meat discoloration. After oxymyoglobin is oxidized to metmyoglobin, the presence of lipid aldehydes decreases the ability of metmyoglobin to be enzymatically reduced, and pro-oxidant activity of metmyoglobin was enhanced in the presence of these aldehydes (Faustman et al. 1999).

4.7 SANITATION

Sanitation is a crucial part of a shelf-life program. According to Marriott et al. (2018), there are many reasons that sanitation needs to be a top priority in a meat plant. The major reason is to decrease bacterial contamination for both food quality and safety purposes. Furthermore, good sanitation programs reduce product waste and financial loss due to discoloration ad spoiled products. Processing and storage at a sufficiently cold temperatures will reduce spoilage and microbial growth, not only on products but also on equipment, supplies, and other areas. Under unsanitary conditions with improper temperature control, certain species of *Pseudomonas* can proliferate every 20 minutes. Meat and poultry are generally expected to avoid spoilage at least twice as long at 0°C (32°F) than at 10°C (50°F).

4.8 CONCLUSIONS

Fresh pork sausage is a great value-added product that increases the value of pork trimmings and pre-rigor sow meat. It is important to understand the meat ingredients, non-meat ingredients, and factors influencing commercial shelf life as the sausage is frozen and refrigerated under a retail display light. It is also important to have excellent sanitation at the plant and temperature control at both meat plants and retail establishments, and employ ingredient technology to minimize microbial growth and lipid and protein oxidation, which contribute to product discoloration.

REFERENCES

Allais, I. 2010. Emulsification. In *Handbook of meat processing*, ed. F. Toldrá, 143–168. Ames: Wiley-Blackwell.
Andersen, H. J., and L. H. Skibsted. 1991. Oxidative stability of frozen pork patties. Effect of light and added salt. *Journal of Food Science* 56(5):1182–1184.
Bakker, W. A. M., J. H. Houben, P. A. Koolmees, U. Bindrich, and L. Sprehe. 1999. Effect of initial mild curing, with additives, of hog and sheep sausage casings on their microbial quality and mechanical properties after storage at difference temperatures. *Meat Science* 51(2):163–174.
Bentley, D. S., J. O. Reagan, and M. F. Miller. 1988. The effects of hot-boned fat type, pre-blending treatment and storage time on various physical, processing and sensory characteristics of nonspecific luncheon loaves. *Meat Science* 23(2):131–138.
Berdahl, D. R., R. I. Nahas, and J. P. Barren. 2010. Synthetic and natural antioxidant additives in food stabilization: current applications and future research. In *Oxidation in foods and beverages and antioxidant applications*, ed. E. A. Decker, R. J. Elias, and D. J. McClements, 272–320. Oxford: Woodhead Publishing.
Bradley, E. M., J. B. Williams, M. W. Schilling et al. 2011. Effects of sodium lactate and acetic acid derivatives on the quality and sensory characteristics of hot-boned pork sausage patties. *Meat Science* 88(1):145–150.
Brown, P. M. 2009. Spices, seasonings, and flavors. In *Ingredients in meat products: properties, functionality and applications*, ed. R. Tarté, 199–210. New York: Springer.

Canadian Food Inspection Agency. 2018. Safe foods for Canadians regulations, Canadian Standards of Identity Volume 7 – Meat Products.

Code of Federal Regulations (CFR). 2015. *Animals and animal products. 9 CFR §318, 9 CFR § 319.141, 9 CFR § 319.6, 21 CFR §170.3, 21 CFR § 182.3169, 21 CFR § 182.3173.* Washington: U.S. Government Printing Office.

Crist, C. A., J. B. Williams, M. W. Schilling, A. F. Hood, B. S. Smith, and S. G. Campano. 2014. Impact of sodium lactate and vinegar derivatives on the quality of fresh Italian pork sausage links. *Meat Science* 96(4):1509–1516.

Decker, E. A., L. C. Faustman, and C. J. Lopez-Bote. 2000. *Antioxidants in muscle foods: nutritional strategies to improve quality.* New York, NY: Wiley-Interscience.

Decker, E. A., and D. J. McClements. 2001. Transition metal and hydroperoxide interactions. *Inform* 12(3):251–262.

Erickson, M. C. 1997. Lipid oxidation: flavour and nutritional quality deterioration in frozen foods. In *Quality in frozen food*, ed. M. C. Erickson, and Y.-C. Hung, 141–173. New York, NY: Chapman and Hall.

Erickson, M. C. 2008. Lipid oxidation of muscle foods. In *Food lipids: chemistry, nutrition, and biotechnology*, ed. C. C. Akoh, and D. B. Min, 322–364. Boca Raton, FL: CRC Press.

Eskin, N. A. M., and R. Przybylski. 2000. Antioxidants and shelf life of foods. In *Food shelf life stability: chemical, biochemical, and microbiological changes*, ed. N. A. M. Eskin, and D. S. Robinson, 175–209. Boca Raton, FL: CRC Press LLC.

Estévez, M., S. Ventanas, R. Ramírez, and R. Cava. 2005. Influence of the addition of rosemary essential oil on the volatiles pattern of porcine frankfurters. *Journal of Agricultural and Food Chemistry* 53(21):8317–8324.

European Union. 2011. Regulation (EU) No 1169/2011 of the European Parliament and of the Council of 25 October 2011 on the provision of food information to consumers.

Faustman, L. C., D. C. Liebler, T. D. McClure, and Q. Sun. 1999. α,β-Unsaturated aldehydes accelerate oxymyoglobin oxidation. *Journal of Agricultural and Food Chemistry* 47:3140–3144.

Gray, J. I., E. A. Gomaa, and D. J. Buckley. 1996. Oxidative quality and shelf life of meats. *Meat Science* 43:S111–S123.

Gray, J. I., and A. M. Pearson. 1994. Lipid-derived off-flavours in meat-formation and inhibition. In *Flavor of meat and meat products*, ed. F. Shahidi, 116–143. London: Chapman and Hall.

Grün, I. U. 2009. Antioxidants. In *Ingredients in meat products: properties, functionality and applications*, ed. R. Tarté, 291–300. New York: Springer.

Harel, S., and J. Kanner. 1985. Hydrogen peroxide generation in ground muscle tissues. *Journal of Agricultural and Food Chemistry* 33(6):1186–1188.

Hazell, T. 1982. Iron and zinc compounds in the muscle meats of beef, lamb, pork and chicken. *Journal of the Science of Food and Agriculture* 33(10):1049–1056.

Hazen, C. 2005. Antioxidants "meat" needs. *Food Product Design* 15(1):61–68.

James, S. S., and C. James. 2012. Quality and safety of frozen meat and meat products. In *Handbook of frozen food processing and packaging*, ed. D.-W. Sun, 2nd ed., 303–323. Boca Raton, FL: CRC Press, Taylor & Francis Group.

Jay, J. M., M. J. Loessner, and D. A. Golden. 2005. *Modern food microbiology* (7th ed.). New York, NY: Springer Science + Buisiness Media, Inc.

Ji, H.-F., H.-Y. Zhang, and L. Shen. 2006. Proton dissociation is important to understanding structure–activity relationships of gallic acid antioxidants. *Bioorganic & Medicinal Chemistry Letters* 16(15):4095–4098.

Johnson, P. B., and G. V. Civille. 1986. A standardized lexicon of meat WOF descriptors. *Journal of Sensory Studies* 1(1):99–104.

Kinsella, J. E. 1989. Flavor perception and binding to food components. In *Flavor chemistry of lipids*, ed. D. B. Min, and T. H. Smouse, 376–403. Champaign: American Oil Chemists' Society.

Legislação: Instrução Normativa SDA - 4, de 31/03/2000. 2000.

Leygonie, C., T. J. Britz, and L. C. Hoffman. 2012. Impact of freezing and thawing on the quality of meat: review. *Meat Science* 91(2):93–98.

Lynch, M. P., and L. C. Faustman. 2000. Effect of aldehyde lipid oxidation products on myoglobin. *Journal of Agricultural and Food Chemistry* 48(3):600–604.

Marriott, N. G., M. W. Schilling, and R. Gravani. 2018. *Principles of Food Sanitation*, 6th ed., 425p. New York: Springer.

Martínez-Tomé, M., A. M. Jiménez, S. Ruggieri, N. Frega, R. Strabbioli, and M. A. Murcia. 2001. Antioxidant properties of Mediterranean spices compared with common food additives. *Journal of Food Protection* 64(9):1419–1419.

McClements, D. J., and E. A. Decker. 2000. Lipid oxidation in oil-in-water emulsions: impact of molecular environment on chemical reactions in heterogeneous food systems. *Journal of Food Science* 65(8):1270–1282.

Min, D. B., and H. O. Lee. 1999. Chemistry of lipid oxidation. In *Flavor chemistry: thirty years of progress*, ed. R. Teranishi, E. L. Wick, and I. Hornstein, 175–187. New York, NY: Kluwer Academic/Plenum Publishers.

Moorhead, S. 2006. Quality and safety of frozen meat and meat products. In *Handbook of frozen food processing and packaging*, ed. D.-W. Sun, 311–324. Boca Raton, FL: CRC Press, Taylor & Francis Group.

Nychas, G.-J. E., E. H. Drosinos, and R. G. Board. 1998. Chemical changes in stored meat. In *Microbiology of meat and poultry*, ed. A. Davies, and R. G. Board, 288–326. London: Blackie Academic and Professional.

Ockerman, H. W., and L. Basu. 2014a. Carcass chilling and boning. In *Encyclopedia of meat sciences*, ed. M. Dikeman, and C. Devine, 2nd ed., 142–147. San Diego: Academic Press.

Ockerman, H. W., and L. Basu. 2014b. Chemistry and physics of comminuted products: other ingredients. In *Encyclopedia of meat sciences*, ed. M. Dikeman, and C. Devine, 2nd ed., 296–301. San Diego: Academic Press.

Offer, G., and P. Knight. 1988. The structural basis of water-holding capacity in meat. Part 1: general principles and water uptake in meat processing. In *Developments in meat science* (Vol. 4), ed. R. Lawrie, 61–171. New York: Elsevier Applied Science.

Panya, A., M. Laguerre, C. Bayrasy et al. 2012. An investigation of the versatile antioxidant mechanisms of action of rosmarinate alkyl esters in oil-in-water emulsions. *Journal of Agricultural and Food Chemistry* 60(10):2692–2700.

Pearson, A. M., and T. A. Gillett. 1996. *Processed meats*. Gaithersburg: Aspen Publishers, Inc.

Pisula, A., and A. Tyburcy. 1996. Hot processing of meat. *Meat Science* 43(1):125–134.

Renerre, M. 2000. Oxidative processes and myoglobin. In *Antioxidants in muscle foods: nutritional strategies to improve quality*, ed. E. A. Decker, L. C. Faustman, and C. J. Lopez-Bote, 113–135. New York: Wiley-Interscience.

Rhee, K. S., L. M. Anderson, and A. R. Sams. 1996. Lipid oxidation potential of beef, chicken, and pork. *Journal of Food Science* 61(1):8–12.

Richard, H. M. J. 1991. Spices and condiments I. In *Volatile compounds in foods and beverages*, ed. H. Maarse, 411–447. New York, NY: Marcel Dekker, Inc.

Romans, J. R., W. J. Costello, C. W. Carlson, M. L. Greaser, and K. W. Jones. 2001. *The meat we eat* (14th ed.). Danville, IL: Interstate Publishers, Inc.

Ross, C. F., and D. M. Smith. 2006. Use of volatiles as indicators of lipid oxidation in muscle foods. *Comprehensive Reviews in Food Science and Food Safety* 5(1):18–25.

Rust, R. E. 1987. Sausage products. In *The science of meat and meat products*, ed. J. F. Price, and B. S. Schweigert, 3rd ed., 457–485. Westport, CT: Food and Nutrition Press, Inc.

Savic, I. V. 1985. *Small-scale sausage production*. Rome, Italy: Publications Division, Food and Agriculture Organization of the United Nations.

Schilling, M. W. 2019. Emulsified meat products. In *Food emulsifiers and their applications*, ed. R. Hartel, 4th ed., 347–377. Springer: New York.

Sebranek, J. G. 2010. Natural vs. artificial casings: evaluating which is best for your product. American Association of Meat Processors. Retrieved from http://www.aamp.com/documents/NaturalvsArtificialCasings.pdf

Sebranek, J. G. 2016. An overview of functional non-meat ingredients in meat processing: The current toolbox. In 68th Annual Reciprocal Meat Conference (Vol. 3), 42–46. American Meat Science Association. https://doi.org/10.4315/0362-028X-69.9.2176

Shahidi, F., and P. Ambigaipalan. 2015. Phenolics and polyphenolics in foods, beverages and spices: antioxidant activity and health effects–a review. *Journal of Functional Foods* 18(b):1–78.

Shahidi, F., P. K. Janitha, and P. D. Wanasundara. 1992. Phenolic antioxidants. *Critical Reviews in Food Science and Nutrition* 32(1):67–103.

Shan, B., Y.-Z. Cai, M. Sun, and H. Corke. 2005. Antioxidant capacity of 26 spice extracts and characterization of their phenolic constituents. *Journal of Agricultural and Food Chemistry* 53(20):7749–7759.

Sindelar, J. 2014. Methods to extend the shelf-life of fresh sausages. Meatingplace. Retrieved from http://www.meatingplace.com/Industry/TechnicalArticles/Details/46450

Smith, J. L., and J. A. Alford. 1968. Action of microorganisms on the peroxides and carbonyls of rancid fat. *Journal of Food Science* 33(1):93–97.

Srinivasan, S., Y. L. Xiong, P. Blanchard, and J. H. Tidwell. 1997. Physicochemical changes in prawns (*Machrobrachium rosenbergii*) subjected to multiple freeze-thaw cycles. *Journal of Food Science* 62(1):123–127.

Suman, S. P., and P. Joseph. 2014. Chemical and physical characteristics of meat: color and pigment. In *Encyclopedia of meat sciences*, ed. M. Dikeman, and C. Devine, 2nd ed., 244–251. San Diego, CA: Academic Press, Inc.

Undeland, I., H. O. Hultin, and M. P. Richards. 2002. Added triacylglycerols do not hasten hemoglobin-mediated lipid oxidation in washed, minced cod. *Journal of Agricultural and Food Chemistry* 50(23):6847–6853.

Wanous, M. P., D. G. Olson, and A. A. Kraft. 1989a. Oxidative effects of meat grinder wear on lipids and myoglobin in commercial fresh pork sausage. *Journal of Food Science* 54(3):545–548.

Wanous, M. P., D. G. Olson, and A. A. Kraft. 1989b. Pallet location and freezing rate effects on the oxidation of lipids and myoglobin in commercial fresh pork sausage. *Journal of Food Science* 54(3):549–552.

Witte, V. C., G. F. Krause, and M. E. Bailey. 1970. A new extraction method for determining 2-thiobarbituric acid values of pork and beef during storage. *Journal of Food Science* 35(5):582–585.

Xia, X., B. Kong, Q. Liu, and J. Liu. 2009. Physicochemical change and protein oxidation in porcine longissimus dorsi as influenced by different freeze-thaw cycles. *Meat Science* 83(2):239–245.

Xiong, Y. L. 1997. Structure-function relationships of muscle proteins. In *Food proteins and their applications*, ed. S. Damodaran, and A. Paraf, 341–392. New York, NY: Marcel Dekker, Inc.

5 Chorizo and Chouriço de Carne

Varieties, Composition, Manufacturing Process and Shelf Life

Luz H. Villalobos-Delgado,[1] Irma Caro,[2]
Bettit K. Salvá,[3] Alexandra Esteves,[5]
Daphne D. Ramos-Delgado,[4] Sergio Soto,[6]
Enrique A. Cabeza-Herrera[7]
Roberto González-Tenorio,[8] and Javier Mateo,[9]

[1]Instituto de Agroindustrias, Universidad Tecnológica de la Mixteca, Carretera a Acatlima Km. 2.5, 69000, Huajuapan de León, Oaxaca, Mexico

[2]Departamento de Pediatría, Inmunología, Obstetricia-Ginecología, Nutrición-Bromatología, Psiquiatría e Historia de la Ciencia. Universidad de Valladolid Avenida de Ramón y Cajal, 7, 47005, Valladolid, Spain

[3]Facultad de Industrias Alimentarias. Universidad Nacional Agraria La Molina. Av. La Molina s/n, La Molina, Lima, Peru

[4]Laboratorio de Salud Pública y Salud Ambiental, Facultad de Medicina Veterinaria, Universidad Nacional Mayor de San Marcos. Av. Circunvalación cdra. 28 s/n San Borja, Lima, Peru

[5]Veterinary and Animal Research Centre (CECAV) and Department of Veterinary Sciences, School of Agrarian and Veterinary Sciences, Universidad de Trás-os-Montes e Alto Douro (UTAD), 5000-801Vila Real, Portugal

[6]Instituto de Ciencias Agropecuarias. Universidad Autónoma del Estado de Hidalgo. Avenida Universidad s/n km 1. CP. 43600. Tulancingo, Hidalgo. Mexico

[7]Departamento de Microbiología, Facultad de Ciencias Básicas, Universidad de Pamplona Ciudad Universitaria Km 1 Vía Bucaramanga Pamplona–Norte de Santander CP 543050, Colombia

[8]Instituto de Ciencias Agropecuarias. Universidad Autónoma del Estado de Hidalgo. Avenida Universidad s/n km 1. Tulancingo, Hidalgo. CP. 43600. Mexico
[9]Departamento de Higiene y Tecnología de los Alimentos, Universidad de León, Campus Vegazana, s/n, 24007, León, Spain

CONTENTS

5.1   INTRODUCTION

In general terms, the names *chorizo* in Spanish and *chouriço/chouriça de carne* in Portuguese (derived from the Latin term *salsicĭum*), are used to denominate a group of raw sausage varieties. They are usually made with minced pork, air-dried for either a short (hours-days; fresh or semidried sausages) or long period (weeks-months, dry-cured sausages), and characterized by the use of red paprika (from dried *Capsicum annuum* fruits), named *pimentón* in Spanish and *páprica* or *pimentão* in Portuguese, at approximate levels of 2–3% (Martín-Bejarano, 2001). However, there are a few exceptions to the above-mentioned definition. On the one hand, there are sausages that can be classified as *chorizo/chouriço* (chorizo-like sausages) named *longaniza* in some regions of Spain, or *linguiça* in Portugal, where the sausage shape is long and narrow (Ministerio de la Presidencia, 2014). On the other hand, the name *chorizo/chouriço* is also used for sausages that do not comply with the above-mentioned definition, such as the Spanish dry-cured sausage *chorizo blanco*, which does not contain paprika, or the Portuguese *chouriço de sangue*, which is a blood sausage.

Paprika, with its high carotenoid content, gives the sausage a distinctive red colour while also contributing to its flavour. The use of paprika as a food colorant in Spain, and thus its use in sausages, became commonplace at the beginning of the 19th century (Abad-Alegría, 2001). In chorizo/chouriço it is common to use powdered dried–smoked, oven-dried or sun-dried *C. annuum* fruits, although a paprika paste made from dried *C. annuum* fruits and water is sometimes used. The flavour of Spanish paprika is classified as sweet, semi hot (*okal*) or hot. The hot variety is seldom used, but when it is, the chorizo is called *chorizo picante*. The main paprika used in Spain is from fruits cultivated and prepared in two main areas (Palacios-Morillo et al., 2014): Extremadura (which is a smoked paprika and granted La Vera Protected Origin Designation; PDO) and Murcia (oven- or sun-dried and granted Murcia PDO). In general, Spanish paprika is highly coloured, showing values of American Spice Trade Association (ASTA) units higher than 100 (Gómez et al., 2008; Pereira et al., 2019).

The basic process for preparing chorizo, chouriço or chorizo-like sausages consists of mincing pork (or occasionally the meat from other animals from slaughtering or hunting) and pork back fat, which are then mixed with salt and the characteristic recipe of species and condiments, and occasionally other allowed ingredients, with the sausage batter stuffed into natural or artificial (water-permeable) casings. After this stage, these sausages can be prepared as fresh/semidried or as dry-cured sausages. In Portugal, chouriços are practically prepared only as dry-cured sausages.

Fresh/semidried chorizos are slightly air-dried, for a time period ranging from a few hours to a few days, usually at a cool temperature (c.a. 2–15°C), before being commercialized and can be subject to further drying during storage (hanging in storage chambers). Thus, the drying of these chorizos can be intense enough to make them change from a not ripened (fresh) to a ripened (semidried) sausage (Gómez et al., 2008; González-Tenorio et al., 2013a; Ministerio de la Presidencia, 2014). Regardless of the drying conditions, chorizos must be stored under re-frigeration, because their final activity water (a_w) is considered to be insufficiently low for microbial stability (Camerati and Garcés, 2017), and must be cooked before consumption. In contrast, dry-cured chorizos are dried for longer periods of time in order to produce shelf-stable and ready-to-eat sausages.

Fresh/semidried varieties are produced in butcheries and small and large meat industries, while dry-cured varieties are produced in small and large meat in-dustries. A small amount of chorizo/chouriço (and of other typical meat products, such as dry-cured ham, salchichones, blood sausages, etc.) is prepared in rural households in winter for self-consumption, and this is considered in some cases as an important sociocultural event for families and communities (Fonte, 2008). In Spain, the estimated total annual consumption of chorizo is approximately 0.7 million kg (MERCASA, 2019).

In Latin America, a variety of typical chorizos or chorizo-like sausages in-troduced by European immigrant colonizers, usually resembling the Spanish/Portuguese-style fresh/semidried chorizos, are also prepared and consumed (González-Schnake and Nova, 2014; Hajmeer et al., 2006). These can be considered creole meat products that have evolved over time into traditional American varieties (Mateo et al., 2009). At present, they constitute an important part of local econo-mies and traditions in Latin America. However, their production has increasingly become more industrialized (González-Schnake and Nova, 2014). Moreover, due to the historical Spanish influence, varieties of chorizo and langgonisang (chorizo-like sausage) are found in the Philippines (Briones et al., 2013), and Portuguese-influenced varieties of chouriços are found in African (Mozambique, Cape Verde) and Asian (Goa, Timor) regions. Due to more recent migration, chorizos/chouriços can be easily found and are popular in countries around the world, classified as ethnic food, such as in the USA or South Africa.

Chorizo sausages in Latin America are normally fresh/semidried and their definition seems to be more flexible than in Spain or Portugal. Mexico is probably the country with the chorizo most similar to the Spanish fresh/semidried chorizo. In Argentina, Bolivia, Colombia and Uruguay, chorizo is used as a generic name for a wide range of fresh/semidry sausages made from coarsely minced meat (Allen, 2015; Gobierno Autónomo Departamental de Santa Cruz, 2013). On the other hand, some Latin American chorizo-like sausages are called salchicha or longaniza, for example the Chilean longaniza de Chillán (Camerati and Garcés, 2017). In Peru, the names chorizo and salchicha appear to be used for regional fresh/semidried chorizo-like sausages, depending more on the regional practices than on technology. In Brazil, what is usually referred to as chorizo in other Latin American countries is called linguiça, while Brazilian chouriço refers to a blood sausage (Allen, 2015).

5.2 FRESH/SEMIDRIED CHORIZO

Fresh sausages are sold raw (without prior heat treatment), refrigerated or (less commonly) frozen, and need to be cooked before consumption (Feiner, 2006). Among these products, there are a lot of fresh/semidried chorizo and chorizo-like sausage varieties (specialties) around Spain, Latin America and other countries, which differ in formulation, manufacturing process and shape. They are usually 2.5–4 cm diameter red-coloured sausages, prepared with minced meat and fat, salt, spices, and condiments and allowed additives (Allen, 2015; González-Tenorio et al., 2013b; Kuri et al., 1995). They are normally dried for a short time and, depending on the length of the drying time, they can be classified as fresh or slightly ripened (semidry) sausages. Some industries also produce pre-cooked, and (optionally) smoked, chorizos. Different chorizos or longanizas from Spain and Latin America, fresh, semidried and pre-cooked are shown in Figures 5.1 and 5.2.

For consumption, fresh/semidried chorizos and chorizo-like sausages can be fried or grilled (as a whole sausage, sliced or crumbled) and consumed with bread or combined with other foods (Mexican tacos, fried eggs with chorizo, chorizo with potatoes, chorizo tapas, etc.). They can also be boiled with wine or cider, for example, *chorizo a la sidra*, or cooked with other foods to prepare regional dishes such as legume stew, for example, *cocido, fabada, feijoada.* etc., rice, for example *paella con chorizo*, or pasta, for example *macarrones con chorizo*.

5.2.1 RAW MATERIALS AND ADDITIVES

In general, fresh/semidried chorizo is manufactured with meat, pork fat and non-meat ingredients such as salt, spices and condiments as well as other allowed ingredients. Pork is the preferred meat, however, beef, lamb, poultry or even game are used in some specialties. When pork is used, the anatomical regions more frequently used are shoulder, neck, belly, trims, and pork back fat. Traditional chorizo contains a relatively high amount of fat, that is, at least 30%, contributing to the

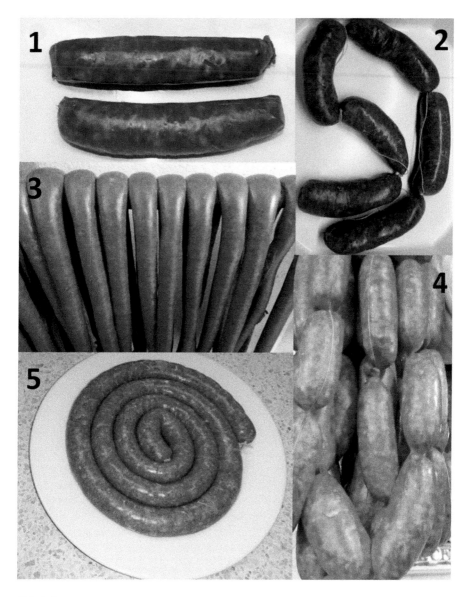

FIGURE 5.1   Fresh/semidried chorizos and chorizo-like sausages: 1–semidried chorizo, Spain; 2–fresh chorizo, Spain; 3–longaniza, Mexico; 4–chorizo, Mexico; 5–salchicha huachana, Peru.

sausage flavour and juiciness (Carmona-Escutia et al., 2019; Pacheco-Pérez et al., 2011).

In Spain, chorizo formulation typically includes pork meat (the use poultry or beef are less frequent), pork fat, salt, Spanish paprika (powdered), garlic (*Allium sativa*) and oregano (*Origanum vulgare*), although other ingredients, such as

FIGURE 5.2 Fresh/semidried chorizos: 1–pre-cooked Antioqueño, Colombia; 2–not co-lored, Lima market, Peru; 3–industrial, stuffed in a semi-permeable plastic casing, Mexico.

vinegar, brandy, etc. can be used. Industrial chorizos can contain allowed additives, that is colorants (cochineal, caramels, paprika extract, red beetroot, etc.), and or-ganic acids and salts (citrate, lactate, acetate, ascorbate), the latter in prepackaged chorizos (SANCO, 2022). In Latin America fresh/semidried chorizos or chorizo-like sausages are typically made with pork, beef, or pork and beef. For example, beef is either the sole ingredient or is combined with pork in specialties such as Antioqueño (Colombia) (Pacheco-Pérez et al., 2011), Cusqueño (Peru, personal communication, butchers at the Mercado Central de San Pedro, Cusco), or Argentinian chorizos (Allen, 2015). Latin American varieties tend to be more spiced than their Spanish counterparts, thus containing a wider variety of in-gredients, which results in different tastes and appearances. A list of non-meat ingredients used in different fresh/semidried chorizo and chorizo-like sausages from Spain and Latin America is shown in Table 5.1. Furthermore, although not fre-quently, some fresh Latin American chorizos can contain water in the formulation, for example *salchicha de Huacho* can contain up to 10% water (Municipalidad Provincial de Huaura, 2017).

One of the main reasons for the differences among Latin American chorizos is the variety, type and amount of dried *Capsicum* fruits used. Mexican varieties of traditional chorizos are especially popular in the states of Hidalgo, Veracruz,

TABLE 5.1

Ingredients Used in Different Fresh/Semidried Chorizo and Chorizo-Like Varieties in Spain and Latin America

Variety, Country	Non-meat Ingredients/Additives	Reference
Butcher's chorizo, Spain	Paprika (okal) from La Vera region (2.3%), common salt (1.7%), brandy, garlic, oregano	Personal communication, G. Mateo, local butcher, Valladolid, Spain
Industrial chorizo, Spain	Paprika, common salt (1.8%), dextrose, dextrin, sodium lactate, garlic, other spices	Label information: Pimursa, Cabezo de Torres, Murcia; and Embutidos Arrieta, Zubiri, Navarra; Spain.
Industrial, Mexico	(Iodized) common salt, potassium chloride, texturized soy proteins, mix of chili peppers, garlic, spices, vinegar, lactate, acetate, diacetate, sorbate, erythorbate, nitrite, colorant additives, artificial flavours, citrate	Label information: FUD, San Pedro, Sinaloa; Aurrera, Ciudad de México; Kir, Qualtia Alimentos, Monterrey, Nuevo León; Mexico
Toluca Green chorizo, Mexico	Fresh green capsicum fruits (cuaresmeño, jalapeño), common salt, garlic, onion, ginger, coriander, spinach, species, vinegar	Quintero-Salazar (2017)
Butcher's chorizo chapin, Guatemala	Common salt, phosphates, dextrose, annatto, nutmeg, minced garlic, thyme, chili pepper, oregano	Personal communication, local butcher, Guatemala
Traditional, El Salvador	Common salt, black pepper, cumin, garlic, oregano, thyme, vinegar, fresh capsicum fruits, annatto	Allen (2015)
Antioqueño, industrial, Colombia	Water, garlic, onion, cumin, pepper, oregano, coriander, common salt, liquid smoke, polyphosphates, nitrites, nisine, soy protein isolate, starch, annatto	Pacheco-Pérez et al. (2011)
Butcher's criollo or campesino chorizo, Colombia	Pork, bacon, common salt (2%), potassium chloride, sodium acetate, parsley, garlic, onion, pepper	Personal communication, Luis M. García, local butcher, Guatemala
Chorizo parrillero, Colombia	Fresh garlic (1%), onion (3.4%), parsley (3%), Paprika (3%), common salt (2%), phosphate (0.1%), sodium nitrite (0.2%)	Personal communication, G.E. Laguado-Corredor, Universidad de Pamplona, Colombia

(Continued)

TABLE 5.1 (Continued)

Variety, Country	Non-meat Ingredients/Additives	Reference
Salchicha de Huacho, industrial, Peru	Common salt, achiote, turmeric, pepper, cumin, nutmeg, nitrites, phosphates, garlic, glutamate, sugar, paprika, fresh *aji amarillo* fruits (*Capsicum baccatum*)	Personal communication, B. Salvá, Universidad Nacional Agraria La Molina Lima, Perú
Salchicha de Huacho, traditional, Peru	Common salt, orange juice, black pepper, cumin, garlic, annatto oil	Municipalidad Provincial de Huaura (2017)
Parrillero, traditional, Peru	Common salt, black pepper, cumin, garlic, oregano, vinegar, carmine (cochineal)	Chirinos and Salvá (2000)
Chuquisaqueño, traditional, Bolivia	Common salt, cumin, pepper, oregano, garlic, red paprika, chili pepper fresh fruits, parsley, peppermint, salt, clove, cinnamon, sugar, green onion, monosodium glutamate	Gobierno Autónomo Departamental de Santa Cruz (2013)

Oaxaca, and Mexico (Becerril-Sánchez et al., 2019), and some of them can be hot-spiced (Allen, 2015). Mexican-style chorizos are characterized by mixtures of dried chili peppers (capsicum fruits), that is *chile guajillo, chile ancho, chile pasilla* and *chile de árbol* (all *Capsicum annuum L. var. annum*), at levels of 2–3% of the weight of the sausage batter. The pepper mixtures are usually added as thick capsicum paste, for example with 50% dried Capsicum fruit and 50% water (Carmona-Escutia et al., 2019; González-Tenorio et al., 2012). In South America, producers use other *Capsicum annuum* varieties in chorizo, such as *trompa de elefante, nora de Murcia, bolita*, papri-king, papri-queen, Sonora, etc. (Pickersgill, 2003). They can be added as paprika paste or as powdered paprika, and in some chorizo varieties, dried *Capsicum* fruits are roasted before being used as an ingredient.

Moreover, apart from the *Capsicum annuum* varieties, all over Latin America chorizos can be made using dried fruits from *Capsicum chinense* varieties, for example, the chorizo commercialized in Cusco city by small-scale butchers contains the Peruvian red pepper named *ají panca* or *ají colorado*. In some chorizo varieties, dried red *Capsicum* spp. fruits can be accompanied by ground annatto (*Bixa Orellana*) seeds or annatto oil to reinforce the red colour (Carmona-Escutia et al., 2019). This is the case for the popular Peruvian salchicha de Huacho also called norteña, huachana or colorado salchicha (González-Schnake and Nova, 2014; Municipalidad Provincial de Huaura, 2017).

Annatto seeds or carmine (cochineal) generally provide the typical red colour of other Latin American chorizo varieties. Varieties of chorizo in different South America countries glow with annatto (Allen, 2015; Ramos et al., 2014), which can be added as annatto oil prepared by frying annatto seeds in vegetable oil. Apart from natural colorants (semi-)synthetic additives, for example amaranth or carmine, can also be used in industrial chorizos.

Furthermore, in Latin America not all chorizos show the characteristic red colour. In Toluca (Mexico), which is considered a chorizo-producing city, both red and green chorizos are made (Becerril-Sánchez et al., 2019). The popular green chorizo, chorizo verde, is made with pork meat and fat, salt and different fresh green vegetables such as fresh coriander, roasted jalapeño chili peppers, tomatillos (*Physalis ixocarpa*), chard, spinach and parsley (Allen, 2015; Ramírez-Muñoz et al., 2015). In Argentina, the popular chorizo criollo (a version of this chorizo is also popular in Spain) is seasoned with garlic, nutmeg, red wine, sugar and pepper (Allen, 2015).

Another key difference in composition among chorizos, one that is especially evident in some regions of Latin America where legislation on additives can be more permissive than in the European Union, is related to both the intended market and the traditions of the manufacturing process. Industrial, non-traditional low-cost Mexican chorizos intended to be commercialized in wholesale markets in big cities can have added mechanically recovered meat and fillers, such as carbohydrates (gums, starches or dextrin), calcium silicate, non-meat proteins (milk proteins, gluten), and textured soy flour, which is one of the most-used ingredients (González-Tenorio et al., 2012). Additionally, by comparing industrial chorizos with traditional varieties, the industrial varieties appear to contain a higher amount of common salt than the traditional varieties, and a range of additives, such as antioxidants (erythorbate or ascorbate), preservatives (nitrite, sorbate, benzoate, parabens) or others (phosphates, glutamate, dextrin, starch, powdered milk, etc.), which were not present in traditional sausages (González-Tenorio et al., 2012; González-Tenorio et al., 2013b). The use of soy protein isolate to prevent drip loss has been suggested for industrial Argentinian chorizos (Porcella et al., 2001).

The chemical and nutrient composition of fresh/semidried chorizos would mainly depend on the proportion of muscle and fat used in the formulation, the use of extenders for reducing costs, and the degree of drying to which the chorizo is subjected. The composition of different types of chorizos or chorizo-like sausages obtained from literature is shown in Table 5.2.

5.2.2   MANUFACTURING PROCESS

Meat and fat are minced together with a mincing machine through sieves with different hole diameters (5–12 mm) and then mixed with the rest of the ingredients. The sausage batter is next tightly stuffed into casings, which are either natural (from pork or beef) or artificial (usually collagen or specific semi-permeable barrier plastic in the case of high technology industries) casings with c.a. 28–38 mm diameter. When the mincing and stuffing equipment works without vacuum, the sausage batter has to be pressed down and stuffed as tightly as possible to avoid air being trapped inside. Before stuffing, the batter can be kept for a day at a cool temperature, leading to a decrease in the redox potential. In Spain, chorizo batter (non-stuffed) can be commercialized under the name of *picadillo de chorizo*. After being stuffed, the filled casings can be tied or left untied, depending on the region or

TABLE 5.2

Composition of Different Types of Chorizo or Chorizo-like Sausages

Variety	Country	Moisture	Protein	Fat	CH	NaCl	pH	a$_w$
Industrial[a]	Spain	49.5	16.7	29.5	~1.5	1.9	N/A	N/A
Butchers' chorizo[b]	Mexico	42.9	16.3	34.9	3.2	2.1	5.1	0.96
Industrial chorizo[b]	Mexico	38.7	13.1	40.2	4.7	2.6	5.1	0.95
Chorizo verde Toluca[c, d]	Mexico	54.8	16.9	15.1	8.6	~2	4.7	0.91
Chorizo Antioqueño	Colombia	59.7	15.5	17.2	N/A	N/A	5,7	N/A
Salchicha de Huacho[e]	Peru	38.2	12.9	44.0	2.4	N/A	5.5	N/A
Tumbes sausage[f]	Peru	51.9	16.7	23.8	N/A	1.7	5.6	0.970

Notes

[a] Label information: Embutidos Arrieta, Zubiri, Navarra, España.

[b] González-Tenorio et al. (2013b).

[c] Ramírez-Muñoz et al. (2015).

[d] Pacheco-Pérez et al. (2011).

[e] Reyes-García et al. (2017).

[f] Ramos et al. (2014)

CH: Carbohydrate, includes lactate, may be obtained by difference or by analysis.

N/A: not available

the specific manufacturer's practices. The sausages that are tied are done so in order to form either strings with portions of different length, such as the approximately 5 cm short spherical portions for the *chorizo bola* or *de bolita* (Allen, 2015), the most common 10–15 cm portion, or individual horseshoe shaped portions of about 50 cm total length. Thread is normally used for tying, while metal clips are less commonly used to fasten chorizos. Other traditional tying material is used in a few varieties such as chorizo Salvadoreño, where sausage strings can be formed with corn husks.

Stuffed chorizos are then suspended from rods or hocks (hanging) for a short air-drying period (from a few hours to a few days, for example, 10 days) and further storage. During drying and storage, moisture loss (up to 20%) occurs, and depending on the length of the time elapsed, the initially fresh sausages can become semidried (ripened) sausages (Panagou et al., 2013), with moisture content between 40–60%, moisture/protein ratios between 2.0–3.5, and water activity from 0.96 to 0.92 (González-Tenorio et al., 2013b). Drying in the long run would also result in lactic acid fermentation and pH decrease (González-Tenorio et al., 2013a).

The drying of fresh/semidried chorizos is normally carried out at room temperature. Spanish chorizos are usually dried for up to 24 hours in refrigeration chambers or at room temperature (12–22°C). Drying for periods longer than 1 day should be carried out under controlled time–temperature combinations, considering that pathogenic bacteria can grow in these products, thus posing a risk to consumer health. In order to control *Staphylococcus aureus*, in particular its ability to produce a thermostable toxin, the drying period should be no longer than 60 hours at 23°C , or 30 hours at 30°C. See degree–hours combinations at Sebranek (2004).

As already mentioned, depending on the length or degree of drying, chorizos are fresh or semidry (slightly ripened) sausages. The moment when the transition from one category to the other occurs seems not to be clearly established. This circumstance may represent a problem for fresh/semidried chorizo classification into the *meat preparation* or *meat product* categories inscribed in the European Union legislation (European Parliament and the Council, 2004). Annex I states that for meat preparations, "meat has undergone processes which insufficiently modify the internal muscle fibre structure of the meat and therefore eliminate the characteristics of fresh meat," and specify that "if the characteristics of fresh meat are completely eliminated after processing, it should no longer be considered a meat preparation, but should fall within the definition of "meat products." Therefore, a slightly dried chorizo should be a meat preparation and a semidried chorizo a meat product. However, the transition point at which a chorizo changes category on the basis of visual or histological characteristics of fresh meat, or not as may be the case, would be difficult to establish (Ros-Berruezo et al., 2016).

Some industrial chorizos, produced mainly in Latin American countries, are cooked (optionally smoked during cooking) after being stuffed (Cabeza-Herrera et al., 2019; Pacheco-Pérez et al., 2011; Romero et al., 2013). In these cases, chorizos are prepared and commercialized as heat-treated meat products (precooked or ready to eat), and are usually vacuum packaged for commercialization.

5.2.3 Storage and Shelf Life

The storage of fresh/semidried chorizos is usually carried out by hanging the sausages in refrigeration chambers with an appropriate relative humidity (for example 75–80%) to avoid the growth of moulds on the chorizo surface, while preventing excessive moisture loss, affecting the yield and economic performance. When chorizos are not dried (some Latin American varieties) or barely dried (a few hours), due to their pH values of approximately 5.5 and a_w equal to or higher than 0.97, they are highly perishable (for example 10 days at 4°C) (Hugo and Hugo, 2015). However, they improve their stability during storage because hung chorizos gradually lose humidity through evaporation thus decreasing their a_w, the extent of which depends on relative humidity and time, as well as retarding microbial growth. This drying process explains the small increments in total microbial counts (lower than 1 logarithmic unit of colony forming units; CFU) found by Perales-Jasso et al. (2018) after 1 week of storing hung chorizos. Over time, the growth of indigenous psychrotrophic lactic acid bacteria (LAB) could take place in refrigerated stored chorizo, increasing the LAB concentration and reducing chorizo pH (González-Tenorio et al., 2013a). This growth presumably would develop deterioration when counts become high, for example <8 log CFU/g; however, the precise effect of this lactic acid fermentation on the spoilage of fresh meat is not clear (Pothakos et al., 2015); neither is it in fresh/semidried chorizos.

Although fresh/semidried chorizo require time–temperature control during drying and storage (IFT/FDA, 2003), there are studies reporting uncontrolled room-temperature storage or display practices in manufacturing and retail establishments, that is, chorizos being stored or displayed at room temperatures, without proper cool chain management. In these cases, producers rely on pH decreases by spontaneous lactic fermentation or the addition of acidifiers such as vinegar (acetic acid), and a_w reduction by drying during storage to achieve microbial stability and the safety of the chorizo (Kuri et al., 1995; Hew et al., 2006).

However, these practices could result in temperature abuse (Escartín et al., 1999; Ingham et al., 2009). In specific markets, chorizos have shown a relatively high prevalence of pathogenic enterobacteria such as *Escherichia coli* O157:H7 or *Salmonella* spp., and *Listeria monocytogenes* (Escartín, Castillo, Hinojosa-Puga, and Saldaña-Lozano 1999; Hew et al., 2005; Gamboa-Marín et al., 2012), which are able to grow in fresh chorizos. In semidried chorizos, internal conditions, such as pH of approximately 5.3 and water activity <0.97, would cause the gradual death of pathogens, although death rates would not be sufficiently quick to ensure a safe destruction during the shelf life of the product (Hew et al., 2006). A pathogen of particular concern is the food-borne *Staphyloccocus aureus* (Normano et al., 2007; Pinto et al., 2005). Due to its resistance to low a_w in chorizos with pH values higher than 5.3 and under temperature abuse conditions, that is <15°C, this pathogen is capable of growing and producing heat-resistant toxins.

Industrial chorizos can be prepacked for refrigerated storage and displayed in bags or trays under vacuum or modified atmospheres (such as mixtures of N_2 and CO_2). Modified atmosphere packaging for display extends chorizo shelf life when compared with chorizo wrapped in air permeable plastic film (Ruiz-Capillas et al., 2012). The shelf life of modified atmosphere or vacuum packaged fresh sausages largely depends on the quality of raw materials, presence of preservatives, atmosphere composition (redox potential), chorizo pH and a_w, as well as the storage temperature (Ruiz-Capillas et al., 2012; González-Tenorio et al., 2013a; Raimondi et al., 2018). Fresh sausages with pH values >5.5 and a_w ≥0.97 are highly perishable due to the presence of microbe-generated off odors and discoloration, with shelf-life periods of up to 20 days (Cocolin et al., 2004). However, semidried chorizos frequently have lower values of both pH and a_w, which would lead to longer shelf life.

Lactic acid fermentation occurs during storage of modified atmosphere or vacuum packaged chorizos (González-Tenorio et al., 2013a). In fresh sausages under refrigerated conditions, the growth of lactic acid bacteria (mainly psychrotrophic *lactobacillus*, such as *L. sakei*) together with *Enterobacteriaceae* (for example *Serratia* spp.), and *Brochotrix thermosphacta* (all facultative anaerobic bacteria) is responsible for spoilage due to the production of off odours (Benson et al., 2014; Dias et al., 2013; Raimondi et al., 2018). Moreover, microbial production of biogenic amines in modified-atmosphere-packaged fresh/semidried chorizos has been found during long storage periods, that is, more than 2 weeks, which might represent a health risk (Ruiz-Capillas, Pintado, and Jiménez-Colmenero 2012; González-Tenorio et al., 2013a). In order to control the growth of pathogenic or spoilage bacteria in packaged chorizo during storage, several studies have suggested the use of different natural antimicrobial approaches such as the use of *Carnobacterium maltaromaticum* strains (González et al., 2013), ethanolic propolis extracts (Gutiérrez-Cortés and Suárez-Mahecha, 2014), or oregano essential oil (Perales-Jasso et al., 2018).

5.3   DRY-RIPENED CHORIZO/CHOURIÇO

Dry-cured Portuguese *chouriços de carne* or *chouriça de carne*, depending on the geographical area of the country, and Spanish chorizos are among the traditional dry-cured sausages from Mediterranean Europe, such as Italian salami or French *saucisson sec*. In some regions of Spain, some chorizo-like sausage varieties are called longaniza, chistorra or morcon. All these are considered shelf stable; that is, they do not require refrigeration during their storage (mainly due to a low a_w) and are a ready-to-eat food (Ordoñez and de la Hoz, 2007). The term *cured* for these sausages means that the sausage has undergone a long ripening process lasting weeks or months, which tends to be slow (Flores, 1997). During ripening, dry-cured sausages undergo lactic fermentation, drying, enzymatic proteolysis and lipolysis, amino acid breakdown, etc., which extension is controlled by external humidity, temperature and air velocity regulation in natural or automatically controlled facilities (Feiner, 2006; Mariansky and Mariansky, 2009).

Most of the traditional sausages in Mediterranean Europe are slow or low-acid fermented sausages, subjected to spontaneous fermentation at low temperatures, final pH from 5.0 to 6.2, and dried to an a_w point that inhibits the growth of the organisms, that is <0.9 (Demeyer et al., 2000; Spaiciani et al., 2009). However, albeit less frequently, some industries can use starters and sugars to accelerate and intensify the sausage fermentation to a pH near to or lower than 5.0. The main quality characteristics of both traditional and industrial types of sausages are a strong and stable colour, succulent typical flavour, and good sliceability (Feiner, 2006), of which the latter tends to be better in industrially produced sausages than in traditional sausages (Comi et al., 2005).

Dry cured chorizos/chouriços are produced all over Spain and Portugal with a great diversity among the different regions. The differences are related to pig breed and feeding systems, as well as specific sausage manufacturing processes, for example, formulation, sieve diameter, sausage shape, ripening conditions, etc., which are influenced by climatic conditions and cultural traditions. A few examples of chorizos/chouriços from Spain and Portugal are shown in Figure 5.3.

In Portugal, chouriços are generally smoked, while in Spain smoking only takes place in some regions, normally in the northwest. As dry-cured chorizos/chouriços belong to the Mediterranean zone sausages, they have a pH value normally higher than 5.0 and aw <0.90 (Ordoñez and de la Hoz, 2007). According to Menéndez et al. (2018), the mean a_w (standard deviation) of a sample of commercial Spanish chorizos was 0.86 (0.05) and the mean pH was 5.12 (0.28). In general, chorizos/chouriços are firm, with a reddish colour and bright cut, where lean and fat are macroscopically visible in fragments (Marcos et al., 2016).

In general, high quality dry-cured chorizos/chouriços are mainly consumed raw, sliced, in baguettes, tapas, etc. Moreover, there are lower-quality chorizos, with a higher content of collagen and/or containing offal, that are used in main dishes, such

FIGURE 5.3 Dry-cured chorizo: 1–Iberian, sliced, Spain; 2–stuffed in *cular* casing, Spain; 3–traditional from Alentejo, Portugal.

as lentil, chickpea, or dried-bean stews, or cooked with vegetables such as cabbage and potatoes (Marcos et al. 2016).

It is worth mentioning the relevance of native pig breeds in the chorizo industry (Marcos et al. 2016). In Portugal and Spain an important part of high-value chorizos are prepared from the meat of an Iberian breed pig called Ibérico in Spain and Alentejano in Portugal. This is a rustic, slow-growing native breed, deeply bound to the *Dehesa* Mediterranean ecosystem (López-Bote, 1998; Marcos et al., 2016). The meat from the Portuguese Bisaro Celtic-origin pig breed is also used in the north of Portugal to prepare premium chouriços, such as the chouriça from Trás-os-Montes (Álvarez-Rodríguez and Texeira, 2019; Marcos et al., 2016). In general, the meat of the animals of this type of breed, raised in extensive or semi-extensive production systems, has greater subcutaneous and intramuscular fat content, associated with a superior sensory quality.

There are traditional varieties of chorizos, chouriços or linguiças that have been granted European quality Protected Geographical Indications designations: Chorizo de Cantimpalos, Chorizo Riojano, Chouriço de Carne de Estremoz e Borba, Chouriço Groso de Estremoz e Borba, Chouriça de Carne de Barroso-Montalegre, Chouriça de Carne de Vinhais ou Linguiça de Vinhais, Chouriço de Abóbora de Barroso-Montalegre, Chouriço Azedo de Vinhais, Chouriço de Portalegre, Linguiça de Portalegre, and Linguíça do Baixo Alentejo ou Chouriço de carne do Baixo

Alentejo. Moreover, many other varieties have national or regional registered quality labels (MERCASA, 2019).

Dry-cured chorizo is also produced in Latin American countries, most notably *chorizo seco* from Argentina, which resembles Spanish chorizo, although their production and popularity is significantly lower than in Spain and Portugal. In Latin American markets the fresh/semidry chorizo varieties predominate.

5.3.1 RAW MATERIALS AND ADDITIVES

The most common meat ingredients used for preparing chorizos/chouriços are pork meat and back fat, although a mix of beef and pork is also common (Feiner, 2006). Moreover, beef (alone) or game (deer, boar) (Soriano et al., 2006), together with pork fat are used for specific varieties or industries. In pork chorizo/chouriço production, lean and fatty meat is used that usually comes from the anatomical regions of the neck, shoulder, ham and belly (Martín-Bejarano, 2001; Ockerman and Basu, 2007). Some traditional chorizos, or their industrial equivalents, include other meat ingredients in their formulation such as blood, rinds and/or offal. For example, the formulation of homemade Galician *chorizo de cebolla* (onion chorizo; northwest region of Spain) includes bloody trimmings, lungs and cooked rind at average percentages of 12%, 4% and 3%, and that of chouriça from Trás-os-Montes (north of Portugal) can include blood, or boiled or brined rind (Marcos et al., 2016).

Fat is the second most important ingredient for chorizos, and comes from the meat pieces and subcutaneous lard (pork back fat) due to its firmness (Elias et al., 2006). The amount of fat in chorizo formulation in the sausage batter is approximately 30% (Feiner, 2006), with common ratios of muscle/adipose from 80/20 to 70/30 (Elias et al., 2006). Lower amounts of fat would produce a lack of succulence, fast drying and hard texture; higher amounts can be penalized by quality standards. According to Spanish standards, a fat content higher than 57 g per 100 g dry matter downgrades the chorizo's commercial category from *extra* to *non-extra* (average), with the exception of Pamplona chorizo and Iberian chorizo varieties, where the maximum fat for the extra category is 65 g of fat/100 g dry matter. Other compositional limits for *extra* chorizos are ≥30 g protein/100 g dry matter, a maximum ratio of collagen/total protein of 16%, with both limits being more permissible for Pamplona and Iberian chorizo. In Portugal, according to their fat/protein ratio, chouriços are classified as *extra* (fat content <2 times the total protein content), and *corrente* (higher fat content) (Marcos et al., 2016). The chemical composition in different high-quality chorizos/chouriços is shown in Table 5.3.

Apart from meat ingredients, there are a few varieties of chorizo that contain relatively large amounts of vegetables in their formulation, mainly pumpkin, onion or potatoes. In *chorizo gallego*, for example, producers add a mixture of onion and pumpkin at levels of 20% of sausage batter (Salgado et al., 2006), and in *chorizo patatero rojo*, they add 25% of boiled potato (Martín-Bejarano, 2001).

Common salt is used at levels of approximately 2% (Feiner, 2006;Martín-Bejarano, 2001), which results in levels near to 4% in the ripened sausage (Zurera-Cosano et al., 2011). The common spices and condiments used to prepare chorizo are powdered paprika at levels of 2–3.5% (used as previously explained for both fresh

TABLE 5.3
Composition of Different Types of Dry-Cured Chorizos/Chouriços

Variety, Country	Moisture (%)	Protein (%)	Fat (%)	NaCl (%)	pH	a_w
Chorizo de Pamplona, Spain[a]	30.0–33.8	18.6–20.4	34.6–36.8	3.0–3.5	4.6–4.9	N/A
Chorizo de León, Spain[b]	28.8	26.6	36.3	3.1	4.9	0.85
Chorizo de Zamora, Spain[c]	21.8–28.7	25.1–29.8	36.8–43.7	3.6–4.4	5.2–5.8	0.80–0.84
Chorizo Gallego de cebolla (with onion), Spain[d]	32.2	13.8	47.4	2.1	4.7	0.84
Chorizo riojano, Spain[e]	26.4	49.5	29.4	4.0	5.23	0.83
Chouriço de Abóbora, Portugal[f]	28.2	19.4	39.1	2.5	4.8	0.87
Chouriço de Melgaço, traditional, Portugal[g]	31.4–34.1	24.4–42.3	33.8–37.1	2.6–2.8	5.3–5.4	N/A

Notes
[a] Gimeno et al. (2000).
[b] Domínguez et al. (1988) and Mateo et al. (1996).
[c] Castro-Alfageme et al. (2004)
[d] Salgado et al. (2006).
[e] Alastrué-Naval et al. (2015).
[f] Patarata et al. (1998).
[g] Brito et al. (2010).
N/A: not available.

and semidried chorizo), and fresh garlic or garlic paste at levels of c.a. 0.5–2 g/kg of sausage batter (Elias et al., 2007; Martín-Bejarano, 2001). Moreover, the use of oregano (up to 1 g/kg) is also common, and some varieties can use small amounts of additional condiments such as pepper, white wine, olive oil, honey, etc. (Garriga and Aymerich, 2007; Marcos et al., 2016; Martín-Bejarano, 2001).

Species and condiments provide necessary organoleptic characteristics, contributing to the typicality of the products, while paprika imparts the characteristic redness to chorizo. The L*, a* and b* colour values range from 35 to 54, 14 to 27, and 9 to 18, respectively (Gimeno et al., 2000; Gómez et al., 2001; Pagán-Moreno et al., 1998), with the levels and colour of the meat and paprika used, and the drying temperature and ripening time being relevant variability factors. The paprika also contributes to the chorizo with regard to flavour compounds, fermentable sugars, natural antioxidants and minerals of technological interest (Mateo et al., 1996; 1997; Aguirrezábal et al., 2000). The amount of fermentable sugars in the sausage batter originating from the addition of 3% of paprika is expected to be of approximately 2 g/kg of meat batter (Aguirrezábal et al., 1998), which would significantly contribute to reducing the sausage pH during ripening.

Sugars, that is dextrose or lactose, and LAB starter cultures are not usually added to the meat mixtures in traditional chorizo/chouriço (Roncalés, 2007; Roseiro et al., 2010), although they can be used in industrial chorizo to obtain a higher and faster pH decrease during fermentation. In these cases, mixed starter cultures of LAB, that is *Lactobacillus sake, L. curvatus, L. plantarum, Pediococcus pentosaceus* and/or *P. acidilactici*, normally accompanied by adjunct cultures composed of *Staphylococcus xylosus* and/or *Staphylococcus carnosus* are added to the batter (Bañón et al., 2011; Flores & Toldra, 2011; González-Fernández et al., 1997; Petaja-Kanninem & Puolanne, 2007). The level of LAB to be added to chorizo could be around 10^7 cfu/g and those of *S. xylosus and/or S. carnous* aroud 10^6 cfu/g (González-Fandos et al., 1999).

Proteins, normally milk proteins, can be added to industrial chorizo to improve cohesiveness. In Spain, the permitted amount of milk proteins in the batter of "extra" chorizos is less than 1% (Presidencia del Gobierno, 2014). Many chorizos, both homemade and industrial varieties have no additives (Roncalés, 2007). However, when additives are used, the most used are nitrite, nitrate, ascorbate and phosphates, which (phosphates) increase firmness and cohesiveness (Fonseca et al., 2011). The composition of different dry-cured high quality chorizos/chouriços is shown in Table 5.3.

5.3.2 MANUFACTURING PROCESS

The production of dry-cured chorizo/chouriço comprises the following main steps (Feiner, 2006): meat and fat selection, meat and fat mincing/chopping, preparation of the sausage mix (batter), optionally keeping the mix at cool temperature for one or two days, filling and tying of casings, and dry-curing. During this last step, sausages can be smoked for a short period.

The meat and fat are minced together though sieve plates, which are usually around a 6–12 mm mesh, although in specific varieties, such as Spanish chorizo *cular,* the sieves can be wider (that is, 18 mm) or the cutting is carried out manually with knives (Martín-Bejarano, 2001). The minced meat and fat are mixed with the remaining ingredients and additives, preferably with vacuum mixers and traditionally by manual kneading, to form a homogeneous and cohesive mass. Optionally, although very common in the traditional manufacturing processes, the batter is compacted and subjected to a resting period of 24 and 48 hours at a temperature of approximately 2–6°C (Elias et al., 2006; Ordoñez and de la Hoz, 2007). This step can improve the activity of nitrate reductase bacteria (Sanz et al., 1997) and contribute to the reduction of the batter redox potential. The mass is commonly inserted into 30–60 mm-diameter hog or beef casings (Elias et al., 2006; Ordoñez and de la Hoz, 2007), although some industries use artificial collagen casings. Sausages are tied normally with thread, resulting in different shapes: horseshoe (named in Spain *sarta* or *herradura),* string (*ristra*) or stick *(vela* and *cular)* (Figure 5.3). The name of *cular* chorizo is given to a chorizo variety stuffed into a *cular* casing, which is obtained from the pig's large intestine, that is, the rectum.

In some small industries and rural homes, the dry-curing of chorizos/chouriços is carried out in conditioned facilities at room temperatures, making standardisation difficult (Elias et al., 2014; Roseiro et al., 2010). However, most of the industrial production involves dry-curing in ripening chambers with control of temperature, humidity and air speed. The environmental conditions vary from place to place and the type of industry. Many traditional chorizos/chouriços are slowly dry-cured, at temperatures usually below 15°C (5–15°C) and relative moisture ranging between 65 and 85%. During the dry-curing step, a spontaneous lactic acid fermentation takes place in the sausage at a moderate rate by the natural occurring microbiota. In contrast, industrial chorizos can include a short initial period of up to 3 days with high moisture (for example 90%) and temperature 18–25°C to promote fermentation; then the sausages continue the dry-curing process at temperatures close to 15°C and relative humidity of 70–80%.

The product can also be slightly smoked (Elias et al., 2006) during the dry-curing. In Portugal, chouriços are frequently smoked (Instituto Português da Qualidade, 2008), while in Spain smoking is only common in the northwest regions, for example Galicia or León (Mateo and Zumalacárregui, 1996). Smoking is commonly carried out with holly firewood and occurs at temperatures ranging from 10 to 40°C for a variable period of time, which can range from 3 to 7 days (this period may be even longer and last up to 42 days). The smoking conditions have influence on the polycyclic aromatic hydrocarbon content in these products (Lorenzo et al., 2011).

Apart from fermentation, a series of biochemical phenomena, lipolysis, proteolysis, amino acid breakdown, etc., determine the distinctive organoleptic characteristics (taste, aroma, texture) (Mateo and Zumalacárregui, 1996; Roseiro et al., 2011). The length of the dry-curing largely depends on the environmental conditions, the fat content and the diameter of the chorizo/chouriço. A 3-week period is a normal length for industrial chorizos stuffed in 35–40 mm casings. In contrast, the dry-curing can last 3–6 months in slowly dry-cured traditional chorizos, or *cular* chorizos stuffed in 50–60 mm casings (Martín-Bejarano, 2001).

The development and type of microbiota present in chorizos/chouriços depends on specific formulations, the microbiology of raw materials and the technological practices (Silva et al., 2003). The microbial population initially present in the meat batter of traditional chorizos/chouriços, that is, without the use of cultures, includes mainly gram-negative bacteria (for example *Pseudomonas* spp. and *Enterobacteriaceae*) and gram-positive bacteria (for example *Bacillus* spp., *Lactobacillus* spp. and *Enterococcus* spp.) as well as yeast (Ordoñez and de la Hoz, 2007).

During fermentation, LAB becomes the main microflora intrinsically present in traditional chorizos/chouriços, followed by gram-positive and catalase-positive *cocci*. LAB increases from about 3–5 log CFU/g to 6–9 log CFU/g and coagulase positive cocci from 3–5 log CFU/g to 5–8 CFU/g have been reported (Garriga and Aymerich, 2007; González-Fandos et al., 1999; Castaño et al., 2002; Fonseca et al., 2013; Prado et al., 2019; Gonzales-Barron et al., 2015). Yeasts are usually present in the batter of chorizos/chouriços at a level that typically ranges from 2.0–4.0 log CFU/g (Ordoñez and de la Hoz, 2007; González-Fandos et al., 1999; Encinas et al., 2000: Prado et al., 2019). These microorganisms can exhibit slight growth during the ripening of the chorizo, resulting in an increase in pH and flavour development (Encinas et al., 2000). On the other hand, *Enterobacteriaceae* counts, which are normally lower than 3 log cfu/g of initial batter under adequate hygienic conditions, might grow and increase to 1–2 log CFU/g at the beginning of the ripening. However, at the end of this process they tend to decrease to values lower than 2 log CFU/g. *Pseudomonas* also tend to disappear during the ripening process (Castaño et al., 2002; Fonseca et al., 2013; Prado et al., 2019).

Fungi can develop on the surface of the dry fermented sausage. A few chorizo varieties, which are non-smoked and dry-cured under appropriate temperature and humidity to favour a slow growth of moulds on their surface, present a desirable homogeneous white mould covering, which is the case of the Cantimpalos chorizo. The microbiota in the surface of these chorizos seems to be dominated by *Penicilium* spp. such as *P. commune* and *P. olsonii* (López-Díaz et al., 2001). *Penicillium* is one of the genera that most often develops in mould-ripened sausages (Bruna et al., 2003) because it has the ability to grow at lower temperatures and a_w than other moulds present in the product.

The LAB communities of chorizo/chouriço have been studied by few authors (Aymerich et al., 2006; Benito et al., 2007; Fonseca et al., 2013; Quijada et al., 2018). When molecular methods were used, the main species of LAB isolated in the chorizo were, in order of abundance, *Lb. sakei*, *Lb. plantarum*, *Lb. curvatus* followed by *P. acidilactici*, *Lactococcus lactis*, and *Enterococcus faecium*. *Enterococcus* spp. develop during the initial fermentation process (reaching 4–6 log CFU/g) and remain constant (or decrease) until the end of the ripening process. Main species of the *Staphylococcus* genus isolated from chorizo have been *S. xylosus, S. carnosus, S. equorum, S. sciuri, S. warnery, S. epidermidis, and S. condiment* (Fonseca et al., 2013; Quijada et al., 2018).

Regarding the pathogenic *S. aureus*, González-Fandos et al. (1999) found that it could grow and produce toxins in chorizo depending on environmental factors, such as temperature and competing microflora. To avoid the growth of this microorganism, temperature control has been suggested during ripening so that the product of "ripening temperature (°C) minus 15°C" multiplied by "ripening time (hours) until pH 5.3" must be lower than 20°C × hours (Vignolo et al., 2010).

5.3.3 STORAGE AND SHELF LIFE

Although dry-cured chorizos/chouriços are shelf stable, their appearance and edible characteristics gradually change during storage, affecting their quality. The speed of these changes depend on the chorizo composition and environmental conditions. In non-packaged chorizos/chouriços stored in warm hanging rooms, the most deleterious changes are spoilage by undesirable growth of moulds and yeast on the surface, when the air humidity on the sausage surface is high, or an excessive hardness when the humidity is low due to evaporation losses. Oxidative spoilage is also possible, especially when the raw materials, that is fat and natural casings, were already oxidized, the oxygen concentration inside the sausage is high or the sausage is highly exposed to light during display. Oxidation can result in rancid flavour development and/or in discoloration, in this case mainly due to the degradation of paprika carotenes. Colour stability depends also on paprika characteristics. The use of smoked-dried paprika (containing antioxidant compounds from smoke) appears to result in a more stable colour when compared with the use of oven-dried paprika (Pereira et al., 2019). Moreover, the colour of pasteurized paprika, the use of which can be recommendable by its low microbial load, can be unstable during chorizo/chouriço storage, causing discoloration (Gómez et al., 2008).

All the above-mentioned problems can be avoided with vacuum packaging or storage under anoxic atmospheres, for example 100% nitrogen or 20% carbon dioxide and 80% nitrogen. This packaging considerably extends the chorizos'/chouriços' shelf life, reaching periods of several months, for example 5–10 months for the Gallego chorizo studied by Fernández-Fernández et al. (2002; 2005). Deterioration in these cases has been attributed to chemical changes and the subsequent physical and taste changes, such as progressive lipolysis, excessive taste persistence and a decrease in cohesiveness and juiciness, which can be slowed with low (refrigeration) storage temperatures (Fernández–Fernández et al., 2002). Another limiting factor in the shelf-life of packaged chorizos is the production of biogenic amines (tyramine), levels of which can increase during storage, especially when the chorizo a_w is high, that is close to 0.89 (González-Tenorio et al. 2013a). The best storage procedure regarding shelf life, according to Fernández–Fernández et al. (2005), should be packaging and frozen storage.

Chorizo/chouriço can be displayed in pieces or in slices on trays packaged under vacuum or anoxic modified atmosphere. In these cases, special attention has to be paid to the survival of *Listeria monocytogenes*, although these bacteria quickly decrease through storage (Menéndez, Rendueles, Sanz, Santos, and García-Fernández, 2018) if present. Where cross-contamination occurs during slicing and packaging, a significant amount of *L. monocytogenes* cells would be able to survive, which are the function of chorizo hurdles (pH, aw, nitrites), microbial composition, storage temperature and time (García-Díez and Patarata, 2017; Possas et al., 2019).

REFERENCES

Abad-Alegría, F. 2001. *Color rojizo en nuestra historia culinaria. El especiado con azafrán y pimentón en las cocinas hispanas*. Zaragoza: Institución Fernando el Católico/Diputación de Zaragoza. https://ifc.dpz.es/recursos/publicaciones/20/47/_ebook.pdf.

Aguirrezábal, M. M., J. Mateo, M. C. Domínguez, and J. M. Zumalacárregui. 1998. Spanish paprika and garlic as sources of compounds of technological interest for the production of dry fermented sausages. *Sciences del Aliments* 18:409–414.

Aguirrezábal, M. M., J. Mateo, M. C. Domínguez, and J. M. Zumalacárregui. 2000. The effect of paprika, garlic and salt on rancidity in dry sausages. *Meat Science* 54:77–81.

Alastrué-Naval, Y., S. Sanz-Cervera, C. Olarte-Martínez, and E. Romero-Melgosa. 2015. Caracterización del Chorizo Riojano. I. Composición físico-química y compuestos volátiles. *Eurocarne* 239:86–92.

Allen, G. 2015. *Sausage, a Global History*. The Edible Series. London: Reaktion Books, Ltd.

Álvarez-Rodríguez, J., and A. Texeira. 2019. Slaughter weight rather than sex affects carcass cuts and tissue composition of Bisaro pigs. *Meat Science* 154:54–60.

Aymerich, T., B. Martín, M. Garriga, M. C. Vidal-Carou, S. Bover-Cid, and M. Hugas. 2006. Safety properties and molecular strain typing of lactic acid bacteria from slightly fermented sausages. *Journal of Applied Microbiology* 100:40–49.

Bañón, S., A. Martínez, and M. López. 2011. Maduración de chorizo y salchichón de Chato Murciano con diferentes cultivos iniciadores (bacterias ácido-lácticas y estafilococos). *Anales de Veterinaria de Murcia* 27:101–108.

Becerril-Sánchez, A. L., G. O. Dublán, A. Domínguez-López, C. D. Arizmendi, and B. Quintero-Salazar. 2019. La calidad sanitaria del chorizo rojo tradicional que se comercializa en la ciudad de Toluca, Estado de México. *Revista Mexicana de Ciencias Pecuarias* 10:172–185.

Benito, M. J., A. Martín, E. Aranda, F. Pérez-Nevado, S. Ruiz-Moyano, and M. G. Córdoba. 2007. Characterization and selection of autochthonous lactic acid bacteria isolated from traditional Iberian dry-fermented Salchichón and Chorizo sausages. *Journal of Food Science* 72:193–201.

Benson, A. K., J. R. D. David, S. E. Gilbreth, G. Smith, J. Nietfeldt, R. Legge, J. Kim, R. Sinha, C. E. Duncan, J. Ma, and I. Singh. 2014. Microbial successions are associated with changes in chemical profiles of a model refrigerated fresh pork sausage during an 80-day shelf life study. *Applied and Environmental Microbiology* 80:5178–5194.

Briones, D. M. S., R. M. Cueto, R. S. Ocampo, J. M. Aballa, and B. Festijo. 2013. Lucban specialty foods as culinary attraction in Quezon Province, Philippines. *International Journal in Management and Social Science* 1(3):7–14.

Brito, N. V., D. Santos, A. P. Vale, I. M. Afonso, E. Mendes, S. Casal, and M. B. P. P. Oliveira. 2010. Chemical characterization of a traditional Portuguese meat sausage, aiming at PGI certification. *British Food Journal* 122(5):489–499.

Bruna, J. M., E. M. Hierro, L. de la Hoz, D. S. Mottram, M. Fernández, and J. A. Ordoñez. 2003. Changes in selected biochemical and sensory parameters as affected by the superficial inoculation of *Penicillium camemberti* on dry fermented sausages. *International Journal of Food Microbiololgy* 85:111–125.

Cabeza-Herrera, E. A., G. E. Laguado-Corredor, and W. H. Suárez-Quintana. 2019. Modeling the growth of lactic acid bacteria in cooked chorizo vacuum-packed under isothermal cooling conditions. *Revista de la Facultad de Ciencias Básicas* 17(1):124–138.

Camerati, P., and R. Garcés. 2017. Determinación histológica y planimétrica de la composición de longanizas comercializadas en la provincia de Arauco y Concepción, región del Bbío-bío, Chile. *Revista Electrónica Veterinaria* 18:8. http://www.veterinaria.org/revistas/redvet/n080817.html.

Carmona-Escutia, R. P., J. E. Urías-Silvas, M. D. García-Parra, E. Ponce-Alquicira, S. J. Villanueva-Rodríguez, and H. B. Escalona-Buendía. 2019. Influence of paprika (*Capsicum annuum L*) on quality parameters and biogenic amines production of a ripened meat product (*chorizo*). *Revista Mexicana de Ingeniería Química* 18:949–966.

Castaño, A., M. C. G. Fontán, J. M. Fresno, M. E. Tornadijo, and J. Carballo. 2002. Survival of *Enterobacteriaceae* during processing of Chorizo de cebolla, a Spanish fermented sausage. *Food Control* 13:107–115.

Castro-Alfageme, S., M. T. Osorio, E. A. Cabeza-Herrera, J. Mateo, and J. M. Zumalacárregui. 2004. Aportaciones a la caracterización del chorizo elaborado en la provincia de Zamora. *Eurocarne* 125:151–162.

Chirinos, E. R., and B. Salvá. 2000. *Elaboración de embutidos*. Lima, Perú: Editorial Bookexpress.

Cocolin, L., K. Rantsiou, L. Iacumin, R. Urso, C. Cantoni, and G. Comi. 2004. Study of the ecology of fresh sausages and characterization of populations of lactic acid bacteria by molecular methods. *Applied and Environmental Microbiology* 70:1883–1894.

Comi, G., R. Urso, L. Iacumin, K. Rantsiou, P. Cattaneo, C. Cantoni, and L. Cocolin. 2005. Characterisation of naturally fermented sausages produced in the North East of Italy. *Meat Science* 69:381–392.

Demeyer, M., A. Raemakers, A. Rizzo, A. Holck, B. De Smedt, B. Brink, C. Hagen, E. Montel, E. Zanardi, F. Murbrek, F. Leroy, K. Vandendressche, K. Lorentsen, K.

Venema, L. Sunesen, L. Stahnke, L. De Vuyst, R. Talon, R. Chizzolini, and S. Eerola. 2000. Control of bioflavour and safety in fermented sausages: first results of a European proyect. *Food Research International* 33:171–180.

Dias, F. S., C. L. Ramos, and R. F. Schwan. 2013. Characterization of spoilage bacteria in pork sausage by PCR–DGGE analysis. *Food Science and Technology Campinas* 33:468–474.

Domínguez, M. C., C. Ferre, and J. M. Zumalacarregui. 1988. Aportaciones a la caracterización del chorizo elaborado en la provincia de León: parámetros químicos y fisicoquímicos. *Alimentaria* 198:19–23.

Elias, M., A. C. Santos, and B. Raposo. 2007. Caracterização de matérias-primas subsidiárias usadas no fabrico de paio de porco Alentejano. *Revista de Ciências Agrárias* 30:424–438.

Elias, M., M. E. Potes, L. C. Roseiro, C. Santos, A. Gomes, and A. C. Agulheiro-Santos. 2014. The effect of starter cultures on the Portuguese traditional sausage "Paio do Alentejo" in terms of its sensory and textural characteristics and polycyclic aromatic hydrocarbons profile. *Journal of Food Research* 3:45–56.

Elias, M., M. J. Fraqueza, and A. Barreto. (2006). Typology of traditional sausage production from *Alentejo*. *Revista Portuguesa de Zootécnia* 1(Ano XIII):2–12.

Encinas, J. P., T. M. López-Díaz, M. L. García-López, A. Otero, and B. Moreno. 2000. Yeast populations on Spanish fermented sausages. *Meat Science* 54:203–208.

Escartín, E. F., A. Castillo, A. Hinojosa-Puga, and J. Saldaña-Lozano. 1999. Prevalence of *Salmonella* in chorizo and its survival under different storage temperatures. *Food Microbiology* 16:479–486.

European Parliament and the Council. 2004. Regulation (EC) No 853/2004 of 29 April 2004 laying down specific hygiene rules for food of animal origin. *Official Journal of the European Union*. Publications Office of the European Union: Luxemburgo. https://eur-lex.europa.eu/legal-content/EN/TXT/PDF/?uri=CELEX:32004R0853&from=EN.

Feiner, G. 2006. *Fresh Sausages. Meat Products Handbook: Practical Science and Technology*. Cambridge: Woodhead Publishing Limited/CRC Press.

Feiner, G. 2006. *Meat Products Handbook: Practical Science and Technology*. Cambridge: Woodhead Publishing.

Fernández–Fernández, E., M. L. Vázquez–Odériz, and M. A. Romero-Rodríguez. 2002. Sensory characteristics of Galician chorizo sausage packed under vacuum and under modified atmosphere. *Meat Science* 62:67–71.

Fernández–Fernández, E., M. L. Vázquez–Odériz, and M. A. Romero-Rodríguez. 2005. Changes in sensory properties of Galician chorizo sausages preserved by freezing, oil immersion and vacuum packing. *Meat Science* 70:223–226.

Flores, J. 1997. Mediterranean vs northern European meat products. Processing technologies and main differences. *Food Chemistry* 4:505–510.

Flores, M., and F. Toldra. 2011. Microbial enzymatic activities for improved fermented meats. *Trends in Food Science & Technology* 22(2-3):81–90.

Fonseca, B., V. Kuri, J. M. Zumalacárregui, A. Fernández-Diez, B. K. Salvá, I. Caro, M. T. Osorio, and J. Mateo. 2011. Effect of the use of a commercial phosphate mixture on selected quality characteristics of 2 Spanish-style dry-ripened sausages. *Journal of Food Science* 76(5):S300–S305.

Fonseca, S., A. Cachaldora, M. Gómez, I. Franco, and J. Carballo. 2013. Effect of different autochthonous starter cultures on the volatile compounds profile and sensory properties of Galician chorizo, a traditional Spanish dry fermented sausage. *Food Control* 33:6–14.

Fonte, M. 2008. Knowledge, food and place. A way of producing, a way of knowing. *Sociologia Ruralis* 48(3):200–222.

Gamboa-Marín, A., S. Buitrago, K. Pérez-Pérez, M. Mercado, R. Poutou-Piñales, and A. Carrascal-Camacho. 2012. Prevalence of *Listeria monocytogenes* in pork-meat and other processed products from the Colombian swine industry. *Revista MVZ Córdoba* 17(1):2827–2833.

García-Díez, J., and L. Patarata. 2017. Influence of salt level, starter culture, fermentable carbohydrates, and temperature on the behaviour of *L. monocytogenes* in sliced chouriço during storage. *Acta Alimentaria* 46(2):206–213.

Garriga, M., and T. Aymerich. 2007. The microbiology of fermentation and ripening. In *Handbook of Fermented Meat and Poultry*, ed. F. Toldrá, 125–136. Iowa: Willey-Blackwell.

Gimeno, O., D. Ansorena, O. Astiasarán, and J. Bello. 2000. Characterization of chorizo de Pamplona: instrumental measurements of colour and texture. *Food Chemistry* 69:195–2000.

Gobierno Autónomo Departamental de Santa Cruz. 2013. *Recetario*. Santa Cruz, Bolivia: Secretaría de Educación y Cultura. http://www.santacruz.gob.bo/archivos/AN3009-2013112731.pdf.

Gómez, R., M. Álvarez-Orti, and J. E. Pardo. 2008. Influence of the paprika type on redness loss in red line meat products. *Meat Science* 80:823–828.

Gómez, R., M. I. Picazo, A. Alvarruiz, J. I. Pérez, D. Valera, and J. E. Pardo. 2001. Influencia del tipo de pimentón en la pérdida de color del chorizo fresco. *Alimentaria* 323:67–73.

Gonzales-Barron, U., V. Cadavez, A. P. Pereira, A. Gomes, J. P. Araújo, M. J. Saavedra, L. Estevinho, F. Butler, and T. Dias. 2015. Relating physicochemical and microbiological safety indicators during processing of linguiça, a Portuguese traditional dry-fermented sausage. *Food Research International* 78:50–61.

González, M. I., W. Yien, J. A. Castrillón, and A. Ortega. 2013. Addition of *Carnobacterium maltaromaticum* CB1 in vacuum pack aged chorizo and morcilla, to inhibit the growth of *Listeria monocytogenes*. V*itae, Revista de la Facultad de Química Farmacéutica* 20(1):23–29.

González-Fandos, M. E., M. Sierra, M. L. García-López, M. C. García-Fernández, and A. Otero. 1999. The infuuence of manufacturing and drying conditions on the survival and toxinogenesis of Staphylococcus aureus in two Spanish dry sausages (chorizo and salchichón). *Meat Science* 52:411–419.

González-Fernández, C., E. M. Santos, I. Jaime, and J. Rovira. 1997. Use of starter cultures in dry-fermented sausage (chorizo) and their influence on the sensory properties. *Food Science and Technology International* 3(1):31–42.

González-Schnake, F., and R. Nova. 2014. Ethnic meat products: Brazil and South America. In *Encyclopedia of Meat Science*, ed. M. Dikeman, and C. Devine, 2nd edition, Vol. 1, 518–521. London: Academic Press.

González-Tenorio, R., A. Fernández-Diez, I. Caro, and J. Mateo. 2012. Comparative assessment of the mineral content of a Latin American raw sausage made by traditional or non-traditional processes. In *Atomic Absorption Spectroscopy*, ed. A. F. Muhammad. Rijeka, Croatia: *IntechOpen*. https://www.intechopen.com/books/atomic-absorption-spectroscopy/comparative-assessment-of-the-mineral-content-of-a-latin-american-raw-sausage-made-by-traditional-or.

González-Tenorio, R., A. Totosaus, I. Caro, and J. Mateo. 2013b. Characterization of chemical and physicochemical properties of sausages marketed in the Central region of Mexico. *Información Tecnológica* 24(2):3–14.

González-Tenorio, R., B. Fonseca, I. Caro, A. Fernández-Diez, V. Kuri, S. Soto, and J. Mateo. 2013a. Changes in biogenic amine levels during storage of Mexican-style soft and Spanish-style fry-ripened sausages with different aw values under modified atmosphere. *Meat Science* 94:369–375.

Gutiérrez-Cortés, C., and H. Suárez-Mahecha. 2014. Antimicrobial activity of propolis and its effect on the physicochemical and sensorial characteristics in sausages. *Vitae, Revista de la Facultad de Química Farmacéutica* 2(2):90–96.

Hajmeer, M., N. Basheer, C. M. Hew, and D. O. Cliver. 2006. Modelling the survival of *Salmonella* spp. in chorizos. *International Journal of Food Microbiology* 107:59–67.

Hew, C. M., M. N. Hajmeer, T. B. Farver, J. M. Glover, and D. O. Cliver. 2005. Survival of *Listeria monocytogenes* in experimental chorizos. *Journal of Food Protection* 68:324–333.

Hew, C. M., M. N. Hajmeer, T. B. Farver, H. P. Riemann, J. M. Glover, and D. O. Cliver. 2006. Pathogen survival in chorizos: ecological factors. *Journal of Food Protection* 69:1087–1095.

Hugo, C. J., and A. Hugo. 2015. Current trends in natural preservatives for fresh sausage products. *Trends in Food Science and Technology* 45:12–23.

IFT/FDA. 2003. Analysis of microbial hazards related to time/temperature control of foods for safety. Chapter IV. *Comprehensive Reviews in Food Science and Food Safety* 2(supplement):33–41.

Ingham, S. C., B. H. Inghan, D. Borneman, E. Jaussaud, E. L. Schoeller, N. Hoftiezer, L. Schwartzburg, G. M. Burnham, and J. P. Norback. 2009. Predicting pathogen growth during short-term temperature abuse of raw sausage. *Journal of Food Protection* 72:75–84.

Instituto Português da Qualidade. 2008. *Norma Portuguesa para Chouriço de carne: Definição, classificação, características e acondicionamento (NP-589)*. Lisboa: Instituto Português da Qualidade.

Kuri, V., R. H. Madden, and M. A. Collins. 1995. Hygienic quality of raw pork and chorizo (raw pork sausage) on retail sale in Mexico city. *Journal of Food Protection* 59(2):141–145.

López-Bote, C. J. 1998. Sustained utilization of the Iberian pig breed. *Meat Science* 49(supplement 1): S17–S27.

López-Díaz, T. M., J. A. Santos, M. L. García-López, and A. Otero. 2001. Surface mycoflora of a Spanish fermented meat sausage and toxigenicity of *Penicillium* isolates. *International Journal of Food Microbiology* 68:69–74.

Lorenzo, J. M., L. Purriños, R. Bermúdez, N. Cobas, M. Figueiredo, and M. C. García-Fontán. Polycyclic aromatic hydrocarbons (PAHs) in two Spanish traditional smoked sausage varieties: "Chorizo gallego" and "Chorizo de cebolla". *Meat Science* 89:105–109.

Marcos, C., C. Viegas, A. M. Almeida, and M. M. Guerra. 2016. Portuguese traditional sausages: different types, nutritional composition, and novel trends. *Journal of Ethnic Foods* 3:51–60.

Mariansky, A., and S. Mariansky. 2009. *The Art of Making Fermented Sausages*, 2nd edition. Andhra Pradesh, India: Bookmagic LLC.

Martín-Bejarano, S. 2001. *Enciclopedia de la carne y los productos cárnicos*. Plasencia, Cáceres: Martín & Macías.

Mateo, J., I. Caro, A. C. Figueira, D. Ramos, and J. M. Zumalacárregui. 2009. Meat processing in Ibero-american countries: a historical view. In *Traditional Food Production and Rural Sustainable Development. A European Challenge*, ed. T. N. Vaz, P. Nijkamp, J. L. Rastoin, 135–148. Cornwall: Ashgate, Publishing Limited.

Mateo, J., and J. M. Zumalacárregui. 1996. Volatile compounds in chorizo and their changes during ripening. *Meat Science* 44(3):255–273.

Mateo, J., M. C. Domínguez, M. M. Aguirrezábal, and J. M. Zumalacárregui. 1996. Taste compounds in chorizo and their changes during ripening. *Meat Science* 44(4):245–254.

Mateo, J., M. M. Aguirrezábal, M. C. Domínguez, and J. M. Zumalacárregui. 1997. Volatile compounds in Spanish paprika. *Journal of Food Composition and Analysis* 10: 225–230.

Menéndez, R. A., E. Rendueles, J. J. Sanz, J. A. Santos, and M. C. García-Fernández. 2018. Physicochemical and microbiological characteristics of diverse Spanish cured meat products. *CyTA Journal of Food* 16(1):199–204.

MERCASA. 2019. Alimentación en España. In *Producción, industria, distribución y consumo*. Madrid: MERCASA-Distribución y Consumo. https://issuu.com/tetraktys5/docs/aee_2019_web?fr=sMmFkYzg5NzY2.

Ministerio de la Presidencia. 2014. *Real Decreto 474/2014, de 13 de junio, por el que se aprueba la norma de calidad de derivados cárnicos*. Ref.: BOE-A-2014-6435. Madrid: Boletín Oficial del Estado. https://www.boe.es/buscar/pdf/2014/BOE-A-2014-6435-consolidado.pdf.

Municipalidad Provincial de Huaura. 2017. *Ordenanza que declara la Salchicha Huachana como producto gastronómico y emblemático de la ciudad de Huacho*. Ordenanza municial no. 036-2017/MPH. Norma legal de 3 de febrero de 2018. Lima: Diario El Peruano. https://busquedas.elperuano.pe/download/url/ordenanza-que-declara-la-salchicha-hua-chana-como-producto-ga-ordenanza-n-036-2017mph-1613218-1.

Normano, G., G. La Salandra, A. Dambrosio, N. C. Quaglia, M. Corrente, A. Parisi, G. Santagada, A. Firinu, E. Crisetti, and G. V. Celano. 2007. Occurrence, characterization and antimicrobial resistance of enterotoxigenic *Staphylococcus aureus* isolated from meat and dairy products. *International Journal of Food Microbiology* 115:290–296.

Ockerman, H. W., and L. Basu. 2007. Production and consumption of fermented meat products. In *Handbook of Fermented Meat and Poultry*, ed. F. Toldra, 9–15. Iowa: Willey-Blackwell.

Ordoñez, J.A., and L. de la Hoz. 2007. Mediterranean products. In *Handbook of Fermented Meat and Poultry*, ed. F. Toldrá, 333–347. Iowa: Willey-Blackwell.

Pacheco-Pérez, W. A., D. A. Restrepo-Molina, J. H. López-Vargas. 2011. Evaluación de un extensor graso sobre las propiedades de calidad del chorizo tipo Antioqueño. *Revista Facultad Nacional de Agronomía Medellín* 64(2):6265–6276.

Pagán-Moreno, M. J., M. A. Gago-Gago, J. A. Pérez-Álvarez, M. E. Sayas-Barberá, M. Rosmini, F. Perlo, and V. Aranda-Catalá. 1998. The evolution of colour parameters during chorizo processing. *Fleischwirtschaft* 78:987–989.

Palacios-Morillo, A., J. M. Jurado, A. Alcázar, and F. de Pablos. 2014. Geographical characterization of Spanish PDO paprika by multivariate analysis of multielemental content. *Talanta* 128:15–22.

Panagou, E. Z., G-J. E. Nychas, and J. N. Sofos. 2013. Types of traditional Greek foods and their safety. *Food Control* 29:32–41.

Patarata, L., G. Saraiva, and C. Martins. 1998. Processo de fabrico de produtos de salsicharia tradicional. In *1ª Jornadas Geográficas de Queijos e Enchidos-Produtos Tradicionais*. IAAS 83–86. IAAS. Exponor.

Perales-Jasso, Y. J., S. A. Gámez-Noyola, J. Aranda-Ruiz, C. A. Hernández-Martínez, G. Gutiérrez-Soto, A. I. Luna-Maldonado, R. E. Silva-Vazquez, M. E. Hume, and G. Mendez-Zamora. 2018. Oregano powder substitution and shelf life in pork chorizo using Mexican oregano essential oil. *Food Science and Nutrition* 6:1254–1260.

Pereira, C., M. G. Córdoba, E. Aranda, A. Hernández, R. Velázquez, T. Bartolomé, and A. Martín. 2019. Type of paprika as a critical quality factor in Iberian chorizo sausage manufacture. *CyTA – Journal of Food* 17(1):907–916.

Petaja-Kanninem, E., and E. Puolanne. 2007. Principles of meat fermentation. In *Handbook of Fermented Meat and Poultry*, ed. F. Toldrá, 31–36. Iowa: Willey-Blackwell.

Pickersgill, B. 2003. Peppers and chillies. *In Encyclopedia of Food Sciences and Nutrition*, ed. C. Benjamin, 2nd edition, 4460–4467. Oxford: Academic Press.

Pinto, B., E. Chenoll, and R. Aznar. 2005. Identification and typing of food-borne *Staphylococcus aureus* by PCR-based techniques. *Systematic and Applied Microbiology* 28:340–352.

Porcella, G., M. I. Sánchez, S. R. Vaudagna, M. L. Zanelli, A. M. Descalzo, L. H. Meichtria, M. Gallingera, and J. A. Lasta. 2001. Soy protein isolate added to vacuum-packaged chorizos: effect on drip loss, quality characteristics and stability during refrigerated storage. *Meat Science* 57:437–443.

Possas, A., V. Valdramidish, R. M. García-Gimeno, and F. Pérez-Rodríguez. 2019. High hydrostatic pressure processing of sliced fermented sausages: a quantitative exposure assessment for *Listeria monocytogenes*. *Innovative Food Science and Emerging Technologies* 52:406–419.

Pothakos, V., F. Devlieghere, F. Villani, J. Björkroth, and D. Ercolini. 2015. Lactic acid bacteria and their controversial role in fresh meat spoilage. *Meat Science* 109:66–74.

Prado, N., M. Sampayo, and P. González. 2019. Physicochemical, sensory and microbiological characterization of Asturian Chorizo, a traditional fermented sausage manufactured in Northern Spain. *Meat Science* 156:118–124.

Quijada, N. M., F. de Filippis, J. J. Sanz, M. C. García-Fernández, D. Rodríguez-Lázaro, D. Ercolini, and M. Hernández. 2018. Different *Lactobacillus* populations dominate in "Chorizo de León" manufacturing performed in different production plants. *Food Microbiology* 70:94–102.

Quintero-Salazar, B. 2017. Why Toluca's chorizos are famous? Perception, reputation, identity and culinary tradition of a sausage that is here to stay. *Nacameh* 11(2):33–49.

Raimondi, S., M. R. Nappi, T. M. Sirangelo, A. Leonardi, A. Amaretti, A. Ulrici, R. Magnani, C. Montanari, G. Tabarnelli, F. Gardini, and M. Rossi. 2018. Bacterial community of industrial raw sausage packaged in modified atmosphere throughout the shelf life. *International Journal of Food Microbiology* 280:78–86.

Ramírez-Muñoz, D., C. Ruiz-Capillas, A. M. Herrero, F. Jiménez-Colmenero, T. Pintado, B. de las Rivas, R. Muñoz, A. Pérez-Baltar, M. C. Cueto-Wong, and N. Balagurusamy. 2015. Caracterización del chorizo verde mexicano durante el procesado y conservación a distintas temperaturas: aminas biógenas. *Revista Alimentos Hoy* 23(33):20–41.

Ramos, D., V. San Martín, M. Rebatta, T. Arbaiza, B. Salvá, I. Caro, and J. Mateo. 2014. Características fisicoquímicas de la salchicha de cerdo del departamento de Tumbes, Perú. *Salud y Tecnología Veterinaria* 2:120–128.

Reyes-García, M., I. Gómez-Sanchez Prieto, and C. Espinoza-Barrientos 2017. *Tablas peruanas de composición de alimentos*. Lima: Instituto Nacional de Salud, Ministerio de Salud. https://repositorio.ins.gob.pe/handle/INS/1034.

Romero, M. C., A. M. Romero, M. M. Doval, and M. A. Judis. 2013. Nutritional value and fatty acid composition of some traditional Argentinean meat sausages. *Food Science and Technology, Campinas* 33(1):161–166.

Roncalés, P. 2007. Additives. In *Handbook of Fermented Meat and Poultry*, ed. F. Toldrá, 77–86. Iowa: Willey-Blackwell.

Ros-Berruezo, G., J. A. Santos-Buelga, J. M. Barat-Baviera, M. O. Conchello-Moreno, and J. Simal-Gándara. 2016. Report of the Scientific Committee of the Spanish Agency for Consumer Affairs, Food Safety and Nutrition (AECOSAN) about histological methods for differentiating between meat preparations and meat products. Reference no. AECOSAN-2016-003. *Revista del Comité Científico* 24:11–25. http://www.aecosan.msssi.gob.es/AECOSAN/docs/documentos/seguridad_alimentaria/evaluacion_riesgos/informes_cc_ingles/HISTOLOGICAL_METHODS.pdf.

Roseiro, L. C., A. Gomes, and C. Santos. 2011. Influence of processing in the prevalence of polycyclic aromatic hydrocarbons in a Portuguese traditional meat product. *Food Chemistry and Toxicology* 49:1340–1345.

Roseiro, L. C., A. Gomes, H. Gonçalves, M. Sol, R. Cercas, and C. Santos. 2010. Effect of processing on proteolysis and biogenic amines formation in a Portuguese traditional dry-fermented ripened sausage "Chouriço Grosso de Estremoz e Borba PGI". *Meat Science* 84:172–179.

Ruiz-Capillas, C., T. Pintado, and F. Jiménez-Colmenero. 2012. Biogenic amine formation in refrigerated fresh sausage "chorizo" keeps in modified atmosphere. *Journal of Food Biochemistry* 36:449–457.

Salgado, A , M. García-Fontan, I. Franco, M. López, and J. Carballo. Effect of the type of manufacture (homemade or industrial) on the biochemical characteristics of Chorizo de cebolla (a Spanish traditional sausage). *Food Control* 17:213–221.

SANCO. 2022. *Database on Food Additives, Application.* DG SANTE, Brussels: Directorate General of Health and food Safety. https://webgate.ec.europa.eu/foods_system/main/? sector=FAD&auth=SANCAS.

Sanz, Y., J. Flores, F. Toldrá, and A. Feria. 1997. Effect of pre-ripening on microbial and chemical changes in dry fermented sausages. *Food Microbiology* 14:575–582.

Sebranek, J. G. 2004. Semidry fermented sausages. In *Handbook of food and beverage fermentation technology*, ed.Y. H. Hui, L. Meunier-Goddik, A. S. Hansen, J. Josephsen, N. Wai-Kit, P. S. Sanfield, and F. Toldrá, 385–396. New York: Marcel Dekker Inc.

Silva, M. V., P Teixeira, T. A. Hogg, and J. A. Couto. 2003. *Manual de Segurança Alimentar de Produtos Cárneos Tradicionais e Enchidos e Produtos Curados.* Porto: AESBUC/UCP.

Soriano, A., B. Cruz, L. Gómez, C. Mariscal, and A. García-Ruiz. 2006. Proteolysis, physicochemical characteristics and free fatty acid composition of dry sausages made with deer (*Cervus elaphus*) or wild boar (*Sus scrofa*) meat: a preliminary study. *Food Chemistry* 96:173–184.

Spaiciani, M., M. Del Torre, and M. L. Stechinni. 2009. Changes of physicochemical, microbiological, and textural properties during ripening of Italian low-acid sausages. Proteolysis, sensory and volatile profiles. *Meat Science* 81:75–85.

Vignolo, G., C. Fontana, and S. Fadda. 2010. Semidry and dry fermented sausages. In *Handbook of Meat Processing*, ed. F. Toldrá, 379–398. Iowa: Willey-Blackwell.

Zurera-Cosano, G., A. Otero-Carballeira, E. Carrasco-Jiménez, F. Pérez-Rodríguez, and A. Valero Díaz. 2011. Report of the Scientific Committee of the Spanish Agency for Food Safety and Nutrition (AESAN) in relation to the effect of salt reduction on the microbiological. *Revista del Comité Científico de la AESAN* 13:59–87. http:// www.aecosan.msssi.gob.es/AECOSAN/docs/documentos/publicaciones/revistas_comite_cientifico/comite_cientifico_13.pdf.

6 Pork Sausages in Asia

Chunbao Li,[1] Siyuan Chang,[2] and Cui Zhiyong[1]
[1]Key Laboratory of Meat Processing and Quality Control, MOE; Key Laboratory of Meat Processing, MOA; College of Food Science, Nanjing Agricultural University, Nanjing 210095, PR China
[2]Institute of Food and Nutrition Development, MOA, Haidian, 100089, Beijing; China

CONTENTS

6.1   INTRODUCTION

Over the past 40 years, pork production in Asian countries has increased, from 11.98 million tons in 1978 to 66.72 million tons in 2018. The majority of pork is produced in Eastern and Southeastern countries, including China, Thailand, Viet Nam, South Korea, Japan and Indonesia (FAO, 2020). In these countries, people have a long history of eating pork. Sausage is one of the oldest pork products in these countries, and the first Chinese sausages were produced during the Northern and Southern Dynasties (420 CE to 589 CE). Dry-cured sausage is very popular in Asian countries, and is mainly prepared using lean pork meat, fat and various other ingredients. Traditionally, dry-cured sausage was sun dried for many days. During the process of drying, microbial fermentation took place to produce flavor components. One characteristic of traditional dry-cured sausages is that they have high salt and high fat contents. For the sake of human health, salt and fat have been significantly replaced by other ingredients greatly in recent years, and other kinds of sausages, for example, cooked sausage, fermented sausage and blood sausage, have been developed. The quality attributes and food safety of sausages have received much attention, and some measures have been taken to improve the quality and safety of sausages. Great advances have been achieved in recent decades. Here, several kinds of sausages popular in some Asian countries are introduced.

6.2   CHINESE SAUSAGES

In China, there are more than two hundred types of sausages, of which Cantonese sausage and Harbin sausage are the most popular. The traditional Chinese sausages were prepared mainly by mixing lean pork meat, fat and other ingredients, including soy sauce but not starch, stuffing the mixture into natural casings and naturally drying for a certain time. During long-term drying, the sausages undergo fermentation by microorganisms existing in the air. Such sausages can be considered as both dry-cured sausage and fermented sausage. However, the flavor of sausages varies greatly with locations, from sweetness in South China to spiciness along the Yangtze River and in West China to saltiness in North China. Most traditional Chinese sausages were handmade and their quality was difficult to control. In recent years, however, the conditions of Chinese sausage processing have been improved greatly. For example, the formulations have been standardized and drying procedures can be finished in air-conditioned rooms. Lipid oxidation and protein oxidation can be controlled by shortening the drying period and by vacuum-packaging products. The nutritional value of sausages can be optimized by replacing fat with dietary fiber and reducing salt in the formulations.

However, some Western-style sausages have been introduced from other countries since the 1900s. The earliest Western-style sausage varieties in China were the wiener and the red sausage (Teewurst). In recent years, such sausages have been reformulated according to Chinese preferences, and a large number of new varieties of sausage with Chinese flavors have been developed. Different from traditional Chinese sausage, Western-style sausages can be made using a great variety of meats, for example, pork, beef, lamb, horse, poultry, fish and edible offal, including blood and liver. The raw meat is marinated or salted, twisted or chopped into a meaty shape, mixed with a variety of ingredients and then stuffed into natural or artificial casings. Finally, the sausages are baked, steamed, smoked, cooled or fermented.

6.2.1 CANTONESE SAUSAGE

In Cantonese sausage, dry-cured meat and dry-cured fish are the most famous traditional ingredients in Guangdong, and can be traced back to the Tang and Song Dynasties, and the production of these Cantonese meat products amounts to 300,000 tons. Cantonese sausage has good appearance and flavor. The protein and fat account for 20% and 30% of Cantonese sausage, respectively (Bai, Chen & Liu, 2012).

Cantonese sausage is a semidry meat product that is produced by mixing pork, salt, sugar, wine and other ingredients, then curing, stuffing, drying or baking. The final product has good appearance with bright-red lean meat and creamy white fat. In some cases, Cantonese sausage has pale yellow fat and brownish lean meat, and tastes salty and sweet with a strong wine flavor (Li, 2012).

During drying and storage, bacteria in the air can grow on the surface of, and sometimes inside, the sausage and produce a variety of enzymes degrading lipids, proteins and carbohydrates to generate short-chain fatty acids, aldehydes, ketones, amino acids, nucleotides, hypoxanthine, lactic acid and other nutrients and flavor components (Bai et al., 2012).

6.2.1.1 Flavor of Cantonese Sausage

The flavor components in Cantonese sausage are mainly derived from food ingredients, lipid oxidation and protein degradation. Du and Ahn (2001) identified about 30 volatile compounds and found that 61–62% of total fatty acids in Cantonese sausage were unsaturated and 36–37% were saturated, including octadecenoic acid, hexadecenoic acid, and octadecanoic acid. The high fat content in sausage (31.73%) induced higher lipid oxidation with higher thiobarbituric acid reactive substances values. In that study, the authors also found that ethanol and ethyl esters accounted for 76% of volatile components in Cantonese sausage, which is associated with addition of a large amount of wine (approximately 8% of meat). This indicates that ethanol and its derivatives may make a great contribution to the characteristic flavor of Cantonese sausage. Hexanal, octane, heptane, pentane, heptanone and hexane were detected in the Cantonese sausage, indicating the occurrence of lipid oxidation, degradation of unsaturated fatty acids or amino acids and the Maillard reaction. In subsequent studies, similar results were observed. For example, Sun et al. (2010) identified the volatile compounds in Cantonese sausage at different stages of processing and storage. They identified 104 volatile compounds. Esters and alcohols contributed 16.94–50.12% and 30.01–65.54% of the total area, respectively. The authors also confirmed the importance of ethanol and its alcohol esters and aldehydes to the formation of flavor of Cantonese sausage.

6.2.1.2 Physiochemical Properties of Cantonese Sausage

In traditional Cantonese sausage, the sugar, salt and fat contents were quite high, which not only affected the texture, flavor and lipid oxidation, but also had a great impact on consumers' health. A strategy is to replace some ingredients with plant-based components.

Sugar is not just an important sweeter, but also acts as a preservative. In a study, Qiu et al. (2012) showed that the addition of 12% sugar significantly reduced water activity (a_w), pH, and the degree of denaturation of sarcoplasmic and myofibrillar proteins in Cantonese sausage. In addition, 12% sugar also showed protective effects against proteolysis of water-soluble proteins, improved the textural characteristics and inhibited microbial growth in Cantonese sausage.

During the processing of Cantonese sausage, free fatty acids increase but phospholipids decrease. The total levels of saturated, monounsaturated and poly-unsaturated fatty acids in neutral lipids did not change too much, but the proportion of polyunsaturated fatty acids significantly decreased in the phospholipid fraction. Lipid and protein oxidation was observed during processing (Qiu et al., 2013). To reduce the fat content in Cantonese sausage, some replacements have been tried. For example, Mesona blumes gum/rice starch mixed gels (18% fat) have better emulsifying stability and water-holding capacity but have similar pH, yield, residual nitrite and TBARS when compared with high-fat (28%) and low-fat (18%) sausages (Feng et al., 2013). During 3-week chill storage, fat-substituted sausages showed textural properties similar to high-fat sausages but higher than low-fat sausages (Feng et al., 2013)

To reduce lipid oxidation, some natural plant extracts or artificial antioxidants are recommended for use. For example, a combination of rosemary with Vitamin E or fresh ginger or liquorice showed a good antioxidant activity with lower acidity value, peroxide value (POV), 2-thiobarbituric acid-reactive substances and total free fatty acids in Cantonese sausage, and the better composite is vitamin E and rosemary (Zeng et al., 2017). In fact, adding single D-sodium erythorbate (0.05–0.1%) to Cantonese sausage can reduce lipid oxidation by inhibiting phospholipase, acid lipase and neutral lipase activities and increasing superoxide dismutase activity, and the antioxidant activity effect is dose-dependent (Shang, Qiao and Chen, 2019). Good correlations have been observed between the added D-sodium erythorbate level and TBARS values, free radical content, metmyoglobin levels, heme iron levels, a* and pH in Cantonese sausage, indicating that adding D-sodium erythorbate can reduce the rate of myoglobin and lipid oxidation, and prevent the discoloration of sausage (Shang et al., 2020). In other cases, dried straw mushrooms, fresh and dried *Flammulina velutipe*, and mulberry polyphenol have also been shown to improve the physiochemical properties of Cantonese sausage. The addition of 4% straw mushrooms increased the amount of essential amino acids 8-fold and the P/S ratio of fatty acid to 0.46, but reduced lipid peroxide value 10-fold in Cantonese sausages (Wang et al., 2018). The addition of 2.5% dry *Flammulina velutipe* significantly decreased fat content while increasing free amino acid contents of Cantonese sausages, and reducing lipid and protein oxidation, hardness, chewiness and overall sensory acceptability of Cantonese sausages (Wang et al., 2019). Mulberry polyphenol (1.0 g/kg) has good antioxidant and antimicrobial effects but may reduce the lightness and redness of Cantonese sausages (Xiang et al., 2019).The addition of 0.5% *Spirulina platensis* polysaccharides can also reduce lipid peroxidation and improve sensory properties of dry-cured sausage (Luo et al. 2017).

Evaluation of dry-cured sausage is quite important. In recent years, several non-destructive analytical techniques have been developed to classify Cantonese sausage. One example is hyperspectral imaging, which is a fast and non-destructive method that captures images of an object using a near-infrared camera, and image analysis is quite critical. Gong et al. (2017) used a support vector machine (SVM) and random forests (RF) to classify intact and sliced Cantonese sausages with an accuracy of the calibration and prediction set of over 90%. Wang and He (2019) extracted the hyperspectral information of the near-infrared band of Cantonese sausage using a successive projections algorithm (SPA), and then the multiple linear regression (MLR) and partial least squares regression (PLSR) algorithms were used to classify the sausage grade. The lean pork meat and pork fat in sausage showed different signals in the near-infrared band, and the modeling results have higher accuracy and anti-interference after separating lean meat and fat meat. The accuracy of SPA-MLR model for the fat region is 100%. The other example is an electronic nose that can be used for extracting the flavor fingerprint map of Chinese-style sausage during processing and storage. Gu et al. (2017) developed SVM and artificial neural networks (ANN) regression models to classify samples into their respective quality phases and predict lipid oxidation using full and optimal sensor arrays. The accuracy reached >95% and 82% for sausages during the processing and storage. SVM and artificial neural network (ANN) regression models showed good performance in predicting AV and POV.

6.2.1.3 Proteolysis in Cantonese Sausage

During relatively long drying time and storage, meat proteins undergo degradation by the action of proteolytic enzymes. In Cantonese sausage, proteins can be classified into five fractions on the basis of protein solubility, the majority of which are alkali-soluble. The amount of alkali-soluble proteins may increase during the processing of Cantonese sausage. The salt-soluble myofibrillar proteins and water-soluble sarcoplasmic proteins may decrease through the process, indicating a significant proteolysis (Sun et al. 2011). On the other hand, protein oxidation occurs during the processing of Cantonese sausage to sarcoplasmic, myofibrillar and alkali-soluble proteins, and the degree of protein oxidation affects the proteolysis and in vitro digestibility of meat proteins by forming S–S bonds and carbonyl groups, which could be attributed to increased protein aggregation and secondary structural changes (Sun et al., 2011; Sun et al., 2011; Sun et al., 2011).

6.2.1.4 Microorganisms in Cantonese Sausage

Like other dry-cured meat products, there are not only spoilage microorganisms that deteriorate the products but also beneficial microorganisms for food fermentation in Cantonese sausage. Wu et al. (2010) observed that traditional Cantonese sausage had lower total viable counts ($5.55 \pm 1.48\log_{10}$cfu/g), lactic acid bacteria ($3.59 \pm 0.68\log_{10}$cfu/g), staphylococci/micrococci (3.96 ± 0.54 $\log_{10}$cfu/g) and yeasts/molds ($3.55 \pm 0.47\log_{10}$cfu/g) compared with other dry-cured sausage. They also found that staphylococci/micrococci were the dominant microbial group in Cantonese sausage, but lactic acid bacteria have been reported to be predominant in other sausages (Lin & Chao, 2001; Drosinos et al., 2005).

Based on such findings, two strains, *Staphylococcus condimenti* and *Micrococcus caseolyticus* isolated from Cantonese sausage were tried as starter cultures in the same type of sausage, and the two strains were shown to accelerate the degradation and oxidation of lipids and proteins, and improve the flavor characteristics of Cantonese sausage. In addition, efforts have been made to improve the texture and reduce the nitrite residue of Cantonese sausage by artificially adding a mixed starter culture of *L. plantarum*, *P. pentosaceus* and *D. hansenii* (Cheng, Liu & Zhang, 2018). A similar study also showed that an artificial inoculation of *Pediococcus pentosaceus* (P), *Staphylococcus xylosus* (S), and a combination of *P. pentosaceus* and *S. xylosus* (P-S) accelerated pH decline and inhibited the growth of enterobacteria in a traditional dry-cured sausage (Du et al., 2019). *Staphylococcus simulans* and *S. saprophyticus* have been isolated from dry-cured sausages, which are coagulase-negative and resistant to 10% of salts and 150 mg/kg nitrites (Sun et al., 2019). These two staphylococci strains can express proteolytic enzymes to degrade both sarcoplasmic and myofibrillar proteins, nitrate reductase activity and lipolytic activity to produce flavor components or their precursors. These strains also exhibited antibiotic susceptibilities and negative hemolytic activity, DNase activity and biofilm formation ability (Sun et al. 2019).

However, the lipid degradation and oxidation were increased, as well. Thus, a balance should be considered between hygiene control and quality control when such starter culture is applied. A high-throughput sequencing comparison of bacterial communities among salami, Chinese dry-cured sausage and Chinese smoke-cured sausage indicates that Chinese dry-cured sausage and Chinese smoked-cured sausage have more diverse bacterial communities than salami. *Staphylococcus* spp. accounted for 97.45% of operational taxonomic units (OTUs) in salami, while *Staphylococcus* spp., *Lactobacillus* spp., *Weissella* spp., *Pediococcus* spp., and *Lactococcus* spp. were the dominant bacteria in two Chinese sausages. For sake of food safety, *Enterococcus* spp., *Photobacterium* spp., *Brochothrix* spp. and *Pseudomonas* spp. were identified in Chinese-style sausages, indicating a poor hygienic status (Wang et al., 2018).

6.2.1.5 Nitrite in Cantonese Sausage

Chen et al. (2007) tracked the changes in nitrite in Cantonese sausage and they found that three factors affected nitrite content during the oven-heating process. In the first 6 hours, moisture loss due to evaporation caused an increase of nitrite content from 10 mg/kg to 60 mg/kg. From 6 hours to 50 hours, the chromogenic reaction caused a reduction in nitrite content from 60 mg/kg to 30 mg/kg. After 50 hours, the flavor-forming reaction and oxygenolysis further reduced the nitrite content from 30 mg/kg to 15 mg/kg.

6.2.1.6 Bioactive Peptides in Cantonese Sausage

During the drying process of dry-cured sausage, proteolysis may occur to produce free amino acids and polypeptides. Some polypeptides could be further degraded by peptidyl peptidases and aminopeptidases into smaller peptides that have antioxidant activity. Sun et al. (2009) isolated two peptide fractions from Cantonese sausage and found that histidine content in the fraction greater than 5kDa increased from 56.67% to 60.23% within the first 6 hours of drying, but decreased to 48.66% at 36 hours.

However, the histidine content in the fraction lower than 5kDa decreased from 60.90% to 56.16% during the whole drying process. Histidine has been shown to have radical scavenging activity because the imidazole group in histidine has the proton-donation ability. The peptide fractions isolated from the Cantonese sausage also showed high antioxidant activity. However, the >5kDa fraction had a greater antioxidant activity than the <5kDa fraction within the first 18 hours and a lower antioxidant activity afterward (Sun et al., 2009).

6.2.2 HARBIN RED SAUSAGES

Harbin red sausage is very famous in North China after being introduced from Litaue about 140 years ago. Harbin sausage is prepared with beef, pork lean, pork back fat and cornstarch. It is also a type of dry-cured sausage. The red color is a characteristic of Harbin sausage. To realize color stability, nitrite is usually added in practice. However, nitrite can be transformed into N-nitrosoamines, which are harmful for human health and thus some efforts have been made to reduce nitrite usage in Harbin sausage. However, challenges are encountered in realizing nitrite reduction and maintaining quality assurance of Harbin sausage.

6.2.2.1 Nitrite Reduction

Nitrite is widely used in meat products for color formation, flavor, and inhibition of oxidation and microbial growth. However, nitrite can react with a secondary amine to produce toxic N-N-nitroso-compounds. Several kinds of N-nitroso-compounds have been found in cured meat products, including N-nitrosodimethylamine (NDMA), N-Nitrosodiethylamine (NDEA), N-Nitrosodi-n-propylamine (NDPA), N-nitrosodiisobutylamine (NDEA), N-nitrosodiphenylamine (NDphA), N-nitrosopyrrolidine (NPYR), Nitrosopiperidine (NPIP) and Nitrosomorpholine (NMor) (Crews, 2014; Mey et al., 2014; Wei et al., 2009).

Sun et al. (2017) investigated whether inoculation of three strains of lactic acid bacteria, including *Lactobacillus pentosus*, *Lactobacillus curvatus*, and *Lactobacillus sake*, could reduce N-nitrosoamine formation in Harbin dry sausage and related quality characteristics. The authors observed that *L. curvatus* showed a good inhibitory impact on NDEA, NDPA, NDPhA and NPIP, but *L. pentosus* only affected the formation of NDPA and NDphA. This could be associated with the fast decline in pH value and water activity. Lactic acid bacteria produce lactic acid that contributes to pH decline and the formation of N-nitrosoamine. Coagulase-negative staphylococci species has also been observed to produce nitric oxide synthase that catalyzes the formation of nitrosomyoglobin in a dry sausage model without nitrite addition (Huang et al., 2020). If this is the case, coagulase-negative staphylococci may a good application for dry-cured and fermented meat products.

A mixture of *Pediococcus pentosaceus* and *Staphylococcus xylosus* with *Lactobacillus curvatus* or *Lactobacillus sakei* significantly decreased moisture content and water activity, and increased redness, hardness and springiness of Harbin sausages (Hu et al., 2019). Such a mixed starter also increased the abundances of volatile compounds, for example, 3-hydroxy-2-butanone, 3-phenylpropanol, 2,3-butanediol, phenylethyl alcohol,

2-methyl-propanal, 2-methyl-butanal, 3-methyl-butanal, 2-pentanone, 2-heptanone and 2-nonanone and ethyl esters (Hu et al., 2019).

6.2.2.2   Fat Replacement, Antimicrobial and Antioxidant Agents

In Harbin sausage, fat content may amount to 30% of the sausage. To avoid the negative impact of high fat, a 50% replacement of pork back fat by camellia oil gels did not affect texture (cohesiveness, resilience, springiness) of sausage, but it increased moisture content, lightness and yellowness. In addition, lipid oxidation was reduced and the fatty acid profile was optimized (Wang et al., 2018).

Spice extracts, for example, cinnamon, clove and anise, have been shown to have good inhibitory effects on the growth of enterobacteriaceae, formation of biogenic amines (cadaverine, putrescine, tyrosamine, 2-phenylethylamine, histamine, and tryptamine) in Harbin sausage, and to improve the sensory quality of the dry sausage. Of the tested spice extracts, cinnamon extract was shown to be better than the other two extracts (Sun et al., 2018).

Replacement of NaCl by a mixture of 70% NaCl, 20% KCl, 4% lysine, 1% alanine, 0.5% citric acid, 1% Ca-lactate and 3.5% maltodextrin was reported to reduce lipid oxidation and protein oxidation and to promote the formation of volatile compounds originated from carbohydrate and amino acid metabolism, beta-lipid oxidation and esterification (Wen et al., 2019). Such a treatment also reduced microbial diversity and *Staphylococcus spp.* and *Lactobacillus spp.* became the dominant genera accompanied by higher brightness and lower hardness and chewiness (Chen et al., 2019).

6.2.2.3   Microorganisms in Harbin Sausage

Several strains of lactic acid bacteria have been isolated from Harbin sausages and their biochemical characteristics have been evaluated. For example, *Lactobacillus fermentun R6* and *Lactobacillus brevis R4* may produce enzymes that degrade myofibrillar and sarcoplasmic proteins, in particular, myosin heavy chain, paramyosin, phosphorylase and creatine kinase-M type (Sun et al., 2020; Sun et al., 2019). An enzyme isolated from *Pediococcus pentosaceus* shows similar biochemical characteristics (Sun et al., 2019).

6.3   SAUSAGES IN OTHER ASIAN COUNTRIES

6.3.1   Japanese Sausages

In Japan, most sausages are prepared using beef, pork or other meat. Fish-based sausage, called kamaboko, could be considered a type of sausage. Normally kamaboko is made with cured ground fish paste called *surimi* (Rourssel and Cheftel, 1988).

6.3.1.1   Wiener Sausage

Wiener sausage is the most popular sausage in Japan, and has a similar texture to traditional fish-paste products like kamaboko and chikuwa. Wiener is actually a kind of emulsion sausage. Pork and beef are finely chopped and seasoned with white pepper, ginger, mace, paprika, coriander and salt. After chopping, the batter is stuffed into casings and the sausages are smoked. This kind of sausage has a relatively long shelf life if frozen. In the formulations, sodium erythorbate is a crucial additional antioxidant that ensures no lipid oxidation occurs in wieners during a 10-month storage at −20°C (Coronado et al., 2002).

6.3.1.2 Arabiki Sausage

Arabiki is a Japanese word meaning *coarsely ground*. Arabiki sausage is coarsely ground smoked pork sausages in natural casings. Arabiki sausage is larger than other cocktail sausages. Arabiki sausage originated in Germany, but unlike the German-style pork sausage, Arabiki sausages are juicy and succulent with slightly sweet and mild smoky flavor.

6.3.2 KOREAN SUNDAE SAUSAGE

Sundae is a traditional Korean blood sausage, which is prepared using lean meat, vegetables, porcine blood, rice, glass noodles and spices such as onion, garlic, ginger and black pepper. It was introduced from China in the Northern and Southern Dynasties or the Yuan Dynasty. The cooked lean meat and vegetables are generally ground and then mixed with other ingredients. The mixture is stuffed into natural casings and cooked for 50 to 60 minutes at 80 to 85°C. At present, Korean blood sausage may be prepared with various functional materials, including natural antioxidants and gelatin. A study indicates that the addition of swollen pig skin with different natural vinegars (Bokbunja, brown rice, cider and lemon vinegars) will increase moisture content, cooking loss, L*, and a* values, but decrease hardness and thiobarbituric acid-reactive substance values in the sausage (Choe and Kim, 2016).

6.3.3 THAI SAUSAGES

Thai sausages are mainly prepared with pork, sometimes beef, chicken, or even fish, and other ingredients, for example, galangal, lemon grass, garlic, coriander, chili, kaffir lime leaves, white pepper and fish sauce.

6.3.3.1 Sai Krok E-san

Sai Krok means *sausage* in Thai. Sai Krok E-san is a popular fermented sausage in northeast Thailand. It is also called Sai Krok Prew because of its sour taste, resulting from meat fermentation by lactic acid bacteria (LAB). Sai Krok E-san is prepared using ground pork, fat, cooked sticky rice, garlic, salt and pepper. Sometimes the sticky rice is replaced by glass noodles or vermicelli. The sticky rice can provide energy for microbial fermentation as well as specific textures and sourness in the sausages. The sausage is usually fermented or dried for several hours to 2–3 days in the sunshine. Such flavorful sausages are usually grilled or fried and served with fresh cabbage leaves, raw chilies and ginger.

Different from Nham sausage, as shown in next section, Sai Krok E-san sausage has some differences in the recipes and manufacturing processes. Sai Kork

E-san sausage contains 30% fat while the amount in Nham sausage is only 8% or less. Thus, some trials have been done to replace fat by konjac gel in formulations of Sai Krok E-san sausage (Sorapukdee et al., 2019). The results indicate that replacement with 15% of konjac gel was the optimum formulation; this produces higher sensorial scores of sourness and overall acceptability. Moreover, it provided 46% fat reduction and 32% energy reduction compared to the control (Sorapukdee et al., 2019).

6.3.3.2 Nham

Nham is an uncooked semi-dry sausage, which is prepared with ground pork, pork skin, cooked sticky rice, garlic, salt, sugar and chilies. Different from Sai Krok E-san, Nham is fermented in ceramic pots for several days. The ingredients are mixed and tightly packed in banana leaves or plastic casings and fermented at 30°C for 3–5 days (Visessanguan, Benjakul, Riebroy, & Thepkasikul, 2004). Nham has a shelf life of less than a week, but its life can be extended by keeping it under refrigeration. In a study, biogenic amines and bacterial communities in Nham were dynamically tracked during processing and storage (Santiyanont et al., 2019).Too-lengthy fermentation and inferior conditions increased the risk of producing high levels of biogenic amines. *Lactobacillus, Lactococcus, Pediococcus* and *Weissella* comprised 90% of total bacteria during fermentation. *Weissella* was predominant in the samples with low biogenic amines while *Lactobacillus* and *Pediococcus* were the dominant genera in the samples with higher biogenic amines (Santiyanont et al., 2019).

In fact, some natural bacteria existing in Nham can produce bacteriocin to inhibit other bacteria growth. A strain of *Lactococcuslact* (WNC 20) was shown to produce a bacteriocin that had an inhibitory effect on relevant lactic acid bacteria, *Listeria monocytogenes, Clostridium perfringens, Bacillus cereus* and *Staphylococcus aureus*. This bacteriocin was heat stable and could be inactivated by a-chymotrypsin and proteinase K, and had identical sequences to nisin Z (Noonpakdee et al., 2003).

In recent years, artificial starter cultures have been introduced to Nham. Sriphochanart and Skolpap (2018) compared the effectiveness of six starter cultures to improve the flavor of Thai fermented sausage, including *Pediococcus pentosaceus, Pediococcus acidilactici, Weissellacibaria, Lactobacillus plantarum, Lactobacillus pentosus,* and *Lactobacillus sakei*. They observed that lactic-acid-bacteria-inoculated sausage had higher levels of 3-methyl-butanoicacid, which may contribute to flavor formation.

6.3.3.3 Other Thai Sausages

Sai Qua, also called Chiang Mai sausage, is processed using ground pork shoulder and many local herbs and spices, for example, lemongrass, galangal, kaffir lime, coriander and red curry paste.

Moo Yor is similar to the Western-style sausage. Moo Yor has a simpler filling and mild flavor. It is spiced with white or black pepper and mixed with tapioca powder. Traditionally, Moo Yor is wrapped in a banana leaf and then boiled in hot water.

Yam Kun Chiang in Thai, originating from China, is a sweet and salty dried sausage. It´s usually pan-fried and sliced with bird´s eye chilis, shallots, cucumbers, tomatoes, cilantro leaves, celery and onions.

6.3.4 INDONESIAN SAUSAGE

Urutan is one of the indigenous fermented foods in Bali, Indonesia. The sausage can be dried (dendeng) and fermented (urutan and bebontot).

Urutan is a traditional spicy sausage. Urutan has a different microbial ecology from other fermented sausages. Urutan is prepared with lean pork, fat, garlic, turmeric, aromatic ginger, chili, pepper, salt and sugar. The sausage is sun-dried for 2–5 days, during which fermentation occurs and the product quality largely depends on bacteria from ingredients. In Urutan, 77.5% of bacteria are *Lactobacilli* and the other 22.5% are *Pediococci* (Antara et al., 2002).

6.3.5 VIETNAMESE SAUSAGES

Nem chua is a unique uncooked fermented sausage in Vietnam, which is prepared using pork, porcine skin, pepper and salt, and wrapped in banana leaves. Nem chua has the flavor of fresh pork, the spicy sensation of pepper and a salty taste. Its formulations and processes vary greatly with locations. Much attention has been paid to microbial ecology in such a fermented sausage.

Tran et al. (2011) isolated 74 lactic acid bacteria from Nem chua. The majority of LAB isolates were *Lactobacillus plantarum* (67.6%) and *Pediococcus pentosaceus* (21.6%). The remaining isolates were *Lactobacillus brevis* (9.5%) and *Lactobacillus farciminis* (1.4%).

Golneshin et al. (2015) isolated one strain of *L. plantarum* (B21) from Nem chua and performed a complete genome sequencing. The genome sequence has 3,284,260 bp with a GC content of 44.47%, containing 3,117 genes with 2,766,912 bp including 2,930 coding sequences, 51 pseudogenes, and 18 frame shifted genes. There were 17 rRNA, 65 tRNA, 2 sRNA genes, and a number of natural plasmids in this strain. When compared with the *L. plantarum* WCFS1 reference strain, 2,668 out of 3,117 genes (80.17%) of B21 were well aligned with those of WCFS1. In addition, 449 genes were unique for the B21 strain.

Golneshin et al. (2017) isolated another strain of *L. plantarum* (A6). The strain was sequenced and 3,218 genes were matched, including 3,028 coding sequences, 92 RNAs, and 98 pseudogenes. Several native plasmids were present in the strain. Compared with the above strain B21, 2,954 out of 3,261 genes (90.6%) of A6 were aligned well with those of B21. A total of 307 genes were unique to the strain A6.

Doan et al. (2012) applied (GTG)5-PCR fingerprinting and MALDI/TOF/MS to identify 119 lactic acid bacteria strains from Nem chua. Five species were identified by both (GTG)5-PCR and in MALDI-TOF MS. Three species showed a great heterogeneity of MALDI-TOF MS profiles from their (GTG)5-PCR profiles (Figure 6.1).

In Nem chua, high abundances of yeasts were identified (5×10^4 to 4×10^5 CFU g^{-1}). The yeast ecosystem produces the C10-fatty acid-derived aroma compound γ-decalactone. Eighty-four different species were identified with *Candida sake* and *Candida haemulonii*. Six strains, including *Yarrowia lipolytica*, can produce γ-decalactone (Le Do et al., 2014).

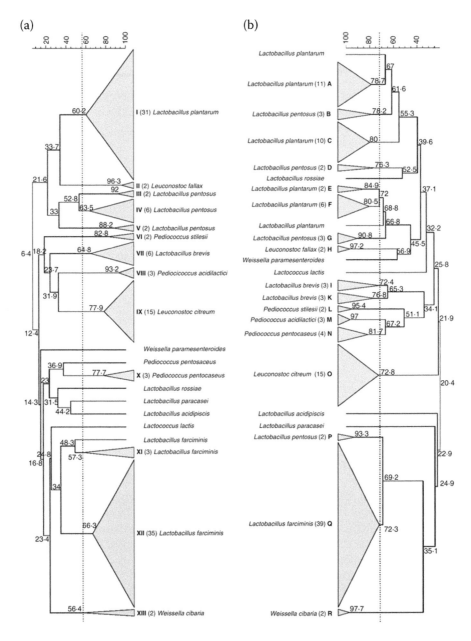

FIGURE 6.1 Dendrogram generated after cluster analysis of the digitized (GTG)5-PCR (a) and MALDI-TOF MS (b) fingerprints of 119 lactic acid bacteria isolates from Nem chua. The dendrogram was constructed using the unweighted pair group method using arithmetic averages with correlation levels expressed as percentage values of the Pearson correlation coefficient. The thirteen clusters in (GTG)5-PCR are designated with the Roman numbers I–XIII. The seventeen clusters in MALDI-TOF MS are labeled A to R. Number of isolates in each cluster (). The dashed vertical lines represent the cluster cut-off values for (GTG)5-PCR (56.4%) and MALDI-TOF MS (72.3%) clustering. (Doan et al., 2012).

6.4 CONCLUSION

In this paper, we introduced the basic procedures and quality control of several typical sausages in China, South Korea, Japan, Thailand, Vietnam, and Indonesia. In Asian countries, the majority of traditional sausages are dry-cured and naturally fermented. Much attention has been paid to microbial ecology in these traditional dry-cured sausages. In addition, lipid oxidation, protein oxidation and other aspects have been studied in Chinese dry-cured sausages. In the future, technological modernization and inoculation of commercial starter culture should be studied.

REFERENCES

Antara NS, Sujaya IN, Yokota A, Asano K, Aryanta WR, Tomita F. Identification and succession of lactic acid bacteria during fermentation of "urutan", a Balinese indigenous fermented sausage. *World Journal of Microbiology & Biotechnology*, 2002, 18: 255–262.

Bai WD, Chen Y, Liu LW. Study progress on flavors of cantonese curingmeat and sausage. *Chinese Food Additives*, 2012, 3: 208–212.

Chen JX, Hu YY, Wen RX, Liu Q, Chen Q, Kong BH. Effect of NaCl substitutes on the physical, microbial and sensory characteristics of Harbin dry sausage. *Meat Science*, 2019, 156: 205–213.

Chen WZ, Rui HM, Yuan HT, Zhang L. Analysis of dynamic chemical changes in Chinese cantonese sausage: Factors influencing content of nitrite and formation of flavor substances. *Journal of Food Engineering*, 2007, 79(4): 1191–1195.

Cheng JR, Liu XM, Zhang YS. Characterization of Cantonese sausage fermented by a mixed starter culture. *Journal of Food Processing and Preservation*, 2018, 42(6): e13623.

Choe JH, Kim HY. Effects of swelled pig skin with various natural vinegars on quality characteristics of traditional Korean blood sausages (Sundae). *Food Science Biotechnology*, 2016, 25(6): 1605–1611.

Coronado SA, Trout GR, Dunshea FR, Shah NP. Antioxidant effects of rosemary extract and whey powder on the oxidative stability of wiener sausages during 10 months frozen storage. *Meat Science*, 2002, 62(2): 217–224.

Crews C. Processing contaminants: N-nitrosamines A2-Motarjemi, *Yasmine Encyclopedia of Food Safety*, Academic Press, Waltham, 2014, 409–415.

Doan NTL, Van Hoorde K, Cnockaert M, De Brandt E, Aerts M, Le Thanh B, Vandamme P. Validation of MALDI-TOF MS for rapid classification and identification of lactic acid bacteria, with a focus on isolates from traditional fermented foods in Northern Vietnam. *Letters in Applied Microbiology*, 2012, 55(4): 265–273.

Drosinos, EH, Mataragas M, Xiraphi N, Moschonas G, Gaitis F, Metaxopoulos J. Characterization of the microbial flora from a traditional Greek fermented sausage. *Meat Science*, 2005, 69: 307–317.

Du M, Ahn DU. Volatile substances of Chinese traditional Jinhua ham and Cantonese sausage. *Journal of Food Science*, 2001, 66(6): 821–831.

Du S, Cheng H, Ma JK, Li ZJ, Wang CH, Wang YL. Effect of starter culture on microbiological, physiochemical and nutrition quality of Xiangxi sausage. *Journal of Food Science and Technology*, 2019, 56(2): 811–823.

FAO, http://www.fao.org/faostat/en/#data/QL. Available on 31 March, 2020.

Feng T, Ye R, Zhuang HN, Rong ZW, Fang ZX, Wang YF, Gu ZB, Jin ZY. Physicochemical properties and sensory evaluation of Mesona Blumes gum/rice starch mixed gels as fat-substitutes in Chinese Cantonese-style sausage. *Food Research International*, 2013, 50(1): 85–93.

Golneshin A, Adetutu E, Ball AS, May BK, Van TTH, Smith AT. Complete genome sequence of *Lactobacillus plantarum* strain B21, a bacteriocin-producing strain isolated from Vietnamese fermented sausage nemchua. *Genome Announcement*, 2015, 3(2): e00055–15.

Golneshin A, Gor M-C, Van TTH, May B, Moore RJ, Smith AT. Draft genome sequence of Lactobacillus plantarum strain A6, a strong acid producer isolated from a Vietnamese fermented sausage (nemchua). *Genome Announcement*, 2017, 5: e00987–17.

Gong AP, Zhu SS, He Y, Zhang C. Grading of Chinese Cantonese sausage using hyperspectral imaging combined with chemometric methods. *Sensors*, 2017, 17(8): 1706.

Gu XZ, Sun Y, Tu K, Pan LQ. Evaluation of lipid oxidation of Chinese-style sausage during processing and storage based on electronic nose. *Meat Science*, 2017, 133: 1–9.

Hu, YY, Chen Q, Wen RX, Wang Y, Qin LG, Kong BH. Quality characteristics and flavor profile of Harbin dry sausages inoculated with lactic acid bacteria and Staphylococcus xylosus. *LWT-Food Science and Technology*, 2019, 114: 108392.

Huang P, Xu B, Shao X, Chen C, Wang W, Li P. Theoretical basis of nitrosomyoglobin formation in a dry sausage model bycoagulase-negative staphylococci: Behavior and expression of nitric oxidesynthase. *Meat Science*, 2020, 161: 108022.

Le Do TT, Thanh Vu N, Phan-Thi H, Cao-Hoang L, Ngoc Ta TM, Wache Y, Nguyen THT. Traditional fermented sausage 'Nem chua' as a source of yeast biocatalysts efficient for the production of the aroma compound gamma-decalactone. *International Journal of Food Science and Technology*, 2014, 49, 1099–1105.

Li WX. Advances of research on flavor of guang- shi sausage. *Meat Industry*, 2012, (4): 49–51.

Lin KW, Chao JY. Quality characteristics of reduced-fat Chinese-style sausage as related to chitosan's molecular weight. *Meat Science*, 2001, 59: 343–351.

Luo A, Feng J, Hu B, Lv J, Oliver Chen CY, Xie S. Polysaccharides in *Spirulina platensis* improveantioxidant capacity of Chinese-style sausage. *Journal of Food Science*, 2017, 82(11): 2591–2597.

Mey ED, Klerck KD, Maere HD, Dewulf L, Derdelinckx G, Peeters MC, et al. The occurrence of N-nitrosamines, residual nitrite and biogenic amines in commercial dry fermented sausages and evaluation of their occasional relation. *Meat Science*, 2014, 96(2): 821–828.

Noonpakdee W, Santivarangkna C, Jumriangrit P, Sonomoto K, Panyim S. Isolation of nisin-producing *Lactococcuslactis* WNC 20 strain from Nham, a traditional Thai fermented sausage. *International Journal of Food Microbiology*, 2003, 81: 137–145.

Qiu CY, Sun WZ, Cui C, Zhao MM. Effect of sugar level on physicochemical, biochemical characteristics and proteolysis properties of Cantonese sausage during processing. *Journal of Food Quality*, 2012, 35(1): 34–42.

Qiu CY, Zhao MM, Sun WZ, Zhou FB, Cui C. Changes in lipid composition, fatty acid profile and lipid oxidative stability during Cantonese sausage processing. *Meat Science*, 2013, 93(3): 525–532.

Rourssel H, Cheftel JC. Characteristics of surimi and kamaboko from sardines. *International Journal of Food Science and Technology*, 1988, 23(6): 607–623.

Santiyanont P, Chantarasakha K, Tepkasikul P, Srimarut Y, Mhuantong W, Tangphatsornruang S, Zo Y, Chokesajjawatee N. Dynamics of biogenic amines and bacterial communities in a Thai fermented pork product Nham. *Food Research International*, 2019, 119: 110–118.

Shang XL, Qiao J, Chen ZX. Changes in lipase and antioxidant enzyme activities during processing of Cantonese sausage with D-sodium erythorbate. *Journal of Food Safety*, 2019, 2019: 1671603.

Shang XL, Zhou ZG, Jiang SH, Guo HZ, Lu YX. Interrelationship between myoglobin oxidation and lipid oxidation during the processing of Cantonese sausage with d-sodium erythorbate. *Journal of the Science of Food and Agriculture*, 2020, 100(3): 1022–1029.

Sorapukdee S, Jansa S, Tangwatcharin P. Partial replacement of pork backfat with konjac gel in Northeastern Thai fermented sausage (SaiKrok E-san) to produce the healthier product. *Asian-Australas J Anim Sci*, 2019, 32: 1763–1775.

Sriphochanart W, Skolpap W. Modeling of starter cultures growth for improved Thai sausage fermentation and cost estimating for sausage preparation and transportation. *Food Science and Nutrition*. 2018, 6: 1479–1491.

Sun F, Kong B, Chen Q, Han Q, Diao X. N-nitrosoamine inhibition and quality preservation of Harbin dry sausages by inoculated with Lactobacillus pentosus, Lactobacillus curvatus and Lactobacillus sake. *Food Control*, 2017, 73: 1514–1521.

Sun FD, Hu YY, Chen Q, Kong BH, Liu Q. Purification and biochemical characteristics of the extracellular protease from Pediococcus pentosaceus isolated from Harbin dry sausages. *Meat Science*, 2019, 156: 156–165.

Sun, FD, Hu YY, Yin XY, Kong BH, Qin LG. Production, purification and biochemical characterization of the microbial protease produced by Lactobacillus fermentum R6 isolated from Harbin dry sausages. *Process Biochemistry*, 2020, 89: 37–45.

Sun FD, Li QX, Liu HT, Kong BH, Liu Q. Purification and biochemical characteristics of the protease from Lactobacillus brevis R4 isolated from Harbin dry sausages. *LWT - Food Science and Technology*, 2019, 113: UNSP 108287.

Sun J, Cao C, Feng M, Xu X, Zhou G. Technological and safety characterization of coagulase-negative staphylococci with high protease activity isolated from Traditional Chinese fermented sausages. *LWT - Food Science and Technology*, 2019, 114: 108371.

Sun Q, Zhao X, Chen H, Zhang C, Kong B. Impact of spice extracts on the formation of biogenic amines and the physicochemical, microbiological and sensory quality of dry sausage. *Food Control*, 2018, 92: 190–200.

Sun WZ, Cui C, Zhao MM, Zhao QZ, Yang B. Effects of composition and oxidation of proteins on their solubility, aggregation and proteolytic susceptibility during processing of Cantonese sausage. *Food Chemistry*, 2011, 124(1): 336–341.

Sun WZ, Zhao HF, Zhao QZ, Zhao MM, Yang B, Wu N, Qian YL. Structural characteristics of peptides extracted from Cantonese sausage during drying and their antioxidant activities. *Innovative Food Science & Emerging Technologies*, 2009, 10(4): 558–563.

Sun WZ, Zhao MM, Yang B, Zhao HF, Cui C. Oxidation of sarcoplasmic proteins during processing of Cantonese sausage in relation to their aggregation behaviour and in vitro digestibility. *Meat Science*, 2011, 88(3): 462–467.

Sun WZ, Zhao QZ, Zhao HF, Zhao MM, Yang B. Volatile compounds of Cantonese sausage released at different stages of processing and storage. *Food Chemistry*, 2010, 121(2): 319–325.

Sun WZ, Zhao QZ, Zhao MM, Yang B, Cui C, Ren JY. Structural evaluation of myofibrillar proteins during processing of Cantonese sausage by Raman spectroscopy. *Journal of Agricultural and Food Chemistry*, 2011, 59(20): 11070–11077.

Sun WZ, Zhou FB, Zhao MM, Yang B, Cui C. Physicochemical changes of myofibrillar proteins during processing of Cantonese sausage in relation to their aggregation behaviour and in vitro digestibility. *Food Chemistry*, 2011, 129(2): 472–478.

Tran KTM, May BK, Smooker PM, Van TTH, Coloe PJ. Distribution and genetic diversity of lactic acid bacteria from traditional fermented sausage. *Food Research International*, 2011, 44: 338–344.

Visessanguan W, Benjakul S, Riebroy S, Thepkasikul P. Changes in composition and functional properties of proteins and their contributions to Nham characteristics. *Meat Science*, 2004, 66(3): 579–588.

Wang Q, He Y. Rapid and nondestructive classification of Cantonese sausage degree using hyperspectral images. *Applied Science-Basel*, 2019, 9(5): 822.

Wang X, Xie Y, Li X, Liu Y, Yan W. Effects of partial replacement of pork back fat by a camellia oil gel on certain quality characteristics of a cooked style Harbin sausage. *Meat Science*, 2018, 146: 154–159.

Wang X, Zhang Y, Ren H, Zhan Y. Comparison of bacterial diversity profiles and microbial safety assessment of salami, Chinese dry-cured sausage and Chinese smoked-cured sausage by high-throughput sequencing. *LWT - Food Science and Technology*, 2018, 90: 108–115.

Wang XP, Xu MY, Cheng JR, Zhang W, Liu XM, Zhou PF. Effect of Flamrnulina velutipes on the physicochemical and sensory characteristics of Cantonese sausages. *Meat Science*, 2019, 154: 22–28.

Wang XP, Zhou PF, Cheng JR, Chen ZY, Liu XM. Use of straw mushrooms (*Volvariella volvacea*) for the enhancement of physicochemical, nutritional and sensory profiles of Cantonese sausages. *Meat Science*, 2018, 146: 18–25.

Wei F, Xu X, Zhou G, Zhao G, Li C, Zhang Y, Chen L, Qi J. Irradiated Chinese Rugao ham: Changes in volatile N-nitrosamine, biogenic amine and residual nitrite during ripening and post-ripening. *Meat Science*, 2009, 81(3): 451–455.

Wen RX, Hu YY, Zhang L, Wang Y, Chen Q, Kong BH. Effect of NaCl substitutes on lipid and protein oxidation and flavor development of Harbin dry sausage. *Meat Science*, 2019, 156: 33–43.

Wu YT, Cui C, Sun WZ, Yang B, Zhao MM. Effects of Staphylococcus condimenti and Micrococcus caseolyticus on the volatile compounds of Cantonese sausage. *Journal of Food Processing Engineering*, 2019, 32(6): 844–854.

Wu YT, Zhao MM, Yang B, Sun WZ, Cui C, Mu LX. Microbial analysis and textural properties of Cantonese sausage. *Journal of Food Process Engineering*, 2010, 33(1): 2–14.

Xiang R, Cheng JR, Zhu MJ, Liu XM. Effect of mulberry (Morus alba) polyphenols as antioxidant on physiochemical properties, oxidation and bio-safety in Cantonese sausages. *LWT - Food Science and Technology*, 2019, 116: 108504.

Zeng XF, Bai WD, Lu CH, Dong H. Effects of composite natural antioxidants on the fat oxidation, textural and sensory properties of Cantonese sausage during storage. *Journal of Food Processing and Preservation*, 2017, 41(2): e13010.

7 Traditional Pork Sausages in Serbia

Manufacturing Process, Chemical Composition and Shelf Life

Tomašević Igor,[1] Simunović Stefan,[2] Đorđević Vesna,[2] Djekic Ilija,[3] and Tomović Vladimir[4]

[1]Department of Animal Source Food Technology, University of Belgrade, Faculty of Agriculture, Nemanjina 6, Belgrade, 11080, Serbia

[2]Institute of Meat Hygiene and Technology, Kaćanskog 13, Belgrade, Serbia

[3]Food Safety and Quality Management Department, University of Belgrade, Faculty of Agriculture, Nemanjina 6, Belgrade, 11080, Serbia

[4]Faculty of Technology, University of Novi Sad, Bulevar cara Lazara 1, Novi Sad, Serbia

CONTENTS

7.1 INTRODUCTION

As early as the 12th century, William of Tyre, medieval chronicler, documented that "the Serbs are rich in domestic small and large livestock, milk, meat, honey and

wax" (Ćorović, 1989). By the first decades of the 19th century, Serbia was profi-
cient in domestic animal breeding, especially pigs, and its agriculture was oriented
towards exporting livestock, mainly to the Austro–Hungarian Empire and
Dubrovnik (Ćorović, 1989). When the Austro–Hungarian Empire prohibited the
import of Serbian livestock in 1881, the records show that the annual export reached
its peak of 350,000 pigs (Romano, 1969). The only way for the Serbian economy to
resolve the problem of the ban was to build facilities for pig slaughtering and meat
processing, and this was done for the first time in 1895 in Belgrade (Romano,
1969). Today, we mark this event as the founding year of the Serbian meat processing industry. We can also claim, based on historical facts, that pig breeding and
pig meat production is deeply rooted in Serbian tradition and has played an important role in the national economy for centuries.

Traditional Serbian pig breeds are Šumadinka, Moravka, Mangulica, Šiška and
Resavka, and they are generally characterized by high fat content. In 1890, the
number of pigs in Serbia was around 3.5 million and it varied throughout the next,
century (Romano, 1969) decreasing to 2.9 million in 2017 (Figure 7.1). Following the
development of the meat industry, sausage production increased from 3,000 tonnes in
1947 to 63,088 t in 2017, while in the same period pork meat production increased
from 77,000 t to 307,000 t (Statistical Office RS). As for the average consumption of
pork per capita per annum, it increased from 12 kg in 1978 to 20.3 kg in 2017.

Average consumption of dried and processed meat, mostly made of pork, also
increased from 12.4 kg in 2006 to 14.4 kg in 2017 (Statistical Office RS). In accordance with the long history of pig breeding, consumption of pork sausages plays
an important role in a typical Serbian diet. The most famous traditional Serbian
sausages fully or partially made of pork meat are Kulen, Sremska, Pirotska,
Petrovska, Užička and Čajna.

It was 1990 when the first Serbian meat products received Protected Designation
of Origin (PDO) status, according to Yugoslav laws at that time. Apart from many
dried meat products, the first sausages that were recognized as PDO were Sremski
kulen and Sremska sausage. Many years later, Petrovská Klobása and Lemeški
kulen were also also included in the list. However, in spite of the fact that Serbia has

FIGURE 7.1 Number of pigs and sausage production in Serbia from 1890 to 2017 (data
from Statistical Office RS).

considerable numbers of meat products that are produced by traditional means, following the recipes inherited from our ancestors, none receive a PDO, Protected Geographical Indication (PGI) or Traditional Speciality Guaranteed (TSG) designation under EU law.

7.2   ČAJNA SAUSAGE

The creation of Čajna sausage was closely related to the beginning of the modern meat industry in Serbia. In fact, Čajna sausage originates from the meat industry Mesopromet (Zemun) where it was made in the early 1960s based on European- and American-type fast-fermenting semi-dry sausages (Milojević, 1996; Pećanac et al., 2017; Radetic, 1997; Vuković et al. 1986). Before production expanded to other industrial facilities, the drying period was extended and Čajna became a dry-fermented sausage with relatively high fat content. In its original form, it was made from the most valuable meat cuts (mainly pork and, to a lesser extent, beef) and pork back fat, and differed from Teewurst produced in Germany (Radetic, 1997). Nowadays, it is normally made from pork meat, pork trimmings and pork back fat, although a small number of producers still use up to 35% beef. As for the other ingredients, only salt (2.5%), black pepper (0.3–0.5%), nitrite and sugars were used until recently, when glucono-delta-lactone (GDL) and starter cultures became used routinely.

Pork back fat and pork trimmings are frozen while the pork meat is chilled (Table 7.1). Freezing the meat reduces water-binding capacity and improves salt diffusion (Lücke, 2007), which accelerates the drying, i.e., reduces the time required for drying of the sausage. The frozen part is cut into small pieces by guillotines and then comminuted in bowl-cutter to particle sizes of approximately 2 mm

TABLE 7.1
Ingredients, Smoking and Ripening Conditions of Čajna Sausage

Sausage Dimensions and Weight	Casing	Ingredients	Smoking	Ripening
ø 35–38 mm diameter, 30 cm in length, weight 300 g	collagen or pork small intestine	chilled pork (50%), frozen pork (25%), frozen back fat (20%), nitrite salt (2%), black pepper (0.5%), spices (0.5%), dextrose (0.4%)	1–3 days, 16–25°C, 0.3 m/s, beach or oak	14–30 days, 22°C,[*] 90–98% RH,[*] 16°C,[**] 74–82% RH[**]

Notes
  [*] start of the drying/ripening;
 [**] end of the drying/ripening
(data taken from Pribiš et al. 1996; Stamenković et al. 1996; Stamenković et al. 1984)

along with salt, nitrites, ascorbic acid, sugars, starter cultures and spices. Starter cultures are first dissolved in water and afterward added to the bowl. At the same time, chilled meat is minced through a plate with 4 mm diameter holes and then added to the bowl-cutter to be mixed with the already comminuted part of the meat batter. Friction occurs during the comminution, and consequently the temperature rises, so special care must be taken in order for the temperature not to exceed $-2°C$.

The batter is stuffed, under vacuum, into collagen casings approximately 38 mm in diameter and 30 cm in length (Milojević, 1996; Pribiš et al., 1996; Stamenković et al., 1996; Stamenković, Hromiš, & Janković, 1988; Vuković et al., 1986). In the beginning, natural casings were used, but soon after they were replaced by artificial ones due to their better mechanical and technological properties as well as for economic reasons (Stamenković et al., 1988). After stuffing, sausages are left for tempering when water condensed on surface should be removed in order to prepare the sausages for smoking. Smoking is usually performed at 20–22°C and 80–88% RH and at air velocity of about 0.3 m/s over 1 to 4 days (Pribiš et al., 1996; Radetic, 1997). Sometimes it can last up to 18 days, depending on the amount of smoke applied (Djinovic, Popovic, & Jira, 2008). The most commonly used woods for smoking are beech and oak. An important food safety risk in smoked meats is the content of polycyclic aromatic hydrocarbons (PAHs), especially if a traditional smoking chamber is used and smoke is obtained by combustion in an open fireplace and not by friction, as is usually the case in industrial production. However, the investigation into the contents of the 16 EU priority PAHs in six different meat products from Serbia (Djinovic et al. 2008), examined during the process of smoking, concluded that the content of all 16 PAHs was lowest in Čajna sausage samples. The maximum level for benzo[a]pyrene (BaP) of 5µg/kg in smoked meat products was not exceeded in any of the Čajna sausage samples (Figure 7.2), confirming the chemical safety of the product in this regard.

When it comes to fermentation temperature, in industrial surroundings it can reach 26–27°C inducing fast fermentation and reducing the production time to only

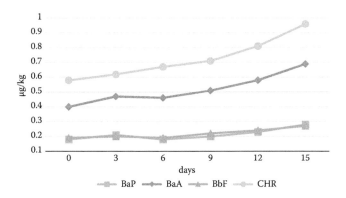

FIGURE 7.2 Content of benzo[a]pyrene (BaP), benzo[a]anthracene (BaA), benzo[b] fluoranthene (BbF) and chrysene (CHR) in Čajna sausage during smoking in a traditional smoking chamber (adapted from Djinovic et al. 2008).

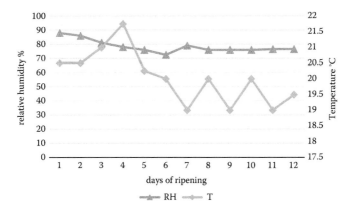

FIGURE 7.3 Temperature and relative humidity conditions during smoking and ripening of Čajna sausage (adapted from Stamenković et al. 1984).

14 days. During the drying period, the temperature and relative humidity (RH) are gradually decreased to 16–19°C and 70–75% (Pribiš et al., 1996; Radetic, 1997; Stamenković et al., 1984; Vuković et al., 1986) (Figure 7.3). Production time usually ranges from 14 to 21 days, depending on the sausage diameter and the extent of fermentation. Apart from temperature, pH decline depends on whether or not starter cultures, sugars or GDL are added. If not, pH does not drop below 5.25, while when sugars and starter cultures are added, minimum pH during ripening is usually around 5.1 (Šutić, Milovanović, & Svrzić, 1990; Vuković et al., 1986). On the other hand, the addition of 0.5% of GDL affects pH, which drops to around 4.9 (Vuković et al., 1986). Generally, the addition of GDL is not desirable from the consumer point of view, because this overemphasizes the sour taste that these products tend to have (Stamenković et al. 1990). When used, the content of GDL should not be higher than 0.2% of meat batter in order to avoid this quality defect (Stamenković et al., 1990). Weight loss ranges from 32.6 to 42.39% (Joksimović, Čavoški, & Fridl, 1981; Stamenković et al., 1990; Stamenković et al., 1996; Stamenković et al., 1988; Šutić et al., 1990; Vuković et al., 1986).

Čajna sausage contains a high fat content, which ranges between 37% and 60% (Džinić et al., 2016; Oluški et al. 1974; Stamenković et al., 1988). According to Stamenković et al. (1991) it should not contain more than 55% of fat at the end of the ripening in order to maintain good sensorial properties. In other words, pork back fat should be added up to 30%. In addition, thiobarbituric acid (TBA) and peroxide values of sausages rise with increased fat content (Soyer 2005). Due to the high fat content and small granulation (2 mm) of meat and fat, its cross-section colour is pale red. Values for lightness (L* value) and yellowness (b* value) decrease during the entirety of the ripening process, while values for redness (a* value) increase until the 18th day of production (Figure 7.4). Recently, the study of Džinić et al. (2016) showed that L* values for Čajna sausages from different producers were between 44.33 and 53.64. The same study reported that values for hardness and springiness of Čajna sausage were slightly lower than those found for

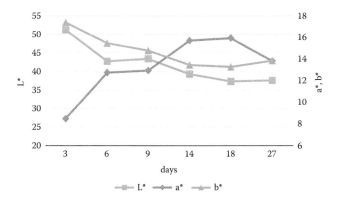

FIGURE 7.4   Values of colour parameters (L*, a*, b*) during 27. days of ripening (adapted from Pribiš et al. 1996).

chorizo with a high fat content (Gómez & Lorenzo, 2013), while cohesiveness was similar to the values reported for salchichón (Fonseca et al. 2015) but higher than those reported for fuet and chorizo (Herrero et al., 2007).

Depending on the amount of fat, protein content ranges from 17 to 26% while moisture content ranges from around 25% to 35% (Dzinic et al., 2015, October; Stamenković et al., 1988). National Regulation (2019) requires that water content in all dry fermented sausages must not exceed 35% to ensure proper drying of sausages. At the same time, the meat industry tends to fulfil the regulatory requirement by increasing fat content of the product while reducing the cost of production. Another common way to make the production process cheaper is to add cooked pork skins up to 10%. Special care must be taken not to exceed the limit of 15% of collagen in total meat proteins set by the same regulation. At the end of the production, the sausage is characterized by a bright surface, relatively dark colour of meat particles on the cross section, odour of smoke and sour taste (Stamenković et al., 1990; Stamenković et al., 1996). However, because of its lower production costs and cheaper price, compared to other fermented sausages on the market, Čajna sausage was and still is the most popular dry fermented sausage in Serbia (Pećanac et al., 2017; Stamenković et al., 1996).

7.3   SREMSKA SAUSAGE

Sremska sausage was named after the Srem region, which lies between the Sava and the Danube Rivers, and its production began in the middle of the 18th century. The assumption is that the forerunner of today's Sremska sausage was a spicy and smoked Lucanica sausage that Roman soldiers carried in their backpacks to consume before and after fierce battles (Stevanovic et al., 2016). In the past, production of Sremska sausage usually required the meat of domestic breeds, such as Mangalitsa, which are characterized by a high amount of fatty tissue, but these were progressively replaced by modern pig breeds in the second half of the 20th century.

In contrast to Čajna sausage, which is produced only industrially, Sremska sausage was traditionally produced in Serbian households and by artisanal producers. However, due to the high demand for Sremska sausage, its production has spread to the meat industry, which has influenced the great diversity in production processes and quality characteristics between traditionally and commercially produced sausages (Tojagić, 1996). In addition, there is little uniformity, even between sausages produced by homemade producers. Sremska sausage was one of the first products that was granted the PDO mark under Serbian law (Radovanović et al., 1992b).

Production is carried out during the winter months, usually in late November and December, when the ambient temperature is relatively low (0°C). Traditionally, this sausage is made from the meat of late-maturing pig breeds slaughtered at the age of 12 months and older because their meat is characterized by lower moisture content (Živković et al., 2011). In households, it is usually produced from pork trimmings with the addition of pork shoulder and pork back fat (Table 7.2). Meat is minced to a particle size of 8–13 mm and mixed by hand with only salt, ground hot and sweet paprikas, garlic and black pepper (Radovanović et al., 1992b). The mixture is then stuffed into small pig intestines of around 32 mm in diameter, making sausages of about 40 cm in length. Immediately after stuffing, sausages are placed to hang on wooden sticks to drain and the surface must be dried in order to prepare them for

TABLE 7.2

Ingredients, Smoking and Ripening Conditions of Sremska Sausage

	Dimensions and Weight	Casing	Ingredients	Smoking	Ripening
Traditional	ø 28–32 mm, 25–60 cm in length, 150–300 g	pig's small intestine	pork (65–75%), back fat (25–35%), salt (2.5–2.7%), hot paprika (0.3–1.3%), sweet paprika (0.3–1.3%), garlic (0.15–0.5%), black pepper (0.3–0.5%)	3–5 days, 18–20°C, 58–65% RH, 0.1 m/s beach	14–15°C[*], 90% RH,[*] 0.5 m/s, 14–15°C,[**] 78% RH,[**] 0.1–0.3 m/s
Industrial	ø 28–36 mm, 25–60 cm in length, 150–300 g	pig's small intestine or collagen casings	pork trimmings (100%), or lean pork meat (65–70%), pork back fat (30–35%), red hot and sweet paprika, garlic, black pepper, nitrate, nitrite, sugar (0.4%), ascorbic acid	2–3 days, 18–22°C, 94% RH, 0.1 m/s beach	16–21°C,[*] 90% RH,[*] 0.5 m/s, 14–16°C,[**] 75–80% RH,[**] 0.1–0.3 m/s[*]

Notes

[*] start of the drying/ripening;

[**] end of the drying/ripening

(Data taken from Oluški et al., 1974; Radovanović et al., 1992b; Rašeta et al., 2010; Tojagić, 1996, 1997; Turubatović & Tadić, 2005; Živković et al., 2011)

smoking (Tojagić, 1997). Cold smoke, usually obtained by pyrolysis of beech, is applied for 6–7 days and sometimes even longer (Djinovic et al., 2008; Tojagić, 1996). Sausages are placed at least 2 m above the fireplace and smoked for few hours each day, or they can be smoked every two days, and in that case a larger amount of smoke is applied. They are then dried in a traditional smoking/ripening chamber for 12–60 days.

The investigation of Tojagić (1996) revealed that the best sensory score was obtained for sausages that had ripened for 12–21 days. However, in households, sausages can sometimes be dried for as long as few months because they are traditionally consumed throughout the whole year. Therefore, its quality also depends on storage period and atmospheric conditions in the chamber (Tojagić, 1996). In addition, atmospheric conditions, which depend on the external temperature and chamber design, also have a great influence on the fermentation process which is anything but fast. When using fermentation and drying as the main safety mechanisms, most fermented meat products are microbiologically stable when a pH of 5.3 and an a_w of 0.89 or lower are obtained within a relatively short period of time.

The traditionally made Sremska sausage has a final pH ≤ 5.6 which, coupled with an a_w ≤ 0.88, achieved during drying, inhibits growth of *L. monocytogenes* and *St. aureus* on this product (Figure 7.5) (Frece et al., 2014; Tomašević & Đekić, 2017; Vesković-Moračanin et al., 2011; Živković et al., 2012). The most dominant lactic acid bacteria (LAB) strains isolated from the sausage are *Lactobacillus delbrueckii ssp. delbrueckii* (26%), *Lactobacillus curvatus* (13.3%), *Lactobacillus plantarum* (10.6%), and *Pediococcus pentosaceus* (10%), while *Lactobacillus fermentum*, *Lactobacillus cellobiosus* and *Leuconostoc mesenteroides ssp. mesenteroides* are present in smaller numbers (Borović, Vesković, Velebit, Baltić, & Spirić, 2009).

At the end of the ripening, the product has a distinctive spicy flavour, dark red colour and firm consistency (Stanišić et al., 2012, May). The red intensity (a* value) and yellow intensity (b* value) of Sremska sausage (Stanišić et al., 2016; Živković et al., 2012) are higher than those reported for salchichón (Fonseca et al., 2015) but are

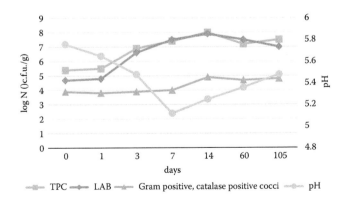

FIGURE 7.5 Changes in total plate count (TPC), lactic acid bacteria (LAB) and gram-positive, catalase positive cocci number during ripening of Sremska sausage (adapted from Živković et al. 2012).

lower than those found for chorizo (Gómez & Lorenzo, 2013), while values for lightness (L*) are similar to those found in chorizo. As already mentioned, Sremska sausage can sometimes ripen for a few months, so its texture varies. However, after optimum ripening time (21 days), values for springiness and cohesiveness are similar to those found in salchichón (Fonseca et al., 2015). In terms of Sremska sausage safety, the investigation of Tasic et al. (2015) showed that content of tryptamine, histamine and tyramine was 34.2 mg/kg, 6.42 mg/kg and 45.2 mg/kg, respectively. Durin theg 21st day of ripening, peroxide values increased from an initial value of 0.5 mEq O_2/kg to 2 mEq O_2/kg while thiobarbituric acid (TBA) values increase from 0.02 mg MDA/kg to 0.12–0.18 mg MDA/kg (Stanišić et al., 2014).

As for industrial production, meat trimmings are most often added to the list of usual ingredients along with nitrites and sugars, and the use of starter cultures is not common. The minimum pH value, when sugars are added (up to 0.3%), is about 5.0, which is slightly lower than traditionally made Sremska sausages (Ducic, Vranic, & Baltic, 2018; Vesković-Moračanin et al., 2011; Živković et al., 2011). Temperature and relative humidity are controlled and they are often set to higher values in the first few days of ripening, which should, in conjunction with added sugar, speed up fermentation and thereby ensure sufficient decrease of pH value. As a result, the production process is shorter and these products tend to have a slightly more sour taste compared to those traditionally produced. At the end of the ripening process, weight loss is around 35% (Radovanović et al., 1992b) while in traditional production it can be even higher depending on the ripening time.

Sremska sausage, when packed in cellophane or plastic foils at 7–10°C, shows no changes for 30 days, while product packed in vacuum at the same temperature can last up to 4 months (Turubatović & Tadić, 2005). A shelf-life study of Sremska sausage packed under vacuum confirmed that it could be stable for 110 days at 4°C (INMES, 2019). As for the oxidative stability, TBARS values showed an increase from 0.36 to 0.54 mg MDA/kg during 120 days of storage. At the same time, peroxide values showed an increase from 0.47 mmol/kg to 0.94 mmol/kg. Although the product was packed under vacuum and stored at 4°C, pH value dropped from 5.17 to 4.94, while the number of lactic acid bacteria (LAB) increased from 4.2 to 7.2 log CFU/g. Total bacterial count (TBC) increased from 4.9 log CFU/g on the first day to 7.6 log CFU/g on the last day of storage (INMES, 2019). When Sremska sausage is packed in a modified atmosphere of 70% N_2 and 30% CO_2, shelf life can be extended up to 210 days (Turubatović & Tadić, 2005).

7.4 PETROVSKÁ KLOBÁSA (PETROVAČKI KULEN)

In the 18th century, Slovaks started settling in what is now Serbia. More precisely, they settled in an area around Bački Petrovac and the province of Vojvodina in northern Serbia in 1745. They passed their cultural heritage and tradition to the local inhabitants, and the manufacturing of the dry fermented sausage Petrovská Klobása being a part of it. This product has been granted PDO under Serbian law since 2007. (Petrović et al. 2007). Each year, a famous festival, Klobásafest, is held in Bački Petrovac, attracting a great number of sausage manufacturers and visitors. Petrovská Klobása is made only from meat of the Landrace pig breed raised in local households. It is made from the meat of pigs that are 9–12 months old and weigh 135–200 kg (Petrović et al. 2011). It is produced only in local households in late autumn and early winter, when the outside temperature is around 0°C (Ikonic et al., 2015, October; Ikonić et al., 2013; Krkić et al., 2013). After slaughtering, carcasses are left to chill during the night and deboned the next day. Nevertheless, some producers use warm meat for sausage production, right after slaughter and deboning has been completed. When it comes to the choice of meat, every producer has its own recipe. Usually, the ratio of meat to fat is around 80:20 (Ikonić et al., 2012; Ikonić et al., 2013; Krkić et al. 2013; Škaljac et al., 2017; Tasić et al., 2012) or, to a lesser extent, 85:15 (Danilovic et al., 2011; Šojić et al., 2015; Šojić et al., 2014).

Meat cuts that are used in production are ham, shoulder, loin, neck, belly, trimmings and back fat. At the beginning, meat and fat are cut by knife into 4–5 cm^3 cubes and then minced through a 10 mm plate (Škaljac et al., 2018; Tasić et al., 2012). Minced meat is transferred to a table where mixing with spices by hand is also performed. In the production of Petrovská Klobása, the usual ingredients, aside from table salt, are garlic, caraway, sugar (sucrose), and hot and sweet ground paprika (Table 7.3). However, the amount of spice is not defined, so it varies among individual producers. This is where the greatest variations between sausages come from. A large amount of hot ground paprika (up to 2.5%) is added, which gives this product its characteristic intense red colour and rather hot taste. In addition, some manufacturers also choose to add sweet paprika, which even further enhances the red colour of this product. According to Škaljac (2014) and Petrović et al. (2007), due to bad relations between the Ottoman and Austro–Hungarian empires during the Ottoman–Habsburg wars, the former stopped the export of black pepper that was and still is traditionally used in sausage production. Consequently, peasants in the south of the Serbia, who were most affected with a lack of the highly

TABLE 7.3
Ingredients, and Smoking and Ripening Conditions of Petrovská Klobása

Sausage Dimensions	Casing	Ingredients	Smoking	Ripening
ø 45–65 mm, 35–50 cm in length	pig's large intestine (rectum) or collagen casings	lean pork (80–85%), back fat (20–25%), hot red paprika (2.5%), salt (1.8%), raw garlic (0.2%), caraway (0.2%), sucrose (0.15%)	10–15 days, 4–8 hours per day, 2–13°C, beech, cherry and apricot	3–4 months, 3–11°C,[*] 65–93% RH,[*] 6–12°C,[**] 43–87% RH[**]

Notes
[*] start of the drying/ripening;
[**] end of the drying/ripening
(Data taken from Ikonić et al., 2013; Petrović et al., 2011; Petrović et al., 2007; Škaljac et al., 2018)

appreciated spice, started using hot ground paprika, readily available in their households. From that point on, red ground paprika became an indispensable ingredient of Petrovská Klobása.

Mixing is also done by hand, using a specific traditional technique of tipping over and squashing the batter for 10–30 minutes (Petrović et al., 2011; Šojić et al., 2014). Well-mixed batter is traditionally stuffed into pigs' large intestines (rectums) or, lately, into collagen casings of similar dimensions and placed to hang and dry on wooden sticks. After a day of resting, sausages are transferred to a traditional smoking/drying room where they are kept for a period of 3–4 months. Interestingly, for the smoking process of Petrovská Klobása, among other types of wood, combustion of cherry and apricot wood is also used (Petrović et al., 2007). This definitely makes its distinctive colour and flavour. This is why Petrovská Klobása, smoked in traditional conditions and using cherry and apricot wood, had significantly lower red intensity (a* value) at the end of the drying period than the sausage smoked in industrial conditions, where smoke was obtained using beech (Škaljac et al. (2018). As for the lightness (L* value), it is lower than those reported for chorizo and Sremska sausage, while yellow intensity (b* value) is similar to that found for Sremska sausage (Gómez & Lorenzo, 2013; Ikonić et al., 2010; Škaljac et al., 2018; Stanišić et al., 2016; Živković et al., 2012).

The traditional drying/ripening process relies on local environmental conditions and temperature, and RH during the process cannot be controlled. However, Šojić et al. (2014) reported that temperature and relative humidity during 12 days of smoking were 5–10°C and 75–85%. In addition, Petrović et al. (2007) reported average daily temperatures of 2.2°C and RH of 67.4% in December, when the smoking of Petrovská sausages traditionally occurs.

The investigation of Danilovic et al. (2011) showed that dominant LAB species isolated from Petrovská Klobása are *Lactobacillus sakei* (36.4%), *Leuconostoc mesenteroides* (37.1%) and *Pediococcus pentosaceus* (18.4%), while *Lactobacillus curvatus* (6%), *Enterococcus caseliflavus* (6%) and *Enterococcus durans* (1%) are present in much lower percentages. Although accelerated by the addition of sucrose, the fermentation process is slow due to the low external temperatures (Danilovic et al., 2011; Ikonic et al., 2015, October; Ikonić et al., 2013) and in most cases pH does not fall below 5.1 (Figure 7.6). At the iso-electric point (pI) of meat proteins (5.4 for myosin) water binding capacity is the lowest (Huff-Lonergan, 2009), so the pH value above 5.1 ensures proper drying of the Petrovská sausage. However, since drying/ripening usually lasts for 4 months, it is common for the water content to fall to around 25% while weight loss ranges from 40–45% (Ikonić et al., 2010).

Traditional manufacturing of Petrovská sausage excludes the addition of contemporary additives such as GDL, nitrates and nitrites, and therefore it is highly appreciated by consumers. The investigation of Ikonić et al. (2010) revealed that the best sensory scores were obtained after 120 days of ripening and that the product is characterized by aromatic and spicy-hot flavour, dark red colour and hard consistency. A long ripening period affects its consistency, during which firmness increases from around 5 N on day 50 up to 14 N on day 120 of production. At the same time, hardness and chewiness increase from 0.48 kg and 0.04 kg to 6.82 kg and 0.89 kg, respectively (Ikonic et al., 2015, October; Ikonić et al., 2010). During

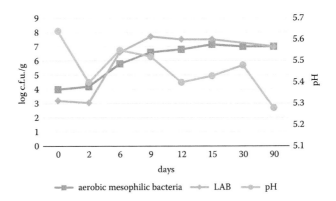

FIGURE 7.6 Changes in number of aerobic mesophilic bacteria, lactic acid bacteria (LAB) and pH of Petrovská Klobása during production (adapted from Danilovic et al. 2011).

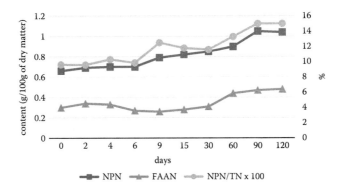

FIGURE 7.7 Changes in non-protein nitrogen (NPN), free amino acids nitrogen (FAAN) and proteolysis index (%) during the ripening of Petrovská sausage (adapted from Ikonić et al. 2013).

the entirety of the ripening process, the content of non-protein nitrogen (NPN) and free amino acids nitrogen (FAAN) increases (Figure 7.7) (Ikonić et al., 2013). Depending on the duration of ripening and the amount of meat and fatty tissue used, protein content at the end of the production varies from 23–30% while fat content ranges from 31–46% (Ikonić et al., 2010; Šojić et al., 2014). During storage, free fatty acid content increases from 14.62 mg KOH/g of lipids to 26.02 mg KOH/g of lipids and 35.09 mg KOH/g of lipids after 2 and 7 months, respectively (Šojić et al., 2014). At the same time, propanal and hexanal content increase from 1.86 μg/g and 0.12 μg/g to 32.59 μg/g and 1.67 μg/g after 7 months of storage. In addition, TBARS value increases during 5 months of storage from 0.8 mg MDA/kg to 3 mg MDA/kg (Krkić et al., 2013). In terms of safety, Tasić et al. (2012) reported maximum values of tryptamine (75.1 mg/kg), phenylethylamine (51.6 mg/kg), putrescine (11.7 mg/kg) and spermine (101 mg/kg) in different samples of Petrovská sausage while histamine was not detected in any sample.

7.5   SREMSKI KULEN

Budmani (1903) describes Kulen as the large intestine filled with blood, fat, flour and offal, similar to Blutwurst, and states that the word *kulen* was mentioned for the first time in the 15th century in Dubrovnik. In addition, it was mentioned in the second edition of the Serbian language dictionary in 1851, and it was described as a big sausage made from offal, similar to Magenwurst (Karadžić, 1852). Today, Kulen is a dry fermented sausage traditionally produced in the countries of the Western Balkan region, including Serbia and Croatia. Depending on the origin of the product, it is called Sremski kulen, Lemeški kulen, Petrovački kulen (Serbia), and Baranjski kulen or Slavonski kulen (Croatia). The word *kulen* is probably derived from the Latin word *colon*, which means large intestine, likely because the intestine serves as a natural casing for this kind of product (Vuković et al. 2012). What is common in all types of Kulen is that they are stuffed into wide-diameter casings and seasoned with plenty of hot red paprika powder.

Sremski kulen is also produced in the Srem region, and its production history is closely related to that of Sremska sausage, as already described. These two were the first meat products granted PDO mark under Serbian law (Radovanović et al., 1992a). In the last century, great demand for this sausage influenced partial transition of its production from household and artisanal manufacturing to the larger industrial meat producers.

For the production of Sremski kulen, only pork ham meat is used. It is made from the meat of late-maturing pig breeds that are slaughtered at around 120–150 kg (Radovanović et al., 1992a). The meat has a hard consistency, a more intense red colour and contains less moisture (Suvajdzic et al., 2018). When it comes to the choice of fat tissue, it is usually made from jowl or back fat (15%) (Radovanović et al., 1992a). Firstly, meat and fat are chopped to 10 cm^3 sized cubes and placed on a perforated plate where they are strained for 24 hours. Afterward, the meat is partially frozen to around −6°C while fat is chilled to 0°C. In case of artisanal production in winter, meat and fat are left outside overnight during which external temperature usually ranges from −5°C to 5°C. Because of the low temperatures, the occurrence of frost is very common during these months.

When it comes to mixing the meat and fat with spices, there are two principle methods. One is to add spices to chopped meat, mix it in a mixing machine and mince it together afterward. The other way is to mince the meat and fat to a particle size of around 8–13 mm and then mix it with salt and paprika, which is more common in industrial production (Radovanović et al., 1992a; Suvajdzic et al., 2018). As mentioned previously, hot paprika is an irreplaceable part of every Kulen and up to 1.5% is added, while salt is added to around 2.6% (Table 7.4). The addition of only two ingredients, in addition to meat and fat, leaves enough space for aromatic compounds, which are derived from proteins and fatty acids, to come to the fore at the end of the ripening. However, some producers also add smaller quantities of homemade red wine in order to improve the colour and flavour of the sausage.

TABLE 7.4
Ingredients, Smoking and Ripening Conditions of Sremki Kulen

Sausage Dimensions, and Weight	Casing	Ingredients	Smoking	Ripening
⌀ 10–11 cm, 15–20 cm in length, 1–1.8 kg	pig's cecum	partially frozen pork ham, –6°C (85–95%), chilled back fat (5–15%), salt (2.6%), hot red paprika (1.5%)	5 days, 18°C, beech (cherry and plum)	2–9 months, 12°C,[*] 85–88% RH,[*] 12°C,[**] 72–75% RH[**]

Notes
[*] start of the drying/ripening;
[**] end of the drying/ripening
(Data taken from Radovanović et al., 1992a; Suvajdzic et al., 2018)

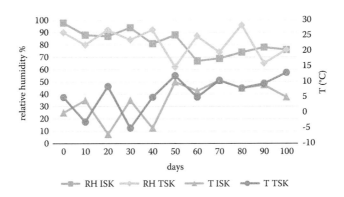

FIGURE 7.8 Relative humidity and temperature conditions during ripening of traditionally (TSK) and industrially (ISK) made Sremski kulen (adopted from Suvajdzic et al. 2018).

What is unique about the production of Sremski kulen is that the batter is stuffed into the pig's cecum, weighing about 1–2 kg after stuffing (Radovanović et al., 1992a; Suvajdzic et al., 2018). The filled cecum is then tied with a rope and then placed into a netting roll. Sremski kulen of very large diameter require longer drying/ripening periods. Suvajdzic et al. (2018), reported values of T and RH during ripening for both traditionally and industrially produced Kulen (Figure 7.8).

The centre of the sausage is characterized by a light red colour, which is the result of uneven moisture distribution. Along with its unique shape, this is exactly what makes this sausage special. This characteristic is especially obvious for traditionally produced Sremski kulen with a moisture content that is higher than the same product produced in industrial conditions. It usually ranges from 30–35% (Radovanović et al., 1992a; Suvajdzic et al., 2018), although sometimes it can be up to 44.2% (Vuković et al. 1988). Because of this, values of hardness and chewiness for traditionally made Kulen are around 20% lower compared to those found for Kulen produced in industrial conditions (Suvajdzic, 2018). The differences in colour between sausages made by the two types of production were also observed (Suvajdzic, 2018), with a lightness (L* value) for traditionally made Kulen of 35.71 and a value for industrially produced Kulen of 32.65. In addition, the red intensity (a* value) and the yellow intensity (b* value) for traditionally made Sremski kulen were also significantly higher.

Smoking of kulen usually lasts around 5 days, during which the temperature should be close to 18°C, while relative humidity and air velocity depend on the external conditions (Radovanović et al., 1992a). For this process, mostly beech but in some cases cherry or plum wood is used (Suvajdzic et al., 2018). Due to the low temperatures during ripening and no addition of sugars or starter cultures, fermentation is relatively slow and pH usually does not drop under 5.25 (Figure 7.9) (Suvajdzic et al., 2018), which is enough to ensure proper drying of the product.

The study of Suvajdzic (2018) also revealed that dominant LAB species were *Lactobacillus sakei* (89.64%) and *Lactobacillus curvatus* (7.22%). During 150 days of ripening, the LAB number was the highest, followed by *Micrococcaceae* and *Enterococcus*, while the number of *Enterobacteriaceae* and *Pseudomonadaceae* was under 2 log CFU/g in the second half of the ripening period (Figure 7.10). Its long drying/ripening period results in big decrease in a_w value, which, coupled with its low pH, is enough to ensure microbiological safety of the product. The content of biogenic amines such as histamine, tryptamine and tyramine was 16.1 mg/kg, 47.6 mg/kg and 95.1 mg/kg, respectively (Tasic et al., 2015).

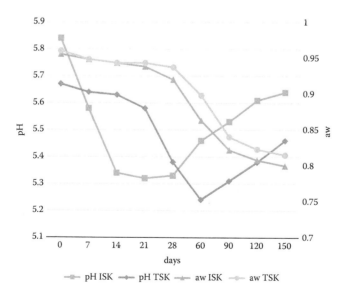

FIGURE 7.9 pH and a_w decrease during production of traditional (TSK) and industrial (ISK) Sremski kulen (adopted from Suvajdzic et al. 2018).

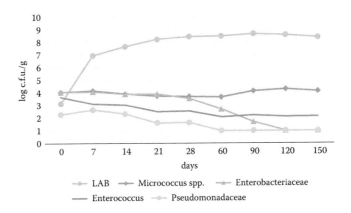

FIGURE 7.10 Changes in microbial count during ripening of traditionally made Sremski kulen (adopted from Suvajdzic et al. 2018).

According to Radovanović et al. (1992a), in industrial production, ripening is done under controlled conditions for at least 2 months, during which RH is gradually decreased while temperature stays at 12°C during the whole period. In contrast, if the sausage is produced in households, the ripening period can last up to 9 months. Then sausages must be transferred to a traditional ripening chamber, which is located in the home's basement, similar to a wine cellar, where they are protected from harmful external influences during the summer months, providing darkness and relatively constant T and RH. Because the ripening time varies, weight loss ranges from 35–53% (Radovanović et al., 1992a; Suvajdzic et al., 2018). The reason for lower moisture content in industrial Kulen is its higher fat content. It is well known that the moisture content of fat is much lower than that of muscle, therefore sausage containing a higher amount of fat will have lower initial water content and vice versa (Incze, 2007; Simunovic et al., 2019). At the end of the ripening, content of protein, fat and salt are around 36%, 26% and 4.95%, respectively (Suvajdzic et al., 2018). The content of NPN increases during ripening from an initial content of 10.15–15.66%, which is lower than some Spanish fermented sausages such as chorizo and salchichón (Lizaso, Chasco, & Beriain, 1999; Salgado et al. 2005). As for the lipid oxidative changes, peroxide, acidity and the TBARS value rise to 0.46 mmol/kg, 2.39 mg KOH/g and 0.11 mg MDA/kg, respectively, on day 150 of ripening (Suvajdzic et al., 2018; Vuković et al., 2011). At the end of the ripening, the product is characterized by a dark red colour, hard consistency, aromatic flavour and hot aftertaste (Suvajdzic, 2018; Vuković et al., 2011).

7.6 LEMEŠKI KULEN

Lemeški kulen is traditionally produced in the village of Svetozar Miletić (Lemeš) located in northwest Serbia near the city of Sombor. The village was founded in 1752 under the rule of the Austro–Hungarian empire. In historical records from 1900 and a monograph about the city of Lemeš, Kulen was described as a "national

dish made from pork seasoned with plenty of paprika and with very hot taste."
(Beljanski, 1984). Lemeški kulen is stuffed either in the cecum or the rectum.
Lemeški kulen is different from other types of kulen sausages because the meat is
minced through plates with smaller diameter holes, a larger amount of hot red
paprika powder is added, and it contain no pork back fat. Unlike Sremski kulen, it is
only produced in households by traditional means. The estimated annual production
of Lemeški kulen is about 3 tons representing only 0.005% of Serbian sausage
production (Vuković, Vasilev, & Saičić, 2014). This places it on the top of the list
of Serbian artisanal meat products in general.

For the production of Lemeški kulen, dry red meat from mature pigs of above
150 kg is preferred. Mainly shoulder, ham and neck are used (Table 7.5). The pigs'
diet is based on products such as corn, barley, alfalfa and soy, which are grown in
the same geographical region (Vuković et al., 2014). Similarly, the red paprika used
in production, also known as Lemeška paprika, is traditionally produced in local
households. According to Vuković et al. (2011), Lemeška paprika contains
9.6–13.2% of sugars and bacteria that are capable of fermenting it to lactic acid. In
addition, except for its pigments, carotene and capsorubin, which give this sausage
an intense red colour, paprika contains high amounts of ascorbic acid and flavonoids
such as quercetin and luteolin (Lee, Howard, & Villalón, 1995). Like most other
vegetables, paprika uses nitrates of the soil that accumulate in variable amounts
depending on the cultivar or cultivation system (Colavita et al. 2014). According to
Vuković et al. (2012), lemeška paprika contains from 38 to 108 mg/kg of nitrates.

The meat is chopped into pieces, then minced to a particle size of 6–8 mm, and then
salt and hot red paprika powder are added. After mixing, the batter is stuffed firmly into
casings, making sure that there is no residual air (Vuković et al., 2011). Although the
batter should be stuffed into the cecum, some producers are partly using the large
intestine (rectum) which allows faster drying due to its smaller diameter. Consequently,
the production process when using this casing is shorter and it is usually finished after
3 months. In contrast, kulen in the cecum must be dried for at least 6 months. After

TABLE 7.5
Ingredients, Smoking and Ripening Conditions of Lemeški Kulen

Dimensions and Weight	Casing	Ingredients	Smoking	Ripening
ø 10–12 cm, 15–20 cm in length, 1–2 kg	pig's cecum	pork ham, shoulder or neck (95%), hot red paprika (3%), salt (2–2.2%)	7 days, beech (oak and black locust)	6 months, 2–8°C,[*] 8% RH,[*] 10–15°C,[**] 5% RH[**]

Notes
[*] start of the drying/ripening;
[**] end of the drying/ripening
(Data taken from Vuković et al., 2014; Vuković et al., 2012)

stuffing, a mash of rope is placed around the kulen cecum, usually between natural intestine creases, giving kulen its distinctive appearance and at the same time pre-venting the casing from bursting, especially in the case of larger products (Vuković et al., 2011). Sausages are hung on wooden sticks, transferred to a smoking/drying chamber and then smoked after a resting day. Lemeški kulen is cold-smoked up to 7 days, usually using beech wood. Because of low temperatures during ripening, fer-mentation is slow and the pH value reaches its minimum of 5.5 after 3 months of drying (Vuković et al., 2012). The investigation of Vasilev et al. (2015) revealed that at the end of the ripening, the most dominant LAB species were *Lactobacillus brevis* (57.9%), *Lactobacillus curvatus ssp. curvatus* (15.8), *Lactobacillus paracasei ssp. paracasei* (10.5%), *Lactobacillus plantarum* (10.5%) and *Lactobacillus fermentum* (5.3%). In addition, the number of LAB is the highest (7.18 log c.f.u./g), followed by *Micrococcaceae* (5.40 log c.f.u./g) and *Enterococcaceae* (5 log c.f.u./g).

Average external temperatures from December to February, when the sausage is made, are always below 10°C, while RH is around 70% (Vuković et al., 2014). In the next 3 months of ripening temperatures are higher, allowing enzymes to get more involved in proteolytic and lipolytic reactions. Because of the liberation of peptides, amino acids and amines during ripening, pH can rise to 5.42–6.12 (Vuković et al., 2014). Water activity value after 3 months of ripening is 0.90, while after 6 months it reaches 0.86. Low water activity and low pH value coupled with the antimicrobial activity of capsaicin from the paprika ensures microbiological safety. The shelf life of Lemeški kulen is around 9 months, after which colour and aroma begin to change while consistency becomes hard. Protein content at the end of the ripening ranges from 28.32–37.79%, while collagen content in total proteins ranges from 5.22–7.26% (Vuković et al., 2014).

7.7   UŽIČKA SAUSAGE

In contrast to the plains of the Vojvodina region, central Serbia is characterized by its mountains and hills, encompassing Zlatibor and Tara national parks. The adminis-trative centre of the Zlatibor district is the city of Užice, for which the sausage was named. In addition to representing a famous tourist attraction, Zlatibor has a long tradition in livestock farming and the production of dry meat and sausages. The production of Užička sausage takes place in winter months when the temperature conditions are perfect for outdoor chilling of carcasses after slaughtering. Only in the case of an unusually warm winter weather is meat deboned immediately after slaughtering. Very low temperatures and occurrence of frost are desirable and very common in this mountainous region. Average daily temperatures from November to February when the sausage is made range from –2.2–3.5°C (Radovanović et al., 1990), instigating a low extent of meat fermentation. For the production of the Užička sausage, pork, beef and pork back fat are used. There is big diversity when it comes to the choice of meat between producers, and therefore quality and sensory properties of sausages vary. While some producers make sausages only from pork or beef, most manufacturers use both kinds of meat. Since pork has a lower price compared to beef, and cost is an important criterion, the share of pork in this traditional meat product

TABLE 7.6

Ingredients, Smoking and Ripening Conditions of Užička Sausage

Dimensions and Weight	Casing	Ingredients	Smoking	Ripening
ø 40 mm, 30–41 cm in length, 700 g	bovine small intestine	pork (20–50%), beef (10–70%), pork back fat (10–20%), salt, sugar, garlic and pepper	21–30 days, beech	0.16 m/s 7–9°C,[*] 72–82% RH,[*] 6–8°C,[**] 70–80% RH[**]

Notes
[*] start of the drying/ripening;
[**] end of the drying/ripening
(Data taken from Borovic et al., 2017, September; Savić & Savić, 1962; Vesković Moračanin et al., 2013; Vesković Moračanin et al., 2010)

has steadily increased in the last decades. Nonetheless, meat and fat are minced to a 3–4 mm particle size and then mixed with 1.5–2.5% of salt (Table 7.6).

Spices such as black and white pepper, hot paprika powder and garlic are distinctive characteristics of every producer. Meat batter is then placed in a cool chamber where it rests up to one day (Savić & Savić, 1962). Afterwards, the batter is stuffed into bovine small intestines in units of approximately 30 cm in length and tied up with a rope at both ends. If two ropes are connected, the sausage acquires the shape of a horseshoe (Savić & Savić, 1962). Sausages are then transferred to a traditional smokehouse where they are kept for 1 month. After a resting day, they are smoked throughout the whole ripening process, which is why the final product is characterized by strong smoky flavour (Borovic et al., 2017, September; Vesković Moračanin et al., 2013). Hot smoking prevents casing bursts due to freezing by increasing the temperature during cold winter months in this mountainous region in which minimum temperature can be as low as –23°C (Radovanović et al., 1990). As for the temperature and relative humidity conditions during the smoking/ripening period, they are usually around 6–9°C and 69–82%, while air velocity is 0.16 m/s (Vesković Moračanin et al., 2010). The added sugar promotes fermentation of the sausage, resulting in a pH decrease from 5.25 on the first day to 4.97 on the seventh day of production and then remaining unchanged until the end of production (Vesković Moračanin et al., 2013). At the beginning of the ripening, dominant bacteria species are *Leuconostoc mesenteroides* and *Lactococcus lactis*, while after 4 days of the process and continuing until the end of production, the most dominant are *Leuconostoc mesenteroides*, *Lactobacillus brevis* and *Lactobacillus sakei* (Borovic et al., 2017).

At the end of the production, Užička sausage is characterized by the light brown colour of the surface, while cross sections are red to light brown. Reported colour values are similar to those of Turkish-type dry fermented sausages such as Sudzuk (Ercoşkun & Özkal, 2011; Stajić et al. 2013). Because of moisture loss, the content of protein and fat rise during ripening while weight loss is usually 42–55%, depending

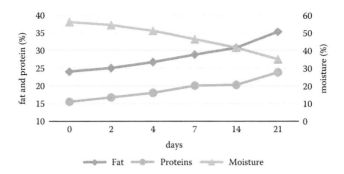

FIGURE 7.11 Changes in chemical composition during production of Užička sausage (Vesković Moračanin et al., 2013).

on the meat and fat ratio (Figure 7.11). Final product contains 4.47–5.12% salt, while the content of collagen in total proteins is around 10% (Vesković Moračanin et al., 2013; Vukašinović, Kurćubić, Kaljević, Maskovic, & Petrović, 2011).

7.8 CONCLUSION

Recent growing demand for meat-free products that suit consumers who are adopting vegan, vegetarian or flexitarian diets has resulted in a proliferation of meat analogues or meat substitutes hitting the markets all over Europe, except for Serbia. The Serbian consumer is a meat eater, and pork meat is the most popular with the masses. Pigs reproduce rapidly and eat everything. Pork meat is the best for curing, smoking and sausage-making. For centuries, Serbian pork sausage recipes were not publicized. They were passed from father to son, and otherwise kept closely guarded. Sausage making has been well known for centuries to Serbs and almost everybody in Serbia knows "what goes inside." However, we are still waiting for the Serbian government to adopt proper standards for traditionally made sausages. This would help to promote these products internationally with specific characteristics, particularly those coming from less-favoured or rural areas. Benefits would include the improvement of the income of farmers in return for a genuine effort to improve quality. This would also support retention of population in rural areas that are the true keepers of every tradition, including sausage-making.

This chapter is a review of available scientific literature regarding production of traditional pork sausages in Serbia, including their manufacturing process, chemical composition and shelf life, which should assist in adopting such standards.

REFERENCES

Beljanski, M. (1984). *Nemeš Militič, Svetozar Miletić (1752–1984)* (A. Kaić, Ed.). Sombor: Prosveta.

Borovic, B., Velebit, B., Vesković, S., Lakicevic, B., & Baltic, T. (2017, September). *The characterization of lactic acid bacteria isolated during the traditional production of*

Užička sausage. Paper presented at the 59th International Meat Industry Conference MEATCON2017, Zlatibor.

Borović, B., Vesković, S., Velebit, B., Baltić, T., & Spirić, D. (2009). Dominant microflora isolated from traditionally fermented "Sremska" sausage. *Tehnologija mesa*, *50*(3–4), 227–231.

Budmani, P. (1903). *Dictionary of croatian or serbian language*. Zagreb: Yugoslav Academy of Sciences and Arts.

Colavita, G., Piccirilli, M., Iafigliola, L., & Amadoro, C. (2014). Levels of nitrates and nitrites in chili pepper and ventricina salami. *Italian Journal of Food Safety*, *3*(2), 114–116. https://doi.org/10.4081/ijfs.2014.1637.

Ćorović, V. (1989). *Istorija Srba* (Z. Stefanović, Ed.). Beograd: BIGZ.

Danilovic, B., Joković, N., Petrović, L., Veljovic, K., Tolinački, M., & Savic, D. (2011). The characterisation of lactic acid bacteria during the fermentation of an artisan Serbian sausage (Petrovska Klobasa). *Meat Science*, *88*, 668–674. https://doi.org/10.1016/j.meatsci.2011.02.026.

Djinovic, J., Popovic, A., & Jira, W. (2008). Polycyclic aromatic hydrocarbons (PAHs) in different types of smoked meat products from Serbia. *Meat Science*, *80*(2), 449–456. https://doi.org/10.1016/j.meatsci.2008.01.008.

Ducic, M., Vranic, D., & Baltic, M. (2018). Selected physico-chemical properties of Serbian dry fermented sausages in different meat industries. *Meat Technology*, *59*(2), 120–126. https://doi.org/10.18485/meattech.2018.59.2.7.

Dzinic, N., Ivic, M., Branislav, S., Jokanovic, M., Tomovic, V., Okanovic, D., & Raljic, J. P. (2015, October). *Some quality parameters of dry fermented sausages (Čajna kobasica)*. Paper presented at the International 58th Meat Industry Conference "Meat Safety and Quality: Where It Goes?", Zlatibor.

Džinić, N., Ivić, M., Jokanović, M., Šojić, B., Škaljac, S., & Tomovic, V. (2016). Chemical, color, texture and sensory properties of Čajna kobasica, a dry fermented sausage. *Quality of Life*, *7*(1-2), 5–11. https://doi.org/10.7251/QOL1601005DZ.

Ercoşkun, H., & Özkal, S. G. (2011). Kinetics of traditional Turkish sausage quality aspects during fermentation. *Food Control*, *22*(2), 165–172. https://doi.org/10.1016/j.foodcont.2010.06.015.

Fonseca, S., Gomez, M., Domínguez, R., & Lorenzo, J. M. (2015). Physicochemical and sensory properties of Celta dry-ripened "salchichón" as affected by fat content. *Grasas y Aceites*, *66*, e059. https://doi.org/10.3989/gya.0709142.

Frece, J., Kovačević, D., Kazazić, S., Mrvčić, J., Vahcić, N., Ježek, D., ... Markov, K. (2014). Comparison of Sensory Properties, Shelf-Life and Microbiological Safety of Industrial Sausages Produced with Autochthonous and Commercial Starter Cultures. *Food Technology and Biotechnology*, *52*(3), 307–316.

Gómez, M., & Lorenzo, J. M. (2013). Effect of fat level on physicochemical, volatile compounds and sensory characteristics of dry-ripened "chorizo" from Celta pig breed. *Meat Science*, *95*(3), 658–666. https://doi.org/10.1016/j.meatsci.2013.06.005.

Herrero, A. M., Ordóñez, J. A., de Avila, R., Herranz, B., de la Hoz, L., & Cambero, M. I. (2007). Breaking strength of dry fermented sausages and their correlation with texture profile analysis (TPA) and physico-chemical characteristics. *Meat Science*, *77*(3), 331–338. https://doi.org/10.1016/j.meatsci.2007.03.022.

Huff-Lonergan, E. (2009). Fresh meat water-holding capacity. In J. P. Kerry & D. Ledward (Eds.), *Improving the sensory and nutritional quality of fresh meat* (pp. 147–160). Cambridge: Woodhead Publishing.

Ikonic, P., Jokanovic, M., Tasic, T., Skaljac, S., Sojic, B., Tomovic, V., ... Petrovic, L. (2015, October). *The effect of different ripening conditions on proteolysis and texture of dry-fermented sausage Petrovská klobása*. Paper presented at the International 58th Meat Industry Conference "Meat Safety and Quality: Where It Goes?", Zlatibor.

Ikonić, P., Ljiljana S. P., Tasić, T., Natalija R. D., Marija R. J., & Vladimir M. T. (2010). Physicochemical, biochemical and sensory properties for the characterization of Petrovská klobása (traditional fermented sausage). *Acta periodica technologica, 41,* 19–31. https://doi.org/10.2298/APT10410191.

Ikonić, P., Petrovic, L., Tasić, T., Jokanovic, M., Savatic, S., Ikonić, B., & Dzinic, N. (2012). The effect of processing method on drying kinetics of Petrovská klobása, an artisan fermented sausage. *Chemical Industry and Chemical Engineering Quarterly, 18*(2), 163–169. https://doi.org/10.2298/CICEQ110909058I.

Ikonić, P., Tasić, T., Petrović, L., Škaljac, S., Jokanović, M., Mandić, A., & Ikonić, B. (2013). Proteolysis and biogenic amines formation during the ripening of Petrovská klobása, traditional dry-fermented sausage from Northern Serbia. *Food Control, 30,* 69–75. https://doi.org/10.1016/j.foodcont.2012.06.021.

Incze, K. (2007). European products. In F. Toldrá (Ed.), *Handbook of fermented meat and poultry* (pp. 307–318). Oxford: Blackwell Publishing.

INMES. (2019). *Shelf life of Sremska sausage packed under vacuum.* Belgrade:Institute of Meat Hygiene and Technology.

Joksimović, J., Čavoški, D., & Fridl, T. (1981). Examination of technological tolerances for the reduction of fat quantity in dry sausage stuff. *Tehnologija mesa, 22*(3), 85–87.

Karadžić, V. (1852). *Dictionary of Serbian language.* Vienna: Printing House of Armenian Monastery.

Krkić, N., Šojić, B., Lazić, V., Petrović, L., Mandić, A., Sedej, I., & Tomovic, V. (2013). Lipid oxidative changes in chitosan-oregano coated traditional dry fermented sausage Petrovska klobasa. *Meat Science, 93,* 767–770. https://doi.org/10.1016/j.meatsci.2012 .11.043.

Krkić, N., Šojić, B., Lazić, V., Petrović, L., Mandić, A., Sedej, I., ... Džinić, N. (2013). Effect of chitosan–caraway coating on lipid oxidation of traditional dry fermented sausage. *Food Control, 32,* 719–723. https://doi.org/10.1016/j.foodcont.2013.02.006.

Lee, Y., Howard, L. R., & Villalón, B. (1995). Flavonoids and antioxidant activity of fresh pepper (*Capsicum annuum*) cultivars. *Journal of Food Science, 60*(3), 473–476. https:// doi.org/10.1111/j.1365-2621.1995.tb09806.x.

Lizaso, G., Chasco, J., & Beriain, M. J. (1999). Microbiological and biochemical changes during ripening of salchichón, a Spanish dry cured sausage. *Food Microbiology, 16*(3), 219–228. https://doi.org/10.1006/fmic.1998.0238.

Lücke, F.-K. (2007). Quality assurance plan. In F. Toldrá (Ed.), *Handbook of fermented meat and poultry* (pp. 535–543). Oxford: Blackwell Publishing.

Milojević, J. (1996). Examination of fatty acid composition changes of free fatty acid fraction during ripening of Čajna sausage with different additives. *Tehnologija mesa, 37*(1), 29–31.

Oluški, V., Miloševski, V., Ćirić, M., & Marinković, S. (1974). Influence of raw material composition and glucono-delta-lactone addition on aging rate and quality of dry sausages. *Tehnologija mesa, 15*(6), 177–182.

Pećanac, B., Tomić, R., Janjić, J., Ćirić, J., Baltić, B., Grbić, S., & Baltić, M. Ž. (2017). Tea sausage-safety and quality. *Veterinarski zurnal Republike Srpska, 17*(1), 81–92. https:// doi.org/10.7251/vetj1701081p.

Petrović, L., Džinić, N., Ikonić, P., Tasić, T., & Tomović, V. (2011). Quality and safety standardization of traditional fermented sausages. *Tehnologija mesa, 52*(2), 234–244.

Petrović, L., Džinić, N., Tomović, V., Ikonić, P., & Tasić, T. (2007). Code of practice-registered geographical indication Petrovská klobása. Intellectual Property Office, Republic of Serbia, Decision No. 9652/06 G-03/06.

Pribiš, V., Svrzić, G., Polić, M., & Đorđević, M. (1996). Color changes of "čajna sausage" during production using various additives. *Tehnologija mesa, 37*(6), 235–240.

Radetic, P. (1997). *Raw sausages.* Beograd: Ilijanum.

Radovanović, R., Čarapić, G., Arbutina, Z., Bojović, P., Čavoški, D., Obradović, D., ... Mojašević, M. (1990). Code of practice-registered geographical indication Užička pork ham. Intellectual Property Office, Republic of Serbia, Decision No. G-1990/05690.

Radovanović, R., Tubić, M., Savić, S., Čavoški, D., Veličković, D., Bojović, P., ... Mojašević, M. (1992a). Code of practice-registered geographical indication Sremski kulen. Intellectual Property Office, Republic of Serbia, Decision No. 2650/92-2.

Radovanović, R., Tubić, M., Savić, S., Čavoški, D., Veličković, D., Bojović, P., ... Mojašević, M. (1992b). Code of practice – registered geographical indication Sremska sausage. Intellectual Property Office, Republic of Serbia, Decision No. 2651/92-2.

Rašeta, M., Vesković Moračanin, S., Borović, B., Karan, D., Vranic, D., Trbović, D., & Lilić, S. (2010). Microclimate conditions during ripening of traditionally produced fermented sausages. *Tehnologija mesa*, *51*(1), 45–51.

Regulation (2019). Rulebook on quality of meat products. *Official Gazette of the Republic of Serbia*, 2019, 50, 100–116.

Romano, J. (1969). First plants for cattle slaughtering and meat processing in Serbia at the beginning of 20th century. *Tehnologija mesa*, *10*(3), 86–88.

Salgado, A., Fontán, M., Franco, I., López, M., & Carballo, J. (2005). Biochemical changes during the ripening of Chorizo de cebolla, a Spanish traditional sausage. Effect of the system of manufacture (homemade or industrial). *Food Chemistry*, *92*(3), 413–424. https://doi.org/10.1016/j.foodchem.2004.07.032.

Savić, T., & Savić, N. (1962). Meat products of the Užice region. *Tehnologija mesa*, *3*(2), 4–6.

Statistical Office of the Republic of Serbia. (1950-2018). *Statistical yearbook of the Republic of Sebia*. Retrieved from https://www.stat.gov.rs/sr-cyrl/publikacije/?d=2&r=#.

Statistical Office of the Republic of Serbia. (2007-2018). *Household budget survey*. Retrieved from https://www.stat.gov.rs/en-US/publikacije/?d=2&r=.

Simunovic, S., Đorđević, V., Bogdanović, S., Dimkic, I., Stankovic, S., Novakovic, S., & Tomašević, I. (2019). *Changes in chemical attributes during ripening of traditional fermented sausage "Pirot ironed"*. Paper presented at the 60th International Meat Industry Conference "Safe Food for Healthy Future", Kopaonik.

Škaljac, S. (2014). *The influence of different technological parameters during standardiza-tion of safety and quality on the color formation of traditional fermented sausage (Petrovačka kobasica)*. (Doctoral dissertation, University of Novi Sad, Novi Sad, Serbia). Retrieved from http://nardus.mpn.gov.rs/handle/123456789/1829.

Škaljac, S., Jokanović, M., Tomović, V., Ivić, M., Tasić, T., Ikonić, P., ... Petrović, L. (2018). Influence of smoking in traditional and industrial conditions on colour and content of polycyclic aromatic hydrocarbons in dry fermented sausage "Petrovská klobása". *LWT - Food Science and Technology*, *87*, 158–162. https://doi.org/10.1016/j.lwt.2017.08.038.

Škaljac, S., Petrović, L., Jokanović, M., Tomović, V., Tasić, T., Ivić, M., ... Dzinic, N. (2017). *The influence of smoking in traditional conditions on content of polycyclic aromatic hydrocarbons in Petrovská klobása*. Paper presented at the 59th International Meat Industry Conference MeatCon2017, Zlatibor.

Šojić, B., Dzinic, N., Tomovic, V., Jokanovic, M., Ikonić, P., Tasić, T., ... Ivic, M. (2015). *Effect of the addition of Staphylococus Xylosus on the oxidative stabilitty of traditional sausage (Petrovská klobása)*. Paper presented at the International 58th Meat Industry Conference "Meat Safety and Quality: Where It Goes?", Zlatibor.

Šojić, B., Petrovic, L., Mandić, A., Sedej, I., Dzinic, N., Tomovic, V., ... Ikonić, P. (2014). Lipid oxidative changes in traditional dry fermented sausage Petrovská klobása during storage. *Hemijska Industrija*, *68*(1), 27–34. https://doi.org/10.2298/hemind130118024S.

Soyer, A. (2005). Effect of fat level and ripening temperature on biochemical and sensory characteristics of naturally fermented Turkish sausages (sucuk). *European Food Research and Technology*, *221*, 412–415. https://doi.org/10.1007/s00217-005-1192-6.

Stajić, S., Perunović, M., Stanišić, N., Žujović, M., & Živković, D. (2013). Sucuk (turkish-style dry-fermented sausage) quality as an influence of recipe formulation and inoculation of starter cultures. *Journal of Food Processing and Preservation, 37,* 870–880. https://doi.org/10.1111/j.1745-4549.2012.00709.x.

Stamenković, T., Dević, B., Đurković, A., Hromiš, A., & Vlaisavljević, M. (1990). Effect of glucono-delta-lactone on pH and sensory properties of tea sausage. *Tehnologija mesa, 31*(5), 177–183.

Stamenković, T., Đurković, A., Janković, D., & Orlić, Z. (1996). Influence of drying and smoking on the formation of sensory properties of tea sausage. *Tehnologija mesa, 37*(6), 256–260.

Stamenković, T., Đurković, A., Orlić, Z., Hromiš, A., Vlaisavljević, M., & Janković, D. (1991). Quality of tea sausage dependent on amount of additional fatty tissue. *Tehnologija mesa, 32*(3), 105–108.

Stamenković, T., Hromiš, A., & Janković, D. (1988). Fundamental characteristics of production and quality of sausage intended for export. *Tehnologija mesa, 29*(11), 333–336.

Stamenković, T., Miloševski, V., Aničić, V., Ćirić, M., & Stamatović, S. (1984). Contribution to study of temperature and relative humidity importance for tea sausage smoking and drying in semi-automatic equipment. *Tehnologija mesa, 25*(5), 139–141.

Stanišić, N., Lilić, S., Petrović, M. D., Živković, D., Radović, Č., Petričević, M., & Gogić, M. (2012, May). *Proximate composition and sensory characteristics of Sremska sausage produced in a traditional smoking house.* Paper presented at the 6th Central European Congress on Food, CEFood2012, Novi Sad.

Stanišić, N., Parunović, N., Petrović, M., Radovic, C., Slobodan, L., Stajić, S., & Gogić, M. (2014). Changes in chemical and physicochemical characteristics during the production of traditional Sremska sausage. *Biotechnology in Animal Husbandry, 30*(4), 705–715. https://doi.org/10.2298/BAH1404705S.

Stanišić, N., Petrović, M., Radovic, C., Petricevic, M., Stanojković, A., & Gogić, M. (2016). *Characteristics of dry fermented "Sremska kobasica" produced in traditional smoking house.* Paper presented at the 51st Croatian & 11th International Symposium on Agriculture, Opatija.

Stevanovic, J., Okanović, Đ., Stevanetic, S., Mirilović, M., Karabasil, N., & Pupavac, S. (2016). Traditional products – base for the sustainable development of Serbian animal origin products. *Food and Feed Research, 43*(2), 127–134. https://doi.org/10.5937/FFR1602127S.

Šutić, M., Milovanović, R., & Svrzić, G. (1990). Influence of additives and sugars on new starter-cultures in production of tea sausage. *Tehnologija mesa, 31*(4), 136–141.

Suvajdzic, B. (2018). *Microbiota and quality parameters of Sremski kulen produced in industrial and traditional conditions.* Doctoral dissertation, University of Belgrade, Belgrade, Serbia. Retrieved from https://uvidok.rcub.bg.ac.rs/handle/123456789/3066.

Suvajdzic, B., Petronijevic, R., Teodorović, V., Tomovic, V., Dimitrijevic, M., Karabasil, N., & Vasilev, D. (2018). Quality of fermeted sausage Sremski Kulen produced under traditional and industrial conditions in Serbia. *Fleischwirtschaft - Frankfurt, 98*(6), 93–99.

Tasic, T., Ikonic, P., Jokanovic, M., Mandic, A., Tomovic, V., Sojic, B., & Skaljac, S. (2015). *Content of vasoactive amines in Sremski Kulen and Sremska Kobasica traditional dry fermented sausages from Vojvodina.* Paper presented at the International 58th Eat Industry Conference "Meat Safety and Quality: Where It Goes?", Zlatibor.

Tasić, T., Ikonić, P., Mandić, A., Jokanović, M., Tomovic, V., Savatić, S., & Petrović, L. (2012). Biogenic amines content in traditional dry fermented sausage Petrovská klobása as possible indicator of good manufacturing practice. *Food Control, 23,* 107–112. https://doi.org/10.1016/j.foodcont.2011.06.019.

Tojagić, S. (1996). The household production as the precursor of industrial production of Sremska sausage. *Tehnologija mesa, 37*(6), 261–265.

Tojagić, S. (1997). Necessity of Sremska sausage standardization as the national product. *Tehnologija mesa, 38*(6), 265–267.

Tomašević, I., & Đekić, I. (2017). HACCP in fermented meat production. In N. Zdolec (Ed.), *Fermented meat products: Health aspects* (pp. 513–534). Boca Raton: CRC Press.

Turubatović, L., & Tadić, R. (2005). Production parameters, packing and critical points in distribution chain if traditional fermented sausages and standard operating procedures. *Tehnologija mesa, 46*(3–4), 212–217.

Vasilev, D., Aleksic, B., Tarbuk, A., Dimitrijevic, M., Karabasil, N., Cobanovic, N., & Vasiljevic, N. (2015). *Identification of lactic acid bacteria isolated from Serbian traditional fermented sausages Sremski and Lemeski Kulen.* Paper presented at the International 58th Meat Industry Conference "Meat Safety and Quality: Where It Goes?", Zlatibor.

Vesković-Moračanin, S., Karan, D., Okanović, Đ., Jokanović, M., Džinić, N., Parunović, N., & Trbović, D. (2011). Colour and texture properties of traditionally fermented "Sremska" sausage. *Tehnologija mesa, 52*(2), 245–251.

Vesković Moračanin, S., Karan, D., Trbović, D., Okanović, D., Džinić, N., & Jokanović, M. (2013). Colour and texture characteristics of "Užička" fermented sausage produced in the traditional way. *Tehnologija mesa, 54*(2). https://doi.org/10.5937/tehmesa1302137V.

Vesković Moračanin, S., Rašeta, M., Đorđević, M., Turubatović, L., Stefanović, S., Janković, S., & Škrinjar, M. (2010). Specificies of "Uzicka" sausage produced in traditional way of manufacture. *Economics of Agriculture, International Scientif Meating Multifunctional Agriculture and Rural Development (V)* (Special issue-2), 211–218.

Vukašinović, M., Kurćubić, V., Kaljević, V., Maskovic, P., & Petrović, M. (2011). Examinations of certain chemical characteristics of fermented dry sausage quality parameters. *Veterinarski glasnik, 66*, 73–84. https://doi.org/10.2298/vetgl1202073v.

Vuković, I., Bunčić, O., Babic, L., Radetić, P., & Bunčić, S. (1988). Investigation of some important physical, chemical and biological changes during ripening of kulen. *Tehnologija mesa, 29*(2), 34–39.

Vuković, I., Saičić, S., & Vasilev, D. (2011). Contribution to knowledge of major quality parameters of traditional (domestic) kulen. *Tehnologija mesa, 52*(1), 134–140.

Vuković, I., Vasilev, D., & Saičić, S. (2014). Code of practice-registered geographical indication Lemeški kulen. Intellectual Property Office, Republic of Serbia, Decision No. 990 2014/3797-G-2013/0004/9.

Vuković, I., Vasilev, D., Saičić, S., & Ivanković, S. (2012). Investigation of major changes during ripening of traditional fermented sausage Lemeški kulen. *Tehnologija mesa, 53*(2), 140–147. https://doi.org/10.5937/tehmesa1202140v.

Vuković, I., Vidaković, B., Bunčić, O., Babić, L., & Milovanović, R. (1986). Examination of the important changes occuring in the course of preparing of tea sausages containing different kinds of additives. *Tehnologija mesa, 27*(10), 297–300.

Živković, D., Radulovic, Z., Aleksic, S., Perunović, M., Stajić, S., Stanišić, N., & Radovic, C. (2012). Chemical, sensory and microbiological characteristics of Sremska sausage (traditional dry-fermented Serbian sausage) as affected by pig breed. *African Journal of Biotechnology, 11*, 3858–3867. https://doi.org/10.5897/AJB11.3363.

Živković, D., Tomovic, V., Perunović, M., Stajić, S., Stanišić, N., & Bogićević, N. (2011). Sensory acceptability of "Sremska" sausage made from meat of pigs of different ages. *Tehnologija mesa, 52*(2), 252–261.

8 Development of Innovative Dry-Cured Product from Pork Shoulder—The *Cuore di Spalla* Case Study
Salting Time Optimization According to the Crossbreed Used

Francesco Sirtori, Chiara Aquilani, and Carolina Pugliese
Department of Agriculture, Food, Environment and Forestry, Section of Animal Sciences, University of Firenze, 50144 Firenze, Italy

CONTENTS

8.1   INTRODUCTION

8.1.1   Prime Cuts in the Italian Heavy Pig Industry

The Italian pork industry is mainly based on heavy pig production, which accounts for 86% of the animals slaughtered in the country (Cozzi 2003). The slaughtering weight and the genotype selected for rearing are greatly influenced by the rules that regulate manufacturing and production of PDO Italian dry-cured ham. The latter is the most valuable product of the Italian heavy pig industry. It is estimated that hind legs of heavy pigs account for almost half of the whole value of the carcass at the slaughtering house. After the production process starts, the ham increases its value almost four times (from 80 euro for fresh product up to 330 euro for the cured ham); according to this, the total remaining parts of the carcass account for the remaining 50% of the value (Belletti and Petrocchi 2008). Among them, the shoulders might represent an untapped value. Pork shoulders can undergo, with proper adjustments, a curing process similar to dry-cured hams, or it can be suitable for manufacture by different processes, including the production of deboned products such as cured loin. The optimization of the shoulder curing process should consider its different characteristics in terms of size: thinner than a ham, it requires special attention during the salting procedure. Furthermore, its chemical composition, including moisture, lean/fat ratio of the different muscles, and its specific physical traits should be taken into account when defining new processing protocols for innovative products.

8.1.2   Effect of Genotype on Dry-Cured Products

The Italian heavy pig chain primarily produces dry-cured products. Several genetic lines, pure breeds and crossbreeds, are used for this purpose. Moreover, a small, but important aspect of production is linked to the use of local pig breeds, which are characterized by several traits that distinguish them from improved genetic lines. Eventually, even if not for PDO chains, crossbreeds between improved and local breeds are also commonly used, since local breeds are often characterized by slow growth rates and low productive and reproductive performance. The cross of local breeds with the improved ones is a widely applied breeding technique aimed to link good productivity with some quality characteristics of local breeds (Franco, Vazquez, and Lorenzo 2014).

In Italy, the local breed with the largest population is the Cinta Senese (Franci and Pugliese 2007). For the above-mentioned crosses, Cinta Senese sows were historically crossed with Large White boars (Pugliese and Sirtori 2012) to overcome some limits of autochthonous pigs related both to *in vita* performance and on carcass and meat quality traits (Sirtori et al. 2011). Some examples can be found also outside Italy, such as Duroc × Iberian in Spain (Fuentes et al. 2014) and Large White × Corsican pigs in France (Coutron-Gambotti 1999). This sets up an extremely diversified framework for the production of dry-cured products. The involvement of local breeds, in particular, has a series of technological implications that have to be addressed in order to obtain high-quality products.

Several studies have outlined differences between local and improved breed products (Franci et al. 2007; Sirtori et al. 2011; Pugliese and Sirtori 2012); when local breeds are involved in dry-cured meat production, as pure breeds or as crossbreeds, some modifications in qualitative characteristics of the raw material should be considered. Firstly, the greater adipogenicity of local breeds influences the final characteristics of the products (Sirtori et al. 2011) as well as the production process, including the diffusion of salt. Salt uptake is important in view of product stability in the initial processing phases, and it influences aroma and texture development by affecting proteolytic, lipolytic and oxidative processes (Toldrá and Flores 1998). Salt penetration in meat is linked to several factors: intramuscular fat (IMF), moisture and product size were the most important, as reported by Škrlep et al. (2016). Indeed, subcutaneous fat prevents hams from excessive moisture loss, which would lead to dehydration and, at the same time, it mediates salt diffusion inside the cured meat cut, helping prevent excessive saltiness. And IMF is an important trait to ensure the development of appropriate aroma and juiciness (Čandek-Potokar and Škrlep 2012). Bermúdez et al. (2017) observed the highest IMF content in hams from the Duroc x Celta crosses (13.92%), whereas pure-bred Celta pigs showed the lowest (8.03%). Moreover, the effects of genotype were especially marked in the first stages of ham manufacturing (fresh ham, and salting and post-salting phases mainly) for most of the examined parameters. Secondly, the higher moisture and lower size, which characterized local breeds, promote salt penetration. Low water holding capacity has been associated with lower seasoning yields as well as higher salt uptake (Čandek-Potokar and Škrlep 2012). These lead to dry-cured products characterized by high salt content (Hersleth et al. 2011).

Concerning alternative cuts to ham, such as shoulder or loin, few studies have been conducted on quality characteristics and salt kinetics as affected by different genotypes. Comparing Celta (C), Celta x Duroc (CxD) and Celta x Landrace (CxL) pigs, Lorenzo and Fonseca (2014), observed lower moisture in CxL shoulders compared to C and CxD, while, at the end of the curing process, final moisture was higher in C samples. As concerns IMF, CxD showed the highest content. The Duroc breed is largely known for its fatness and it is widely employed in crossbreeds to increase IMF. Eventually, salt content was affected by genotype immediately after salting and at the end of processing. In both phases, Celta pure breed products showed the lowest content. The same authors also evaluated the volatiles profile of cured shoulders. Genotype affected several volatile compounds and, in the final products, the amounts of alcohols, acids and esters were significantly different between genotypes. Similar results were also observed in dry-cured loins of Chato Murciano when compared to a commercial crossbred. The local breed showed lower moisture and higher IMF (Salazar et al. 2013). Moreover, dry-cured loins differed for their fatty acids profile, with Chato Murciano products showing the highest content of MUFA, especially oleic acid. Local breeds have been hypothesized to have a higher predisposition for depositing oleic acid, due to differences in *de novo* lipids synthesis and turnover (Pugliese and Sirtori 2012). Moreover, authors who compared local genotypes with crossbreed and commercial genotypes have found such differences to be less manifest in products obtained by crossbreeding between local and commercial genotypes (Elías et al. 2000; Carrapiso et al., 2003).

8.1.3 SALTING PHASE IN THE MANUFACTURING PROCESS OF DRY-CURED PRODUCTS

Dry-cured products, ham in particular, are highly consumed products around the world (Pérez-Santaescolástica et al. 2018). The curing process to which these products are subjected consists of characteristic phases through which many lipid and protein conversion reactions lead to the production of particular flavours (Bermúdez et al. 2015). The process is characterized by a stabilization phase at low temperature that includes the curing and resting steps, and a further phase of dry ageing at increasing temperatures. As for the salting phase, it involves adding salt to control the endogenous enzymatic activity, so it is responsible for the salty taste and typical texture of dry-cured ham (Flores et al. 2012).

Even if the dry-curing process is quite similar in the southern European pork chain, regional differences and differences in the manufacturing of each type of dry-cured product are common. Among several factors that can be changed, the amount of salt applied is one of the most variable. The amount of salt can be undetermined or exact. The first is mainly seen in Spain, in some parts of France and in Italy, where the product is completely covered by salt for several days. In the second procedure, the exact amount of salt per kg of fresh ham is placed on the lean surface and hand-rubbed (Toldrà 2002). The main goal of salt is to stabilize the product, from a safety point of view.

Salt as a food preservative has been used for millenia. The mechanisms by which salt preserves the product are many, including reducing water activity (a_w) and decreasing water content (Garcia-Gil et al. 2012). In recent years there has been increasing consumer attention to the nutritional composition of food. In 2003, the World Health Organization (WHO), recommended the limits for the number of calories that should be supplied by different compounds, including salt. In 2012, the WHO recommended a reduction of salt intake as this is considered one of the main dietary risk factors for cardiovascular disease and various other diseases (Inguglia et al. 2017). Recently, the European Union (EU) was forced to implement salt reduction initiatives because of its negative effect on hypertension (Corral et al., 2013). Especially in southern Europe, where the traditional seasoning methods are characterized by the use of large amounts of salt (Pugliese and Sirtori 2012). This concern has led to the development of new recipes to reduce the quantity of salt in products.

Over the years, salt reduction strategies have been many and varied. At first, complete replacement with other chloride salts was attempted, resulting in negative effects on texture, flavour, appearance and conservation (Inguglia et al. 2017). In recent years, producers have tried to reduce the quantity of salt by changing the parameters of the technological process, modifying the length of the entire process and, at the same time, using animals with particular characteristics of the raw material (amount of fat, moisture, weight of product). Salt penetration in meat is affected by many characteristics of the raw product: size (whole or portioned), moisture, fat, etc. (Sánchez-Molinero and Arnau 2008; Gou et al. 2008). These factors affect the diffusion process where the salt absorption is strictly linked to water loss caused by differences in concentration and osmotic pressures (Raoult-Wack 1994). Diffusion is a passive transport system, (no energy is requested) in which a solute (salt) contained in a solvent (water) passes from an area of high concentration to an area of low concentration. In the opposite direction, water comes out due to osmosis, and this phenomenon is influenced by several factors (Sirtori et al. 2019). Temperatures during the entire seasoning process play an important role in the penetration of the salt, and a high temperature before salting and the consequent dryness of surface can limit the salt diffusion. On the other hand, a low temperature in the salting phase leads to the crystallization of the salt, which then hardly diffuses. Other affecting factors are moisture, fat content and size. Since water acts as a carrier for salt, higher levels of moisture promote salt diffusion while a greater amount of fat (characterised by low moisture) hinders salt diffusion (Sirtori et al. 2019).

8.1.4 Influence of Salt on Final Product Characteristics

Given the multiple functions of NaCl, the reduction of salt, required by consumers and health institutions, must take into account its influence on quality characteristics, both sensorial and technological (Corral et al. 2014). Salt has been proved to have an important role in developing optimal texture through proteolysis (Purriños et al. 2012; Harkouss et al. 2015), as well as in participating in the formation of flavours and aromas due to its lipids' pro-oxidant role during curing (Andrés, Cava, and Ruiz 2002; Purriños et al. 2012). Salt must therefore be reduced without modifying the chemical–physical processes since it is an important inhibitor of muscle proteases. An excessive reduction can lead to a high proteolysis and, therefore, to an extremely soft product (Jiménez-Colmenero et al. 2010). At the same time, an excessive reduction can lead to the formation of an unacceptable aromatic profile for the consumer. The last effect is linked to the pro-oxidant capacity of the salt which, in the right doses, leads to the formation of characteristic volatile compounds. The possible cause of the pro-oxidant action of salt is attributed to many factors: capacity to disrupt cell membrane integrity facilitating the access of oxidizing agents to lipid substrates; ability to liberate iron ions; and the ability to inhibit the activity of antioxidant enzymes (Mariutti and Bragagnolo 2017). In dry-cured products, besides its function in developing sensory and technological properties, salt penetration is pivotal to ensure product safety and stability over time (Martuscelli et al. 2017). As mentioned above, the primary antimicrobial action of the salt is the reduction of water and, consequently, of a_w. Another inhibited function is given by the sodium ion of the salt that leads to: a) water efflux through the semipermeable membrane of bacteria, causing an osmotic shock which may result in the injury or death of bacterial cells; b) limited oxygen solubility; and c) interference with cellular enzymes. It also forces bacterial cells to expend energy to exclude Na^+ ions (Inguglia et al. 2017).

Salt content is the major factor determining product acceptability for consumers. The reduction of salt must be linked to maintaining product acceptance, sometimes masking the reduction with the addition of agents such as spices or other types of salts that do not contain the sodium ion (Kloss et al. 2015).

8.2   CUORE DI SPALLA CASE STUDY

8.2.1   INTRODUCTION TO INNOVATION

Research has already focused on processing pork shoulder to obtain a cured product similar to dry-cured ham. Hallenstvedt et al. (2012) examined the quality characteristics of fresh shoulder intended for cured production. Also, Delgado et al. (2002) and Sárraga et al. (2007) investigated the quality of pork products obtained from dry-cured shoulder which, in this case, was manufactured as whole piece, including the bone (similarly to dry-cured ham production). However, products obtained by deboning and stuffing of shoulder muscles require more investigation, even though, in the Italian tradition, some successful products obtained by this technique are present. For instance, the Culatello di Zibello, made from the hindquarters' internal muscles of ham, has obtained a PDO (CE n°1263/96) and accounts for 73,000 pieces/year produced by both small manufacturers and industrial producers (Tregear et al. 2007). The innovation proposed for Cuore di Spalla manufacturing lies in the employment of the muscles from the, forelimb which have humerus and scapula bones. These muscles, once trimmed of most of the external fat, are stuffed into synthetic cases. Salting is carried out using a mixture of salt, black pepper and other minor spices. The introduction of this kind of new pork product is intended to have multiple positive effects on local pork chains, offering complementary production to the manufacturers, a high-quality product without synthetic curing agents and with a reduced fat content to consumers, and the opportunity to diversify their offerings to retailers.

8.2.2   MANUFACTURING AND ANALYSIS

In the specific case of study discussed in this chapter, two different crossbreeds were used. Of 28 castrated males, 14 belonged to the Duroc x Large White (DxLW) genotype, and 14 were crossbred between Cinta Senese and Large White (CxLW). Cinta Senese was tested in crossbreeding in order to add a further value trait to the innovative product, strengthening its link with the territory by using a local breed. Eventually, two salting times were tested in order to properly consider their variability due to the different sizes of the products (linked to different genotypes) as well as the effects of fat trimming and of stuffing on salt diffusion.

The animals were reared in the same conditions and they were slaughtered at the same age, but at different live weight, due to the lower growth performance of the Cinta Senese crossbreed (143±15 kg for DxLW and 122±15 kg for CSxLW). Shoulders were deboned, trimmed and stuffed. Products obtained from the right shoulders were salted for 3 days (Low salt: L), while the left shoulders were salted for 5 days (High salt: H). After salting, the products were washed, stuffed into synthetic casings and rested in a chamber for 4 weeks (T 4–6°C, RH 70–85%). After resting, products were dried at 14–20°C and 60–80% RH for 5 months (Figure 8.1).

Parameters related to product quality were assessed on the final products: instrumental color parameters (L*, a*, b*) and texture profile analysis (TPA) were measured on a slice of each product. For TPA, the following parameters were calculated: hardness (N), adhesiveness (J), cohesiveness (dimensionless), springiness

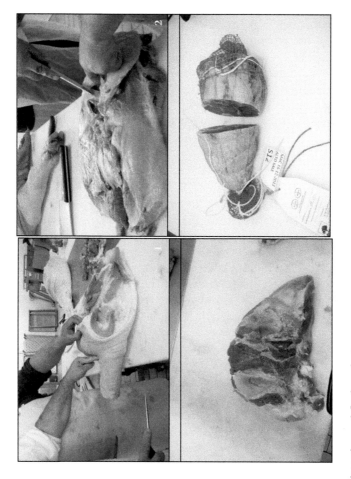

FIGURE 8.1   Shoulder dissection and final product.

(mm) and chewiness (J) (Ruiz-Ramírez et al. 2005). Chemical characterization was performed through the assessment of moisture, intramuscular fat (IMF) and protein content. The fatty acid profile was also examined (Folch et al., 1957; Morrison and Smith 1964). Results were expressed as grams of FAs per 100 g of dry product.

Volatile compounds (VOCs) were analyzed using the SPME-GC-MS technique. VOCs were identified by matching EI mass spectra against NIST 05 or the Wiley 07 spectral library and Kovats indices. A detailed description of the method can be found in Pugliese et al. (2010).

The obtained products were finally evaluated by a trained panel of 10 judges, who tried 2 slices (2 mm thick) of each Cuore di Spalla. Panelists scored 12 sensory traits using a 10 cm unstructured scale (ranging from *very low* to *very high*) according to a quantitative–descriptive method. The 12 parameters were grouped by appearance (visual intramuscular fat, color intensity and oiliness), texture (hardness), aroma (aroma and after-flavor) and taste (saltiness, after-taste, rancidity, chewiness, juiciness and persistence).

Data were analyzed by analysis of variance using the GLM procedure (SAS 2012).

8.2.3 Results

8.2.3.1 Changes in Quality Traits Due to Genotype Effect

Deboned and stuffed products belong to the Italian tradition; however, traditional manufacturing processes have been progressively lost through the decades. In fact, however, they might be hardly applicable today, clashing with new technologies, production ideas and market demand. Following the latest requests for low-fat, low-salt and preservative-free pork products, the pork processing industry has become interested in new products able to address multiple requirements, but also able to maintain high quality standards. According to the ever-changing concept of quality, if some decades ago very lean products were strongly preferred, today the fat tissue, within healthy limits, is regaining interest. Today, fat is recognized for its importance in the qualitative characterization of products, especially seasoned ones. This led to a resurgence interest in crossbreeding in several regions, especially in southern Europe. There local breeds are crossed with selected breeds to take advantage of additive and non-additive genetic variances to counterbalance some of the limits of both autochthonous pigs and of improved pigs regarding the fat amount (Pugliese and Sirtori 2012). Likewise, the salt content of processed pork products has raised concerns about their consumption, but as with the fat content, salt has important functions in the development of quality traits. So the comparison between two crossbreeds and two salting times has been tested to see if it can meet the current demands of modern consumers who require products that are healthier, such as having lower salt content, but who do not want to lose the sensory profile they are used to.

The genotypic comparison carried out in the production of Cuore di Spalla showed how different breeds can differently contribute to the characteristics of the product. The use of Cinta Senese and Duroc breeds as paternal lines led to a clear distinction between the two products (Table 8.1).

TABLE 8.1

Genotypic Comparison Carried Out in the Production of the "Cuore di Spalla"

	Genotype		
	CSxLW	**DxLW**	*P-value*
Initial weight g	2500.0	3098.1	**
Final weight g	1484.3	1857.1	**
Seasoning loss %	40.85	40.23	ns
Moisture %	47.12	45.25	*
Crude protein %	59.92	58.24	ns
Ether extract %	17.92	24.49	**
Ash %	21.10	16.77	**
Salt %	18.43	13.43	**

The use of Duroc in crossbreeding resulted in a product with greater size and an even distribution of intermuscular fat compare to a local breed, such as Cinta Senese, in pure breeding. The contribution of the Duroc breed on IMF content is well known (Lebret et al. 2011; Sirtori et al. 2011). Concerning product size, using local breeds in crossbreeding often led to a limitation of the animal growth, which resulted in a smaller-size product. Cinta Senese is considered a tardive breed; therefore, at the same age as the improved breeds, it shows lower performance and lesser carcass maturation. The recorded differences in IMF and the size and moisture of products highlighted the importance of these parameters on an important issue for consumers, the salt content. Indeed, salt content is negatively correlated with IMF and positively with moisture, (Škrlep et al. 2016) the first being a barrier to penetration (Martuscelli et al. 2017) and the second parameter representing the carrier via which the salt penetrates. Furthermore, the increase in the product's size causes an extension of the diffusion time.

Consumers' purchase decisions consider some primary visual and tactile factors that always determine their choices. Colour and meat texture in pork meat are influenced by many intrinsic (for example breed, gender, age, type of muscle) and extrinsic (for example feeding, pre-slaughter handling, slaughtering) factors (Tikk et al. 2008). The use of two different breeds resulted in different colorimetric parameters. Indeed, Duroc, thanks to its higher slaughter weight, reached greater values of IMF and myoglobin content with higher values for colorimetric parameters; in other words, Duroc in crossbreeds resulted in redder and brighter products (Table 8.2) (García-Macías et al. 1996; Latorre et al. 2003; Galián, Poto, and Peinado 2009; Suzuki et al. 2005; Gjerlaug-Enger et al. 2010).

The differences between genotypes in raw material characteristics, such as IMF and moisture, could have affected the TPA (Table 8.3), but in the case of hardness, it seems irrelevant.

TABLE 8.2

Color Parameters of "Cuore di Spalla"

	Genotype		
	CSxLW	**DxLW**	**P-value**
L*	36.23	40.02	*
a*	16.10	17.81	*
b*	4.90	6.67	*

TABLE 8.3

Texture Parameters of "Cuore di Spalla"

	Genotype		
	CSxLW	**DxLW**	**P-value**
Hardness (N)	9.53	11.02	ns
Cohesiveness	0.50	0.53	*
Springiness (mm)	6.73	7.12	*
Chewiness (N*mm)	30.02	22.52	ns

The inclusion of the Duroc breed permitted a greater deposit of total lipids with a possible influence also on the fatty acids profile. The different growth rate (Candek-Potokar et al. 1998) and precocity (Franco, Vazquez, and Lorenzo 2014) compared to CSxLW, together with the genetic effect, were linked to total FAs content (Table 8.4), which was higher in the DxLW group, no matter if they were saturated (SFAs), monounsaturated (MUFAs) or polyunsaturated fatty acids (PUFAs).

Despite the differences recorded in raw meat with regard to aromatic profile (Table 8.5), genotypic effect had limited influence on the main families of volatile compounds. This result will be confirmed by the panel test shown in Table 8.6 showing few differences for the aromatic parameters. Many authors indicated aldehydes as the main family and as the major contributors of overall aroma, due to their low thresholds (Huan et al. 2005; Ramírez and Cava 2007; Marušić et al. 2014; Lorenzo and Carballo 2014), but these results were mainly found in dry-cured ham characterized by a long dry-curing process. In this study, aldehydes were not the most abundant family. This result could be attributable to several factors, with the low level of oxidation present in the products probably the main one. Indeed, the present products had a lower maturing time with respect to ones reported in other studies (Sunesen et al. 2001; Huan et al. 2005; Ramírez and Cava 2007; Marušić et al. 2014; Narváez-Rivas, Gallardo, and León-Camacho 2014); a second reason could be the low content of PUFAs from which many aldehydes derive (Ramírez and Cava 2007; Laranjo et al. 2015).

TABLE 8.4

Fatty Acid Profile of "Cuore di Spalla"

	Genotype		
	CSxLW	DxLW	*P-value*
Total lipid	18.32	23.49	**
C16:0	4.40	5.69	**
C16:1	0.60	0.77	**
C18:0	1.78	2.47	**
C18:1	6.43	8.64	**
C18:2	1.03	1.46	**
C18:3	0.04	0.07	**
SFA	6.55	8.64	**
MUFA	7.20	9.59	**
PUFA	1.16	1.63	**

TABLE 8.5

Volatile Profile of "Cuore di Spalla"

	Genotype		
	CSxLW	DxLW	*P-value*
Aldehydes	*4.56*	*4.83*	*ns*
- Hexanal	0.28	0.26	ns
Acids	*6.96*	*5.68*	*ns*
- Butanoic acid	6.43	8.64	*
- Hexanoic acid-2-ethyl	1.03	1.46	*
- Octanoic acid	0.04	0.07	*
- n-Decanoic acid	6.55	8.64	*
Alcohols	*10.98*	*19.81*	**
Esters	*8.12*	*7.54*	*ns*
Ketones	*1.44*	*1.28*	*ns*
Thiols-Sulphur compounds	*0.64*	*0.63*	*ns*
Terpenes	*0.25*	*0.07*	*ns*
Furans	*0.14*	*0.18*	*
Alkanes	*0.10*	*0.13*	*ns*

This hypothesis was confirmed by hexanal amount. Indeed, hexanal is one of the major products of oxidations, and in this study, it recorded low values (Marušić et al. 2014). Therefore, hexanal value suggests a low oxidation for both the genotypes (CSxLW and DxLW). Lastly, considering the large amount of alcohol observed, the

TABLE 8.6
Sensory Properties of "Cuore di Spalla"

	Genotype		
	CSxLW	**DxLW**	*P-value*
Appearance			
Intramuscular fat	3.16	3.05	ns
Lean colour	3.98	3.42	*
Oiliness	3.25	2.73	*
Texture			
Hardness	4.20	3.29	*
Aroma			
Aroma	2.90	3.00	ns
After-flavour	2.16	1.53	ns
Taste			
Saltiness	5.07	4.43	*
After-taste	1.68	1.22	ns
Rancidity	0.13	0.20	ns
Chewiness	3.00	2.62	*
Juiciness	4.45	4.02	*
Persistence	5.21	4.84	*

lower level of aldehydes can be explained by their reduction to corresponding alcohols by alcohol dehydrogenase (Sunesen et al. 2001). Although acids can be produced in different ways, such as hydrolysis, degradation of triglycerides and phospholipids, deamination of amino acids, microbial activity (Huan et al. 2005) and aldehydes oxidation (Škrlep et al. 2016), there were limited differences between the studied genotypes. CSxLW showed lower values for butanoic, octanoic and n-decanoic acids and higher value of hexanoic acid-2-ethyl with respect to DxLW. However, the low capacity of long chain acids to influence the aroma may suggest that, between the two genotypes, only the difference found in butanoic acid can lead to different perceivable flavours. Indeed, the main acids that characterize the flavour are short chain ones (<6 carbon atoms) that have an important role on aroma due to their odour as vinegar, cheese or cucumber (Stahnke 1995) together with low threshold values.

Butanoic acid is reported to have a strong cheese flavour (García-González et al. 2008). The main family found in this product was alcohols, commonly observed in low seasoning products, as confirmed by Muriel et al. (2004). The formation of alcohols attributable to aldehydes degradation, microbial fermentation for branched alcohols and PUFAs oxidation for straight chain ones (Muriel et al. 2004) gains importance as the carbon chain of this compounds increases (García-González et al. 2008). In the present study, the total amount was almost double in DxLW.

Esters are another group of important compounds due to their low threshold values (Lorenzo and Carballo 2014), but they did not show any differences between genotypes. They contribute to the development of typical aged-meat aroma (Huan et al. 2005) and fruity odour (Zhao et al. 2017), but this is probably due to the short seasoning time, and few differences between the two experimental groups were observed. Ketones, as well as esters, were not affected by genotype and, thus, no difference is expected in the perception. Considering that high concentration of ketones is the signal of low product quality (Pastorelli et al. 2003), no noticeable difference between the genotypes were found for thiols, sulphur compounds, terpenes or alkanes. On the other hand, furans, which are considered important for desirable aroma development due to their low threshold values (Ramírez and Cava 2007), were more present in DxLW. This could be attributed to the higher amount of C18:2 as the formation of these compounds starts from linoleic acid (Muriel et al. 2004).

As displayed in Table 8.6, genotype did not influence panellists' perceptions of either olfactory or taste parameters. The genotype effects on the aromatic profile were minimal, especially for compounds with low perceptive threshold values such as aldehydes (Pugliese et al. 2010). Only saltiness confirmed the higher salt content in CSxLW, in agreement with chemical analysis. As already reported in the literature (Carrapiso, Jurado, and García 2003), some instrumental measurements often cannot be correlated with differences perceived by panellists; indeed, colour and IMF seem in contrast with the results of instrumental determination. The higher values of juiciness and persistence in CSxLW versus DxLW are attributable to the higher salt content that accentuated these perceptions for panellists (Lorido et al. 2015; Lorido et al. 2016).

8.2.3.2 Changes in Quality Traits Due to Different Salting Times

The trial suggested that is feasible to further reduce the salt content to meet consumers' demand and EU requirements. Indeed, no differences between the two salting times, either for physical–chemical or texture parameters, were observed (Table 8.7).

TABLE 8.7
Physical-Chemical and Texture Parameters of "Cuore di Spalla"

	Salting Time		
	H	**L**	*P-value*
Initial weight g	2,833.0	2,766.2	ns
Final weight g	1,688.6	1,652.8	ns
Seasoning loss %	40.66	40.42	ns
Moisture %	45.80	46.57	ns
Crude protein %	58.72	59.44	ns
Ether Extract %	20.93	21.48	ns
Ash %	19.58	18.29	*
Salt %	16.52	15.33	*
L*	37.99	38.45	ns
a*	16.51	17.40	ns
b*	5.84	5.73	ns

TABLE 8.8
Texture Parameters of "Cuore di Spalla"

	Salting Time		
	H	L	*P-value*
Hardness (N)	11.37	9.17	ns
Cohesiveness	0.51	0.52	ns
Springiness (mm)	6.96	6.89	ns
Chewiness (N*mm)	28.35	24.20	ns

The changes in texture (Table 8.8), usually caused by an increasing salt content, are correlated to dehydration, which causes meat hardening (Ruiz-Ramírez et al. 2005), and to proteolysis inhibition (Monica Flores et al. 1997; Martín et al. 1999). Since products showed the same seasoning loss for both salting times, the lack of significant differences is likely due to the high salt percentage in both products and to the reduced gap between H and L products for these parameters.

Considering the main compounds' families, the aromatic profile (Table 8.9) revealed few differences between salting times, also in this case due to shorter curing time if compared to a product such as ham. Taking into consideration the individual compounds, the few that showed differences recorded higher values in H than in the L group, which are attributable to salt's pro-oxidizing and solubilising role, as reported in other studies (Purriños et al. 2012).

Despite these differences, panellists only perceived a change in saltiness between the two product types (Table 8.10), as reported also in other research (Corral, Salvador, and Flores 2013; Škrlep et al. 2016).

TABLE 8.9
Volatile Profile of "Cuore di Spalla"

	Salting Time		
	H	L	*P-value*
Aldehydes	4.86	4.53	ns
Acids	5.86	6.76	ns
Alcohols	19.00	11.79	*
Esters	7.08	8.64	ns
Ketones	1.45	1.26	ns
Thiols-sulphur compounds	0.75	0.51	ns
Terpenes	0.27	0.06	ns
Furans	0.17	0.15	ns
Alkanes	0.11	0.12	ns

TABLE 8.10

Sensory Properties of "Cuore di Spalla"

	Salting Time		
	H	**L**	*P-value*
Appearance			
Intramuscular fat	2.95	3.31	ns
Lean colour	3.71	3.62	ns
Oiliness	3.05	2.95	ns
Texture			
Hardness	3.85	3.53	ns
Aroma			
Aroma	2.78	3.09	ns
After-flavour	1.82	1.88	ns
Taste			
Saltiness	5.15	4.34	*
After-taste	1.37	1.53	ns
Rancidity	0.17	0.17	ns
Chewiness	2.79	2.79	ns
Juiciness	4.26	4.21	ns
Persistence	5.09	4.95	ns

8.3   CONCLUSION

The study found that is possible to valorize pork shoulder despite its being considered as a secondary product of the heavy pig production chain. This is made possible by developing an innovative manufacturing process, which takes into account the effects of different genotypes on product characteristics. Optimization of processing also considered two possible salting times, according to the expected differences in size and IMF of products obtained from two different crossbreeds, one of which involved a local breed. Indeed, even if the use of Cinta Senese in crossbreeding did not affect the aromatic profile of the new product, it affects the physical–chemical parameters, size and IMF content. On the other hand, genotype was irrelevant to the chosen salting times. Salting time affected only the saltiness attribute in sensory evaluation, and the volatile profile between high-salt- and low-salt-level products showed similar results. However, both the tested salting times led to high salt content in both products. Thus, it seems feasible to produce the Cuore di Spalla with further reduced salting times, resulting in a positive influence on health providing that product safety is guaranteed.

REFERENCES

Andrés, A. I., R. Cava, and J. Ruiz. 2002. "Monitoring Volatile Compounds during Dry-Cured Ham Ripening by Solid-Phase Microextraction Coupled to a New Direct-Extraction Device." *Journal of Chromatography A* 963 (1–2): 83–88. https://doi.org/1 0.1016/S0021-9673(02)00139-5.

Belletti, M., and R. Petrocchi. 2008. "La Filiera Delle Carni Suine in Italia: Aspetti Produttivi e di Mercato." *Progress in Nutrition* 10 (2): 109–133.

Bermúdez, R., D. Franco, J. Carballo, and J. M. Lorenzo. 2015. "Influence of Type of Muscle on Volatile Compounds throughout the Manufacture of Celta Dry-Cured Ham." *Food Science and Technology International* 21 (8): 581–592. https://doi.org/10.1177/1082 013214554935.

Bermúdez, R., D. Franco, J. Carballo, and J. M. Lorenzo. 2017. "Sensory Properties and Physico-Chemical Changes in the Biceps Femoris Muscle during Processing of Dry-Cured Ham from Celta Pigs. Effects of Cross-Breeding with Duroc and Landrace Pigs." *Italian Journal of Food Science* 29 (1): 123–137. https://doi.org/10.14674/112 0-1770/ijfs.v436.

Čandek-Potokar, M., and M. Škrlep. 2012. "Factors in Pig Production That Impact the Quality of Dry-Cured Ham: A Review." *Animal* 6 (2): 327–338. https://doi.org/10.101 7/S1751731111001625.

Candek-Potokar, M., L. Zlender, L. Lefaucheur, and M. Bonneau. 1998. "Effect of Age at Slaughter on Chemical Traits and Sensory Quality of Longissimus Lumborum Muscle in the Rabbit." *Meat Science* 48 (1–2): 181–187. https://doi.org/10.1016/S0309-174 0(97)00088-0.

Carrapiso, A. I., A. Jurado, and C. García. 2003. "Note: Effect of Crossbreeding and Rearing System on Iberian Ham Volatile Compounds." *Food Science and Technology International* 9 (6): 421–426. https://doi.org/10.1177/1082013203040396.

Corral, S., A. Salvador, C. Belloch, and M. Flores. 2014. "Effect of Fat and Salt Reduction on the Sensory Quality of Slow Fermented Sausages Inoculated with Debaryomyces Hansenii Yeast." *Food Control* 45: 1–7. https://doi.org/10.1016/j.foodcont.2014.04 .013.

Corral, S., A. Salvador, and M. Flores. 2013. "Salt Reduction in Slow Fermented Sausages Affects the Generation of Aroma Active Compounds." *Meat Science* 93 (3): 776–785. https://doi.org/10.1016/j.meatsci.2012.11.040.

Coutron-gambotti, C, and G Gandemer. 1999. "Lipolysis and Oxidation in Subcutaneous Adipose Tissue during Dry-Cured Ham Processing" *Food Chemistry* 64: 95–101.

Cozzi, Giulio. 2003. "Meat Production and Market in Italy." *Agriculturae Conspectus Scientificus (ACS)* 68 (2): 71–77.

Delgado, G. L., C. S. Gómez, L. M. S. Rubio, V. S. Capella, M. D. Méndez, and R. C. Labastida. 2002. "Fatty Acid and Triglyceride Profiles of Intramuscular and Subcutaneous Fat from Fresh and Dry-Cured Hams from Hairless Mexican Pigs."*Meat Science* 61 1), 61–65. https://doi.org/10.1016/S0309-1740(01)00163-2.

Elías, M., J. Tirapicos Nunes(Universidade de Evora (Portugal). Area Departamental de Ciências Agrárias), and C. Sanabria. 2000. "Effect of the Genotype and of the Feeding Regime on Some Chemical and Physical Characteristics of 'Palaio' and 'Barrancos Ham.'" In *Options Méditerranéennes. Série A: Séminaires Méditerranéens (CIHEAM)*. CIHEAM-IAMZ.

Flores, M., M. C. Aristoy, T. Antequera, J. M. Barat, and F. Toldrá. 2012. "Effect of Brine Thawing/Salting on Endogenous Enzyme Activity and Sensory Quality of Iberian Dry-Cured Ham." *Food Microbiology* 29 (2): 247–254. https://doi.org/10.1016/j.fm.2011 .06.011.

Flores, Monica, Daphne A. Ingram, Karen L. Betti, Fidel Toldrá, and Arthur M. Spanierip. 1997. "Sensory Characteristics of Spanish 'Serrano' Dry-Cured Ham." *Journal of Sensory Studies* 12 (3): 169–179.

Folch, J., M. Lees, and G. H. Sloane Stanley. 1957. "A Simple Method for the Isolation and Purification of Total Lipides from Animal Tissues." *The Journal of Biological Chemistry* 226 (1): 497–509.

Franci, Oreste, and C. Pugliese. 2007. "Italian Autochthonous Pigs: Progress Report and Research Perspectives." *Italian Journal of Animal Science* 6 (Suppl. 1): 663–671. https://doi.org/10.4081/ijas.2007.1s.663.

Franci, Oreste, Carolina Pugliese, Anna Acciaioli, Riccardo Bozzi, Gustavo Campodoni, Francesco Sirtori, Lara Pianaccioli, and Gustavo Gandini. 2007. "Performance of Cinta Senese Pigs and Their Crosses with Large White 2. Physical, Chemical and Technological Traits of Tuscan Dry-Cured Ham." *Meat Science* 76 (4): 597–603. https://doi.org/10.1016/j.meatsci.2007.01.020.

Franco, Daniel, José Antonio Vazquez, and José Manuel Lorenzo. 2014. "Growth Performance, Carcass and Meat Quality of the Celta Pig Crossbred with Duroc and Landrance Genotypes." *Meat Science* 96 (1): 195–202. https://doi.org/10.1016/j.meatsci.2013.06.024.

Fuentes, Verónica, Sonia Ventanas, Jesús Ventanas, and Mario Estévez. 2014. "The Genetic Background Affects Composition, Oxidative Stability and Quality Traits of Iberian Dry-Cured Hams: Purebred Iberian versus Reciprocal Iberian×Duroc Crossbred Pigs." *Meat Science* 96 (2): 737–743. https://doi.org/10.1016/j.meatsci.2013.10.010.

Galián, M., A. Poto, and B. Peinado. 2009. "Carcass and Meat Quality Traits of the Chato Murciano Pig Slaughtered at Different Weights." *Livestock Science* 124 (1–3): 314–320. https://doi.org/10.1016/j.livsci.2009.02.012.

Garcia-Gil, N., E. Santos-Garcés, I. Muñoz, E. Fulladosa, J. Arnau, and P. Gou. 2012. "Salting, Drying and Sensory Quality of Dry-Cured Hams Subjected to Different Pre-Salting Treatments: Skin Trimming and Pressing." *Meat Science* 90 (2): 386–392. https://doi.org/10.1016/j.meatsci.2011.08.003.

García-González, Diego L., Noelia Tena, Ramón Aparicio-Ruiz, and Maria T. Morales. 2008. "Relationship between Sensory Attributes and Volatile Compounds Qualifying Dry-Cured Hams." *Meat Science* 80 (2): 315–325. https://doi.org/10.1016/j.meatsci.2007.12.015.

García-Macías, J. A., M. Gispert, M. A. Oliver, A. Diestre, P. Alonso, A. Muñoz-Luna, K. Siggens, and D. Cuthbert-Heavens. 1996. "The Effects of Cross, Slaughter Weight and Halothane Genotype on Leanness and Meat and Fat Quality in Pig Carcasses." *Animal Science* 63 (3): 487–496. https://doi.org/10.1017/S1357729800015381.

Gjerlaug-Enger, E., L. Aass, J. Ødegård, and O. Vangen. 2010. "Genetic Parameters of Meat Quality Traits in Two Pig Breeds Measured by Rapid Methods." *Animal* 4 (11): 1832–1843. https://doi.org/10.1017/S175173111000114X.

Gou, P., R. Morales, X. Serra, M. D. Guàrdia, and J. Arnau. 2008. "Effect of a 10-Day Ageing at 30 °C on the Texture of Dry-Cured Hams Processed at Temperatures up to 18 °C in Relation to Raw Meat PH and Salting Time." *Meat Science* 80 (4): 1333–1339. https://doi.org/10.1016/j.meatsci.2008.06.009.

Hallenstvedt, E., N. P. Kjos, M. Øverland, and M. Thomassen. 2012. "Changes in Texture, Colour and Fatty Acid Composition of Male and Female Pig Shoulder Fat Due to Different Dietary Fat Sources." *Meat Science* 90 (3): 519–527. https://doi.org/10.1016/j.meatsci.2011.08.009.

Harkouss, Rami, Thierry Astruc, André Lebert, Philippe Gatellier, Olivier Loison, Hassan Safa, Stéphane Portanguen, Emilie Parafita, and Pierre Sylvain Mirade. 2015. "Quantitative Study of the Relationships among Proteolysis, Lipid Oxidation, Structure and Texture throughout the Dry-Cured Ham Process." *Food Chemistry* 166: 522–530. https://doi.org/10.1016/j.foodchem.2014.06.013.

Hersleth, Margrethe, Valérie Lengard, Wim Verbeke, Luis Guerrero, and Tormod Næs. 2011. "Consumers' Acceptance of Innovations in Dry-Cured Ham: Impact of Reduced Salt Content, Prolonged Aging Time and New Origin." *Food Quality and Preference* 22 (1): 31–41. https://doi.org/10.1016/j.foodqual.2010.07.002.

Huan, Yanjun, Guanghong Zhou, Gaiming Zhao, Xinglian Xu, and Zengqi Peng. 2005. "Changes in Flavor Compounds of Dry-Cured Chinese Jinhua Ham during Processing." *Meat Science* 71 (2): 291–299. https://doi.org/10.1016/j.meatsci.2005.03.025.

Inguglia, Elena S., Zhihang Zhang, Brijesh K. Tiwari, Joseph P. Kerry, and Catherine M. Burgess. 2017. "Salt Reduction Strategies in Processed Meat Products – A Review." *Trends in Food Science and Technology* 59: 70–78. https://doi.org/10.1016/j.tifs.2016.10.016.

Jiménez-Colmenero, F., J. Ventanas, and F. Toldrá. 2010. "Nutritional Composition of Dry-Cured Ham and Its Role in a Healthy Diet." *Meat Science* 84 (4): 585–593. https://doi.org/10.1016/j.meatsci.2009.10.029.

Kloss, Loreen, Julia Dawn Meyer, Lutz Graeve, and Walter Vetter. 2015. "Sodium Intake and Its Reduction by Food Reformulation in the European Union – A Review." *NFS Journal* 1: 9–19. https://doi.org/10.1016/j.nfs.2015.03.001.

Laranjo, Marta, Ana Cristina Agulheiro-Santos, Maria Eduarda Potes, Maria João Cabrita, Raquel Garcia, Maria João Fraqueza, and Miguel Elias. 2015. "Effects of Genotype, Salt Content and Calibre on Quality of Traditional Dry-Fermented Sausages." *Food Control* 56: 119–127. https://doi.org/10.1016/j.foodcont.2015.03.018.

Latorre, M. A., P. Medel, A. Fuentetaja, R. Lázaro, and G. G. Mateos. 2003. "Effect of Gender, Terminal Sire Line and Age at Slaughter on Performance, Carcass Characteristics and Meat Quality of Heavy Pigs." *Animal Science* 77 (1): 33–45. https://doi.org/10.1017/S1357729800053625.

Lebret, B., A. Prunier, N. Bonhomme, A. Foury, P. Mormède, and J. Y. Dourmad. 2011. "Physiological Traits and Meat Quality of Pigs as Affected by Genotype and Housing System." *Meat Science* 88 (1): 14–22. https://doi.org/10.1016/j.meatsci.2010.11.025.

Lorenzo, José M., and J. Carballo. 2014. "Changes in Physico-Chemical Properties and Volatile Compounds throughout the Manufacturing Process of Dry-Cured Foal Loin." *Meat Science* 99: 44–51. https://doi.org/10.1016/j.meatsci.2014.08.013.

Lorenzo, José M., and Sonia Fonseca. 2014. "Volatile Compounds of Celta Dry-Cured 'lacón' as Affected by Cross-Breeding with Duroc and Landrace Genotypes." *Journal of the Science of Food and Agriculture* 94 (14): 2978–2985. https://doi.org/10.1002/jsfa.6643.

Lorido, Laura, Mario Estévez, Jesús Ventanas, and Sonia Ventanas. 2015. "Salt and Intramuscular Fat Modulate Dynamic Perception of Flavour and Texture in Dry-Cured Hams." *Meat Science* 107: 39–48. https://doi.org/10.1016/j.meatsci.2015.03.025.

Lorido, Laura, Joanne Hort, Mario Estévez, and Sonia Ventanas. 2016. "Reporting the Sensory Properties of Dry-Cured Ham Using a New Language: Time Intensity (TI) and Temporal Dominance of Sensations (TDS)." *Meat Science* 121: 166–174. https://doi.org/10.1016/j.meatsci.2016.06.009.

Mariutti, Lilian R.B., and Neura Bragagnolo. 2017. "Influence of Salt on Lipid Oxidation in Meat and Seafood Products: A Review." *Food Research International* 94: 90–100. https://doi.org/10.1016/j.foodres.2017.02.003.

Martín, L., J. J. Córdoba, J. Ventanas, and T. Antequera. 1999. "Changes in Intramuscular Lipids during Ripening of Iberian Dry-Cured Ham." *Meat Science* 51(2): 129–134. https://doi.org/10.1016/S0309-1740(98)00109-0.

Martuscelli, Maria, Laura Lupieri, Giampiero Sacchetti, Dino Mastrocola, and Paola Pittia. 2017. "Prediction of the Salt Content from Water Activity Analysis in Dry-Cured

Ham." *Journal of Food Engineering* 200: 29–39. https://doi.org/10.1016/j.jfoodeng.2
016.12.017.

Marušić, Nives, Sanja Vidaček, Tibor Janči, Tomislav Petrak, and Helga Medić. 2014.
"Determination of Volatile Compounds and Quality Parameters of Traditional Istrian
Dry-Cured Ham." *Meat Science* 96 (4): 1409–1416. https://doi.org/10.1016/j.meatsci.2
013.12.003.

Morrison, W. R., and L. M. Smith. 1964. "Preparation of Fatty Acid Methyl Esters and
Dimethylacetals from Lipids with Boron Fluoride–Methanol." *Journal of Lipid
Research* 5 (4): 600–608. http://www.ncbi.nlm.nih.gov/pubmed/14221106.

Muriel, E., T. Antequera, M. J. Petrón, A. I. Andrés, and J. Ruiz. 2004. "Volatile Compounds
in Iberian Dry-Cured Loin." *Meat Science* 68 (3): 391–400. https://doi.org/10.1016/
j.meatsci.2004.04.006.

Narváez-Rivas, Mónica, Emerenciana Gallardo, and Manuel León-Camacho. 2014.
"Chemical Changes in Volatile Aldehydes and Ketones from Subcutaneous Fat during
Ripening of Iberian Dry-Cured Ham. Prediction of the Curing Time." *Food Research
International* 55: 381–390. https://doi.org/10.1016/j.foodres.2013.11.029.

Pastorelli, G., S. Magni, R. Rossi, E. Pagliarini, P. Baldini, P. Dirinck, F. Van Opstaele, and
C. Corino. 2003. "Influence of Dietary Fat, on Fatty Acid Composition and Sensory
Properties of Dry-Cured Parma Ham." *Meat Science* 65 (1): 571–580. https://doi.org/1
0.1016/S0309-1740(02)00250-4.

Pérez-Santaescolástica, C., J. Carballo, E. Fulladosa, José V. Garcia-Perez, J. Benedito, and
J. M. Lorenzo. 2018. "Effect of Proteolysis Index Level on Instrumental Adhesiveness,
Free Amino Acids Content and Volatile Compounds Profile of Dry-Cured Ham." *Food
Research International* 107 (December): 559–566. https://doi.org/10.1016/j.foodres.2
018.03.001.

Pugliese, C., F. Sirtori, L. Calamai, and O. Franci. 2010. "The Evolution of Volatile
Compounds Profile of 'Toscano' Dry-Cured Ham during Ripening as Revealed by
SPME-GC-MS Approach." *Journal of Mass Spectrometry* 45 (9): 1056–1064. https://
doi.org/10.1002/jms.1805.

Pugliese, Carolina, and Francesco Sirtori. 2012. "Quality of Meat and Meat Products
Produced from Southern European Pig Breeds." *Meat Science.* https://doi.org/10.1016/
j.meatsci.2011.09.019.

Purriños, Laura, Daniel Franco, Javier Carballo, and José M. Lorenzo. 2012. "Influence of
the Salting Time on Volatile Compounds during the Manufacture of Dry-Cured Pork
Shoulder 'Lacón.'" *Meat Science* 92 (4): 627–634. https://doi.org/10.1016/j.meatsci.2
012.06.010.

Ramírez, M. Rosario, and Ramón Cava. 2007. "Effect of Iberian × Duroc Genotype on Dry-
Cured Loin Quality." *Meat Science* 76 (2): 333–341. https://doi.org/10.1016/
j.meatsci.2006.11.017.

Raoult-Wack, A.L. 1994. "Recent Advances in the Osmotic Dehydration of Foods." *Trends
in Food Science & Technology* 5 (8): 255–260. https://doi.org/10.1016/0924-2244(94)
90018-3.

Ruiz-Ramírez, J., X. Serra, J. Arnau, and P. Gou. 2005. "Profiles of Water Content, Water
Activity and Texture in Crusted Dry-Cured Loin and in Non-Crusted Dry-Cured Loin."
Meat Science 69 (3): 519–525. https://doi.org/10.1016/j.meatsci.2004.09.007.

Salazar, E., J. M. Cayuela, A. Abellán, A. Poto, B. Peinado, and L. Tejada. 2013. "A
Comparison of the Quality of Dry-Cured Loins Obtained from the Native Pig Breed
(Chato Murciano) and from a Modern Crossbreed Pig." *Animal Production Science* 53
(4): 352–359. https://doi.org/10.1071/AN12237.

Sánchez-Molinero, F., and J. Arnau. 2008. "Effect of the Inoculation of a Starter Culture and
Vacuum Packaging during the Resting Stage on Sensory Traits of Dry-Cured Ham."
Meat Science 80 (4): 1074–1080. https://doi.org/10.1016/j.meatsci.2008.04.029.

Sárraga, C., M. D. Guàrdia, I. Díaz, L. Guerrero, J. A. García Regueiro, and J. Arnau. 2007. "Nutritional and Sensory Quality of Porcine Raw Meat, Cooked Ham and Dry-Cured Shoulder as Affected by Dietary Enrichment with Docosahexaenoic Acid (DHA) and α-Tocopheryl Acetate." *Meat Science* 76 (2): 377–384. https://doi.org/10.1016/j.meatsci.2006.12.007.

Sirtori, Francesco, Riccardo Bozzi, Oreste Franci, Luca Calamai, Alessandro Crovetti, Antonio Bonelli, Doria Benvenuti, Chiara Aquilani, and Carolina Pugliese. 2019. "Effects of Genotype and Salting Time on Chemical, Physical and Sensorial Traits of a New Pig Seasoned Meat Product 'Cuore di Spalla.'" *Italian Journal of Animal Science* 18 (1): 898–909. https://doi.org/10.1080/1828051X.2019.1597645.

Sirtori, Francesco, Alessandro Crovetti, David Meo Zilio, Carolina Pugliese, Anna Acciaioli, Gustavo Campodoni, Riccardo Bozzi, and Oreste Franci. 2011. "Effect of Sire Breed and Rearing System on Growth, Carcass Composition and Meat Traits of Cinta Senese Crossbred Pigs." *Italian Journal of Animal Science* 10 (4): 188–194. https://doi.org/10.4081/ijas.2011.e47.

Škrlep, Martin, Marjeta Čandek-Potokar, Nina Batorek Lukač, Maja Prevolnik Povše, Carolina Pugliese, Etienne Labussière, and Mónica Flores. 2016. "Comparison of Entire Male and Immunocastrated Pigs for Dry-Cured Ham Production under Two Salting Regimes." *Meat Science* 111: 27–37. https://doi.org/10.1016/j.meatsci.2015.08.010.

Stahnke, L. H. 1995. "Dried Sausages Fermented with Staphylococcus Xylosus at Different Temperatures and with Different Ingredient Levels – Part II. Volatile Components." *Meat Science* 41 (2): 193–209. https://doi.org/10.1016/0309-1740(94)00069-J.

Sunesen, L. O., V. Dorigoni, E. Zanardi, and L. Stahnke. 2001. "Volatile Compounds Released during Ripening in Italian Dried Sausage." *Meat Science* 58 (1): 93–97. https://doi.org/10.1016/S0309-1740(00)00139-X.

Suzuki, K., M. Irie, H. Kadowaki, T. Shibata, M. Kumagai, and A. Nishida. 2005. "Genetic Parameter Estimates of Meat Quality Traits in Duroc Pigs Selected for Average Daily Gain, Longissimus Muscle Area, Backfat Thickness, and Intramuscular Fat Content." *Journal of Animal Science* 83 (9): 2058–2065. https://doi.org/10.2527/2005.8392058x.

Tikk, Kaja, Gunilla Lindahl, Anders H. Karlsson, and Henrik J. Andersen. 2008. "The Significance of Diet, Slaughter Weight and Aging Time on Pork Colour and Colour Stability." *Meat Science* 79 (4): 806–816. https://doi.org/10.1016/j.meatsci.2007.11.015.

Toldrà, Fidel. 2002. "Dry-Cured Meat Products." In *Publications in Food Science and Nutrition* (3rd ed.). Trumbull, CT: Food & Trition Press, Inc.

Toldrá, Fidel, and Mónica Flores. 1998. "The Role of Muscle Proteases and Lipases in Flavor Development during the Processing of Dry-Cured Ham." *Critical Reviews in Food Science and Nutrition* 38 (4): 331–352. https://doi.org/10.1080/10408699891274237.

Tregear, Angela, Filippo Arfini, Giovanni Belletti, and Andrea Marescotti. 2007. "Regional Foods and Rural Development: The Role of Product Qualification." *Journal of Rural Studies* 23 (1): 12–22. https://doi.org/10.1016/j.jrurstud.2006.09.010.

Zhao, Jian, Meng Wang, Jianchun Xie, Mengyao Zhao, Li Hou, Jingjing Liang, Shi Wang, and Jie Cheng. 2017. "Volatile Flavor Constituents in the Pork Broth of Black-Pig." *Food Chemistry* 226: 51–60. https://doi.org/10.1016/j.foodchem.2017.01.011.

9 Traditional Pork Sausages in Bulgaria
Composition and Shelf Life

Teodora Popova
Agricultural Academy, Institute of Animal Science, 2232
Kostinbrod, Bulgaria

CONTENTS

9.1   INTRODUCTION

Pork is one of the most preferred meats in Bulgaria. The consumption of pork has doubled during the last decade, and it dominates the production of red meat in

Bulgaria—71% of the total quantity produced in 2016—due to its affordable price. Pork is an important component in many processed meat products, such as sausages. The types of sausages produced in Bulgaria are as numerous as the flavors of the countries that have influenced Bulgarian cuisine.

A quick overview of the history of sausage-making in Bulgaria dates back to the 19th century, when sausages were home-made and traditionally dry-cured. At that time there were no meat grinders; the meat was chopped by cleavers on logs and manually stuffed into the casings. Only much later, during the first half of the 20th century, under the influence of foreign sausage-makers—usually Austrian, Czech and Hungarian attracted by the cheap raw materials in the country and the investments of enterprising Bulgarians—sausage-making has evolved into an industry and new European products have been introduced.

An important step in the development of the meat industry in Bulgaria, including sausage-making in the second half of the 20th century, was the introduction of a state standard (a Bulgarian state standard covering all the industrial branches) to control and guarantee the production of high-quality food including meat products.

Over the years, especially after 1990, due to drastic changes in political and economic conditions in Bulgaria, the meat industry underwent changes, as well. Often the State standard was replaced by technological documentation developed by each manufacturer. This process considerably affected meat products and their quality; a lot of new additives were introduced and applied in the meat products, which sometimes, though very rarely, led to deviations in the quality of the meat products. In an attempt to minimize the detrimental effects sometimes observed in the quality of meat products, a new standard called *Stara Planina* was developed in 2010 for the production of products made of beef, pork and poultry meat, including certain types of sausages. The technological requirements described therein were developed by the Expert Food Safety Counsel of the Bulgarian Food Safety Agency together with the Association of Meat Processors in Bulgaria. The products made under this standard should not contain soya, potato starch, fibers or mechanically deboned meat. They carry a special label with the colours of the Bulgarian national flag. The standard is voluntary for the manufacturers of meat products and sausages, but it is a serious and successful attempt to dramatically improve the produce quality. Nevertheless, manufacturers try to preserve the sausage-making traditions in Bulgaria, with its extraordinary products that have been known for centuries.

Usually the sausages made in Bulgaria contain both pork and ruminant meat, mostly beef or veal, combined in different ratios. However, some typical sausages exist that are made entirely from pork. This chapter will give an overview of the unique pork sausages that have been made since ancient times in the country, provide a brief description of their historical background and focus on the current methods of their preparation, in industry and at home, composition and shelf life. More recent characteristic pork sausages will also be described to provide sufficient information on the use of pork for high quality meat products in Bulgaria.

FIGURE 9.1   East Balkan pigs (photo credit: Assoc. Prof. Dr. Jivko Nakev).

9.2   PORK MEAT USED IN SAUSAGE-MAKING IN BULGARIA

The pork meat sold as raw meat or used for manufacturing meat products is from various breeds and hybrids. These include the Danube White breed, which is a Bulgarian breed, as well as Danish and Belgian Landrace, Danish Yorkshire and Danish Duroc, Pietrain, Camborough 23, Youna, PIC 410, Titan. The only autochthonous Bulgarian breed is the East Balkan pig, which is traditionally reared on pasture and is found in the eastern part of Bulgaria (Figure 9.1).

The meat of the East Balkan pig possesses a distinctive taste, has a marble-like texture, and is rich in intramuscular fat (Table 9.1 and Figure 9.2).

The colour of the meat is dark, and the fat is grainy and pale yellowish. The meat of the animals is highly suitable for manufacturing of traditional products, especially the dry-cured sausage called Smyadovska lukanka. Originally this sausage

TABLE 9.1

Chemical Composition of the Meat in East Balkan Pigs (Adapted from Marchev et al., 2010)

Trait,	Moisture	Protein	Fat	Ash
%	70.41–71.24	21.57–23.00	4.55–6.16	1.03–1.13

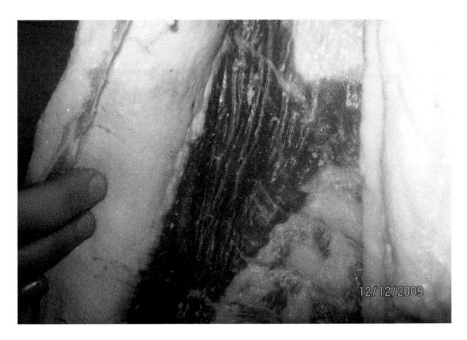

FIGURE 9.2   Meat from the East Balkan pig (photo credit: Assoc. Prof. Dr. Jivko Nakev).

was made of pork and beef meat from another local breed called Bulgarian Iskar cow, or buffalo meat, and is cold-smoked, which makes it different from the other sausages of the lukanka type. Since the East Balkan pigs are traditionally reared on pasture, the meat has a favorable fatty acid composition (Table 9.2) when compared to the intensively reared modern breeds or hybrids.

9.3 PORK CUTS

In order to get full perception of the pork sausages made traditionally in Bulgaria, a brief presentation of the pork cuts and meat used follows (Figure 9.3).

Pork in Bulgaria that will be used for further processing into sausages includes:

1. Lean meat separated from the leg, shoulder and loin; it contains up to 5% fat.
2. Semi-fat meat, containing up to 35% fat, obtained mostly from the ribs, shoulder, leg and neck.
3. Fat meat–contains up to 60% fat and is separated from the leg, shoulder, neck, belly and ribs.
4. Sort meat–includes the muscle and adipose tissue from the whole carcass, with removed bloody parts and tendons, no fat and internal adipose tissue.
5. Meat separated after deboning of the whole neck and removing bloody parts, tendons, nerves and trimmed hanging pieces.
6. Meat separated from loin, cleaned of excessive fat and pieces of bone.

TABLE 9.2

Fatty Acid Composition of Meat and Back Fat of East Balkan Pigs (Adapted From Popova and Marchev, 2015)

Fatty Acids, %	Inner Layer of the Back Fat	Outer Layer of the Back Fat	*m. Longissimus dorsi*
C14:0	1.20	0.95	1.08
C16:0	17.47	16.84	20.35
C16:1n-7	1.81	1.22	3.58
C18:0	3.25	4.41	5.79
C18:1n-9	52.79	53.66	49.81
C18:2	20.27	20.13	14.36
C18:3	2.28	1.84	1.14
C20:2	0.44	0.42	0.34
C20:3	0.10	0.08	0.17
C20:4	0.09	0.20	2.77
C20:5	0.07	0.07	0.23
C22:5	0.20	0.15	0.37
SFA[1]	21.93	22.20	27.22
MUFA[2]	54.61	54.89	53.39
PUFA[3]	23.46	22.90	19.38
PUFA/SFA	1.08	1.04	0.72

Notes
[1] Saturated fatty acids;
[2] Monounsaturated fatty acids;
[3] Polyunsaturated fatty acids

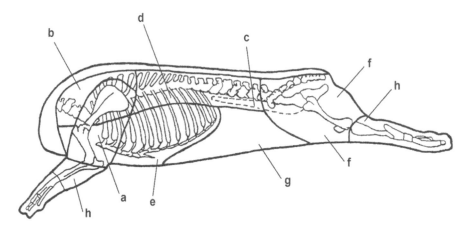

FIGURE 9.3 Pork carcass and cuts (Marinova and Popova, 2011) a–shoulder, b–neck, c–sirloin, d–loin, e–ribs, f–leg, g–belly, h–shank.

7. Ribs and bones for ragout–contain about 45% muscle. Separated from the ribs after deboning of the loin, shoulder and neck.
8. Skin and other wastes–rough tendons, muscle edges and the skin from scalded carcasses.
9. Firm fat–the outer layer of the back fat, the fat on the legs, shoulders and jowls.
10. Semi-firm fat–includes the inner layer of the back fat and thus from the ribs.
11. Soft fat–the fat from the belly as well as the intermuscular fat.
12. Bloody meat–separated at the point of slaughter, includes the bloody trimmings of the jowls and neck as well as trimming from sore tissue.
13. Sort meat with fat–includes the meat and fat from skinned halves without leaf-fat and subcutaneous fat with up to 3 cm without the meat from the legs, loins, fillets, neck or hock, cleaned of bloody parts and tendons.

9.4   BULGARIAN TRADITIONAL PORK SAUSAGES

A large variety of traditional pork sausages exists in Bulgaria. Of particular interest is the group of dry-cured products, which is the largest group and can be further separated into *pressed* and *not pressed*. Pressing is the process that unites the unique traditional Bulgarian dry-fermented sausages and other products. Pressing causes the flat shape of the products, favours drying, and improves and consolidates the links between the meat particles. Bulgarian sausages are usually named after the geographic areas from which they originate. These names have been preserved; however, with industrial manufacturing, the influence of the geographic factor on the quality of the products has been eliminated.

9.5   DRY CURED SAUSAGES LUKANKA

Lukanka is a traditional dry-cured product that has been tightly connected to the life of Bulgarians and has a centuries-long history. Through information from the International Commercial and Industrial Almanac, we learn that in 1909 there were 56 manufacturers and traders of lukanka officially registered (Tsachev et al., 1999). Along with families producing sausages with their own raw materials on tens of thousands of farms in Bulgaria, in the above-mentioned period lukanka has also been produced in established artisans' shops.

During the years, four main regions for lukanka manufacturing have been established, as presented in Figure 9.2 (Tsachev et al., 1999). The first is the northern part of Bulgaria—Troyan, Gabrovo and Kotel, serving the bigger cities such as Lovech and Pleven. The second area includes the Podbalkan valley and Sredna Gora—Koprivshtitsa, Panagyurishte, Karlovo, Kazanlak and Nova Zagora that were oriented to Plovdiv. The third region was the area near the Black Sea and Eastern Bulgaria. The fourth region was Sofia and Kyustendil (Figure 9.4).

In his studies, Marinov (1975, 1976) showed that after 1910, and mostly after the Balkan wars, the production of traditional Bulgarian meat products, including lukanka, increased dramatically. In addition to the shops of the manufacturers, there were also shops from which significant amounts of meat products were delivered

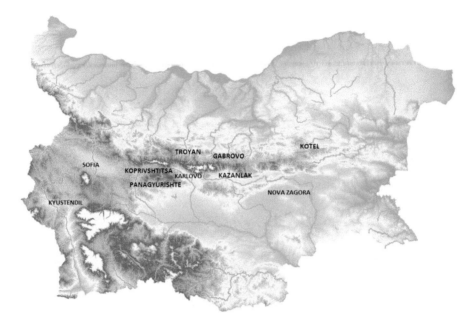

FIGURE 9.4   Main regions for the manufacturing of lukanka.

from artisans in the capital and provinces. As a consequence of the rapid increase in the population of the capital and the need for more meat products, deliveries came from all over the country.

This dry fermented sausage has a flat shape and brownish-red interior in a skin normally covered with a white fungus (Figure 9.5a).

The final product is characterized as having low water activity and a slightly acid taste. It can be stored under refrigeration (during summer) or at environmental temperatures in winter (Dimitrov et al., 2017). The mixture of fine pieces of meat and fat gives the interior of the sausage a grainy and mosaic-like structure (Figure 9.5b).

Lukanka can be made of pork and veal combined in various ratios, but also using only veal or pork and spices (black pepper and cumin). The pork variety has been named Trapezitsa. It originates from the name of the historical hill, Trapezitsa, in the old Bulgarian capital of Veliko Turnovo. This name has acquired popularity. It is specific because it is used and known only in the territory of Bulgaria and has a long history and reputation. Under this name the product is popular and produced throughout the country.

9.6   RESEARCH WORK ON LUKANKA DRY FERMENTED SAUSAGE

Lukanka has drawn the interest of researchers and this is the type of dry-cured sausage that has been mainly studied. Very detailed descriptions of the technology and quality in the production of lukanka are available in a document developed by the Superior Veterinary and Control Institute for Animal Products in 1941. Nowadays, the technology of this kind of dry fermented sausage, as well as the

(a)

(b)

FIGURE 9.5   (a) Lukanka, (b) cut lukanka showing the mosaic structure of the interior.

quality and safety requirements, have been included in various manuals and text-books for students in food technology and engineering, veterinary medicine, etc.

Certain manuals include the technology of home-preparation of lukanka in some regions in Bulgaria. This proves the long-lasting traditions and the characteristics of the composition and the preparation method of this kind of dry fermented sausage in the different geographical regions in Bulgaria. In 1950s–1960s when climatic dryers were introduced in meat processing factories all over Bulgaria, the production of this dry-cured product became year-round and lost its seasonal and regional character. The composition and the quality parameters of the product have been controlled by standards, and it has been produced successfully in Bulgaria without any influence of geographical factors on its traits.

Increasing the quality of lukanka as a desirable and traditional dry-cured product has gained the attention of researchers in meat science. Numerous publications provide thorough information about the importance of factors such as the selection of raw materials; meat production; mechanical treatment; the inclusion of curing materials; stuffing, draining, smoking and drying; pressing; the application of starter cultures; and the possibilities for prolonged shelf life.

Studies on the physicochemical traits and some quality problems date back to the 1960s. In 1964, Zlatev and Dimitrova published a profound analysis of the reasons for quality deterioration of Lukanka, as well as how to minimize them. The authors described the excess of fat that appeared on the surface, the wrinkling of the casings, excessive mould growth, mushy product, very soft texture, etc.

The colours of the exterior and the cut surface are important quality traits of lukanka sausage. It is known that the appearance and the colour of the product considerably affect the perception of the consumer. Colour is the trait on which the product is evaluated. In this regard, in the 1970s, Bulgarian scientists (Lalov and Danchev, 1972; Danchev et al., 1975) carefully studied the possibilities for objective evaluation of the lukanka colour, applying different methods and devices, as well as some technological processes influencing the colour formation.

Studies on the physicochemical processes during manufacturing, and particularly lipolysis and autoxidation, were begun in 1965 by Petrov. Only recently have there been studies describing the possibilities of restricting oxidative processes through various antioxidants applied in lukanka sausages (Dragoev and Balev, 2004; Dragoev et al., 2004; Balev et al., 2005), emphasizing the application of natural antioxidants (herbs and mixtures of herbs). The authors have reported a significant decrease in the formation of primary and secondary products of oxidation in Lukanka-type sausages due to the presence of antioxidants.

The studies on the dynamics of the microbiological changes and the microflora of lukanka sausages cover a vast and interesting complex of factors and processes that affect the quality of the final product. The microorganisms in the meat and spices in the production of lukanka sausage play a decisive role in the formation of the unique flavour and aroma of this product. Borpuzari and Boschkova (1993) studied the distribution of microflora in lukanka emphasizing the *micrococcaceae*, aiming to isolate and select strains of micrococci for possible use as starter cultures in the industrial production of lukanka. Of 354 strains, the authors classified 213 as micrococci and 141 as staphylococci.

Later, Dimov et al. (2010) reported more than 200 isolates of the *Lactobacillus* genera that were isolated during the different stages of the fermentation process in naturally fermented lukanka. The isolates were taxonomically identified by 16S rRNA and the majority were classified as being related to *Lb. plantarum, Lb.brevis and Lb. sakei.* Research into the production of lactic acid, determination of pH and screening for producers of antimicrobials showed that one of the most important roles of *Lactobacillus* strains in this process is their acidifying properties for the preservation of the final product, depending on the total cell number, and of the acidifying properties of the particular strains.

The safety requirements and control during the production of such a delicate product as the lukanka sausage have always been very high. The batches are subjected to strict sensory, physicochemical and microbiological control.

9.7   COMPOSITION AND TECHNOLOGY OF LUKANKA TRAPEZITSA

The raw materials and composition for the manufacturing of lukanka Trapezitsa are presented in Table 9.3.

Selection of the raw materials: The pork used for manufacturing of lukanka Trapezitsa must be acquired from healthy and relaxed animals. Meat from sows and boars is not allowed for the process. The pork should be cooled to 4°C.

Preparation of the meat: The deboned and sorted lean pork and pork from the ribs and belly is cut into pieces of 100–150 g. The lean meat is placed into refrigerator at 0°C–2°C for 24 hours, while the meat from the belly is frozen in freezing tunnels at –36°C for 48 hours.

Preparation of the meat batter: The lean meat is minced (d = 4 mm), whereas the meat from the belly is cut into small pieces to alleviate the work of the bowl

TABLE 9.3 Raw Materials and Composition for the Manufacturing of Lukanka Trapezitsa

1. Raw materials	Quantity
1.1. Pork lean	70%
1.2. Pork from the belly part and ribs	30%
Total:	100%
2. Curing and spices	
2.1. Salt	20g/kg
2.2. Potassium nitrate	0.4 g/kg
2.3. Sugar	4 g/kg
2.4. Black pepper	3 g/kg
2.5. Cumin	4 g/kg
3. Casings–cow large intestines or synthetic	

cutter. The general processing is done in the bowl cutter. First the frozen meat is placed in the bowl and cut to 4–6 mm particles. During cutting the curing agents and spices that have been well mixed are added. At the end, the already minced lean meat is added. When potassium nitrate is used in the process, the meat thus cut and prepared should be left to mature in the refrigerator at 2°C–4°C.

Stuffing: The meat is densely stuffed into casings that might be cow intestines. The sausages are then placed in rooms to drain and dry at 10°C–12°C, and relative humidity 80–85% for 24 hours.

Drying: The drying is carried out in drying chambers at 10°C–12°C, and at a relative humidity of 75%. During drying, lukanka sausages are pressed 3 times—on the 6th, 14th and 21st days. The pressed sausages are dried until the required water content is achieved. Pressing is done between wooden flat surfaces. During the drying process, white fungus might appear and develop on the surface of the product. It is considered to be beneficial and necessary, since it regulates the drying and even in low relative humidity it protects the sausage and slows the fast drying of the peripheral layers. In addition, the fungus positively affects the flavour of the dry fermented sausages.

The parameters of the final product, lukanka Trapezitsa, made entirely of pork, should meet the requirements listed in Table 9.2 (Figure 9.6) (Table 9.4).

9.8 BANSKI STARETS

Another group of traditional dry-cured sausages made exclusively of pork are the so called *starets*, with a famous representative known as *Banski starets*. This dry-cured sausage is produced from lean pork meat and meat acquired from the belly part of the pork carcass. Before processing, the meat should be cooled. Generally, the raw materials and composition of Banski starets is presented below (Table 9.5):

Selection of raw materials: Lean meat, meat from the belly of the carcass. The meat should be previously cooled.

Preparation of the meat batter: The boned and sorted meat with removed tendons is cut into pieces (200 g) and placed in a refrigerator at -5°C for 24 hours. The meat is then cut in a bowl cutter until the meat particles reach 18–20 mm.

Stuffing: The stuffing is densely packed into casings that might be cows' large intestines, sheep caecum or synthetic casings. No air is allowed into the stuffed casings. Any air is removed through a slight pinch immediately after stuffing. The sausages are hung and placed in a room for draining and drying at 10–12°C, at relative humidity 80–85% for 24 hours.

Drying: The drying is done in drying rooms at 10–12°C, relative humidity 75%. During the drying, the product should be pressed 3–4 times. Drying continues until the required water content is achieved (32–38%) (Table 9.6).

The same technology is applied also to a similar dry-cured sausage from the other side of Bulgaria, called *Strandzhansko dyado*. The difference between the two dry-cured sausages comes from the spices used. Usually for Banski starets the spices that are added into the meat are cumin, black pepper and chili pepper, whereas for the Strandzhansko dyado they should be chili pepper, paprika and savory.

(a)

(b)

FIGURE 9.6   (a) Banski starets, (b) Banski starets–cut.

We should note that traditionally, and when they are made at home, these two kinds of sausages are stuffed into the pork stomach or bladder (Figure 9.7).

9.9   BABEK

A typical representative of the pork dry-cured sausages is the so-called *babek*. It is similar to the above-described Banski starets and Strandzhansko dyado, but differs in the cutting of the meat raw materials. It is done through meat mincer and not in a

TABLE 9.4

Quality Characteristics of Lukanka Trapezitsa

Trait	Requirements
Shape and size	The sausages are pressed, straight or slightly bent, 40–55 cm long
External appearance and color	The sausages should have light red to dark brown colour
Cut surface	Evenly distributed pieces of pork (2–5 mm) that are white or pale pink; vno isible tendons or fasciae
Water content (% of the total weight)	28–32
Fat (% of the dry matter)	65
Protein (% of the dry matter)	28
Salt (%)	4.2
Nitrites, mg/100 g	4.5

TABLE 9.5

Composition of Banski Starets

	Quantity
1. Raw materials	
1.1. Pork lean	80%
1.2. Pork from the belly part and ribs	20%
Total:	100%
2. Curing and spices	
2.1. Salt	23 g/kg
2.2. Sodium nitrite	0.1 g/kg
2.3. Cumin	5 g/kg
2.4. Black pepper	3 g/kg
2.5. Red pepper (chili)	0.5 g/kg
3. Casings–cow large intestines, sheep caecum or synthetic	

bowl cutter as described above. Several varieties of babek that are made from pork exist, with the most popular called *Karlovski babek* (Table 9.7).

Selection of raw materials: Lean meat, meat from the belly of the carcass. The meat should be previously cooled.

Processing: The boned and sorted meat with tendons removed is cut into pieces (200 g) and placed in a refrigerator at 2–4°C for 24 hours. The lean meat is minced (grid hole size = 4 mm) and, together with the meat from the belly that has been cut into larger pieces, is put into a mixer to be combined with the curing agent, salt and spices. The meat is then refrigerated at -5–8°C to become firm for a period of 18–20 hours. Furthermore, the meat thus prepared and firm is subjected to mincing through

TABLE 9.6
Quality Characteristics of Banski Starets

Trait	Requirements
Shape and size	The sausages are pressed, straight or slightly bent, 30–50 cm long
Exterior and colour	The sausages should have dark red colour
Cut surface	Meat should have pink to dark red colour, fat should be while or pale pink; the meat and fat particles should be 18–20 mm
Water content (% of the total weight)	32–38
Fat (% of the dry matter)	60
Protein (% of the dry matter)	28
Salt (%)	2.5–3.0

a grid with 10 mm holes and then is stuffed into casings. It should be noted that the typical babek produced in non-industrial conditions is made with meat that is chopped by a cleaver.

Stuffing: The stuffing is densely packed into sheep caecum casings, or sometimes synthetic casings. No air is allowed into the stuffed casings. Any air is removed through a slight pinch immediately after the stuffing. The sausages are hung and placed in a chamber for draining at 13–15°C, relative humidity 75–80%, for 24 hours.

Drying: After draining, the sausages are transferred into drying rooms at 10–12°C, relative humidity 78–85 %. After 1–15 days, the relative humidity is reduced by 2–3%. During the drying, the product should be pressed 3–4 times. Drying continues for 100–120 days (Table 9.8).

There are several types of babek sausages, largely differentiated by the spices used to flavour them. Most popular are babek Madara (it is a fast maturing, dry-cured sausage of the babek type, seasoned with both ground black pepper and black pepper grains, paprika, chili, cumin, savory and garlic, and cold-smoked for 2–3 days at 16–22°C, 70–80% relative humidity) and babek Trakiets (seasoned with cumin and red pepper).

The described dry-cured products should be stored at –1–5°C, RH 75–80% for 60 days.

9.10   COOKED AND COOKED SMOKED SAUSAGES

The cooked and cooked smoked sausages form large groups of sausages and salamis manufactured in Bulgaria. Most of the products include a combination of pork and beef, however some entirely pork representatives are worth mentioning here. The cooked and cooked smoked sausages are not traditional. In comparison with the dry-cured products, their manufacture started relatively late under the influence of foreign sausage makers. Despite this, the cooked and cooked smoked sausages have

(a)

(b)

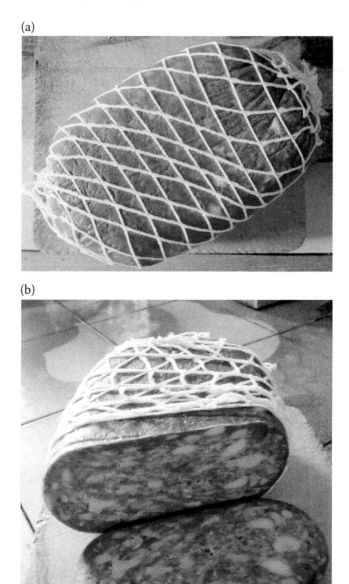

FIGURE 9.7   (a) Babek, (b) Babek–cut.

TABLE 9.7
Composition of Karlovski Babek

1. Raw materials	Quantity
1.1. Pork lean	80%
1.2. Pork from the belly part and ribs	20%
Total:	100%
2. Curing and spices	
2.1. Salt	25 g/kg
2.2. Potassium nitrate	0.4 g/kg
2.3. Sugar	1 g/kg
2.4. B lack pepper	4 g/kg
2.5. Red pepper (chilli)	1 g/kg
2.6. Paprika	4 g/kg
2.7. Fennel seeds	3 g/kg
3. Casings–sheep caecum	

TABLE 9.8
Quality Characteristics of Karlovski Babek

Trait	Requirements
Shape and size	Flat, with the natural shape of the casings
External appearance and colour	The sausages should have red to brown red colour
Cut surface	Meat should have red to light brown colour, with evenly distributed fat pieces
Water content (% of the total weight)	40
Fat (% of the dry matter)	65
Protein (% of the dry matter)	28
Salt (%)	4.6

been produced for a long time in Bulgaria. These kinds of sausages are very popular and very often found on the tables of Bulgarian consumers.

9.11 COOKED SAUSAGES

In this group we may include a large variety of frankfurters, linked sausages and salamis with larger diameter. They have a short shelf life. The frankfurters, despite their German name, are quite popular in Bulgaria. A typical pork variety is called Strandzha after a mountain in the southeastern part of the country. In the past, the

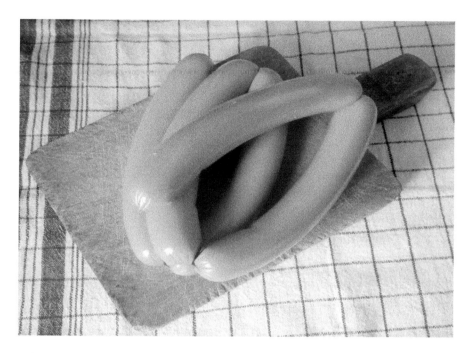

FIGURE 9.8   Krenvirsh Strandzha.

frankfurters, called *krenvirsh* in Bulgarian, were stuffed into sheeps' small intestines; today, however, synthetic casings are mostly used (Figure 9.8).

The components of the sausage are presented below (Table 9.9):

Selection of raw materials: The meat used for preparation might be fresh, cooled or frozen.

TABLE 9.9
Composition of Krenvirsh Strandzha

1. Raw materials	Quantity
1.1. Pork semifat	70%
1.2. Pork lean	30%
Total:	100%
2. Curing and spices	
1. Salt	22 g/kg
2. Sodium nitrite	0.1 g/kg
3. Black pepper	4 g/kg
4. Nutmeg	1 g/kg
3. Casings–sheep small intestine, synthetic	

Processing of the meat and preparation of meat batter: The meat, together with cold water, ice, curing agents and spices, is put into a bowl cutter until a batter with fine texture is obtained. For a better blending of the components of the batter, after the cutting, the batter is processed with a microcutter.

Stuffing: The ready batter is stuffed into sheeps' small intestines or proper synthetic casings and linked through twisting. The sausages are then treated at 80–95°C until the casings are a pale red-brownish colour.

Cooking: The sausages are cooked in tanks with water or at steam at 76–78°C until they reach an interior temperature of 72°C.

Cooling: The sausages are cooled under a cold shower until the temperature of the pieces equals the environmental temperature (Figure 9.9) (Table 9.10 and 9.11).

9.12   SALAMI KAMCHIA

Selection of raw materials: The meat used for preparation might be fresh, cooled or frozen pork acquired from healthy animals.

Processing of the meat and preparation of meat batter: The lean meat, acquired from the leg, is processed in a bowl cutter with the addition of nitrite, salt and spices. The semifat meat is cut into pieces of 500–800 g. The meat is then cut in a bowl cutter until the meat particles reach 10 mm. For improved homogenization, the batter is processed in a microcutter. If there is none available, the batter is cut for 2–3 minutes longer.

Stuffing: The ready batter is stuffed into a large intestine or synthetic casing and left to drain. The sausages are then treated at 80–95°C for 50–70 minutes until the colour of the casing is a pale red-brownish.

Cooking: The sausages are cooked in tanks with water or steam at 76–78°C for 1–1.5 hours until the interior temperature reaches 72°C.

Cooling: The salami are cooled under a cold shower for 15–20 minutes, until the temperature of the pieces equals the environmental temperature (Table 9.12).

The cooked sausages and salami can be stored up to 4 days at -1–5°C (Table 9.13).

Selection of raw materials: The pork should be acquired from healthy animals. The meat might be fresh, chilled or correctly thawed.

Processing: After removing the bones, the meat is cut into 200 g pieces. Half of the lean pork is minced through a grid (hole size = 4 mm) while the other half and the semifat meat are minced though a grid (hole size = 8–10 mm). The meat that is minced through the grid with 4 mm holes is further subjected to the bowl cutter and additionally cut. The processing continues in a mixer in the following sequence: the meat cut into the bowl cutter is put into the mixer, followed by the lean meat and then the semifat meat with the spices (chili pepper, garlic, coriander) and the curing agent. Mixing continues until the mass becomes homogenous.

Stuffing: Stuffing is dense, mainly in synthetic casings; the salami are left to drain and after that subjected to thermal treatment.

Cooking: The salami are placed in a cooking chamber at 80–95°C until the casing turns red-brown in colour and the core temperature reaches 45°. The cooking then is done at 80°C until the temperature in the core of the product reaches 72°C.

FIGURE 9.9 Kamchia salami.

Immediately after cooking the salami are smoked at 40–50°C for 6–8 hours. Smoking can be done after cooling of the product at 22°C for 1–2 days. The smoked salami are dried at 15–25°C and relative humidity of 75–80%, until the water content reaches 42% (Table 9.14).

TABLE 9.10

Quality Characteristics of Krenvirsh Strandzha

Trait	Requirements
Shape and size	Cylinder, straight or slightly curved, with equal pieces linked through twisting, 12–18 cm long and 20–22 mm in diameter
Exterior and colour	The colour of the product should be pale red-brownish
Cut surface	Light pink in colour, without tendons, fasciae or tallow
Water content (% of the total weight)	62
Fat (% of the dry matter)	70
Protein (% of the dry matter)	21.5
Salt (%)	2.2

TABLE 9.11

Composition of Salami Kamchia

	Quantity
1. Raw materials	
1.1. Pork lean	50%
1.2. Pork semifat	50%
Total:	100%
2. Curing and spices	
2.1. Salt	22 g/kg
2.2. Sodium nitrite	0.07 g/kg
2.3. Black pepper	3 g/kg
2.4. Garlic	1 g/kg
3. Casings cow large intestines, sheep caecum or synthetic	

9.13 BULGARIAN BLOOD SAUSAGES

In addition to the above-described traditional and popular sausages, pork might be used for production of blood sausages. Usually, this type of sausages includes mostly by-products such as lungs, hearts and liver, as well as blood or bloody pork meat, and are cooked. They have been known for centuries in Bulgaria and have a unique flavour. Blood sausages are closely connected to Christmas, when the pigs are slaughtered. Typical representatives of Bulgarian blood sausages are karvavitsa and bahur. The blood sausages can be stored up to 2 days at -1–5°C, RH 75–80% (Figure 9.10).

TABLE 9.12

Quality Characteristics of Salami Kamchia

Trait	Requirements
Shape and size	Cylinder, straight
Exterior and colour	The colour of the product should be pale red-brownish
Cut surface	The particles of semifat meat should be evenly distributed, with a size of 10 mm; the meat should have pink to red colour; the fats should be white to pale pink
Water content (% of the total weight)	60
Fat (% of the dry matter)	70
Protein (% of the dry matter)	19
Salt (%)	2.2

TABLE 9.13

Composition of Cooked, Smoked Salami Smyadovo

1. Raw materials	Quantity
1.1. Pork lean	55%
1.2. Pork semifat	45%
Total:	100%
2. Curing and spices	
2.1. Salt	23 g/kg
2.2. Sodium nitrite	0.05 g/kg
2.3. Red pepper (chili)	1 g/kg
2.4. Garlic	4 g/kg
2.5. Coriander	3 g/kg
3. Casings–synthetic	

9.14 KARVAVITSA

Etimologically the name of this sausage comes from the Bulgarian word for blood. Karvavitsa is very popular throughout the country (Table 9.15).

Selection of raw materials: Meat acquired from the heads of the pigs (jowls), lung, tendons, rice and spices, onion

Processing: The rice is soaked in cool water for 2–3 hours, and the tendons are cooked. The cutting is done in a bowl cutter in the following sequence: first the bloody pork, the raw lung, the cooked tendons and the spices. The raw materials are

TABLE 9.14

Quality Characteristics of Cooked, Smoked Salami Smyadovo

Trait	Requirements
Shape and size	Straight, 40–50 cm long and 46–60 mm in diameter
Exterior and colour	The surface should be clean, slightly wrinkled, brown-red colour, no moulds; slight white cover is admissible
Cut surface	No hollows, evenly distributed meat particles with pink-red colour (8–10 mm) and fat with pale yellow to white colour (8–10 mm); the texture should be elastic and dense
Water content (% of the total weight)	42
Fat (% of the dry matter)	60
Protein (% of the dry matter)	27
Salt (%)	3.3

cut until the particles are sized 3–8 mm. After that, the materials are mixed and the rice is added. Mixing continues until the components are evenly distributed.

Stuffing: The batter is stuffed into pork large intestines that are tied at both ends and pricked with a needle.

Cooking: Cooking is done in large tanks filled with water for 60–80 minutes, at 80–85°C until the internal temperature reaches 72°C.

Cooling: The cooked sausages are cooled under a shower and put into a cooling chamber.

As can be seen, the sausage described does not contain blood, however, in other varieties blood is included. It must be obtained from healthy animals and processed appropriately to meet the requirements for the blood for nutritional purposes. In this case, the cooking lasts longer, usually 2–2.5 hours (Table 9.16).

It should be mentioned that karvavitsa might be consumed warm or cold and also is used in Bulgarian cuisine for preparation of various dishes with cabbage or beans (Figure 9.11).

9.15 BAHUR

This is a traditional Bulgarian delicacy that belongs to the sausages. As with kar-vavitsa, it is also made of pork bloody meat, meat acquired from the head of the pig, lungs, spleen and heart that should originate from healthy animals (Table 9.17).

Selection of raw materials: The raw materials used for the preparation of this type of sausage should be fresh, cooled, or frozen and properly thawed meat from cooked pork heads, pork lung, pork liver, pork bloody meat and a variety of spices.

Processing: The pork heads, after being cleaned of, hair and ear epithelium are additionally washed and cooked for 2–3 hours until the meat is easily separated from the bones. The meat is then cleaned from the snout and the pigmented parts of

FIGURE 9.10 Karvavitsa.

the eyes. The lung is cleaned of bronchial branches. The liver is cleaned of bile ducts, fasciae and other undesireable particles. The pork bloody meat is cleaned of clots, and the rice is boiled at 75–80°C until slightly swollen. All the raw materials prepared as mentioned above are mixed with onion, minced in a mincer (hole size =

TABLE 9.15
Composition of Karvavitsa

1. Raw materials	Quantity
1.1. Pork jowls	50%
1.2. Lungs, hearts, spleen	36%
1.3. Tendons	10%
1.4. Rice	4%
Total:	100%
2. Curing and spices	
2.1. Salt	28 g/kg
2.2. Savory	5 g/kg
2.3. Onion	20 g/kg
2.4. Paprika	5 g/kg
3. Casings–pork large intestines	

TABLE 9.16
Quality Traits of Karvavitsa

Trait	Requirements
Shape and size	Curved as the natural shape of the casings
Exterior and colour	Clean, without damage and hollows, with gray to gray-brown colour
Cut surface	The casing should be stuck to the mea; the texture should be crumbly; the flavour should be pleasantly salty and consistent with the spices used
Water content (% of the total weight)	Up to 65
Fat (% of the dry matter)	70
Salt (%)	3

15 mm) and mixed with the salt and spices for 2–3 minutes until a homogenous mass is obtained. The processing might be done also in a bowl cutter.

Stuffing: The mass is stuffed into pork large intestines, tied in the shape of a horseshoe with a size of 20–30 cm. The sausages are pricked with a needle and cooked.

Cooking: The raw sausages are cooked in water at 85–95°C for 2–2.5 hours until the meat is soft and the internal temperature is 76°C.

Cooling: The cooked sausages are immediately cooled in cold water for about 2 minutes and then left at 4°C (Table 9.18).

FIGURE 9.11   Bahur.

TABLE 9.17
Composition of Bahur

1. Raw materials	Quantity
1.1. Pork meat from the heads (cooked)	50%
1.2. Pork lungs	20%
1.3. Pork liver	10%
1.4. Pork bloody meat	15%
1.5. Rice	5%
Total:	100
2. Curing and spices	
2.1. Salt	30 g/kg
2.2. Black pepper	2 g/kg
2.3. Red pepper (chilli)	4 g/kg
2.4. Allspice	1 g/kg
2.5. Onion	30 g/kg
3. Casings–pork large intestines	

TABLE 9.18
Quality Characteristics of Bahur

Trait	Requirements
Shape and size	Curved as the natural shape of the casings
Exterior and colour	Clean, without damage and hollows, with gray to gray-brown colour
Cut surface	Mosaic with size of the particles 10–15 mm, the casing stuck to the meat, the rice evenly distributed and well cooked; the texture should be soft and crumbly, the aroma pleasant and the flavor, slightly salty and seasoned
Water content (% of the total weight)	Up to 65
Salt (%)	3

9.16   COOKED PORK BABE

Babe is a type of cooked sausage that is quite similar to the above described karvavitsa and bahur, but is stuffed into pork stomachs. The latter has given the name of the sausage (Figure 9.12) (Table 9.19).

Selection of raw materials: Meat from pork heads, and also pork stomachs, that may be fresh, cooled, or frozen and thawed. If the meat is acquired from frozen pork heads they can have been stored for up to 6 months and the stomachs for up to 3 months.

Processing: The heads, without tongue and jowls but with the ears left on, are cleaned very well of hair and ear epithelium. The clean heads are additionally washed with clear water, the mouth is rinsed and the heads are put in brine for 2–3 days. After that period the heads are cooked 2–3 hours until the meat is easily separated from the bones. Still warm, the meat is cut to pieces 2–3.5 cm. The spices are added and mixed.

The stomachs are pierced in the wide end of the glandular part, the epithelium is turned out, and they are washed very carefully.

Stuffing: The meat, mixed with the spices, is stuffed into the stomachs and they are sewn up and pricked on the entire surface for the air to be removed.

Cooking: Thee stuffed stomachs are cooked in tanks for 60 minutes at 90°C.

Cooling: The products are placed on tables in cooling rooms at 0°C. The pieces are pressed for 8–10 hours until they are entirely cooled (Table 9.20).

9.17   TRADITIONAL PORK SAUSAGES IN SOME REGIONS OF BULGARIA

Puska is a unique pork delicacy that is made in winter and is traditional for the southern part of Bulgaria. It is made with raw pork meat stuffed into a pork bladder that is called *puska* in this part of the country, which is how it got its name. The

FIGURE 9.12 Pork babe.

method of making puska dates back a few centuries and has been passed down from generation to generation. Today, unfortunately, fewer people make their own puska since it is time- and labour-consuming. However, everyone who has tasted the sausage even once will never forget its unique taste and specific aroma. Sometimes puska can be compared with the above-described babek, Banski starets and

---

TABLE 9.19
Composition of Cooked Babe

1. Raw materials	Quantity
1.1. Pork acquired from the heads	100%
Total:	100%
2. Curing and spices	
2.1. Salt	30 g/kg
2.2. Black pepper	3 g/kg
2.3. Red pepper (chili)	3 g/kg
2.4. Garlic	2 g/kg
3. Casings–pork stomach or caecum	

TABLE 9.20
Quality Characteristics of Babe

Trait	Requirements
Shape and size	Slightly curved kidney or oval flat shape
Exterior and colour	Clean, without damage and hollows, with gray to gray-brown colour
Cut surface	Mosaic structure with size of the fat particles 10 mm, no bone particles or broth under the casing is allowed; the texture should be elastic, the aroma pleasant and the flavour slightly salty and seasoned
Water content (% of the total weight)	Up to 55
Fat (% of the dry matter)	58
Salt (%)	3

Strandzhansko dyado, however, the opinion of the experts is that puska is saltier. Some even say that this is one of the saltiest sausages in the entire world, since 40 g of salt are necessary per kg of meat when making puska, whereas in other popular sausages the quantity of salt rarely exceeds 25 g/kg of raw meat.

The typical spices used in puska are black pepper and cumin, however some manufacturers add paprika, leeks, and even brandy. Usually the meat used is semifat (30%) and lean (70%). It should be well drained and left for one night in a cool place. The next day the meat is cut into small pieces (5 cm), put into a wooden tray and salt is added. This is left overnight and stirred several times. The spices are then added and the meat is stuffed into the bladder that has been previously soaked in water and dried, The stuffed bladder should not contain air or hollows. When the bladder is stuffed, its opening is stitched and it is hung to dry in a cold and airy place for 3–4 months (Figure 9.13).

(a)

(b)

FIGURE 9.13 (a) Puska drying, (b) Puska ready for consumption (both photos: Mr. Bozhidar Kolev).

FIGURE 9.14 Dry karvavitsa.

In some cases, when stuffed, the sausage might be pricked with a needle and pressed for 3–4 days to remove the air. Usually when the weather gets warmer, the sausage is put into a chest and covered with wood ash where it might be stored several months (Figure 9.14).

FIGURE 9.15   Pacha.

9.18   KARVAVITSA (DRY)

This variety of karvavitsa is not cooked but is left to dry. It is typical in the southwestern part of Bulgaria. It is prepared from lungs, kidneys, heart, spleen and a small amount of bloody meat, acquired from the neck and belly. The seasoning per kg of meat is 25 g salt and 3 g black pepper. Additional spices that are used are mint, savory, cumin and anisette seeds, for a total of 10 g/kg meat. The meat is well mixed with the salt and spices and stuffed into pork large intestines, about 1 kg per sausage. The sausages are tied at both ends, pricked and hung to dry in an airy place. They might be consumed raw or cooked (Figure 9.15).

9.19   PACHA

Pacha is not a typical Bulgarian product, however it is a traditional Balkan dish and among the Turkish people. The word has Persian origins but comes to Bulgaria from the Turkish language (*paça*). Usually, Turkish people prepared pacha from a lamb's head and feet, and in the Caucasus it is called *hash* and prepared from beef. In Bulgaria pacha is prepared from the meat that is separated from the head of the pig after cooking. The meat is seasoned with salt, garlic, black pepper and bay leaf. The mixture is stuffed into pork large intestines and then further cooked. The ready pieces are hung for storage.

9.20 CONCLUSION

The pork sausages described above are only a small part of the meat products that are traditional for Bulgaria, however they are representative of the main groups of sausages that are manufactured in the country. Most of them have existed for centuries, especially the dry-cured sausages. Today a great variety of products has been maintained by the manufacturers, however the traditions have been well preserved and so have some of the meat products, among them products made entirely of pork (File Elena dry-cured pork loin, Trapezitsa roll or Role Trapezitsa–a pressed, dry-cured product made of pork from the neck) have been recognized by the EU and registered as traditional specialities under Regulation No 835/2014. It should be noted that the flavours of the pork sausages and products nowadays might not be exactly the same as it was in the past, but this is due to the introduction of modern breeds of pigs with different meat traits and different nutrition when compared to the old breeds. Nevertheless, they will remain preferred and traditionally consumed by the Bulgarian people.

ACKNOWLEDGEMENTS

The author would like to gratefully acknowledge Associate Professor Dr. Damyan Katsarov for his help with information about the traditional sausages, and Mr. Bozhidar Kolev, producer of and award winner for traditional puska sausage.

REFERENCES

Balev, D., T. Vulkova, S. Dragoev, M. Zlatanov, S. Bahtchevanska, 2005. A comparative study on the effect of some antioxidants on the lipid and pigment oxidation in dry-fermented sausages. *International Journal of Food Science & Technology*, 40(9), 977–983.
Borpuzari R.N., K. Boschkova, 1993. Isolation and characterization of micrococcaceae from Bulgaria raw-dried sausage. In Proceedings of the 39th International Congress of Meat Science Technology, August 1–6, Calgary, Canada, File No S6PO1. WP.
Danchev, S., M. Lalov, K. Kostov, S. Dzhevizov, 1975. Influence of some technological operations on the colour formation in dry cured sausages. *Food Industry*, 1, Sofia (BG).
Dimitrov Todorov, S., S. Stojanovski, I. Iliev, P. Moncheva, L.A. Nero, I.V. Ivanova, 2017. Technology and safety assessment for lactic acid bacteria isolated from traditional Bulgarian fermented meat product "lukanka". *Brazilian Journal of Microbiology*, 48, 576–586.
Dimov, S.G., S. Stojanovski, R. Stoyanova, N. Kirilkov, S. Antonova-Nikolova, I. Ivanova, 2010. Molecular Typing of Lactobacilli isolated from dry sausage "Lukanka": comparison of whole cell protein (WCP) versus dna-based methods. *Biotechnology & Biotechnological Equipment*, 24, sup1, 598–601.

Dragoev, S., D. Balev, D. Dontchev, M. Diltcheva, 2004. Optimization of the composition of natural antioxidants suitable for dry-fermented meat products. *Bulgarian Journal of Agricultural Science (Bulgaria)*, 10(1), 79–88.

Dragoev, S., D. Balev, 2004 An application of blend of natural antioxidants in dry-fermented sausages. In Proceedings of 50th International Congress of Meat Science and Technology, August 8–13, Helsinki, Finland, Session 3. Microbiology and Safety, pp. 59–602.

Lalov, M., S. Danchev, 1972. Possibilities for objective evaluation of the colour in dry cured sausages. *Food Industry*, 3, Sofia (BG).

Marchev, J., R. Nedeva, Zh. Nakev, S. Ivanova-Peneva, 2010. Quality and fatty acid profile of meat from pigs from the East Balkan breed reared in different habitats. *Journal of Animal Science*, 47(5), 48–56.

Marinov, M., 1975. How Bulgarian lukanka was created. Food Industry, 1, Sofia (BG).

Marinov, M., 1976. Thus the sausage industry in Bulgaria was born. *Food Industry*, 9, Sofia (BG).

Marinova, P., T. Popova, 2011. Meat and meat products. *Eniovche*, Sofia, 167.

Petrov, P., 1965. Some physicochemical processes during manufacturing of dry cured products. *Food Industry*, 2, Sofia (BG).

Popova, T., J. Nakev, Y. Marchev, 2015. Fatty acid composition of subcutaneous and intramuscular adipose tissue in East Balkan Pig. *Biotechnology in Animal Husbandry*, 31(4), 543–550.

Tsachev, H., S. Yoncheva, M. Mladenova, 1999. *From the top of XX century: past, present and future of the meat processing industry in Bulgaria*, Sofia, Amadeus company LTD, (BG).

Vasilev, K., 2003. Technology of meat products. *Matkom*, Sofia.

Zlatev, I., N. Dimitrova, 1964. On some unacceptable changes in quality of lukanka. *Food Industry*, 4, Sofia (BG).

10 Traditional Pork Sausages in Brazil
Manufacturing Process, Chemical Composition and Shelf Life

Yana Jorge Polizer Rocha, Juliana Cristina Baldin,
Heloísa Valarine Battagin, and
Marco Antonio Trindade
Universidade de Sao Paulo Faculdade de Zootecnia e
Engenharia de Alimentos, 225, Duque de Caxias Norte

CONTENTS

10.1 INTRODUCTION

According to the Brazilian Animal Protein Association (ABPA), Brazilian pork production in 2018 was 3,974 thousand tons, 16% of which was destined for export and 84% for the domestic market, with 15.9 kg/inhabitant of consumption per capita. The consumption of pork in Brazil occurs mainly in the processed forms, with 89% being through meat products (Associação Brasileira de Proteína Animal ABPA, 2019), in which are included the traditional pork sausages. Among the typical Brazilian meat products made exclusively from pork, some sausages are very traditional, including mainly Tuscan sausage (known in Brazil as *Toscana* sausage), *Calabresa* sausage and smoked loin, whose ingredients and additives, chemical composition, manufacturing processes and shelf life will be detailed in this chapter.

The creation of meat products started in approximately 1500 BCE, coming from the Mediterranean region which has a favourable climate for their maturation (Ordóñez et al., 2005). By that time, even though people did not know about microorganisms, the deterioration of food that was not consumed immediately was already known. One of the ways to avoid spoilage was to extend the meat's useful life, by mixing with salt and aromatic herbs, stuffing the minced meat into casings, and dehydrating, providing a product with a more pleasant flavour.

Meat products are defined as products prepared with meat, giblets or fats, and edible by-products from slaughter animals, plus condiments, spices and authorized additives. The quality of a meat product can be influenced by the selection of raw materials and ingredients in order to prevent damage in the processing steps and ensure the quality of the final product (Alencar, 1994; Ordóñez et al., 2005).

The main objectives of industrialized meat processing are to increase shelf life, develop different flavours, and take advantage of parts of the animal's carcass that would not be easily commercialized in raw form. Due to the high nutritional value and a large amount of water available in meat, they are easy targets for deterioration-causing and pathogenic microorganisms, making it necessary to use additives, heat, cold, and good manufacturing practices to obtain non-harmful meat derivatives for consumers (Terra, 1998). In meat products, it is necessary to use preservatives, usually synthetic, to control the oxidation and microbial proliferation that is due to the grinding (comminution) of meat causing an increase in the exposed area, and the natural presence of microorganisms. The use of additives is important for the safety and quality of the products, making them attractive to consumers and contributing to the increase of shelf life, allowing the products to remain on the market for a longer time and all year round. As well, additives need to meet some requirements for their use in food: to be necessary for manufacturing technology, to be registered with the competent organ, to be used in correct quantities, to respect the maximum required limit, to be harmless, to preserve the characteristics of the products, and to preserve the nutritional values of food. (Evangelista, 2001).

The reduction of lipid oxidation, through the addition of antioxidants, synthetic or natural, is also very important for preserving the shelf life of sausages. Some of the factors that directly influence the speed of the chemical reaction and favour lipid oxidation are the amount of oxygen present in the food (it is necessary to use vacuum packaging); the fat composition (the amount used and the type of unsaturated fatty acid influences the reaction); exposure to light (avoid direct light exposure on food displayed in refrigerated counters); storage temperature (do not leave food refrigerated at room temperatures, nor frozen for an excessive time); water activity (products with high water activity are more prone to oxidation); and the presence of pro-oxidants (metals, heme groups of myoglobin molecules, and enzymes such as lipo-oxidases can act as catalysts for reactions) (Ordóñez et al., 2005).

In general, Brazilian legislation (Brasil, 2000) allows the addition of the following ingredients and additives to traditional sausages made from pork:

Pork meat: Must represent the largest portion of the product, added in pieces or ground.

Lard: Pork fat adds flavour and changes the texture of the product.

Salt (NaCl): Its functions are to solubilize myofibrillar proteins, intensify the flavour, dehydrate and reduce water activity, exercise bacteriostatic action.

Water: Participates in the solubilization of proteins, limits the rise in the temperature of the meat mass (that is, ice water must be used) and promotes homogenization of the ingredients.

Sugar: Has bacteriostatic action and participates in the group of ingredients that gives the characteristic flavour of the products; also helps in the development of colour (caramelization) during the cooking of sausages.

Sodium nitrite and sodium nitrate: Work as preservatives and act against anaerobic microorganisms. They participate in the creation of the characteristic reddish colour of cured meat products since they attach to myoglobin, forming nitrosomyoglobin. When it comes to a cured and cooked meat product, nitrosomyoglobin is transformed into nitrosohemochrome, which gives it the typical pinkish colour. Sodium nitrate is added only to formulations of uncooked products, as it requires bacterial action to be transformed into nitrite and only thentakes part in the cure. The amount established by the National Health Surveillance Agency (ANVISA) for the additives sodium nitrite and sodium nitrate is 0.015/100 g and 0.03/100 g, respectively; the sum of both, determined as the maximum residual amount of sodium nitrite, must not exceed 0.015/100 g.

Sodium erythorbate: This antioxidant does not have a stipulated maximum-use amount.

Stabilizers: The most common stabilizers used in meat products are sodium polyphosphate, sodium tripolyphosphate and tetrasodium pyrophosphate. The maximum amount of addition allowed is 0.5/100 g.

Condiments: Aromatic vegetable substances, such as herbs and spices, with or without nutritional value, that modify the sensory characteristics of the products. They can be used whole, ground or in the form of oils and extracts.

Monosodium glutamate: A food additive approved by ANVISA, most used and known as a flavour enhancer. The maximum-use quantity allowed is not specified, according to the RDC 272/2019, meaning the necessary amount that the product requires is permitted (Brasil, 2019). This additive enhances a specific flavour, considered the fifth basic taste, called umami (Jinap; Hajeb, 2010; Alves et al., 2013).

Thickeners: Legislation (Brasil, 2019) allows the use and establishes the maximum addition of the following thickening agents: 0.5/100 g of carrageenan and 0.3/100 g of xanthan gum and carob bean gum.

Extenders: the most commonly used are cassava starch and soy proteins. The soy protein in isolated, concentrated, and texturized forms, is added to the meat products to enhance the functional characteristics and reduce costs (Feng, Xiong and Mikel, 2003; Hu et al., 2013). The cassava starch is largely used in the meat industry due to its high water-holding capacity during cooking, low gelatinization temperature, absence of odour and pigments that may interfere negatively in the functional proprieties of starch, and its low cost (Seabra et al., 2002; Peroni, 2003; Bourscheid, 2009; Zhu, 2015; Romero-Bastida et al., 2016).

Casings: Can be natural or artificial. Among natural casings, it is common to use those of porcine, bovine or ovine origin, which vary in size (diameter). Artificial ones can be of cellulose, collagen or plastic origin.

10.1.1 TOSCANA SAUSAGE

A class of meat products widely appreciated in Brazil, especially for family barbecues, is the raw sausage, known as fresh sausage, and these are the most-consumed processed meat products in Brazil (Associação Brasileira de Proteína Animal ABPA, 2019).

In Brazil, the standards for the identification and quality of sausages are defined by Normative Instruction No. 4, of March 31, 2000 (Brasil, 2000), for national or international commercialization. According to this Normative Instruction, sausages are meat products obtained from comminuted meat of different animal species, seasoned, with or without adipose tissue, with or without the addition of ingredients, stuffed into a natural or artificial casing and subjected to a specific technological process. The classification varies according to the manufacturing technology and can be grouped as fresh, dry, cured and/or matured, or cooked product. Sausages are also classified according to the composition of the raw materials and manufacturing techniques, with Toscana sausage classed as a raw and cured meat product, obtained exclusively from pork meat and fat, and with the addition of other ingredients that modify properties such as texture, conservation and flavour.

10.1.2 INGREDIENTS AND CHEMICAL COMPOSITION

According to the technical regulation of identity and quality (Brasil, 2000), fresh sausages must have the following chemical composition: 70% maximum moisture, 30% maximum fat, 12% minimum protein and 0.1% maximum calcium. Mandatory ingredients are meat from a butchery animal species, exclusively pork for *Toscana sausages*, and salt. Optional ingredients include fat, water, vegetable and/or animal protein, sugars, plasma, intended additives, aromas, spices and condiments. The addition of non-meat proteins, as aggregate protein, is allowed with a maximum content of 2.5%; however, their addition is not allowed in the Tuscan sausage (BRASIL, 2000). The use of mechanically separated meat (CMS) in all fresh sausages (raw and dried) is prohibited. For packaging, they can use natural or artificial wraps, plastic, or similar packaging, and boxes. The wraps may be protected by strange, which must be approved by the competent public agency (Brasil, 2000).

Most manufacturers of Toscana sausage in Brazil use water, salt, onion, garlic, sugar, coriander, white pepper, parsley, red pepper, oregano, nutmeg, rosemary and beet extract. They use sodium lactate or sodium carbonate as an acidity regulator. They use sodium tripolyphosphate or disodium pyrophosphate as stabilizers. They use monosodium glutamate as a flavour enhancer. The antioxidants used are sodium erythorbate and ascorbic acid. And, as a preservative, they use sodium nitrite. They use cochineal carmine as a natural colouring.

10.1.3 Manufacturing Process

Some cuts, such as pork tenderloin, ham or pork loin, and pork fat, are used for the manufacture of Tuscan sausage.

The first stage of processing is cutting the meat and fat to an appropriate size, then placing them in the grinder. Grinders are essentially an endless screw, with perforated fixed plates and knives. The meat-based raw materials break under pressure, causing deformation of the connective tissue and crushing of the muscle and fat tissues. The raw materials can be used fresh, frozen, or in both forms, depending on the type of grinder. However, the ingredients are usually ground at a low temperature, but not frozen, making them easier to cut, and reducing wear and tear on the equipment, and reducing the loss of liquid and contamination in the meat. For the sausage processing, the pork meat and fat are milled in a grinder, generally using a disc with 8 mm holes, then the ingredients are mixed according to the manufacturer. Condiments, curing salts and other additives are weighed separately and incorporated into the mixer, where the pieces of fat or lean meat, resulting from the previous step, are mixed to form a paste with uniformly distributed fragments of fat and meat.

Next is the curing process. Curing is the addition of salts to preserve, enhance the flavour, and improve the colour of the product. Usually sodium chloride and sodium nitrite and nitrate are used. The meat mass then rests for a period of 5–10 hours at a refrigerated temperature. It is customary, both in industrial and artisanal manufacturing, that the meat mass with added salts rest overnight before being stuffed into casings. Prior to this the natural casings are washed, and the mass is compacted to prevent air bubbles from forming, which contributes to oxidation during storage. *Toscana* sausages are usually stuffed into natural porcine casings of around 28 mm, followed by the stringing of segments into links of 15 cm. After these procedures, fresh sausages are vacuum-packed and stored under refrigeration at 0–7°C, or frozen at −12°C or colder.

10.1.4 Shelf Life

Tuscan sausages must be stored at refrigerated or frozen temperatures because this is a fresh sausage that has not undergone a pasteurization process and has high water activity. These sausages' shelf life is defined at the discretion of each manufacturer; in general, the recommended storage temperature is up to 7°C for refrigerated sausages, with shelf lives ranging from 15 up to 38 days. The shelf life for frozen sausages is approximately four months if kept at −12°C or colder.

10.2 COOKED AND SMOKED SAUSAGES

Among the traditional cooked and smoked sausages in Brazil made exclusively from pork are Calabrese (*Calabresa*), paio and Portuguese (*Portuguesa*). The first is a Brazilian sausage whose name originates from the name of the pepper commonly used in its composition, of the genus *Capsicum*, which, in dehydrated form, in flakes with seeds, is known as calabrese pepper. Portuguese sausage has a horseshoe shape and a pronounced garlic flavour. Paio, in turn, has a shorter length and, a larger diameter and may or may not be produced exclusively with pork. All these products are easily found in butcher shops and supermarkets in all regions of Brazil and used especially in pizzas, sandwiches, as appetizers, and in the typical Brazilian dish *feijoada*.

According to Brazilian legislation (Brasil, 2000), Calabresa sausage is defined as "the product obtained exclusively from pork, cured, with the addition of ingredients, which must have the spiced flavour characteristic of calabrese pepper, submitted or not to the process of stuffing or similar for dehydration and/or cooking, the smoking process is optional." The product is presented in two common forms: the traditional and the thin one, which has a smaller calibre. One can find it seasoned with additional ingredients, such as olives, provolone, fine herbs, arugula, parmesan and cheese. When produced without the cooking step, the Calabresa sausage is called *fresh Calabresa*, and its processing is similar to that of Tuscan sausages, varying only in the types of condiments, mainly the addition of calabrese pepper, and the length of the links (larger, or even no links, in the case of fresh Calabresa).

Portuguesa sausage is also processed exclusively with pork, cured, and with added ingredients; however, it must be cooked and smoked. The traditional characteristics of this product are the horseshoe shape, the joined ends of the buds, and the pronounced garlic flavour.

Paio, in turn, is not a sausage necessarily produced exclusively with pork, and may or may not have up to 20% of beef in its composition. It is cured, with added ingredients, and smoked. It is one of the main ingredients found in the traditional Brazilian dish called *feijoada*.

Two other very common and more accessible products, in terms of price and availability, are the sausages of the Calabresa "type" and Portuguesa "type," which are similar to the products of the names to which they refer; however, they are not produced only with pork. They must go through the cooking process, like the paio, and the use of up to 20% of mechanically separated meat (MSM) is allowed, which is generally from chicken.

10.2.1 INGREDIENTS AND CHEMICAL COMPOSITION

Brazilian legislation (Brasil, 2000) defines pork meat and salt as mandatory ingredients for Calabresa and Portuguesa sausages. The casings used can be natural or artificial. As for additional ingredients, for Calabresa and Portuguesa sausages, the addition of non-meat proteins, such as aggregate protein, is not allowed. For paio, Calabresa-type and Portuguesa-type sausages, this addition can be made, respecting the maximum content of 2.5%. In all cases, the optional ingredients are as follows:

TABLE 10.1
Physico-Chemical Characteristics of Sausages

	Fresh	Cooked	Dry
Moisture (maximum)	70%	60%	55%
Fat (maximum)	30%	35%	30%
Protein (minimum)	12%	14%	15%
Ash–dry basis (maximum)	0.1%	0.3%	0.1%

Source: Data from Brazil, "Normative Instruction No. 4, of March 31, 2000."

fat, water, sugars, spices and condiments. Dyes, flavourings, flavour enhancers, acidity regulators and stabilizers can also be used and are widely used in the production of sausages to enhance the specific sensory characteristics of each product.

Regarding physical–chemical characteristics, the maximum levels of moisture, fat and calcium and minimum levels of protein must adhere to the standards showed in Table 10.1, according to the mentioned legislation.

Products of the most traditional brands in Brazil, which are sold throughout Brazil, vary in their composition when compared to each other, due to the different ingredients used and the forms of processing, but the levels of protein and fat are around the following:

Fresh Calabresa sausage: 16% protein and 20% fat
Cooked and smoked Calabresa sausage and Portuguesa sausage: 15% protein and 20% fat
Paio, Calabresatype" and Portuguesa-type sausages: 17% protein and 28% fat.

10.2.2 MANUFACTURING PROCESS

The described cured sausages are easily found in any supermarket or butcher shop, and even in bakeries in Brazil, and artisanal manufacturing is also conventional. Processing varies little when comparing industrial and artisanal production, which follow a generic pattern (description follows). The grinding, mixing and stuffing steps are similar to the previously described manufacturing of Toscana sausage. For the stuffing, the level of contents in the casings needs to be controlled according to the capacity, to prevent wrinkling or rupture while cooking.

After stuffing, according to the pattern of each type of product, the sectioning of the sausages to be cooked are made by twisting them at regular spaces with different sizes. For the larger Calabresa and Calabresa-type sausages, the links are approximately 15 to 20 cm long, and each link weighs about 180 to 250 g. As for thin Calabresa sausages, a 15 cm link weighs about 80 g. In the case of paio, which has a shorter length (about 10 cm) and a larger diameter, the weight of each link usually ranges from 120–180 g. The characteristic horseshoe shape is also taken into account in Portuguesa and similar sausages, and each link weighs about 300 to 400 g.

The next step is cooking. Domestic ovens are used in artisanal production, but smokehouses are used for cooking and smoking operations when it comes to large-scale production. In cooking, there is a gradual increase in the temperature of the smokehouse until the internal temperature of the sausages approaches 74°C. The required time and temperature are based on the levels sufficient to cook the products without compromising the typical flavour and aroma.

The smoking process is optional for Calabresa, but mandatory for Portuguesa and paio sausages, and can occur concurrently with cooking. Products can be subjected to smoking using the traditional method or using liquid or powdered smoke. The traditional method exposes cooked sausages to products generated by the combustion of wood or pure chemical compounds, applied in proportions that will improve the sensory quality of the food. Liquid or powdered smoke can be added directly to the sausage mass in levels of 0.1–0.4%, but spraying liquid smoke on the sausage or immersing it in liquid smoke is also possible. The advantage of using this type of smoke is that adding smoke flavour takes less time. However, the final colour of the product may be slightly darker compared to that achieved by conventional smoking.

The final stage of the manufacturing process leads to packaging, which is usually plastic and uses cardboard boxes as secondary packaging. Vacuum packaging is not always used for fresh Calabresa, which is kept under refrigeration or frozen, but is used for some other sausages (cooked and smoked) so that they maintain their characteristics at room temperature.

10.2.3 Shelf Life

When compared to other cooked and smoked sausages, fresh Calabresa sausages have higher requirements concerning the mechanisms applied for preservation after packaging. These products are usually packaged in plastic and vacuum-packed. Manufacturers estimate a shelf life of 45–60 days if stored at up to 7°C, or 180 days if frozen at temperatures below –10°C.

Calabresa, Calabresa-type, Portuguesa, Portuguesa-type, and paio sausages are usually cooked and smoked, and these additional non-mandatory processes are ways of attributing a longer shelf life since they reduce the microbial load (pasteurization) and the water activity. These products are also usually vacuum-packed but have different conservation recommendations than fresh ones. Depending on the salt content and water activity, some cooked and smoked sausages can be kept at room temperature (up to 22°C) for up to 90 days (while the vacuum package remains sealed) or, after opening the package, for up to 7 days under refrigeration. However, products with a higher water activity (and less salt), even when cooked/smoked, need to be stored under refrigeration (up to 7°C).

10.3 PORK LOIN SAUSAGES

Currently, a great quantity of cooked, smoked, and raw cured cold meats is available in the national market. Among them, pork ham, smoke-cured sausages, Canadian-style loin sausage and turkey breast are the most popular, especially to prepare pizzas. In Brazil, Canadian-style loin sausage is the main ingredient of the pizza that bears its name, and the most traditional recipe uses tomato sauce, tomatoes, cheese, oregano and, of course, Canadian-style loin sausage. This sausage is also appreciated in several other dishes, such as appetizer/finger foods, snacks, sides for salads, main ingredients or as a complement to other main courses (Pizzas e Massas, 2015; Silva, 2016). Despite being very easy to find in any grocery or supermarket, and the existence of several brands that produce Canadian-style loin sausage, there are no data available about the volume of its production and consumption in Brazil.

The consumption of quality cuts of pork has increased in recent years due to the varied options, and pork tenderloin has become a more common choice in the consumer's daily diet than it used to be, when it was served only on very special occasions. It is a tender and lean kind of meat (do Amaral Souza et al., 2011; Silva, 2016). According to Magnoni; Pimentel (2007) and Falleiros et al. (2008), the pork loin is an option for peopl who need to control high blood pressure since it is low in saturated fat and cholesterol and contains higher levels of potassium than any other kind of meat. As potassium helps regulate sodium levels, which increase the retention of liquids in the organism, health professionals currently recommend pork loin in the diets of patients with hypertension (do Amaral Souza et al., 2011; Silveira, 2013; do Carmo and Rodolpho, 2019). In Brazil, the pork loin has great potential for the development of products (do Amaral Souza et al., 2011; Silva, 2016).

According to the definition given by the Code of Federal Regulations for Identity and Quality of Loin (Regulamento Técnico de Identidade e Qualidade de Lombo) (Brasil, 2000), the loin is understood, followed by any suitable specification, as the industrialized meat product obtained from the lower back of pork, sheep, and goats, with added ingredients and processed with the adequate technological procedure. This same normative instruction, on the quality and identity standards of the loin cut, contemplates the classification of existing types of loin cuts, with processing being based on the production of Canadian-style loin sausage, cooked loin sausage, seasoned loin, dried cured loin, seasoned rack, and others. This classification varies with the technology used in its production, in which the Canadian-style loin and the cooked loin are made into sausages. Both meat products, traditional pork sausages, are largely made in Brazil and will be described next.

10.4 CANADIAN-STYLE LOIN SAUSAGE

Canadian-style loin sausage is made from back bacon, also known as Canadian bacon, a popular Canadian product prepared with the loin of pork. The term *Canadian bacon* is used in the United States to differentiate it from traditional bacon (which is usually made from pork belly). The name Canadian bacon was coined to denote a cured, smoked and cooked pork loin, sold in slices or pieces. Its flavour is described as being very similar to ham when compared to other kinds of meat products, since it is obtained from a lean cut. In Brazil, this cut is made into sausages that are called Canadian-style loin sausages and flavoured to suit Brazilian tastes.

10.4.1 Ingredients and Chemical Composition

According to Brazilian legislation (Brasil, 2000), *Canadian-style loin sausage* is the name of a product obtained from the cut of pork called the loin, in a whole integral or a partial piece with added ingredients, stuffed into natural or artificial casings and processed with adequate technology. Smoking is optional. According to technical regulations for the identity and quality of such products, loin and salt are considered mandatory ingredients, whereas animal- and plant-based protein, sugar, maltodextrin, condiments, flavours and spices, and intentional additives are optional ingredients. The maximum addition of 2% of non-meat-based protein is allowed (Brasil, 2000). Generally, the Canadian-style loin sausages available in Brazil are processed with the following ingredients/additives: curing agent sodium nitrite, stabilizers (most often sodium polyphosphate, sodium tripolyphosphate and tetrasodium pyrophosphate), antioxidant sodium erythorbate, monosodium glutamate as a flavour enhancer, thickeners (often carrageenan, xanthan gum or carob bean gum), sugars, colourants (typically caramel colouring and cochineal carmine), an acidity regulator, sodium lactate, soy protein, cassava starch, salt and smoke aroma.

Brazilian legislation (Brasil, 2000) requires that the product presents the following physico-chemical characteristics (Table 10.2).

The average protein and fat content of the most relevant brands in the national market vary between 16–20% and 3–6%, respectively. Regarding the caloric value, the variation is 92.5 Kcal–125 Kcal. Consumers consider such a product to be a quality sausage, with a soft flavour, and it is perceived as a product with low calories and lower fat content (Pizzas e Massas, 2015).

Because sodium consumption is related to hypertension, heart conditions, diabetes, strokes, and kidney diseases (de Wardener and MacGregor, 2002; Dickinson and Havas, 2007; O'Flaherty et al., 2012), some countries, including Brasil, require information on the sodium content on food labels shown in milligrams (mg). The difference in sodium content in the popular brands is high, and this value varies on average between 670–1,470 mg in portions of 100 g of the product. According to the information described in these product labels regarding sodium content, and

---

TABLE 10.2
Physico-Chemical Characteristics of Canadian-Style Loin Sausage

Characteristics	Composition (%)
Moisture (maximum content)	72
Protein (minimum content)	16
Carbohydrates (maximum content)	1
Fat (maximum content)	8

Source: Data from Brazil, "Normative Instruction No. 4, of March 31, 2000."

---

according to the legislation in force concerning nutrition details, Canadian-style loin sausages available for consumers are classified as a type of food with high levels of sodium; in Brazil a food is considered to have high levels of sodium when it contains 400 mg of sodium or more per 100 g or 100 ml as it is sold. In Brazil, the main strategies proposed by the Health Ministry to reduce sodium consumption are the promotion of healthy eating habits through campaigns in the media, information provided to producers and consumers, and an agreement with the industry for the reformulation of processed foods (Nilson et al., 2012; Guarapuava, 2015).

In 2011, an agreement was made between the National Health Surveillance Agency (ANVISA), the Health Ministry, and the Brazilian Association of Food Industries (ABIA) to reduce the amount of sodium in processed food. After this agreement, several terms of commitment with different reduction goals were signed, establishing sodium reduction for meat products in 2013. Given the situation explained, the reduction of sodium, especially for sausages, was implemented or at least taken into account by the Brazilian food industries (Brasil, 2013; Martins, 2014; Brasil, 2017).

10.4.2 Manufacturing Process

This product's manufacture uses whole pieces or fractions of pork meat (loin). When the product comes from whole pieces of meat, the pieces are trimmed first, and the loin cuts are standardized according to their presentation. After that, the meat is salted and cured in a mix of ingredients called a brine solution or curing saline solution. In this step, the meat pieces are immersed in the brine/curing saline solution. However, to accelerate the curing/salting process, the brine/curing saline solution can be injected into the meat with the aid of a manual or automatic injector. The latter promotes the distribution of the curing agents, additives and ingredients quickly and more thoroughly, since the brine is pumped inside the meat through a needle or several needles. It is recommended that the injection happens in several directions to ensure uniformity and for more brine retention in the pieces of meat. For products obtained from the whole piece of meat, the injection of curing salts through the brine solution is recommended.

After the injection, the pieces must be treated mechanically, which can be carried out using a piece of equipment that disrupts the collagen fibres, and tumbles or massages the cuts to better incorporate the brine and extract the proteins. Subsequently, the pieces remain refrigerated for the curing process. The curing process must be at least 12 hours long, for a good colour distribution, but the established time varies according to the types of curing agents and forms of application (superficial or internal). After that, the pieces are stuffed and, in this case, the casings usually have a larger diameter and may be natural, cellulosic or plastic. After the stuffing, the loin sausages are usually cooked in smokehouses where they are kept for a specified time (average variation from 6 to 8 hours) going through many phases but gradually increasing the temperature until the pieces reach an internal temperature of 72°C. Smoking is optional but, usually, the technological process includes cooking and smoking (Daguer, 2009; Polon, 2010; Ščetar et al., 2010; Dalla Costa and Dalla Costa, 2014; Silva, 2016; Sousa, 2017; Fellows, 2018).

The main objective of smoking is to add the aroma, flavour and colour that are inherent to the product (Pearson and Gillett, 2012; Hui et al., 2014; Fellows, 2018; Previatti, 2019).

According to Fellows (2018), there are four types of smoking techniques: cold, in which aroma and colour are added to the food, but it is not cooked; warm smoking (25–40°C); hot smoking (60–80°C); and the dissolution of smoke composites in water that forms a smoke concentrate called liquid smoke that can be added to foods. However, the preferred techniques for processing this kind of product are hot smoking or liquid smoke. Usually, when the technological process includes cooking and smoking, these phases are divided into three stages: the first consists of drying the product, to add colour and remove surface moisture; in the second step, while the chamber temperature is high, the smoke is applied; and cooking it is the final step (Bressan et al., 2010; Dalla Costa & Dalla Costa, 2014).

The most traditional process employed in the second phase consists of exposuring the Canadian-style loin sausages directly to the smoke that is produced by burning raw wood sawdust (Rocco, 1996). The hot smoking process presents several advantages and its preservative action results of a series of factors and barrier technology, such as reduction of the moisture to decrease the product water activity, and the antioxidant and antimicrobial action of some chemical elements in smoke (Brul et al., 2003; Codex Alimentarius Commission, 2009;Gonçalves et al., 2010; Fellows, 2018). However, aromas and smoke condensates as well as liquid steam are frequenty used instead of burning wood (Álvarez-Ordóñez et al., 2008; Werlang, 2019). Some companies add liquid smoke directly to the product mass, either to ground products, or by spraying or coating whole pieces. However, some of the features of the product surface, such as the colour, may be lost, as is the bacteriostatic effect of hot smoking (Toldrá et al., 2014; Sousa, 2017; Fellows, 2018).

Finally, the pieces are cooled, packaged and stored under refrigeration. Cured products are usually packaged in pre-shaped vacuum bags or under a modified atmosphere. Canadian-style loin sausage can also be sold in vacuum skin packs, making the product look appealing, while also providing a convenient way to separate the parts if commercially offered in slices. However, in Brazil, thermoformed multilayer films are the vacuum packages most used by the food industry (Bressan et al., 2010; Daguer, 2009; Ščetar et al., 2010; Dalla Costa & Dalla Costa, 2014). This product is available to the consumer in different sizes and weights, in whole pieces or in slices.

10.4.3 Shelf Life

Each company determines the expiration dates for its own products, which vary from 60 to 90 days for whole pieces, and 45–60 days when commercialized in slices. These differences are related to the production process, type of package and storage of these products (Wensing, 2018). The microbiological stability and the safety of these products are a result of the synergistic action of several factors (Santos, 2008; Leistner; Gould, 2012). On the whole, the preservation of these products comes from the antimicrobial/antioxidant mixture of nitrites, spices, sodium chloride (salt) and phosphates, the severe heat during the production process, the antimicrobial components in the smoke, the type of package used and storage at low temperatures (Evangelista, 2005; Fellows, 2018). The smoking process associated with the use of salts (sodium chloride, sodium nitrite) and heating reduce and control microorganism growth and improve these products' shelf life (Sousa, 2017).

10.5 COOKED LOIN SAUSAGE

According to Brazilian legislation (Brasil, 2000), cooked loin sausage is a product obtained from the carcass cuts of loin, whole or in pieces, with ingredients added, stuffed in natural or artificial casings and processed with adequate technological cooking, smoked or not. The mandatory and optional ingredients established are the same as described for the Canadian-style loin sausage (Brasil, 2000). There are few differences in the ingredients usually described on the products' labels when compared to the Canadian-style loin sausage. The main differences are the use of maltodextrin, acidulants (most commonly citric acid and lactic acid) and flavourings (often clove, Jamaican pepper, chili, cinnamon and bay leaf). This product in Brazil is usually seasoned with natural spices; some of the most used are cinnamon, nutmeg, coriander, black pepper, red pepper, clove, parsley, salvia, spring onion, basil, garlic, oregano and onion (Bressan et al., 2010; Polon, 2010).

Unlike Canadian-style loin sausage, which is obtained only from the whole loin or a section, cooked loin sausages can also be made from ground meat. The production process is similar to that of Canadian-style loin sausages when composed of whole parts. However, when ground meat is used in the process, the stages differ from the ones described for Canadian-style loin sausage. For the meat industry, using ground meat in the process is more advantageous since the formulations can be developed with differeing quantities and a variety of ingredients, possibly providing a higher yield from the production process (Bonfim et al., 2015; Gadekar et al., 2015).

For the cooked loin sausage obtained from ground meat, the meat is chopped first and then minced using a meat grinder, after which the additives and condiments are added. The mixture (meat mass) is mixed long enough to obtain satisfactory homogenization and uniformity. In the products made from meat chopped or ground in bigger pieces, the primary meat should be cold (−2.2–1.6°C). Furthermore, during the procedure, relatively low temperatures help obtain particles with more defined shapes. The meat mass stays refrigerated for 12–15 hours (curing) after the mixing step. Next, the mass is stuffed into natural, cellulosic or plastic casings with larger diameter (about 7–10 cm). After being put in casings, the sausages are cooked. Even though smoking is optional for this product, most cooked loin sausages available for Brazilian consumers are smoked (Silva, 1995; Rocco, 1996; Daguer, 2009; Gonçalves et al., 2010; Dalla Costa & Dalla Costa, 2014). The cooking and smoking steps for this product, as well as the packaging, are the same as for Canadian-style loin sausage. This product is also available for the consumer as the whole piece in different sizes and weights or sliced portions.

The requirements regarding the physico-chemical characteristics are the same as those established for Canadian-style loin sausage (Table 10.2) (Brasil, 2000). The average protein and fat content of most quality brands in the Brazilian market are from 18–21.25% and 3.75–7.60%, respectively, similar to Canadian-style loin sausages. Caloric value varies from 110–150 Kcal/100 g. Sodium content, as stated on the labelling, also varies (from 972 milligrams up to 1,072 milligrams in each 100 g portion of the product), which is lower than the Canadian-style loin sausage. However, this product is also classified as having high levels of sodium in its composition (Brasil, 2010; 2012).

The expiration dates determined for the cooked loin sausage by the most relevant companies in the market are the same as reported for the Canadian-style loin sausage. In Brazil, the cooked loin sausage is also used to prepare pizza or consumed as appetizers/finger food and snacks. It is also consumed as a side dish with salads, meat stuffing (*farofa*) or as an ingredient or complement to main courses.

REFERENCES

Alencar, N. 1994. Cursos rápidos de tecnologia de produtos cárneos: embutidos finos – série A. Belo Horizonte: SEBRAE-MG, Fundação Centro Tecnológico de Minas Gerais/ CETEC.

Álvarez-Ordóñez, A., Fernández, A., López, M., Arenas, R., & Bernardo, A. 2008. Modifications in membrane fatty acid composition of *Salmonella typhimurium* in response to growth conditions and their effect on heat resistance. International journal of food microbiology. 123(3):212–219.

Alves, L. G. C., da Silveira Osório, J. C., Osório, M. T. M., Fernandes, A. R. M., da Cunha, C. M., da Cunha Cornélio, T., ... & de Souza Fuzikawa, I. H. 2013. Características qualitativas do filé mignon de cordeiros marinados com adição de glutamato de sódio. Pubvet. 7(1):2678–2755.

Associação Brasileira de Proteína Animal (ABPA). 2019. Relatório anual 2019. http://abpa-br.org/wp-content/uploads/2019/07/5350a_final_abpa_relatorio_anual_2019_portu-gues_vesao_web.pdf

Bourscheid, Cristiane. 2009. Avaliação da influência da fécula de mandioca e proteína texturizada de soja nas características físico-químicas e sensoriais de hambúrguer de carne bovina. Trabalho de Conclusão de Curso de Graduação, Universidade do Estado de Santa Catarina, Pinhalzinho.

Bonfim, R. C., Machado, J. D. S., Mathias, S. P., & Rosenthal, A. 2015. Aplicação de transglutaminase microbiana em produtos cárneos processados com teor reduzido de sódio. Ciência rural. 45(6):1133–1138.

Brasil. Ministério da Agricultura, Pecuária e Abastecimento. Instrução Normativa n° 4, de 31 de março de 2000. Aprova os Regulamentos Técnicos de Identidade e Qualidade de Patê, de Bacon ou Barriga Defumada e de Lombo Suíno. Diário Oficial da União, Brasília, v. 21, p. 15–28, 2000.

Brasil. Ministério da Saúde. Resolução RDC n° 360, de 23 de dezembro de 2003. Aprova o Regulamento Técnico sobre Rotulagem Nutricional de Alimentos Embalados, tornando obrigatória a rotulagem nutricional. Disponível em: <http://portal.anvisa.gov.br/wps/wcm/connect/1c2998004bc50d62a671ffbc0f9d5b29/RDC_N_360_DE_23_DE_DEZEMBRO_DE_2003.pdf?MOD=AJPERES>. Acesso em: 21 jan. 2020.

Brasil. Agência Nacional de Vigilância Sanitária. Resolução n° 24 de 15 de junho de 2010. Estabelece critérios para divulgação de produtos alimentícios. Disponível em: <http://portal.anvisa.gov.br/documents/33864/284972/RDC24_10_Publicidade%2Bde%2Balimentos.pdf/c406d0df-e88b-407a-9c0f-30da652f4a44>. Acesso em: 19 jan. 2020.

Brasil. Agência Nacional de Vigilância Sanitária. Portaria n° 54 de 12 de novembro de 2012. Regulamento Técnico sobre Informação Nutricional Complementar. Disponível

em:<http://portal.anvisa.gov.br/documents/%2033880/2568070/rdc0054_12_11_2012
.pdf/c5ac23fd-974e-4f2c-9fbc-48f7e0a31864>. Acesso em: 17 jan. 2020.

Brasil. 2013. Ministério da Saúde. Termo de Compromisso s/no entre o Ministério da Saúde e as Associações Brasileiras das Indústrias de Alimentação, Associação Brasileira das Indústrias de Queijo, Associação Brasileira da Indústria Produtora e Exportadora de Carne Suína, Sindicato da Indústria de Carnes e Derivados e Associação Brasileira dos Produtores e Exportadores de Frangos, de 5 de Novembro de 2013. Brasília: Ministério da Saúde.

Brasil. 2017. Ministério da agricultura. Empresa Brasileira de Pesquisa e Agropecuária. Orientações para a redução do consumo de sódio, açúcar e gorduras. Disponível em:<https://www.embrapa.br/busca-de-publicacoes/-/publicacao/1063849/orientacoes-para-a-reducao-do-consumo-de-sodio-acucar-e-gorduras>. Acesso em: 18 de janeiro de 2020.

Brasil. 2019. Ministério da saúde. Agência nacional de vigilância sanitária - ANVISA - Aprova o regulamento técnico: Aditivos alimentares para uso sem carnes e produtos cárneos, suas respectivas funções, limites máximos e condições de uso. RDC N° 272, DE 14 DE MARÇO DE. Disponível em: http://portal.anvisa.gov.br/documents/101 81/3437262/RDC_272_2019_.pdf/b39e2979-4b68-4f9c-adbd-d8be6c0be543. Acesso em: 18 de janeiro de 2020.

Bressan, M. C., Oda, S. H. I., Faria, P. B., Rodrigues, G. H., Miguel, G. Z., Vieira, J. O., & Martins, F. M. 2010. Produtos cárneos curados e defumados: mais sabor e maior valor agregado. Ed. UFLA: Lavras.

Brul, S., Klis, F. M., Knorr, D., Abee, T., & Notermans, S. 2003. Food preservation and the development of microbial resistance. In: Food preservation techniques. Woodhead Publishing. 1:524–551.

Codex Alimentarius Commission. 2009. Code of practice for the reduction of contamination of food with polycyclic aromatic hydrocarbons (PAH) from smoking and direct drying processes. CAC/RCP, 68.

Daguer, Heitor. 2009. Efeitos da injeção de ingredientes não cárneos nas características físico-químicas e sensoriais do lombo suíno. Tese apresentada como requisito parcial à obtenção do grau de Doutor em Tecnologia de Alimentos ao Programa de Pós-Graduação em Tecnologia de Alimentos, Universidade Federal do Paraná.

Dalla Costa, F. A., & Dalla Costa, O. A. 2014. Industrialização de produtos cárneos. Associação Brasileira dos Criadores de Suínos (Coordenação editorial). Manual de Industrialização dos Suínos. Brasília, DF: Associação Brasileira dos Criadores de Suínos, 176–198.

de Wardener, H. E., & MacGregor, G. A. 2002. Sodium and blood pressure. Current opinion in cardiology. 17(4):360–367.

Dickinson, B. D., & Havas, S. 2007. Reducing the population burden of cardiovascular disease by reducing sodium intake: a report of the Council on Science and Public Health. Archives of internal medicine. 167(14):1460–1468.

do Carmo, A. B., & Rodolpho, D. (2019). Carne Suína e suas possíveis doenças enzoóticas. Revista Interface Tecnológica. 16(2):235–244.

do Amaral Souza, R., Santos, E. L., da Conceição Pontes, E., de Queiroz Costa, J. H., da Silva, S. H. B., Temoteo, M. C., & Lins, J. L. F. 2011. As tendências de mercado da carne suína. Pubvet. 5(1):1157–1164.

Evangelista, J. 2001. Tecnologia de alimentos. São Paulo: Atheneu.

Evangelista, José. 2005. Conservação de alimentos. Tecnologia de alimentos. São Paulo: Editora Atheneu.

Falleiros, F. T., Miguel, W. C., & Gameiro, A. H. 2008. A desinformação como obstáculo ao consumo da carne suína in natura. XLVI Congresso da Sociedade Brasileira de

Economia, Administração e Sociologia Rural. USP – Faculdade de Medicina Veterinária e Zootecnia, Universidade de São Paulo. Pirassununga, SP. Jul 2008.

Fellows, Peter J. 2018. Tecnologia do Processamento de Alimentos. Princípios e Prática. 4º edição. London: Artmed Editora.

Feng, J., Xiong, Y. L., & Mikel, W. B. 2003. Textural properties of pork frankfurters containing thermally/enzymatically modified soy proteins. Journal of food science. 68(4):1220–1224.

Gadekar, Y. P., Sharma, B. D., Shinde, A. K., & Mendiratta, S. K. 2015. Restructured meat products-production, processing and marketing: areview. Indian journal of small ruminants (The). 21(1):1–12.

Gonçalves, C. A. A., Ciabotti, E. D., & Arantes, L. 2010. Efeito do sal de cura e antioxidante eritorbato de sódio na aceitação de lombo suíno defumado. Enciclopédia Biosfera, Centro Científico Conhecer. 6(10).

Guarapuava, P. R. 2015. Análise do teor de sódio em rótulos de mortadelas comercializadas no Brasil. Revista Instituto Adolfo Lutz. 74(3):239–246.

Hu, H., Fan, X., Zhou, Z., Xu, X., Fan, G., Wang, L., ... & Zhu, L. 2013. Acid-induced gelation behavior of soybean protein isolate with high intensity ultrasonic pretreatments. Ultrasonics sonochemistry. 20(1):187–195.

Hui, Y. H., Astiasaran, I., Sebranek, J., & Talon, R. 2014. Handbook of fermented meat and poultry. New York: John Wiley & Sons.

Jinap, S., & Hajeb, P. 2010. Glutamate. Its applications in food and contribution to health. Appetite. 55(1):1–10.

Leistner, L., & Gould, G. W. 2012. Hurdle technologies: combination treatments for food stability, safety and quality. New York: Springer Science & Business Media.

Martins, A. P. B. 2014. Redução de sódio em alimentos: uma análise dos acordos voluntários no Brasil. Série Alimentos, Instituto Brasileiro De Defesa Do Consumidor (IDEC), São Paulo, 1–90.

Magnoni, D., & Pimentel, I. 2007. A importância da carne suína na nutrição humana. São Paulo: UNIFEST.

Nilson, E. A. F., Jaime, P. C., & Resende, D. D. O. 2012. Iniciativas desenvolvidas no Brasil para a redução do teor de sódio em alimentos processados. Revista Panamericana de Salud Pública. 32:287–292.

O'Flaherty, M., Flores-Mateo, G., Nnoaham, K., Lloyd-Williams, F., & Capewell, S. 2012. Potential cardiovascular mortality reductions with stricter food policies in the United Kingdom of Great Britain and Northern Ireland. Bulletin of the World Health Organization. 90:522–531.

Ordóñez, J. A. et al. 2005Tecnologia de alimentos: alimentos de origem animal. Porto Alegre: Artmed.

Pearson, A. M., & Gillett, T. A. 2012. Processed meats. UK: Springer.

Pizzas e Massas. 2015. Editora Insumos. São Paulo. 3(21):34–38. Disponível em: <http://www.insumos.com.br/pizzas_e_massas/materias/415.pdf>. Acesso em: 01 jan. 2020.

Peroni, Fernanda Helena Gonçalves. 2003. Características estruturais e físico-químicas de amido obtido de diferentes fontes botânicas. Dissertação de mestrado - Faculdade de Engenharia de Alimentos, Universidade Estadual Paulista "Júlio de Mesquita Filho", São José do Rio Preto.

Polon, Paulo Eduardo. 2010. Otimização da produção da indústria de embutidos. Tese de Doutorado: Universidade Estadual de Maringá.

Previatti, Izabel Cristina Bergmann. 2019. Análise do poder discriminativo de avaliadores sensoriais submetidos a testes sequenciais utilizando diferentes limpadores de palato. Tese de Doutorado: Universidade de São Paulo.

Rocco, S. C. 1996. Embutidos, frios e defumados. Embrapa – SPI, Coleção Saber, 94p.

Romero-Bastida, C. A., Tapia-Blácido, D. R., Méndez-Montealvo, G., Bello-Pérez, L. A., Velázquez, G., & Alvarez-Ramirez, J. 2016. Effect of amylose content and nanoclay incorporation order in physicochemical properties of starch/montmorillonite composites. Carbohydrate polymers. 152:351–360.

Santos, T. M. 2008. Resistência de microrganismos patógenos (Clostridium, Salmonela, e Listeria) em embutidos crus e cozidos e carnes armazenadas em embalagem com atmosfera modificada. Pubvet. 2(24).

Ščetar, M., Kurek, M., & Galić, K. 2010. Trends in meat and meat products packaging: a review. Croatian journal of food science and technology. 2(1):32–48.

Seabra, L. M., Zapata, J. F. F., Nogueira, C. M., Dantas, M. A., & Almeida, R. D. 2002. Fécula de mandioca e farinha de aveia como substitutos de gordura na formulação de hambúrguer de carne ovina. Ciência e Tecnologia de Alimentos. 22(3):244–248.

Silva, C. A. B. D. (1995). Unidade de processamento de carnes. Série Perfis Agroindustriais. Brasília: Ministério da Agricultura, do Abastecimento e da Reforma Agrária, Secretaria do Desenvolvimento Rural.

Silva, Gabriela de Barros. 2016. Lombo canadense elaborado com diferentes teores de carne PSE e cloreto de sódio. Tese de doutorado, Universidade Federal de Lavras.

Silveira, Tayna Melo. 2013. Relatório de estágio de acompanhamento de rotina de estabelecimento matadouro-frigorífico e processador de produtos cárneos industrializados e in natura de suínos. Monografia apresentada para conclusão do curso de Medicina Veterinária, Faculdade de Agronomia e Medicina Veterinária da Universidade de Brasília.

Sousa, Ana Claudia Montuan de. 2017. Desenvolvimento de lombo canadense defumado produzido com carne de javali. Trabalho de Conclusão de Curso, Universidade Tecnológica Federal do Paraná.

Terra, N. N. 1998. Apontamentos de tecnologia de carnes. São Leopoldo: Universidade do Vale do Rio dos Sinos.

Toldrá, F., Hui, Y. H., Astiasaran, I., Sebranek, J., & Talon, R. (Eds.). 2014. Handbook of fermented meat and poultry. Chichester, UK: John Wiley & Sons.

Wensing, Cristiane Silvano. 2018. Estudo para preservação de presunto cozido superior utilizando conservante natural. Relatório Técnico/Científico apresentado ao Curso de Engenharia Química, Universidade do Sul de Santa Catarina.

Werlang, Gabriela Orosco. 2019. Aplicação de modelo preditivo do comportamento de Salmonella enterica durante a elaboração de salame de carne suína. Tese de doutorado, Universidade Federal do Rio Grande do Sul.

Zhu, F. 2015. Composition, structure, physicochemical properties, and modifications of cassava starch. Carbohydrate polymers. 122:456–480.

11 Indian Traditional Pork Products and Their Quality Attributes

Akhilesh K. Verma,[1] Pramila Umaraw,[1] V. P. Singh,[1] Pavan Kumar,[2] Nitin Mehta,[2] and Devendra Kumar[3]

[1]Department of Livestock Products Technology, College of Veterinary and Animal Sciences, Sardar Vallabhbhai Patel University of Agriculture and Technology, Meerut 250110, (Uttar Pradesh) India

[2]Department of Livestock Products Technology, College of Veterinary Science, Guru Angad Dev Veterinary and Animal Sciences University, Ludhiana 141004, (Punjab) India

[3]Division of Livestock Products Technology, Indian Veterinary Research Institute, Bareilly 243122, Uttar Pradesh, India

CONTENTS

11.1 INTRODUCTION

The livestock sector plays an important role in the economy of India, as it contributes 4.9% of GDP (gross domestic product) and employment to 8.8% of the population (National Accounts Statistics 2019; Basic Animal Husbandry and Fisheries Statistics 2017). Consumption and production of pork rank first among all meat consumption patterns throughout the world (Verma et al., 2019). Pork is very popular in the northeastern part of India and this region contributes more than 63% of the total pig population of India. According to the 20th Livestock Census 2019, India has 9.06 million pigs; most of these are non-descriptive (7.16 million) and only 21% of total pigs are of exotic breeds. Although increasing popularity of and demand for pork products is seen in India, there was a decline of 12.03% of total pig numbers in 2019 (9.06 million) as compared to 2012 (10.29 million). India produced 464 thousand metric tons of pork in 2014–2015, which comes to approximately 8% of the total animal protein source. Globally among all types of meat, pork is one of the most popular; it represents around 37% of the meat consumed in the world (Verma et. al.et al., 2015). In India pork production contributes about 10% and China is first in world pork production (Verma et al., 2016). The average meat yield of India's indigenous breeds is much less (35 kg per animal) than the world average (78 kg per animal).

Traditionally, pigs were reared by the marginal section of society who kept scavenging, non-descriptive/mongrels, and as a result pig farming was not a preferred occupation. These animals have very low meat yield and the quality of meat is very poor: lower juiciness, hardness, toughness, etc. Scavenging further makes these animal very prone to various parasitic infestations, leading to deterioration of animal health and posing serious public health hazards. There is also a sizeable Muslim population in India, for whom pork consumption is prohibited.

A trend has developed that involves importing germplasm from high-performance exotic pig breeds such as the Large White Yorkshire, Landrace, etc. and rearing the pigs in India using modern pig-rearing practices. These animals or their germplasm was used for improving native breeds by crossbreeding. Some native breeds with good body conformation that are better adapted to local climatic conditions were reared and popularized for pork production. Ghungroo, Niang Megha, Ankamali, Agonda Goan and Tany-Vo are examples of some popular pig breeds of India. The unique features of Indian pig farming can be summarized as follows:

1. In India, pig farming is mostly unorganized, barring some commercial farms in Kerala, Goa and Punjab, and up to 70% of pork comes from small-scale producers.
2. The capital investment is very low in small-scale pig rearing, with pigs fed on mostly kitchen/hotel waste, and vegetable and agricultural by-products.
3. Distribution of the pig population throughout India is not uniform and most of the pig population is theconcentrated in tribal belts of India, such as the North Eastern states of Assam, West Bengal, Uttar Pradesh, Jharkhand, Nagaland and Punjab.
4. Still, about 79% of the pig population is made up of native breeds, much of which is inbred due to prevailing non-scientific breeding practices with very little emphasis given to selection.

11.2 TRADITIONAL INDIAN PORK MEAT PRODUCTS

Traditional meat products have high sensory qualities and are produced on a small scale using locally available ingredients and processing techniques handed down from ancient times. These products are prepared to satisfy the nutritional and sensory demands of particular populations according to their unique tastes and the processing conditions. Thus these products are very popular among certain populations in specific areas. These traditional meat products are prepared with locally available raw materials and seasonings, and are processed according to local conditions, making them popular with nearby populations. Traditional meat products are well-regarded in India due to the desirable organoleptic attributes very much preferred by local peoples and the important role played by these products in human nutrition and food security. Meat pickle, meat kabab, meat tikka, meat samosa, meat cutlets, roasted tandoori, biryani, meat curries, enrobed and battered products, etc., are popular traditional meat products found throughout India and are available at various local shops, restaurants, hotels, etc. These are prepared from chicken, pork, sheep and goat.

India is a country comprising several ethnic groups with diverse traditions, food habits and food preferences. These groups prepare some specific meat products by using locally available raw materials, applying old practices that have been preserved and inherited through generations. These products vary from region to region (region specific) and have unique flavours due to specific ingredients and processing practices. In India, especially in North Eastern states, meat is an integral part of human food and there have been several types of traditional meat products available created by diverse ethnic groups. Here, the average expenditure on meat is two to three times higher than the national average (Mahajan et al., 2015). Blood sausage, offal with maize, rice flour, dry meat powder with spices and herbs, a unique preparation of animal fat in dry gourds or bamboo containers, the use of fermented bamboo shoots, soybeans, banana leaves, sesame leaves, etc. are some important traditional meat preparations in teh North East (Lalthanpuii et al., 2015).

Traditionally, meat consumption is very high in the North Eastern states of Indi Mahajan, and this leads to several practices for preparation of ethnic meat products which clearly reflect their socio-cultural and economic life (Mahajan et al., 2015). In this region, rice remains the staple food and for meat eaters, various meat products prepared since time immemorial form an integral part of their lifestyle, customs and food habits. This resulted in the development of several methods of preparation of meat products to best use ocally available resources for sustainable development (Singh et al., 2007). These ingredients, such as soybeans, bamboo shoots, mustard leaves, sesame, etc. are added in different meat preparations in various amounts (Kadirvel et al., 2018).

These products are shelf stable at ambient temperature and their shelf life varies from few days to several months, depending upon the preparation methods as well as preservation technologies adopted by ethnic peoples. These are mostly manufactured on a small scale at household levels and are consumed at community levels. The preparation methods and knowledge of ingredients is passed from generation to generation by practice and listening to experts or old people. These products are prepared by people who don't know about the importance of food hygiene and sanitation, so poor hygiene and lack of sanitation while preparing and packaging these products is a main constraint in commercialization of these products. With increasing urbanization, industrialization, education and better infrastructure facilities, a large population of ethnic and tribal peoples are moving to other places and thus disseminating information about these products to people in new places as well as increasing demand for and popularity of these products.

11.3 THE COMMON TECHNIQUES USED FOR PRODUCTION OF PORK PRODUCTS

Meat processing in the Asian sub-continent is still not very mechanized and depends largely on traditional methods such as fermentation, smoking and drying, drying and pickling. The major portion of meat processing that is done is in the form of thermal processing to produce ready-to-eat curry, roasts or other dishes. The major reasons are the tropical climate, consumer preferences that perceive recently cooked meat as better than stored, lack of large-scale demand, and processing infrastructure. Borah et al. (2018) in a study on ethno-traditional processing of pork in Assam, India, revealed that people in this region preferred home processing, primarily during festive periods, or in winter when the availability of pork is greater. The common processing techniques used are further discussed.

11.3.1 FERMENTATION/FERMENTED PRODUCTS

Fermentation of meat and other food products is a very old practice used for improving nutritive value, preserving shelf life, shortening cooking time and improving organoleptic attributes of the food products by applying beneficial microorganisms such as *Lactobacillus*, *Streptococcus*, *Lactococcus*, *Bifidobacterium*, etc. (Singh et al., 2012). These products, depending upon the type of culture and processing parameters used, produce unique aromas in addition to higher nutritive values. Thus these products are widely accepted in different cultures across the globe. Out of all the microbial cultures, lactic acid bacteria is most commonly used for the fermentation process due to its beneficial effects, higher fermenting/acidifying capacity and safety (Kumar et al., 2017). Fermentation enhances the nutritive value of food, increasing soluble matter by reducing solid matter, reducing the amount of anti-nutritional and toxic compounds such as tannins, phytates and polyphenols to acceptable limits, producing vitamins and essential amino acids, improving digestibility and palatability, saving on transport by reducing bulk volume, saving energy, inhibiting growth of harmful microorganisms, improving sliceability and redness, etc. (Kumar et al., 2017; Gadaga et al., 1999; Simago, 1997).

For fermentation of meat products, a mixture of various species of micro-flora is used to prepare traditional fermented meat products. As these cultures vary greatly in concentration, efficiency and type, they can produce inconsistent and variable results. Commercially this has been overcome by applying starter cultures of selected microorganisms at suitable numbers under controlled conditions.

The people of the North Eastern part of India have inherited various traditional methods developed for preserving locally available meat. These people use meat as well as various kinds of offal, scraps, trimmings, blood and even from unappealing parts such as viscera, stomach, tongue, etc., in the development of various products similar to sausages.

The offal and other leftover parts of animals are rich in nutrients, and the food materials containing offal have nutritive value in hilly regions (Rai et al., 2010b; Tamang et al., 2016).

11.3.1.1   Ingredients

In North East India, bamboo is found in surplus and it is an integral part of the food habits of the people of this region. Various parts of the plants are used as ingredients for preparing a wide variety of products. Banana leaves, sesame leaves, cilantro, ginger, garlic, mustard leaves, chili, black pepper, cardamom, etc. are used for improving organoleptic attributes of traditional products as well as increasing their acceptability and palatability. The use of some typical ingredients and peculiar preparation processes, such as use of fermented bamboo shoots, mustard leaves, sesame leaves, banana leaves, soybean seeds and fermented soybean seeds; lengthy and tedious cooking practices; fermentation by hanging meat over a fireplace; drying of meat for several weeks by hanging in a kitchen, etc. are some reasons for their distinctive flavours.

Fermented bamboo shoot is a very important food item in North Eastern India. For its preparation, fresh and tender bamboo shoots are used. These shoots are washed properly and cut into small pieces after removal of the sheath. The pieces are put into a basket with an inner coating of banana leaves and are covered with banana leaves. A hole is made at the bottom of this basket to drain the sap. The basket is tied to a stick for support and some heavy material or a stone is placed in the basket on the banana leaves. For proper drainage of sap, a hole is created in the centre of the shoots by inserting a stick, which is left in the basket. The stick is rotated periodically to ensure proper removal of sap. The sap is collected and within 15–20 days, the shoots and juice are fermented and cooked for consumption. This juice has an excellent preservative effect and is widely used in various meat preparations in place of vinegar (Mao and Odyuo, 2007). The fermented bamboo shoots are dried under sunlight and stored before being added as an ingredient in various food preparations in different amounts, depending upon the specific product or for peculiar taste, flavour, aroma, etc. Alternatively, wet fermented bamboo shoots are traditionally stored in hollow internodes of bamboo plants with one end covered by banana leaves.

11.3.1.2   Some Popular Traditional Fermented Pork Products of India

i. **Honoheingrain:** Honoheingrain is a popular fermented curry-based meat product widely consumed in Assam, India. It is prepared traditionally by people of the Dimasa tribe using pork either from wild boar or domesticated pigs (Chakrabarty et al., 2014). After slaughtering, the hair and skin are removed. The carcass is boiled in water for some time and then cut into several small pieces and kept for drying and fermentation by s on a bamboo mat 2–3 feet above a fire until completely dry. From these fermented and dried pieces, meat curry is prepared that is usually served during festive occasions such as a wedding, etc. Chakrabarty et al. (2014) noted the presence of *Lactobacillus brevis, Lactobacillus plantarum, Enterococcus faecium, Bacillus cereus, Bacillus circulans, Saccharomyces cerevisiae*, etc. in honoheingrain.

ii. **Kargyong:** Kargyong is an ethnic meat product with a soft to hard texture and brown colour, similar to the sausages of Sikkim and Arunachal Pradesh of India, Bhutan and Tibetans of China. The Sherpa, Lepcha and Bhutia of Sikkim and Thukpa of Bhutan especially like this product, usually consumed from November to December (Rai et al., 2009). Kargyong is usually prepared at home for domestic consumption or for serving at festive occasions, but pork kargyong (faak kargyong) is even available at local restaurants and hotels. It is usually prepared from locally available meat: pork, yak or beef. For its preparation, pork and fat are mixed with garlic and ginger paste and salt, and the mixture is finely chopped. This meat mixture is stuffed into the clean intestine of a pig, yak or goat as natural casing, and both ends of this intestine are closed by tightly tying or twisting. Normally the size of intestine varies from 40 to 60 cm with a diameter of 3–4 cm. The meat-mixture-filled intestine is first cooked in boiling water for 20–30 minutes followed by smoking and drying by hanging on bamboo strips above a kitchen oven for 10–15 days. The product is eaten either by frying in edible oil along with onion, garlic, coriander, and tomato or as curry product by adding seasonings chillies into this or by rather cooked sausage before fermentation and drying (Rai et al., 2009). Microorganisms responsible for the fermentation of kargyong belong to the genera *Lactobacillus, Leuconostoc, Bacillus, Micrococcus* and yeasts like *Debaryomyces hansenii* and *Pichia anomala* (Rai et al., 2010a). Rai et al. (2010b) noted *kargyong* as a very nutritious meat product containing 16% protein, 50% fat, 32% carbohydrate and 2.8% minerals on dry matter basis.

iii. **Noau Soun:** This is an ethnic fermented pork product of the Assam state of India. Pork is cut into small pieces and cooked in boiling water. The water is drained and these cooked pork pieces are collected. Boiled rice is smeared over these cooked pork pieces. The product is tightly packed into locally made bamboo cans. These bamboo cans filled with cooked pork and rice are kept for some time to facilitate fermentation. This product has a long shelf life and distinctive taste due to fermentation (Hazarika, 2013).

iv. **Tungrymbai:** Tungrymbai is a traditional fermented food product of indigenous Khasi tribes in the Meghalaya state of India. Soybean seeds, black sesame and pork are the main ingredients of tungrymbai. It is prepared by fermenting soybeans in bamboo baskets along with leaves of the slamet (*Pymium pubinerve*) plant. For its preparation, soybean seeds are washed with clean water and boiled for 1–2 hours to get the desired softer texture. After boiling, hot water is drained and the product is cooled to ambient temperature. These properly cooked soybean seeds are put into a bamboo basket wrapped in fresh leaves of the slamet plant along with hot charcoal. The product is transferred in a jute bag and kept at ambient temperature for 3–4 days for fermentation, preferably near a fireplace. The soybeans are taken out and ground with a wooden mortar and pestle. This pre-cooked tungrymbai is sold in the local market of Meghalaya, India. The use of water for boiling pork and fermented soybeans significantly affects its nutritive quality and taste (Govindasamy et al., 2018).

v. It is a very nutritious food source, containing the antioxidant properties of sesame as well as high protein and the hypertension-reducing properties of soybeans. It can be consumed in a vegetarian diet by cooking with sesame seeds, garlic, green chilies, turmeric, salt and ginger. It is a cheap source of high-quality protein for local people and has unique taste and flavour due to its intricate, distinctive preparation method and the use of bamboo and slamet leaves. In non-vegetarian tungrymbai, pork fat is heated in a frying pan and mashed garlic, ginger paste, garlic leaves, turmeric, green chilies, black sesame seeds and salt are fried in pork fat until a brown colour appears. Finely ground pork cuts are added into it and mixed properly. The product is heated for some time at low heat, and after that the pre-cooked tungrymbai is added to this mixture. The product is again slowly heated until the desirable texture, colour and flavour are achieved. Sohliya et al. (2009) noted *Bacillus subtilis* as the main fermenting microorganism present in Tungrymbai and the main source of the fermenting microorganisms as fresh leaves of slamet. The use of uncleaned grinders, pestles and other utensils also act as sources of fermenting microorganisms in this product. Tamang (2003) noted the presence of *Bacillus* strains in Kinema, a traditional soybean fermented product of Bhutan similar to tungrymbai. These strains exhibit strong peptidase and phosphatase activity, resulting in spontaneous fermentation and enhancement of nutritive value of Kinema.

vi. **Bagjinam:** A fermented ethnic pork product of Nagaland, prepared by the people of the Sema tribe in Nagaland by using their deep-rooted cultural traditions. It is consumed as curry.

vii. **Sa-um:** Fermented pig fat (lard) collected from the inner abdominal region; however fat obtained from other parts is also used. The pig fat is chopped into small pieces and cooked in a specially designed container (sa-um bur) prepared from the dried fruit of the bottle gourd (*Lagenaria siceraria*), locally known as *um* (Sawmliana, 2013). The cooking and fermentation is done by keeping the container over a fireplace for 3 days or more. This ingredient is used during the preparation of other traditional products (Lalthanpuii et al., 2015).

11.3.2 SMOKED AND DRIED PRODUCTS

Smoking is a very popular ancient technology used for preserving meat as well as preparing a variety of products. It increases the palatability, taste, appearance and colour of the products. The heat treatment rendered at the time of smoking, as well as the presence of organic acids in smoke, decrease the pH of the meat surface, helping form a rind or skin on the surface of meat products. This rind has pre-servating effect by creating a barrier against food pathogens and spoilage organ-isms. During smoking, more than 300 compounds are generated, notably formaldehydes, phenols, acids and gases. Out of these, formaldehyde has bacter-iacidal properties, whereas phenols exert bacteriostatic and antioxidant properties. The practice of smoking meat originated with ancient humans who hung pieces of meat above the fires in their dwelling places.

The most common woods used for smoking meat and meat products in India are pecan, almond, hickory, alder, apple, cherry, mulberry, grape, orange, peach, per-simmon, sassafras, maple, oak and mesquite. Most of these impart a characteristic flavour to the smoked products.

Some common traditional smoked pork products of India are as follows:

i. **Doh tyrkhong:** A smoke-dried pork product of Meghalaya.
ii. **Asan adin:** A product prepared by the Mising/Mishing community of Assam. It is first boiled and then smoke dried for 2–3 days over charcoal. Another variant is as a din wrapped in Ekkum leaves (*Phrynium* spp.). Both these products have good shelf life and can be stored for up to 2 weeks in winter (Borah et al., 2018).
iii. **Tambe-Akom:** Prepared by the Miami tribe of Arunachal Pradesh, it is a smoked product in which pieces of pork are mixed with condiments, stuffed in hollow bamboo containers and smoked.
iv. **Dohsnam:** Dohsnam is sausage made from pork and is a common dish in East Asian countries. During its preparation, the large intestine of a pig is used as casing and into this various ingredients, such as pig blood, ginger, garlic, spices, etc., are stuffed before the sausages are cooked. There are regional and cultural modifications in its preparation, resulting in wide variations in taste and shape. Soon and Koo (2009) observed it as the most popular emulsified meat product in Korea and widely attributed to the aesthetic value of Korean diets (Kwon et al., 2017). During its preparation by Khasi tribal people of Meghalaya, the large intestine of a pig is collected and washed thoroughly with hot water and salt to remove any attached intestinal content and the epithelial layer. The stuffing material is mainly blood, crushed ginger, garlic, onion, salt, and chili that have been cooked separately and are stuffed into the casing using a funnel. Both ends of the casing are then tightly closed using thread and the sausages are cooked in hot water. After cooking, these sausages are rested at room temperature for about an hour and consumed fresh or stored for later use. The presence of blood and the natural casing (large intestine) make it highly nutritious and a rich source of iron and fat.
v. **Sarep:** Sarep is a smoked meat product of Mizoram state of India. For sarep preparation, wild or domesticated meat of various animals such as deer, wild boar, birds, squirrels, rodents, etc. can be used, but pork is generally preferred. The fresh meat is cut into pieces of suitable size and skewered onto thick bamboos sticks. The skewered meat pieces are placed above a

fire at a proper distance from the direct flame for some time so that the meat is both cooked and smoked. The duration of the smoking depends upon the meat type, the size of meat pieces and the length of time until the product is consumed (Lalthanpuii et al., 2015).

vi. **Vawksha Rep:** A smoked meat product of Mizoram state of India, it is prepared from boiling pork pieces with mustard leaves, fermented soybean, oyster mushrooms, ginger and garlic pastes, salt, chili, etc. The smoked pork is fried in a hot pan with oyster sauce, chili, salt, etc. and the mixture is allowed to cook at low heat (Kadirvel et al., 2018).

vii. **Anishi:** Anishi is a traditional product of the Ao Naga tribes of Nagaland, India. It is prepared by wrapping stacked fresh, clean and mature leaves of *Colocasia* and leaving for a week until these leaves turn yellow. The yellow leaves are ground into a paste, and salt, chili, ginger, garlic, etc. are mixed into it. The paste is given the shape of cake and this cake is dried over a kitchen fire. Pork or smoked pork is cut into pieces and dried. This dried pork is cooked along with the cake (Mao and Odyuo, 2007).

viii. **Axone/Akhone:** Prepared from fermented soybean (*Glycine max*) by the Sema tribe of Nagaland, India. Soybeans are boiled in water until their texture becomes soft. The water is drained and the cooked beans are wrapped in banana leaves. These wrapped cooked beans are hung above a fireplace for heat treatment and to facilitate fermentation for a week. These fermented beans are ground, along with spices, salt, tomato, etc. to make chutney. Alternatively, the paste is also given the form of a cake and these cakes are cooked in a fireplace or the beans are sundried to preservie this for longer periods. These cakes or paste or dried beans are cooked with meat and consumed (Mao and Odyuo, 2007).

ix. **Adin:** This smoke-dried pork product prepared with fermented bamboo is a relished delicacy of Arunachal Pradesh.

x. **Ashi Kioki:** This is another dried pork product that is consumed with fermented bamboo shoots and soy beans. It is a local favourite of Naga tribes in Assam and Nagaland.

xi. **Faak Kargyong:** An indigenous sausage-like dish popular in Sikkim, Arunachal and Darjeeling, it is commonly prepared in winter by Bhutia, Lepcha, and Sherpa communities in Sikkim, Thukpa of Bhutan. The finely chopped lean meat is mixed with ginger, garlic, salt and a little water and stuffed into natural casings, preferably with a diameter of 3–4 cm and 40–60 cm long, obtained from pig/yak/cow, locally called *gyuma*. The filled and sealed gyuma are boiled for about half an hour and then placed near a fire for smoking and drying for 10–15 days. The dried kargyong is eaten after rehydration in boiling water for 5–10 minutes or in curry.

xii. **Goan Chouricos** (similar to Spanish chorizo, Goan sausages): These are traditional sausages similar to the Portuguese sausage and are a typical re-flection of Portuguese influence on Indian cuisine. Traditionally prepared by stuffing 1.5 cm diameter casings with pork pieces, seasoning and spices, tied and dried. After drying they are slowly smoked. These were prepared

        originally for the season in which fish are scarce, but are now widely con-
        sumed as a specialty dish. These dried sausages are then eaten plain or as
        curry.

   xiii. **Satchu:** Satchu is shelf stable dried or smoked pork consumed by the
        Himalayan people of India, especially in Tibet and Sikkim. Besides pork,
        satchu is also commonly prepared from locally available beef and yak. This
        is usually a method for preserving raw meat at ambient temperature for long
        durations. The meat is cut into several 60–90 cm long strands and into these
        meat pieces turmeric powder, edible oil or butter and salt are incorporated
        and mixed manually. The meat strands are put on bamboo strips or wooden
        sticks and dried in an oven for 10–15 days. The product can also be smoked
        in an oven along with drying as per preference by consumers (Rai et al.,
        2009). Due to low moisture and smoking, the product has a long shelf life
        and can be kept at ambient temperatures for several weeks without any need
        for refrigeration or cold chain.

Satchu is eaten plain or by preparing gravy after hydrating it with water. Satchu is used for curry preparation after washing with water followed by soaking in water for a short period of time. The water-soaked pieces of meat are fried with butter along with other seasoning agents such as ginger, garlic, onion, chilies and salt, and a final product with thick gravy is formed, often consumed along with noodles or boiled potatoes. Another popular method of satchu consumption is to fry it along with chilies and spices and consume with homemade alcoholic beverages at special occasions or at festive celebrations.

11.3.3 Pickled Products

Pickling is a traditional practice used for long-term preservation of foods and vegetables by using multiple hurdles (hurdle concept): reducing moisture content, decreasing pH, increasing mineral contents, using a high amount of oils, cooking, frying, addition of seasonings, etc. Pickled products are very commonly available in India and very much liked by local populations. Meat pickle is prepared by adding vinegar/acetic acid, edible oil, salt, spices and condiments, and it has a very long shelf life at ambient temperature (Gadekar et al., 2010). Reduced pH and lower water activity are the two important factors that increase the shelf life of meat pickle at ambient temperature. In addition to increasing the shelf life, pickling also has an impact on the sensory quality of meat, as it improves tenderness, flavour, aroma, colour, palatability, etc.

    Organic acids are commonly used to reduce the pH of meat pickles during pickling for preservative effects. These acids are more effective in an undissociate state and this higher acidity affects the temperature required for destroying microorganisms in a food (Gadekar et al., 2010). The pH of meat pickle increases noticeably up to 7 days and after that becomes stable with very little further change in pH value. In some commercial preparation of meat pickles, various preservatives, especially antifungal agents (such as sodium benzoate) or antioxidant compounds such as BHA/BHT (butylated hydroxyanisole/butylated hydroxyl toluene) are added to check the growth of fungi and prevent a rancid odour by inhibiting lipid oxidation.

Some examples of pickled pork products:

i. **Ngam Toonpak (Arunachal Pradesh):** Small pork pieces along with bamboo shoots are kept inside hollow bamboo containers.

ii. **Achar Doh Sniang:** A variety of pork pickle originally prepared by the people of Meghalaya that has slowly become a favourite throughout North Eastern India. Its preparation methods vary among various tribes, but in general, boiled pork pieces are fried and mixed with fried spices in mustard oil. Fresh pork is cut into small chunks and boiled to cook and remove fat. The water is drained off and kept in bottle. The spices and seasoning ingredients such as cumin seeds, coriander, black pepper, cinnamon bark, chilies, garlic, onion, turmeric, etc. are fried in pre-heated mustard oil. Into this fried mixture, salt is added to get the desired taste and flavour. This mixture is added to bottles containing pork pieces for seasoning and mixed well. The bottles are sealed properly to avoid any post-processing contamination. The shelf life of this product is about 2 weeks. The spices have antibacterial effect and help in extending the shelf life of the pickle as well as increasing its functional quality (Govindasamy et al., 2018).

iii. **Pork Pickles:** Pre-salted pork chunks are cooked under pressure for about 20 minutes with a little added water. These chunks are acidified with 2.5% acetic acid and fried in oil until brown. This mixture is then added to other boiled or fried curry and spices and cooked for two minutes. The cooked preparation is then placed in bottles/jar sand kept for direct consumption. Such pickles have very long shelf stability at ambient temperature.

11.3.4 READY-TO-EAT/CURRY PRODUCTS

Ready-to-eat curry-like preparations are very common in India. The Indian palate prefers fresh meat seasoned with local condiments and spices. These preferences can be related to the tropical climatic conditions, lack of infrastructure, processing techniques and storage facilities. Due to these preferences, most of the delicacies are freshly made curries such as Jadoh, Pork vindaloo, Balchauo de porco, Feijoada, Pandi curry, etc. The common method of preparing curry dishes involves cooking the meat with condiments and spices. The common ingredients in curry preparations of India's North Eastern region are bamboo, bamboo shoots, beans, sesame seeds, ginger, garlic and spices such as turmeric, black pepper, chili, cardamom, cinnamon and mustard.

Some examples of traditional curry pork products are as follows:

i. **Dohkhleik:** A delicious salad made from boiled pork and viscera along with onion and chili. In some recipes, vegetables are also added. During its preparation, pork viscera, especially the brain and head, are chopped into small pieces and cooked in boiling water. The water is drained and the cooked pork pieces are transferred to a container. The green chili and onion are chopped into fine pieces and sprinkled over the meat. Salt is added for taste and the mixture is transferred to a container in which vegetables such as tomatoes, beans and carrots are added to give a pleasant appearance. The product is consumed along with salad as a meal. This is consumed in nearly every native household in the Meghalaya state of India, as it is believed to exert positive effects on skin. This product is very tasty and highly nutritious due to the presence of raw vegetables and because it is made without the use of oil for frying or cooking (Govindasamy et al., 2018). Blah and Joshi (2013) calculated the nutritive value of dohkhleik (pork salad) and reported that a typical serving of 40 g contains 7.4% protein, 1.7% fat, 0.1% fibre, 2.2% carbohydrate, 55 Kcal energy, 21.3 mg calcium and 1 mg iron.

ii. **Momo:** Momos are a very common traditional snack in North East India and gaining popularity throughout India due to their typical spicy taste and flavour. It is prepared in different sizes and shapes, and finds a place in the menu of local hotels, restaurants, etc. with different nomenclature. Dough of wheat flour is prepared by adding water, salt, boiled tomato and green chili paste, which is kneaded and rolled into a thin sheet. This thin sheet of dough is cut into a circular shape and moulded into a pinwheel shape filled with meat and seasonings. The incorporation of meat into fillings improved the yield and nutritional quality of momos significantly with higher fat, protein, energy, and organoleptic attributes. These are steam cooked in a locally made steamer called a *moktu* that has oiled racks for holding raw products. The momos are steamed for 20–30 minutes and served hot with meat, soup or sauce. The quality of momos depends upon the quality of flour and filling, and the cooking method.

iii. **Doh Jem:** A popular, delicious pork product of the Meghalaya state of India. It is prepared using the offal/viscera of pig carcasses, such as the stomach, blood, large and small intestines, etc. For its preparation, offal from pig carcasses is collected and cleaned properly. The offal is cooked in boiling water, along with salt, for about 1 hour. The blood, if used, is cooked for 1 hour along with salt. These cooked viscera are cut into small pieces. A ginger, garlic, and onion paste is prepared and fried in mustard oil. During frying, turmeric powder and crushed black pepper are added. The mixture is fried continuously for 2–3 minutes and then sesame oil is added into it and frying continues for another 5 minutes. Into this fried paste, the small pieces of cooked viscera are added and the mixture cooks 45 minutes more. Salt is added and stirred in well for homogenous distribution of all ingredients. In most of these preparations, sesame seed is used. This enhances the functional quality of this product by increasing antioxidant property, mineral contents and important vitamins, leading to beneficial effect on controlling blood pressure, cholesterol and other metabolic diseases (Govindasamy et al., 2018). Blah and Joshi (2013) reported a very high nutritive value for doh jem and estimated that in a serving of 60 g, it had 10.9% protein, 10.6% fat, 1.1% fibre, 3.9% carbohydrate, 151 Kcal energy, 51.3 mg calcium, 6.5 mg iron and 3.3 µg vitamin C. The high content of calcium and iron is very important for maintaining bones and general health.

iv. **Jadoh:** A mixture of pork with red rice, prepared by the Khasi people of Meghalaya. The mixture of pork in red rice gives this product a unique colour and appearance. It is similar to non-vegetarian biryani, and green chili, garlic, onion, etc. are added to make it more delicious. Pork pieces along with fat, garlic, onion, ginger, black pepper, bay leaves, salt, turmeric, vegetable oil, etc. are the main ingredients used to prepare jadoh. A paste of onion, garlic, ginger, turmeric powder and black pepper are fried in pre-heated oil. The burnt tips of bay leaves are also added. Into this fried material, pork pieces are added and fried again until browned. This is followed by the addition of washed red rice and again fried for 2–3 minutes. The product is cooked with hot water. The rice used during the preparation of this product is unpolished, hence it provides the benefits of high fibre, iron and vitamins (Govindasamy et al., 2018). Blah and Joshi (2013) estimated the nutritive value of jadoh sniang (pork pulao) as 15.4% protein, 5.6% fat, 0.3% fibre, 62.5% carbohydrate, and 362 Kcal energy value with 30.7 mg calcium, 0.8 mg iron and 3.3 µg ascorbic acid (vitamin C) in a serving portion of 180 g.

v. **Sawhchiar/Sawchian:** A ready-to-eat porridge prepared from pork or chicken meat in the Mizoram state of India. It is an integral part of the feasts of Mizo peoples and prepared regularly during feasts or celebrations. For its preparation, pork is cut into small pieces and cooked in boiling water for some time. After cooking, rice is added into the cooked meat and stirred until it is a semi-solid texture. Traditionally available ingredients such as lettuce, black pepper, cardamom and spinach are also added to this preparation for enhanced taste and acceptability.

vii. **Bai:** A dish popular in Mizoram, India. It is a pork curry prepared with spinach and bamboo shoots.

viii. **Cheu:** A popular dish among the Deuri community of Assam, India. It is semi-boiled pork, which is smeared with turmeric, red chilies and salt, skewered on bamboo sticks and roasted over fire (charcoal).

ix. **Doh Klong:** A Meghalayan delicacy in which boiled pork pieces are fried with local spices and served hot.

x. **Doh Kpu:** A famous delicacy of Meghalaya in which finely minced pork is mixed with condiments and prepared as a curry.

xi. **Pork Sorpotel:** Famous Goan dish prepared with pork liver, heart and kidneys in a gravy of garlic, chilies, ginger, cinnamon and feni.

xii. **Pork Vindaloo:** Popular hot and sour curry in Goa. Pork is spiced with chilies, ginger and vinegar.

xiii. **Balchauo de porco:** Spicy pork dish popular in Goa. Prepared with rice, spices and tomato sauce

xiv. **Feijoada:** Another curry dish of Goa. It is a stew of salted pork, red beans, spices and coconut milk.

11.4 CONCLUSION

Pork plays a central role in the non-vegetarian diet in North Eastern cuisine, Coorgi cuisine, the Goan diet and in various parts of south India, especially Kerala. The wide use of pork in these regions can be related to the climatic conditions, scare availability of animals, low agricultural produce, etc. in the North Eastern area and to Western colonisation in Southern areas. The traditional recipes or processing techniques are highly environment- and health-friendly, as chemical use is technically nil. Major processing is done at home with limited production, except for a few dishes such as Goan sausages that have been successfully produced on a commercial scale. Most of these recipes have the potential for large-scale production and commercialisation. The factors holding these products back from gaining more popularity is the lack of standardisation, the need to optimize processing conditions, lack of public awareness, and niche production and consumption.

REFERENCES

Basic Animal Husbandry and Fisheries Statistics. 2017. http://dahd.nic.in/Division/statistics/animal-husbandry-statistics-division

Blah, Mandari Mary, and S. R. Joshi. 2013. Nutritional content evaluation of traditional recipes consumed by ethnic communities of Meghalaya, India. *Indian Journal of Traditional Knowledge* 12(3): 498–505.

Borah, B., A. Borgohain, R. Roychoudhury S. et al. 2018. A study on ethno-traditional processing of pork in Assam, India. *International Journal of Chemical Studies* 6(2):2151–2156.

Chakrabarty, J., G. D. Sharma, and J. P. Tamang 2014. Traditional technology and product characterization of some lesser-known ethnic fermented foods and beverages of North Cachar Hills District of Assam. *Indian Journal of Traditional Knowledge* 13(4):706–715.

Gadaga, T. H., A. N. Mutukumira, J. A. Narvhus, and S. B. Feresu. 1999. A review of traditional fermented foods and beverages of Zimbabwe. *International Journal of Food Microbiology* 53:1–11.

Gadekar, Y. P., R. D. Kokane, U. S. Suradkar, R. Thomas, A. K. Das, and A. S. R. Anjaneyulu. 2010. Shelf stable meat pickles—a review. *International Food Research Journal* 17: 221–227.

Govindasamy, K., B. B. Banerjee, A. A. P. Milton, R. Katiyar, and S. Meitei. 2018. Meat-based ethnic delicacies of Meghalaya state in Eastern Himalaya: preparation methods and significance. *Journal of Ethnic Foods* 5(4):267–271.

Hazarika, M. (2013). *Traditional Meat Products of North Eastern Region of India*. Vet Helpline India. https://www.vethelplineindia.co.in/tag/meat/.

Kadirvel, G., B. B. Banerjee, S. Meitei, S. Doley, A. Sen and M. Muthukumar. 2018. Market potential and opportunities for commercialization of traditional meat products in North East Hill Region of India. *Veterinary World* 11:118–124.

Kandeepan, G. 2016. Quality characteristics of ready to eat dumplings (momo) prepared from yak meat. *FleischWirtschaft International* 2:126–131.

Kumar, P., M. K. Chatli, A. K. Verma, et al. 2017. Quality, functionality, and shelf life of fermented meat and meat products: a review. *Critical Reviews in Food Science and Nutrition* 57(13):2844–2856.

Kwon, D. Y., Y. E. Lee, M. S. Kim, et al. 2017. *Tells about Korean diet: culture, history and health.* Seoul, Korea: Elsevier Korea.

Lalthanpuii, P. B., B. Lalruatfela, Zoramdinthara, and H. Lalthanzara. 2015. Traditional food processing techniques of the Mizo people of Northeast India. *Science Vision* 15(1):39–45.

Mahajan, S., J. S. Papang, and K. K. Datta. 2015. Meat consumption in North East India: pattern, opportunities and implications. *Journal of Animal Research* 5(1):37–45.

Mao, A. A., and Odyuo, N. 2007, January. Traditional fermented foods of the Naga tribes of Northeastern, India. *Indian Journal of Traditional Knowledge* 6(1):37–41.

National Accounts Statistics. 2019. Central Statistical Organisation, GoI. https://www.nddb.coop/information/stats/GDPcontrib

Rai, A. K., J. P. Tamang, and U. Palni. 2010a. Microbiological studies of ethnic meat products of the Eastern Himalayas. *Meat Science* 85:560–567.

Rai, A. K., J. P. Tamang, and U. Palni. 2010b. Nutritional value of lesser-known ethnic meat products of the Himalayas. *Journal of Hill Research* 23(1&2):22–25.

Rai, A. K., U. Palni, and J. P. Tamang. 2009. Traditional people of the Himalayan people on production of indigenous meat products. *Indian Journal of Traditional Knowledge* 8(1):104–109.

Sawmliana, M. 2013. *The Book of Mizoram Plants (Includes Wild Animals, Birds, etc.)* (2nd ed., pp. 1–526). Aizawl, Mizoram:P. Zakhuma.

Simango, C. 1997. Potential use of traditional fermented foods for weaning in Zimbabwe. *Social Science Medicine* 44:1065–1068.

Singh, A., R. K. Singh, and A. K. Sureja. 2007. Cultural significance and diversities of ethnic foods of Northeast India. *Indian Journal of Traditional Knowledge* 6(1):79–94.

Singh, V. P., V. Pathak, and A. K. Verma. 2012. Fermented meat products: organoleptic qualities and biogenic amines–a review. *American Journal of food technology* 7(5):278–288.

Sohliya, I., S. R. Joshi, R. K. Bhagobaty, and R. Kumar 2009. Tungrymbai: a traditional fermented soybean food of the ethnic tribes of Meghalaya. *Indian Journal of Traditional Knowledge* 8(4):559–561.

Soon, H. C., and B. C. Koo. 2009. Product characteristics and shelf life effect of low-fat functional sausages manufactured with Sodium Lactate and Chitosans during storage at 10°C. *Korean Journal of Food Science and Animal Resources* 29(1):75–81.

Tamang, J. P. 2003. Native microorganisms in the fermentation of kinema. *Indian Journal of Microbiology* 43(2):127–130.

Tamang, J. P., N. Thapa, T. C. Bhalla, and Savitri. 2016. Ethnic fermented foods and beverages of India. In *Ethnic Fermented Foods and Alcoholic Beverages of Asia*, ed.J.P. Tamang. India:Springer. DOI:10.1007/978-81-322-2800-4_2.

Verma, A. K., M. K. Chatli, N. Mehta, P. Kumar, and O. P. Malav. 2016. Quality attributes of functional, fiber-enriched pork loaves. *Agricultural Research* 5(4):398–406.

Verma, A. K., M. K. Chatli, D. Kumar, P. Kumar, and N. Mehta. 2015. Efficacy of sweet potato powder and added water as fat replacer on the quality attributes of low-fat pork patties. *Asian-Australasian Journal of Animal Sciences* 28(2):252.

Verma, A. K., M. K. Chatli, P. Kumar, and N. Mehta. 2019. Antioxidant and antimicrobial activity of porcine liver hydrolysate in meat emulsion and their influence on physico-chemical and color deterioration during refrigeration storage. *Journal of Food Science* 84(7):1844–1853.

12 Italian Salami

A Comprehensive Analysis

Emanuela Zanardi and Enrico Novelli
Dipartimento di Scienze degli Alimenti e del Farmaco,
Università degli Studi di Parma, Strada del Taglio, 10, 43126
Parma, Italy

CONTENTS

12.1 Introduction 285
12.2 The History of Dry-Fermented Sausage Manufacturing in Italy 286
12.3 The Two Main Ingredients: Salt and Pork 287
     12.3.1 The Availability of Salt (Sodium Chloride) in Italy 287
     12.3.2 Pigs and Pork Production in Italy (with Historical References) 289
12.4 The Manufacturing of Meat Products 290
     12.4.1 The Italian Heavy Pig 291
12.5 Italian Salami: Focusing on Three Types with Particular Traits 293
     12.5.1 Sopressa Vicentina PDO (the Largest Salame) 293
     12.5.2 Ciauscolo (the Spreadable Salame) 298
     12.5.3 Salame Milano (the Most Renowned) 301
12.6 Conclusion 306
References 307

12.1 INTRODUCTION

Salami is the generic Italian term that identifies dry-fermented sausages, a wide and diverse class of meat products. However, other terms such as soppressata, sopressa, and salsiccia define some typologies of dry-fermented sausages that are usually produced in Southern Italy with a few exceptions (e.g., Sopressa Vicentina). In this context, it is important to keep the meaning of the words distinct. *Salume* is the Italian word for *salted and cured meats*, whereas the word *salami* means dry-fermented sausages, and *salame* is the corresponding singular form to indicate a single piece.

The root of the term *salami* can be traced back to the word *salt*, a substance that was kneaded into minced meat. It seems that this term entered the common usage in 1436 through Niccolò Piccinino, a commander of arms employed by the Duke of Milan. He based his operations in Parma where he made the following order for "... porchos viginti a carnibus pro sallamine..." which means "twenty pigs to make the salame." Before then, cured, bagged meats were called, rather generically, *botulus* or *insicia*.

285

Later, in 1581, the word *salame* appeared for the first time in a cooking manual, referring to a pork sausage. This book was created by Vincenzo Cervio, a meat carver of Casa Farnese. Still later, in 1678, Vincenzo Tanara wrote a recipe on how to make salame: a mix of weighed fat or lean meat along with 6% sodium chloride and 12 ounces of pepper for every 100 lb of meat. However, it is interesting to remember that in ancient times, the term *salamen*, even before it referred to salted meat, customarily meant salted fish, a product that would later come to be known as stockfish and salted cod. It is highly probable that, in that same period, the word *salamen* entered into common usage as a means of describing an obtuse and expressionless person, like a codfish (Spisni, 2017).

In this discussion on Italian charcuterie with specific reference to fermented and seasoned sausages, we want to guide the reader through the history of these products with some examination of its most important cultural and social implications. For this purpose, we will describe the historical–cultural significance of the two main ingredients of salami, namely salt and pork, first in the peasant tradition and then in the industrial one. In Italy, all production of salami carrying a mark of origin is carried out using only meat obtained from heavy pigs, in order to have products of distinguishable quality. In addition, salt is so important that in the Vicenza dialect the salame in its traditional size (less than 800 g of weight at the end of ripening) is called *salado*, which means *salty*, to emphasize the fundamental role played by salt in the transformation of meat. The discussion then goes into detail on three Italian salami, specifically the sopressa Vicentina, the ciauscolo and the salame Milano. We've chosen these three salami because there are some specific peculiarities in their production technology that clearly distinguishes them from other bagged and fermented products, and for the wide diffusion and notoriety in the market enjoyed by the products themselves, as is the case of the salame Milano.

12.2   THE HISTORY OF DRY-FERMENTED SAUSAGE MANUFACTURING IN ITALY

The first producers of sausages were the Egyptians, the Phoenicians, the Greeks, the Etruscans and, lastly, the Romans. The Romans, in contrast to the previous groups, made important innovations in the use of salt, pepper and spices. Into Imperial Rome, dried meats (*siccamen*), hams (*perna*), lard (*lardum*), and lucanica were transported, along with wine, oil, dried figs and wheat. Subsequent Germanic populations extended the practice of these typically European food traditions by carrying them on well into the Middle Ages when salami, ham and a kind of mortadella were laid out on poor people's feast day tables, but were rarely seen on the tables of nobles, bishops and popes, who believed these foods too simple and cheap.

From the Renaissance onward, cured meats were elevated to a place of honor on the tables of elegant courts. In Italy, between the 12th and 17th centuries, the crafts related to the transformation of pork were honed, and the profession of *norcino*—named for the inhabitants of Norcia, a city in central Italy, famous for the tradition of salting and bagging of pork—appeared. These workers organized themselves into *Corporazioni* or guilds: the Corporazione dei Salaroli originated in Bologna just as the Confraternita dei Facchini di San Giovanni originated in

Florence. In 1615, Pope Paul V recognized the *Confraternita norcina* dedicated to Saint Benedict and Scholastica. Later, Pope Gregorio XV elevated it to *Arciconfraternita*—a Confraternita that belongs to a network of Confraternitas, which, in 1677, joined the association of Pizzicaroli (*pizzicaroli* means *delicatessen sellers*) of Norcia and Cascia and norcini empirical physicians (Benasaglio, 2017).

Today, in Italy, one can find artisanal salame, which is made with a very simple recipe (usually pork plus salt and pepper and, sometimes, nitrate salt), along with industrial salame—whose recipe also includes sugars, nitrate and ascorbate salts, and a mix of lactic acid bacteria (LAB) as bacterial starters. Artisanal salame is usually available in osteria (taverns) or in trattoria and is sold at itinerant markets. The industrial salame is usually sold at small and large shops, as well as in the supermarkets of large, organized retail outlets. Both artisanal and industrial salame can be served in even the most prestigious starred restaurants.

In 2016, the Italian production of salami accounted for 107,600 tons, representing 9.2% of all the meat products of Italian manufacturing (ASSICA Associazione Industriali della Carne e dei salumi, 2019). These figures include salami protected by quality schemes for protected designations of origin (PDO) and protected geographical indications (PGIs), both laid down by Regulation (EU) n. 1151/2012, and those not included in that legal framework. The production of salami designated by PDO and PGI marks (Table 12.1) accounted for about 10–15% of the total amount of Italian salami produced in 2016. As a matter of fact, in Italy, in addition to the production of salami with identifiable, specific characteristics, in particular those linked to their geographical origin, a wide and diverse range of dry-fermented sausages can be found. Overall, approximately 80 different types of salami have been recorded and included in a collection of all of the Italian meat products, sorted according to region. Some of them are spread throughout the marketplace all over the country and have an important market share, whereas others are limited to local or very small geographical areas and niche markets (INSOR Istituto Nazionale di Sociologia Rurale, 2002).

This culinary success is thanks to the skills and determination of Italian farmers and producers who have kept manufacturing traditions alive, and who have passed on a gastronomic, cultural, social and economic heritage. A survey of the typology of the dry-fermented sausages sector of Northern Italy highlighted that small companies contribute substantially to the production, manufacture and retail of these products (Conter et al., 2007). However, artisan production, in which raw meat is obtained from animals bred on the farm, and local processing units manufacturing traditional products from raw meat coming from different local farms or slaughterhouses, are widespread in Central and Southern Italy, as well (Talon et al., 2007).

12.3   THE TWO MAIN INGREDIENTS: SALT AND PORK

12.3.1   The Availability of Salt (Sodium Chloride) in Italy

The factors that contributed to the development of the manufacture of salted meats in Italy were, firstly, the availability of salt, and, secondly, the possibility of keeping and/or raising pigs. Significant strides in the production and transport of salt

TABLE 12.1

Italian Salami Designated by PDO and PGI Marks

	Production Italian Region	Produced Amount (kg/ year 2016)	Estimated Value (Mln euro)
Protected Designation of Origin (PDO) mark			
Salame di Varzi	Lombardia	497,174	6.9
Salame Brianza	Lombardia	190,159	2.8
Salame Piacentino	Lombardia	1,491,213	15
Salamini Italiani alla Cacciatora	Friuli Venezia Giulia, Veneto, Lombardy, Piedmonte, Emilia-Romagna, Umbria, Toscana, Marche, Abruzzo, Lazio, and Molise	3,571,862	40
Soppressata di Calabria	Calabria	58,368	1.1
Sopressa Vicentina	Veneto	50,211	0.47
Salsiccia di Calabria	Calabria	97,776	1.6
Protected Geographical Indication (PGI)			
Ciauscolo	Marche	294,559	N/A
Finocchiona	Toscana	1,041,732	7.8
Lucanica di Picerno	Basilicata		
Salame Cremona	Lombardia	205,837	2.7
Salama da sugo	Emilia-Romagna	67,112	1.3
Salame d'oca di Mortara	Lombardia	1,726	0.05
Salame S. Angelo	Sicilia	100,332	0.95
Salame Felino	Emilia-Romagna	3,179,366	29
Salame Piemonte	Piedmonte	N/A	N/A

(*Source:* www.qualigeo.eu)

N/A = not available

occurred under the Roman Empire. Salt became a trade good, like gold. The militiamen were paid with salt, hence the term *salario* (wage). Sea water, collected in large artificial basins near the coasts, evaporated naturally, thus enabling the concentration of sodium chloride. Extensive salt pans of this type were known among the Italic populations, but it was certainly the Romans who made salt production a real industry—an industry on which they held a monopoly. Rome exclusively controlled some of the most productive, including the ones called *Campus Salinarum Romanarum*.

The salt trade was so important that it led to the construction of a dense network of roads that connected the most important cities in Europe, the Arab countries and the

Far East. One of the most important was the Via Salaria in Italy, which connected the city of Rome with Porto d'Ascoli on the Adriatic Sea. Beginning in the 10th century, Venice became the center of salt production and trade thanks to a widespread network of salt pans in its territory. In 1200, after two centuries of expansion, salt production had reached its peak, placing the Venice lagoon in first place among Mediterranean producers. A considerable competition for Venice came from Northern Europe where there were many salt mines. In 1232, the opening of Hall's salt mine in Tyrol, near Innsbruck, supplied Bolzano with salt. In 1287, Hall produced 530 wagons of salt per week, mostly traded in the market of Lombardy and the Upper Veneto markets, reaching as far as Vicenza. On the other side of the Mediterranean Sea was Genoa, whose salt monopoly was firmly controlled by the House of San Giorgio, the most powerful financial institution in the medieval West. In the 13th and 14th centuries, this house imported the salt produced in Hyères (France), Cagliari (Italy) and Ibiza (Spain). But the sea was not the only resource for salt production. The Po valley, in the north of Italy, was formed slowly by sedimentation, and contains in its depths, enclosed between layers of impermeable clay, considerable quantities of fossil sea salt, and, for this reason, salty waters and wells were widespread in the low plains, hills, and mountains from which salt had been drawn in the past to meet local demand (Hocquet, 1999; Hocquet, 2003; Laszlo, 2004; Ministero delle Politiche Agricole Alimentari e Forestali, 2020).

12.3.2 Pigs and Pork Production in Italy (with Historical References)

The findings in an Etruscan settlement in Forcello (Bagnolo S. Vito, near Mantua), an Etruscan city of the 5th century BCE, are worthy of note when it comes to pig breeding in Italy. Among the (as many as 50,000) excavated bones, more than 60% probably belonged to the swine species, and, proportionally, many hind limbs were missing. The slaughtering age of the pigs was around two or three years (Ministero delle Politiche Agricole Alimentari e Forestali, 2020). They were small pigs of both sexes (65–75 cm high at the withers at the time of slaughter). This means that, in the Po Valley, Etruscans practiced a stable and specialized type of breeding for the production of pork with a great interest in focusing on the thigh as the area for salting and ripening. Later, during the Roman domination, the Po Valley saw a remarkable extension of oak woods where pigs grazed. According to Strabone (*Geografia*), the Emilia region supplied pork and live pigs to markets all across Italy. According to Columella (*De Re Rustica*), in the Roman period, there were permanent and "rational" pig farms. The great agricultural and demographic crisis of the 3rd and 4th centuries caused a great expansion of uncultivated and wooded areas, and, consequently, relaunched the wild and semi-wild breeding of pigs.

Following the Longobard invasion (569), the culture of a semi-nomadic civilization spread, exploiting what nature spontaneously offered, including the forest with its various fruits and by-products. Among these, the humble pig was one of the most important. According to Baruzzi and Montanari (1981), the medieval Po Valley pigs were thin and slender, with long and thin legs and dark red or blackish in color. The transition from the forest to the pigpen took place with the resumption of agriculture and the associated demographic development that began in the 10th to

11th centuries. In his *Opus ruralium commodorum*, Piero De Crescenzi, a 13th century Bolognese agronomist, wrote that "acorns, chestnuts and similar things must be given to pigs, or the beans, or the barley, or the wheat: because these things not only fatten, but give delightful flavor to meat."

Throughout the Po Valley, pig breeding has always occurred in the flat and hilly parts. Initially, this was because it was covered by oak woods that provided the acorns on which the omnivorous pig was mainly fattened. Subsequently, breeding and fattening were dependent on products derived from the by-products of cow milk (whey) and corn. Pigs were slaughtered at between 1 and 4 years of age. The long farming period was a consequence of the genetic characteristics of the breeds being bred, with high rusticity, low earliness and a diet that was certainly less than adequate. Pigs were slaughtered primarily in November and December, however, always in the winter (Ministero delle Politiche Agricole Alimentari e Forestali, 2020).

A more scientific and structured method of pig farming began in Italy in 1872. At the end of the 19th century, the importation of breeding pigs from England began to improve some characteristics considered important for the production of long-ripened, cured meats. The industry at that time required, just as it does today, heavy carcasses from which to obtain meat with intrinsic technological features suitable for making them into cured meat products, especially dry-cured ham. In the Po Valley, pig breeding was traditionally partnered with the dairy industry, which made large volumes of whey available, from early spring (with pigs who weigh 35–45 kg of live weight) up to the end of the month of November. During these 9–10 months in the piggery, the pigs grew to a live weight of 160–180 kg. Therefore, the breeding registered in the herd books began to augment quickly, and with the support of the genetic control centres established by the Ministry of Agriculture in 1960, a systematic selection program for the Large White and Landrace breeds was launched (Ministero delle Politiche Agricole Alimentari e Forestali, 2020).

12.4   THE MANUFACTURING OF MEAT PRODUCTS

In the Mediterranean area, the technique of salting coupled with dehydration by natural dry air had been a common method for preserving meat since ancient times. In the countries north of the Alps and in those of Eastern Europe, smoking by burning vegetal masses was the traditional method of preserving meat, and this method was also used by populations of the Po Valley during the Roman period. Subsequently, fermentation and cooking techniques began to become more widespread. Fermentation, in particular, was obtained by mixing salt, spices, sugars and wine with minced meat.

The tradition and manufacture of charcuterie products in Italy is not homogeneous, and different local practices and cultural influences can be found in different parts of the country, as described by Ballarini (2003). Probably the oldest Italian salami is that of Sant'Angelo di Brolo in Sicily. The charcuterie tradition began there in the 11th century, following the invasion of the island by the Normans, who ended the Arab domination during which eating pork was prohibited.

12.4.1 THE ITALIAN HEAVY PIG

It is difficult, if not impossible, to talk about the Italian charcuterie tradition without considering its uniqueness, embodied by the use of meat obtained from so-called heavy pigs as they are slaughtered at an average live weight of at least 160 kg. In the regulations for the production of all Italian PDO or PGI marked salami is an obligation to use meat obtained from Italian heavy pigs, as their lean and fat cuts are of higher quality and, as such, are suitable for the production of dry fermented sausages with medium or long ripening times.

In Italy, almost the whole pig production chain is dedicated to the production of dry-cured hams that have the protected designation of origin, such as Parma and San Daniele ham. The farming of the heavy pig is aimed at supplying the industry with thighs of a high technological quality and the capacity for salting and seasoning steps. For the production of the Italian heavy pig, the Consortia for the protection of Parma and San Daniele ham allow only some purebred animals (Italian Large White and Italian Landrace) and crosses with the Italian Duroc breed. In addition, the same breeds coming from other countries or the subjects of other breeds can be used for the production of crossed pigs only according to selection programs coherent with those of the Italian selection process. In general, the effect of genetic types on the quality of green hams may be due to, among other causes, the different frequency of the negative halothane gene and the different degree of adiposity of the carcass of different breeds. The higher the lean meat percentage of the carcass, the higher the loss of weight of the meat during ripening and the more evident the scarce sensorial characteristics of dry-cured ham (Russo and Costa, 1995).

One must keep in mind that an insufficient fat covering of the thigh causes an increase in weight loss during ripening and a lowering of the organoleptic characteristics of the dry-cured ham. The Italian selection scheme for heavy pigs follows the same objectives of other countries but with more attention to the quality of the thigh intended for dry-cured ham. One important goal was that of keeping breeding animals carrying the allele for halothane sensitivity out of selection programs, even if in heterozygous form, so as not to end up with meat that has a lighter color, lower pH 4.5 values, and a lower water holding capacity (Russo et al., 1989). The Consortia of Parma and San Daniele ham fixed a minimum thickness of the subcutaneous fat of the thigh with the aim of limiting the ripening weight loss and of pursuing the sensory quality objectives of the end product.

Nevertheless, the chemical stability of the subcutaneous fat against oxidative stress is not any less important than other technological factors. To avoid defects attributable to the oxidation of fat during ripening, or the unpleasant phenomenon of fat softening in addition to the oiliness and yellow appearance on the surface, the Consortia set a maximum percentage of 15% for linoleic acid in thigh subcutaneous fat, as well as an iodine number that was set at a level of 70 (number of grams of iodine that will react with the double bonds in 100 g of fat). These two parameters for fat composition were introduced to limit the content of polyunsaturated fatty acids (PUFA) that can affect the composition, the consistency and the oxidability of the fat itself. Baldini et al., (1983) and Micossi et al., (1996) found positive correlations between iodine number, linoleic acid percentage and the peroxidability index in subcutaneous fat from hams of heavy pigs. These results justify the limit of 2% linoleic acid—on dry matter basis—in the diet of finishing heavy pigs that was imposed by the Consortia (5% of dietary fat on dry matter basis, as proposed in the new production regulations).

However, selection and improved feeding and breeding techniques have significantly increased the growth rate of modern genetic types. As a consequence, the relationship between maturity and weight is weaker. This is why the Consortia empirically introduced a limit on the slaughtering age of the pigs, which may be of at least 9 months, correlating to an average weight per batch (live weight) of 160 kg (more or less 10%). Linking age and live weight at slaughter was aimed at ensuring an adequate degree of meat maturity and equilibrating the proportions of lean meat and fat in the processing cuts. This balance has been obtained through the feeding practices adopted by pig farmers, whose goals is a pig carcass with an optimal degree of fatness accomplished with a low rate of growth (Mordenti, Rizzitelli, and Cevolani, 1992). Nevertheless, the Consortium of Parma ham recently submitted a revision of the rules of production in which they changed the maximum cold weight of the carcass, raising it to 168 kg, which means that the live weight of the pigs, assuming an average yield in the carcass of 80%, can reach or slightly exceed 200 kg. If we are to accept the rationale behind this choice, we must likewise acknowledge that age is more important than weight for the achievement of an optimal level of the meat's capacity to be submitted to long a ripening process. These live weights are considerably higher than those currently in use in many European countries, whose values vary between 110 and 120 kg of live weight.

About 85% of the pigs slaughtered in Italy belong to the heavy pig category, of which about 50% are in the meat class U (from 50% to less than 55% of lean meat) and about another 30% are in the meat class R (from 45% to less than 50% of lean meat), according to the Regulation (EU) n. 1308/2013. From a territorial point of view, 80% of pig farms are concentrated in four Italian regions: Veneto, Lombardy, Piedmont, and Emilia-Romagna.

The age at slaughter can have an important effect on carcass and meat quality. Virgili et al., (2003) compared the quality of the carcass and meat of pigs slaughtered at 8 and 10 months of age of four different genetic types and both sexes. Dressing percentage was higher ($P < 0.01$) for older and heavier pigs. Pigs of 10 months of age showed higher dressing percentage, but the muscularity index was lower ($P < 0.01$) in 10- vs. 8-month-old pigs, which is in concordance with the greater back fat thickness and higher total fat found in older and heavier pigs (Bittante et al., 1990). The weight of the lean cuts increased ($P < 0.01$) with age, but lean cut yield decreased ($P < 0.01$) in carcasses from older pigs ($P < 0.01$). The semimembranosus muscle (SM) from younger pigs was lighter (higher L* values; $P < 0.05$) than the SM of older pigs, a* (redness) values were not different ($P > 0.05$) between slaughter ages, the SM from pigs slaughtered at 8 months of age was more ($P < 0.01$) yellow (higher b* value) than that from 10-month-old pigs. Other studies found higher a* values in older pigs or heavier pigs (Franci et al., 1994).

It is also interesting to note that ham subcutaneous fat from 10-month-old pigs had greater percentages of palmitic ($P < 0.05$) and oleic ($P < 0.01$) acids and lower ($P < 0.01$) moisture, and linoleic and linolenic acids than subcutaneous fat from 8-month-old pigs. With increasing age and weight, the partition of ingested energy turns from muscular to adipose tissue growth, which increases the ratio between de novo synthesized fatty acids (mainly C18:1; Enser, 1991) and dietary fatty acids (C18:2 and C18:3). Zappa and Pugliese (1991) reported a positive correlation between mono-unsaturated fatty acid and back fat thickness, and a negative correlation between back fat thickness and both back fat moisture and PUFA. Although the fatty acid composition of subcutaneous fat of heavy pigs was found to be affected by several factors—including gender, age, energy intake, dietary fatty acid composition, carcass weight and composition, and genetic background (Della Casa et al., 1999; Zappa and Pugliese, 1991; Piedrafita, Christian, and Lonergan, 2001)—the back fat with lower percentages of PUFA appears more suitable for dry-cured ham production and, generally speaking, for other long ripened meat products, as well.

12.5 ITALIAN SALAMI: FOCUSING ON THREE TYPES WITH PARTICULAR TRAITS

Italian dry fermented sausages have been widely studied in the last few decades. Scientific know-how has become essential for consistent production of safe and high-quality products, and many facets of it have been the subject of excellent reviews and books, mainly referring to industrial production. In the following sections, the authors invite the reader to focus on three different Italian salami and their specific and typical traits.

12.5.1 Sopressa Vicentina PDO (the Largest Salame)

Sopressa Vicentina is a dry-fermented sausage marked PDO and made throughout the territory of Vicenza (North East Italy) (Commission Regulation (EC) n. 492/2003). The production of sopressa Vicentina has its origins in the peasant tradition of the slaughtering of pigs and their transformation into meat products at the beginning of the winter season, which was very widespread in the province of Vicenza. Since Vicenza did not have a tradition of making other cured meat products, such as ham or other cured meats, such as coppa (muscles of the neck region) or pancetta (muscles and fat of the belly region), practically the entire edible part of the carcass was used for the production of sopressa and other sausages to be eaten after cooking (fresh sausage and cotechino). Sopressa is the only bagged, fermented meat produced in Italy that is made using all the cuts of the pork carcass. It is difficult to explain or trace the origin of this completely peculiar custom. In fact, it is known that in the areas of the country with a greater pork and charcuterie tradition, especially Emilia-Romagna but also including Lombardy, Piedmont and Tuscany, just to name a few, the cuts of the pig carcass find well diversified destinations (raw ham with thigh, coppa with the neck, lard seasoned with fat cuts, etc.).

In the foothills of the Veneto, however, there does not seem to be such a wide and varied tradition in the transformation of pork, so most of its meat was used to produce sausages to be fermented and some other sausages for fresh consumption after cooking. Still today, this tradition survives in rural areas of the province of Vicenza, where some families raise one or two pigs to be transformed into delicatessen products exclusively for family consumption. These are pigs that are often brought to a slaughter weight of more than 200 kg with an age close to 15 months, hence the popular custom of saying that such pigs must have crossed *le due lune d'Agosto* (the two moons of August), meaning more than 12 months of age.

It is difficult to get reliable information on the origin of this dry-fermented sausage. The first historical annotation about sopressa comes from the artist Jacopo da Ponte, known as Jacopo Bassano (1515–1592), who, between 1576 and 1577, depicted Martha and Mary Magdalene welcoming Jesus into their home (*Christ in the House of Martha, Mary and Lazarus*; Galleria Palatina, Palazzo Pitti, Florence, Italy). On the table set for this important occasion is a sausage, already partially sliced, with the appearance of a sopressa. It is also reported in the mercurials of the Vicenza Chamber of Commerce which, starting from 1862, includes the item *Salame e sopressa*, an item that is not present in the price lists of other provinces of the Veneto, thus awarding Vicenza first place for this precious dry-fermented sausage.

According to manufacturing regulations, sopressa Vicentina is made using lean meat (ham, shoulder, loin, neck, belly) and back fat obtained from pigs whose genetical types, live weight and age are the same as those described for Italian heavy pigs (at least 9 months of age and 130 kg carcass weight). The cuts used for this meat product, according to the production regulation in force, must come from animals born and raised on farms located in the restricted geographical area of the province of Vicenza. Moreover, the unique quality of this dry-fermented sausage lies in the fact that all cuts of the pig carcass are used for its production. The external appearance is cylindrical. The external binding for improving the adhesion of the casing with the meat mixture is accomplished using an uncolored twine, which is first pulled along the major axis (harness) and then along a series of rings of the same material, placed crosswise above the harness, which cover the entire length of the sopressa (Figure 12.1). The use of nets for this purpose is not allowed.

There are four categories of fresh weight: the lightest is from 1–1.5 kg (for which the minimum ripening time should be at least 60 days; the heaviest is from 3.5–8 kg of fresh weight (at least 120 days of ripening). The length of sopressa Vicentina can vary from 25–30 to 40–50 cm. After cutting, the slice appears to have slightly opaque colors, fat and lean, of medium-large grain, and without well-defined borders (Figure 12.2). The selected cuts of meat are minced using molds with holes that have a diameter of 6–7 mm. The minced meat, once brought to a temperature between +3°C and +6°C, is added to the other ingredients itemized here at their maximum percentages for use: salt (2.7%), cracked pepper (0.3%), mix of ground spices (0.05%), garlic (0.1%), sugars (0.15%) and nitrate salts. Microbial culture preparations, of proven autochthonous origin, may also be added to start fermentation. These ingredients and the minced meat are then mixed until they have a light appearance of fat smearing. The resulting mixture is then bagged using natural casings with a minimum diameter of 8 cm. Drying takes place as follows: for the first 12 hours after stuffing, the product is kept at high temperatures (20–24°C) to allow the initial drip; then, for the next 4–5 days, at descending temperatures from 22–24°C down to 12–14°C.

According to tradition, the production of the sopressa is usually done at the beginning of winter, and the drying step is still conducted today at lower temperatures

FIGURE 12.1 Sopressa Vicentina PDO at 6 months of ripening.

(usually not higher than 15°C), which implies that the length of the ripening phase will be longer. In accordance with the indications prescribed by the production regulations, the protein content of the end product has to be higher than 15%, with fat between 30% and 43%, ash between 3.5% and 5%, residual moisture less than 55%, and with a pH between 5.4 and 6.2. Generally speaking, two types of sopressa exist: those without garlic and those with it, a distinction easily made thanks to the use of coloured twine. Ripening bestows an external whitish color, which is grayish-brown underneath due to the mold that covers it. Upon cutting, it appears reddish, almost rose colored, with a characteristic irregular white marbling due to the component of fat surrounding the muscular part. Flavor should recall that of its spices and the fragrance of aromatic herbs with or without a garlicky odor; the taste is slightly sweet, whereas the consistency is slightly chewy. The sopressa with the characteristics described above is a dry-fermented sausage whose production is quite widespread in the Veneto region. However, only sopressa Vicentina has the PDO mark. Even in the absence of updated statistical data about the sopressa Vicentina PDO production (50,211 kg in 2016), it can be considered as representing the lesser part of the total regional production of this kind of fermented meat product.

The physical–chemical and microbiological features of this product have only been partially investigated. Balzan et al. (2009) carried out a study on the sopressa Vicentina PDO produced by three different industrial manufacturers (two geographically located on the plain, the third one in an area about 400 meters above sea

FIGURE 12.2   Cross section of sopressa Vicentina PDO at 6 months of ripening.

level) according to production guidelines and with the use of a starter culture consisting of *Lactobacillus sakei* and *Staphylococcus xylosus*. The bacterial strains that were employed were isolated from artisanal sopressa, typified, characterized from a technological functionality point of view and in terms of food safety, and then used as technological starters. *Lactobacillus sakei* is a lactic acid bacteria species highly adapted to the meat environment. Belonging to the indigenous microbiota of meat, it is naturally active for the fermentation of artisan dry-cured sausage. Generally, it maintains its viability during the entire ripening period, that is to say, for several months. Its capacity to ferment hexoses to produce acids, also at low temperatures, makes it suitable for those drying and ripening processes conducted at environmental conditions similar to artisanal methods of manufacturing.

   Staphylococcus xylosus (a coagulase-negative staphylococcus) is commonly utilized as a lipolytic starter culture for fermented sausages (Talon and Montel, 1997). It plays an important role in the development of the flavor (through its proteolytic activity) and color of fermented products, as well as in the reduction of nitrate to nitrite. Andrighetto, Zampese, and Lombardi, (2001) studied the microflora of several samples of the so-called salame Veneto and sopressa, mainly produced in the province of Vicenza, all made without the use of starter cultures. The authors pointed out that in the salami (a dry fermented sausage of less than a kilogram with a diameter between 6 and 8 cm ripened for 30–60 days) *Lactobacillus sakei* and *Lactobacillus curvatus* were the most common (39.1% and 34.8%, respectively), whereas *Lactobacillus sakei* was the main Lactobacillus species present in the long-ripened sopressa (92.3% of the isolates). Microbial fermentation during the first week of processing significantly influenced the pH values (Figure 12.3a). All the samples were characterized by a pH value that after 6 days of drying varied between 4.9 and 5.3, while the final pH at 90 days showed

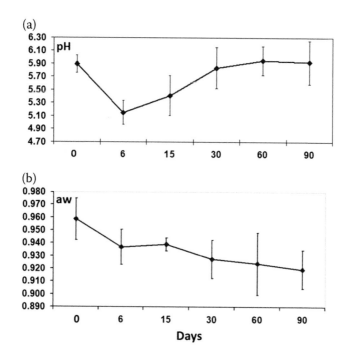

FIGURE 12.3 Variation of pH (a) and water activity (b) of sopressa Vicentina PDO during the ripening process.

values ranging between 5.7 and 6.3 ($P < 0.001$). The weight loss during processing and consequent salt concentration caused the gradual decreasing of water activity to final values between 0.87 and 0.91 ($P < 0.001$) (Figure 12.3b). *L. sakei* showed the capacity to colonize the product with high charges until the end of the seasoning (90 days), while *S. xylosus* exhibited a significant decrease, starting from the second month of ripening, in line with the increase in other species of non-starter lactic acid bacteria, probably favored by the rise in the pH value observed after the second week of ripening (Figure 12.3a).

During the ripening of the sopressa Vicentina PDO, a thick layer of mold formed on the casing surface. The morphological analysis of the colonies isolated from the samples and ripening rooms showed the presence of different molds, mainly green in color, with a velvety texture and cottony center. White or yellowish-white molds were also detected. The subsequent Random Amplified Polymorphic DNA-Polymerase Chain Reaction (RAPD-PCR) analysis made it possible to identify the presence of a large genetic polymorphism. At least 13 different genetic profiles were detected by combining the results of the RAPD-PCR and the analysis of the sequences. A clear prevalence of molds belonging to the *Penicillium* genus was noted, both on the isolates from the sopressa Vicentina PDO and those derived from environmental sampling. *Penicillium nalgiovense* is the clearly dominant species (59.7%) both in the air and in the dry-fermented sausages samples, followed by *Aspergillus candidus* (12.9%), *Penicillium nordicum* (8%) and *Penicillium olsonii*

(6.4%). There was a greater biodiversity in the samples ripened in the factories located on the plaisn compared to those aged in the mountains (data not published). The rise in the pH value during ripening that was evident in the samples (Figure 12.3a) is at least partly due to the metabolic activity of the molds that started to grow on the surface of the casing.

12.5.2 CIAUSCOLO (THE SPREADABLE SALAME)

Ciauscolo is a lightly smoked salame that has the PGI mark. According to the most rigorous tradition, ciauscolo, also known as *ciabuscolo* or *ciavuscolo*, is a delicatessen product made in a rather limited geographical area delineated by the Umbrian–Marche mountainous hinterland (Central Italy). This location has a climate that seems to predispose and allow the ciauscolo to mature in natural conditions. The name derives from the Latin *cibusculum* which means *small meal*, as the salame was consumed sparingly in the interval between breakfast and lunch as well as between lunch and dinner. This eating habit, now no longer practiced, is the reason for the longer ripening times that are observed in household products compared to industrial ones.

There are at least three historical documents that contain references to ciauscolo. The first is an annual communication of the prices of the main food products of the time; the second is an inventory of a Camerino retail shop dated 1737; and the third is a notarial deed containing an inventory of assets, drawn up on the occasion of an inheritance, that dates back to 1750.

The main feature of ciauscolo is its spreadability, and this distinguishes it from other dry-fermented sausages. This characteristic is obtained thanks to the particular mixture's composition and to the processing techniques, including an environment that is very humid. The meat used for its preparation must come from pigs whose average live weight per single batch was between 145 kg and 200 kg. The cuts allowed are belly, up to a maximum of 70%; shoulder, up to a maximum of 40%; and ham and loin trimmings, up to a maximum of 30%. The distinct character of this salami is formed in the grinding phase, which is repeated twice using a continuous meat grinder equipped with two molds, the first of which has large holes, the second of which has 2–3 mm diameter holes. In this way, a very fine and homogeneous mixture is obtained to which salt (2.5–3%), ground black pepper (0.3–0.5%), white wine (0.5–1 l/100 kg), crushed garlic (0.025–0.045%), ascorbate and nitrate salts are added. In the past, the selection of meat cuts included a greater percentage of fatty cuts—hence a color of the end product that was white–pale pink—and the mixture was ground even finer by undergoing, on average, three repeated grinds. Therefore, its spreadability was even greater than it is today.

Starter cultures are not generally used. However, in some production plants they are employed (Trani et al., 2010). The mixture must be stuffed into natural casings from the pig or cow. The binding is usually done with thin hemp twine to form bridles in a variable number of steps, depending on the industrial or homemade production type and on the length of the product (admissible length is 15–45 cm). The product is then transferred to a drying room with an ambient temperature or under controlled conditions of 19–23°C and 65–70% relative humidity (RH) where it remains for a week. On the third or fourth day of drying, it undergoes smoking. In traditional domestic practice, in the same tightly closed room, braziers containing sawdust of softwood, such as beech, poplar, etc., are lit. The smoking room is left to saturate, and after 3 or 4 hours, an air exchange takes place. At the end of the drying process, the product is brought to the ripening room or to a cellar, where in 4–11 weeks, depending on the seasonal variations and market demand, the production cycle is completed (never less than 15 days). In the ripening rooms, the temperature varies from 12°C–14°C in winter, and is 18°C in summer, and the RH is around 80–85%.

Given that the factories are close to the chain of the Sibillini mountains, there are very particular climatic conditions and the temperature, even in the summer months, rarely exceeds 30°C. The ventilation is exclusively natural and the air exchange is performed once a day, in the evening in the warmer months. In domestic production, the slaughtering and subsequent processing of the meat is traditionally concentrated in the winter months, generally between the days immediately preceding Christmas and Epiphany, considered the first "big cold" of the winter period. According to local traditions of home production, the steps of drying/smoking and ripening are slightly different. In homes, braziers are rarely used and smoking comes directly from the fireplace of the kitchen. This operation is performed once a day for a week, after which time the ciauscoli are transferred to the cellar, notoriously the coolest place in the house due to its consistent microclimate, thus making it ideal for the maturation and conservation of food products.

Almost all the products obtained in the home processing of pork meat undergo a period of rest in the kitchen that varies in relation to the type of salami in question. This practice can be traced back to ancient times when the meat products, immediately after their preparation, were left in kitchen—where they were usually also prepared—for a first drying. The chimneys of old houses, over stoves and fires used for both heating and cooking, conducted smoke, and this was an additional important factor for cured meats due to the subsequent acquisition of organoleptic characteristics. As with many other fermented meat products, the tradition of making them at home is still as diffuse as that of salami produced on an industrial level. In home preparation the minced meat is spread on a table, added to the other ingredients and mixed manually (as they say in the local dialect of the Marche, *menata*, which literally means kneaded, something like what takes place in the making of homemade pasta).

In addition to the traditional recipe, there is also a less known variant whose mixture is characterized by the presence of liver (about 8%) and the absence of garlic (Spisni, 45). The distinctive marker of this recipe, especially at the home level, is the use of the meat cuts that are richest in blood, such as those in the throat and the areas surrounding the jugulation point. These meats help give a dark red color, tending to brown, to the end product. Given that this is a product liver-based, and therefore rich in carbohydrates, it has a particularly strong flavor, linked to the high fermentability of the mixture, which tends to be subject of lipid esters hydrolyzation (Cerutti, 1999). The ciauscoli are usually consumed between the end of March and the end of June, since an overly long conservation period can lead to rancidity of the fat component, leading to a complete compromise of the entire product. Industrial production is almost continuous throughout the year.

As for the production of ciauscolo immediately before or during the warm season, the percentage of salt and the quantity of fat vary, with a slight increase of the former and a decrease of the latter. The final weight ranges from 0.4–2.5 kg, and the external appearance is cylindric with a diameter of 4.5–10 cm. The minimum protein content is 15% whereas the fat varies between 32% and 42%, with ash content hovering around 4–5%. The proportion of water to protein must be in a ratio no larger than 3:1 and the ratio of fat/protein must have a maximum of 2:8.

The microbial flora of ciauscolo was investigated analysing samples processed from 15 to 45 days and obtained from 14 artisan and 8 industrial production plants located in different geographical areas of the Marche region (Silvestri et al., 2007). The LAB varied from 6 to over 8 log cfu/g, while coagulase negative cocci (CNC) ranged from less than 2 to over 6 log cfu/g, and yeasts from 2 to a little less than 6 log cfu/g. In almost all the samples, the total coliforms and *E. coli* counts were <10 cfu/g. The analysis of LAB, CNC and yeast highlighted a great variability among the 22 productive sites, suggesting that the composition of the microbial population of ciauscolo is strongly related to the manufacturing techniques employed, as well as the environmental diversity that characterizes different geographical areas. It is interesting to note that, in all the samples, LAB was the dominant group over CNC and yeasts.

Using the culture-independent method of PCR coupled with denaturing gradient gel electrophoresis (DGGE), the authors found that among the samples tested, *Lactobacillus sakei* and *Lactobacillus curvatus* were the most frequently detected. This is in line with what has been shown for other dry fermented sausages made in Italy (Cocolin et al., 2001; Coppola et al., 1998) and also in other European countries (Aymerich et al., 2003). The prevalence of these two psychotropic species could be due to the low temperatures used during the drying and ripening steps (Parente, Grieco, and Crudele, 2001). Contrary to what was expected, Silvestri et al., (2007) did not find any relationship between the site of production or the geographical area and the microbial profile of the samples (bacteria and yeasts) using the unweighted pair group method with arithmetic mean (UPGMA) tree. This was most likely a consequence of the standardisation in the use of technologies both in the artisanal and in the industrial plants that flatten the effect that the natural climate and environment of each productive site could have on the microbial ecology of the product. On the contrary, the same authors found interesting results regarding the yeast ecology in terms of the DGGE results. It seems that the garlic, sometimes used in generous quantities, may have had a repressive effect on the yeast, potentially due to its fungistatic effect (Olesen and Stahnke, 2000). Among others, *Debaryomyces hansenii* was the most frequently detected among the samples, probably due to its high tolerance to salt and its ability to grow at low temperatures and to metabolize organic acids (Osei et al., 2000). At the same time, *D. hansenii* also has a positive effect on the development of the flavor and on the stabilization of the reddening reaction (Rantsiou and Cocolin, 2006).

Two minor species of yeast, namely *Rhodotorula mucillaginosa* and *Trichosporon brassicae*, were isolated from the meat mixture but not from the ciauscolo during fermentation (Aquilanti et al., 2007). While the first type of yeast has been described as occurring in fermented meat products (Gardini et al., 2001), the latter had never been isolated before, probably due to the low competitiveness of this genus in the microbial meat fermenting ecosystem. Federici et al. (2014) examined the composition of LAB strains isolated from 120 samples of ciauscolo produced without starter cultures by 14 artisanal plants located in the Marche region. According to their investigation *Lactobacillus sakei, Pediococcus pentosaceus,* and *Lactobacillus plantarum* were the most common found inside the tested samples using the analytical approach of amplified ribosomal DNA restriction analysis (ARDRA) and RAPD-PCR.

Interesting to note was also the presence of *Lactococcus lactis* ssp. *lactis* that was only apparent in the first 3 days of the ciauscolo's ripening (Aquilanti et al., 2007). In the authors' opinion, this could represent a sort of differentiating trait of this salame, as opposed to other dry fermented sausages, where hetero-fermentative LAB species are dominant from the beginning of processing (Parente et al., 2001). Afterward, *L. curvatus, L. plantarum, S. xylosus, S. saprophyticus, S. equorum* and *D. hansenii* become the prevalent microbial flora during the ripening phase of ciauscolo as evidenced by 16S rRNA gene V1 and V3 regions that were used as targets in the DGGE profiling of the LAB and CNC communities. According to Ranucci et al. (2013) and Trani et al. (2010), the pH and water activity of ciauscolo ripened for 20–30 days varied from 4.7 to 5.5 and from 0.90 to 0.93, respectively. After studying the biochemical traits of ciauscolo collected from 14 different production plants throughout the Marche region, Trani et al. (2010) concluded that, in this kind of salame, lipolysis is of minor importance, as compared to other types of dry-fermented sausages, as far as the development of the chemical features of the end product are concerned. On the contrary, proteolysis, thanks to the fine mincing of the meat cuts, the low pH and the water activity, can exert a strong effect on sensory traits due to the effect of microbial enzymes that reduce the protein fractions, mainly sarcoplasmic proteins, to small peptides and amino acids, which are precursors of aromatic compounds. They also found a significant correlation between gly-pro dipeptide content and LAB load, due to the X-prolyl dipeptidyl aminopeptidase activity of a serine protease (Miyakawa et al., 1991), which had been previously identified in several LAB species by Casey and Meyer (1985). In the same samples, the concentration of methyonine and aspartic acid seem to be a possible molecular marker capable of distinguishing between double and single mincing. The number of mincing steps can favor proteolysis by increasing the contact surface, shortening the muscle fibers, and delivering cellular sarcoplasm along with the enzymes.

12.5.3 SALAME MILANO (THE MOST RENOWNED)

Milano salame is one of the most widely consumed in Italy and its production, initially located in the province of Milan (hence the name), has now extended to the entire country. In spite of this, there is neither a protection mark nor any real production regulation. The Milanese tradition of charcuterie is a long-standing one. For centuries, the Milanese *cervellati*, which literally means brains, have been famous throughout Italy. They are sausages made of pork seasoned with cheese and spices, including saffron, which gives the casing its characteristic, deep yellow color. In 1993, the Italian National Unification (UNI) published a voluntary standard that defines the composition and the characteristics of the Milano salame, which was updated in 1996 with the standard UNI 10268:1996. Consumption of this product is widespread in many other countries and continents, so much so that in Brazil the Ministério de Agricultura, Pecuária and Abastecimento (MAPA) has drawn up a technical *regulation de Identitade and Qualidade de Salame tipo Milano* which sets the criteria for its production.

According to UNI 10268:1996, the salame Milano is cylindrical in shape, consisting of a mixture of pork or pork and beef, the latter in an amount not greater than 20%, obtained from skeletal muscle of the carcass, and hard pork fat, salt, pepper in pieces or minced, and garlic. Ingredients can also include wine; sugar and/or dextrose, and/or fructose, and/or lactose; skimmed milk powder or caseinates; starter cultures; sodium and/or potassium nitrate; sodium and/or potassium nitrite; as well as ascorbic acid and its sodium salt (Cantoni, 2003). A characteristic of the salame Milano is its so-called rice grain pasta texture, in which the thin lean and fat particles maintain their own edges (hence, without fat smearing) and are evenly distributed. In the past, this particular conformation was obtained using a machine called a *finimondo* (which literally means *end of the world*). This is a cutting machine designed in Italy in the early 20th century, and it consists of two sets of circular steel knives (with a diameter of 200–300 mm) rigidly mounted on two axes rotating in opposing directions.

The pork, in large pieces at a very low temperature, enters the *finimondo* and is cut into strips of about 10 mm thickness, which are immediately passed for a second time through the same machine. The pork, thus cut together with already minced beef, and the pork fat, cut into frozen cubes, is then ground through a mold whose holes are 3–3.5 mm in diameter. The meat coming out of the meat grinder is then mixed with the other ingredients using a second special machine, called a sprea-der–mixer. It is unique because it does not stress the meat overly much during mixing, and at the same time, it produces a uniform mixture of fat and meat, keeping the particles divided so as to obtain the characteristic appearance of grains of rice, which is evident when the salame is sliced. It is very hard to obtain this technological result using common kneading machines equipped with spatulas.

These days, the cutter is replacing the *finimondo* almost everywhere. This is a machine that ingeniously takes the place of the meat grinder and the kneading machine combined. Unlike the meat grinder, it does not crush the meat and lard (chilled and frozen), but cuts them cleanly, dryly and uniformly. Thanks to the high cutting speed of the blades or knives, the result is very fine and perfectly homogeneous mixtures, where the particles of lean and fat are immediately and completely amalgamated with other ingredients. (For details about the employment of the bowl cutter for the manufacture of salame, see Feiner, 2016.) The salame Milano is generally made of 70–75% lean pork (mainly shoulder) and 25–30% pork back fat. The use of beef is less frequent nowadays. Salt is added at 2.8–3.0%, sugars at 5–7 g/kg, and spices at around 5 g/kg. The spices most commonly used are white pepper, cardamom and garlic. Red wine is frequently added, as well, at around 1 l/100 kg of the mixture mass. In the absence of stringent manufacturing regulations, other recipes can be found, and different bromatological compositions of the end product are possible, as has been shown by different studies (Casiraghi et al., 1996; Novelli et al., 1998; Zanardi et al., 2000). In Table 12.2, some physical, chemical, and sensory data regarding salame Milano are reported.

TABLE 12.2

Mean, Minimum and Maximum of Determined Variables on 37 Samples of Salame Milano (adapted from Casiraghi et al., 1996)

Variable	Mean	Minimum	Maximum
pH	5.38	4.79	6.11
Moisture (g/100 g)	36.25	24.01	46.42
Protein (g/100 g)	25.76	20.9	35.6
Fat (g/100g)	32.71	21.44	41.6
Soluble nitrogen (g/100 g)	4.42	3.17	5.69
Collagen (g/100 g)	2.23	1.56	3.76
NaCl (g/100 g)	4.7	3.03	6.14
NaCl/moisture	0.13	0.07	0.25
Collagen/protein	0.09	0.06	0.13
Fat/protein	1.29	0.82	1.82
Moisture/protein	1.44	0.67	2.08
Soluble N/protein	0.17	0.14	0.21
Lactic acid (g/100 g)	1.21	0.71	1.96
Acetic acid (g/100 g)	0.13	0.05	0.26
L*	47.23	39.37	52.4
a*	22.09	18.16	23.31
b*	6.33	5.08	7.44
Modulus (N/cm^2)	38.48	10.61	68.61
Elasticity index	0.44	0.25	0.57
"Aged taste" score	4.34	2.71	6.3
Preference score	4.08	2.33	5.46

The mixture is then stuffed into fibrous casings whose diameter can vary from a little less than 50 mm to 110 mm and beyond. The minimum ripening time has been fixed at 4 weeks for sizes not over 50 mm, to 15 weeks for the largest diameters. The voluntary standard (UNI 10268:1996) fixed the following physico-chemical characteristics: total proteins (min. 20%), collagen/protein ratio (max. 0.12), moisture/protein ratio (max. 2.30), fat/protein ratio (max. 2.00), pH ($\geq$ 5.2). Traditionally, natural casings are used, such as the so-called *crespone* (obtained from the large intestines of swine), hence this salame is also known as *crespone*. After bagging, the binding is carried out, which, especially in the case of large size products, is a very laborious operation effectuated by pulling the hemp twine first in the longitudinal direction and then in the crosswise one with very narrow links. Therefore, the proximate composition and quality of salame Milano on the market can vary greatly according to different recipes and the different processing technologies adopted by the numerous producers.

To identify the main quality features of this dry-fermented sausage, Casiraghi et al. (1996) tested n = 37 samples of salame Milano (average weight 3.5 kg; mean age 3.5 months), purchased on the market and produced by the largest Italian manufacturers. Using physical, chemical, and sensorial analyses, the authors explored the data by means of multivariate statistical techniques (principal component analysis, PCA; multiple linear regression; distance-weighted least-squares smoothing algorithm, DWLS; and quantitative descriptive analysis, QDA) to search for correlations between the analytical characteristics and the sensorial score of preference and the attribute "aged taste," given by a panel of judges (Table 12.2). Generally speaking, all the tested samples had chemical and physical data within the thresholds fixed in the voluntary standard (UNI, 10268:1996), with only four samples having pH values lower than the fixed limit, between 4.8 and 5.2.

According to the results of PCA, the variables with explained variance lower than 60% were considered irrelevant in the chemometric model, therefore only six of them were selected: pH, L*, modulus, lactic acid, moisture/protein, and NaCl/moisture were considered, from a statistical point of view, as the more descriptive. The PC analysis showed that texture and color, instrumentally measured by means of modulus and L* (sample luminosity), were found to have great significance. According to QDA graphical display for the ratio moisture/protein and L*, best quality corresponded to a lower than average value. For the ratio NaCl/moisture and modulus, the best quality corresponded to a higher than average value, whereas for pH, best quality was related to high values of the parameter. Finally, for lactic acid, positive correlation with sensory preference was noted for low values. Samples with the highest preference scores showed values slightly higher or lower than the mean for parameters related to ripening (modulus, L*, NaCl/moisture, moisture/protein), but preference decreases as the product becomes too ripe (high value of modulus and NaCl/moisture). This is in line with the general belief of producers that higher preference is attributed to the product that is medium-aged. Higher pH values and less lactic acid are also related to greater degrees of preference.

In the opinion of producers, the optimal texture of salame Milano is moderately hard with a basically dark red appearance, after slicing, including pinkish parts well balanced with the fat parts and a slightly spicy flavor. The flavor of dry-fermented sausage is due to chemical compounds arising from carbohydrates, lipids and proteins through many chemical reactions that happen during ripening (Viallon et al., 1996). The type and quality of fresh meat used in production, the processing conditions, the addition, or not, of starter cultures, and the quantity and type of spices have a decisive effect on the characteristics of the end product (Stahnke, 1994; Montel et al., 1996). Meynier et al. (1999) investigated the chemistry of aroma compounds of salame Milano samples taken from the Italian market that were representative of the main commercial brands. The volatile compounds were studied by chemical (extracting them by dynamic headspace volatile concentration using a purge-and-trap concentrator) and olfactometry analysis, identifying them by Gas-Chromatography-Mass-Spectrometry (GC/MS) and quantifying with an external standard using a flame ionisation detector.

The gas chromatographic effluent was sniffed by a trained panel of eight people to establish the relationships between the odors perceived and the volatile compounds identified by GC/MS. A total of volatile compounds between n = 93 and n = 123 (varying among the different commercial brands) were identified. According to

TABLE 12.3
Results from GC Effluent Sniffing of Purge-and-Trap Extract of Salame Milano (adapted from Meynier et al., 1999)

Odor	Compound Identified [a]	Tentative Identification [b]
Butter, caramel, rancid	2-Methylbutanal	
Chocolate	n.i.[c]	
Gas, onion, unpleasant	1-Propene-3-methylthio	
Musty, fruity		Propyl acetate
Fruity, apple, floral		Ethyl butanoate
Grass, floral	Hexanal	
Butyric, unpleasant		3-3'-thiobis-1-propene
Potatoes	Heptanal	
Mashed potatoes		2,6-Dimethylpyrazine
Potatoes, popcorn	n.i.	
Mushroom	1-Octen-3-ol	
Lemon, fruity, green	Myrcene + α-phellandrene	
Menthol		
Floral, rose		Cineole
Fruity, fresh		o-Cymene/γ-terpinene
Mushroom, floral		Octanol
Fruity, eucalyptus	Terpinolene	
Potatoes, butter		Decanal
Meaty, bread, toasted		Dipropyl trisulfide

Notes
[a] Molecule found at a retention index close to that of the odor and having an odor close to that described in the literature.
[b] Molecule found at a retention index close to that of the odor or having an odor close to that described in the literature.
[c] n.i. = not identified.

GC/MS identification, around 60% of the identified volatile compounds came from spices, nearly 19% from lipid oxidation, a little less than 12% from aminoacid catabolism, and close to 5% from fermentation and/or microbial esterification. In spite of a very long list of identified volatile compounds, the number that concretely could have a key role in the aroma of salame Milano were slightly less than 20 (Table 12.3). Five odors were attributed to terpenes, and three others could be tentatively attributed to terpenes (from the spices). Three odors could be related to non-heterocyclic-sulphur compounds, which play a key role in meat flavor because of their very low threshold of perception (Mottram, 1991). Three more compounds were attributed to lipid oxidation (hexanal, heptanal and decanal), whereas 2-methylbutanal arose from amino acid catabolism. Two odors were attributed to esters: propyl acetate and ethyl butanoate (they were described as fruity)—esters are essential for the typical aroma of fermented sausages. Given that the chemical and biochemical mechanisms that govern the formation of lipid oxidation compounds and protein catabolism are generally common, the authors concluded that some differences in the aromas of different brands of salame Milano could be related to the type and the quantity of spices included in the recipe.

Rebecchi et al. (1998) used the RAPD fingerprinting technique and 16S rDNA sequencing to characterize the dynamics of lactobacilli and staphylococci growth during the 60 days of salame Milano ripening that was effectuated without the use of starter culture. The results demonstrated a strong selective effect on microflora due to the processing conditions. In the fresh mixture, several microbial groups were detected, such as micrococci, coagulase-positive staphylococci, lactobacilli, enterococci, coliforms and fungi. The progressive acidification and the increase in ionic strength after 15 days of ripening showed that lactic acid bacteria reached their highest level, whereas, after 1 month of ripening, the microbial community was mainly represented by *Lactobacillus plantarum*, *Lactobacillus sakei*, *Staphylococcus xylosus* and *Staphylococcus sciuri*. At the same time, most of the initial contaminant microflora had disappeared.

After the first month of ripening and up until the end of the process (60 days), the bacterial community remained unchanged. According to the RAPD fingerprint, at the beginning of the fermentation process more than 30 strains of staphylococci were detected in the meat mixture. After 15 days, only five RAPD profiles were detected, and, after 30 days, only two dominant strains were identified. These latter strains were not detected at the beginning of the process. Furthermore, significant differences were highlighted in the nitrate reductase, in the proteolytic and lipolytic activity of the different staphylococcal strains that took turns during the 60 days of ripening. Papa, Zambonelli, and Grazia (1995) made an experimental batch of salame Milano using 75% pork and 25% lean beef, 3% sodium chloride, 1% powdered milk, 0.35% dextrose, 0.1% cracked black pepper, 0.01% cinnamon, 0.015% garlic, 0.006% cloves, 0.5 l/100 kg white wine, nitrate, nitrite and ascorbate (0.009%, 0.01% and 0.05% respectively). To the mixture, a strain of *Lactobacillus plantarum* P70 previously isolated from other dry-fermented sausages—where it showed a remarkable vigor and acidogenic activity—was added. During the 70 days of processing the pH varied from 5.82 to 5.05 and the a_W from 0.964 to 0.93, both measured in the center of the salame. Higher and lower values for the same variables were respectively measured in the outer part of the salame. On the seventh day of drying, the LAB count reached 7×10^8 cfu/g, always measured from the center of the samples. The authors stressed the fact that the physico-chemical outputs were not homogeneous in the salame Milano of large diameter, with a gradient that occurs from the outside toward the center. The main microbial activity, especially those of LAB, occurs at the outermost section of the salami. Thus, the substantial changes took place here, while at the center everything came to a stop when lactic fermentation had terminated.

12.6   CONCLUSION

The pig has had a fundamental role as a source of noble proteins and more generally of sustenance for families in peasant Italy of the last two centuries. The breeding of pigs for the production of meat and cured meats for family self-consumption is a practice that still exists in the rural areas of many Italian regions. Generally these are animals that are slaughtered at a live weight between 200 and 300 kg and whose meat is almost completely transformed into charcuterie products using salt, spices and potassium nitrate. The pig is slaughtered between the end of November and the first half of December and the salame is ripened according to the natural temperature and humidity of the season. The ripening never ends before 30–40 days for smaller sized salami. Today, this traditional method of processing pork is sparking interest in Italy from the industry that wants to meet the expectations of an increasingly numerous number of consumers attentive to the quality and naturalness of food products. Another challenge that will directly affect the breeding of heavy pigs in Italy concerns efficiency in terms of economic and environmental sustainability of this productive chain, which at the moment, is lagging behind the breeding of light pigs.

Although Italian dry-fermented sausages still rely on traditional and local manufacturing processes, in industrial production their formulations have evolved in the last decades, driven by a healthier nutritional profile. The first factor behind this change is the raw material with higher concentrations of minerals and vitamins than in the past, as well as a progressively reduced fat content and a balanced content of saturates and unsaturated fatty acids. The modulation of the use of ingredients (for example salt, fat, plant extracts) and additives (for example nitrates and nitrites), together with the use of suitably optimized process technologies for the control of drying and curing conditions, has led to the production of salami, especially those produced on an industrial scale, of a composition more aligned with the nutritional recommendations of the scientific community and responding to the demand of some groups of modern consumers for the reduction or elimination of additives and preservatives from the lists of ingredients to satisfy a preference for energy-reduced and minimally processed products.

REFERENCES

Andrighetto, C., Zampese, L., and Lombardi, A. 2001. RAPD-PCR characterization of lactobacilli isolated from artisanal meat plants and traditional fermented sausages of Veneto region (Italy). *Letters in Applied Microbiology* 33:26–30.

Aquilanti, L., Santarelli, S., Silvestri, G., Osimani, A. Petruzzelli, A., and Clementi, F. 2007. The microbial ecology of a typical Italian salami during its natural fermentation. *International Journal of Food Microbiology* 120:136–145.

ASSICA (Associazione Industriali della Carne e dei salumi). 2019. Rapporto Annuale. Analisi del settore e dati economici 2018. https://www.assica.it/it/pubblicazioni/rapporto-annuale/archivio.php (accessed July22, 2020).

Aymerich, T., Martín, B., Garriga, M., and Hugas, M. 2003. Microbial quality and direct PCR identification of lactic acid bacteria and nonpathogenic staphylococci from artisanal low-acid sausages. *Applied and Environmental Microbiology* 69:4583–4594.

Baldini, P., Palmia, F., Pezzani, G., and Lambertini, L. 1983. Indagine sulla composizione del grasso superficiale di prosciutti freschi destinati alla produzione del prosciutto di Parma. *Industria Conserve* 58:219–222.

Ballarini, G. 2003. Piccola storia della grande salumeria italiana. Milano: Edra.

Balzan, S., Busin, F., Alberghini, L., Giaccone, V., and Novelli, E. 2009. Microbiological profile of Soprèssa Vicentina PDO from raw materials to final products. Book of Abstracts of the 55th International Congress of Meat Science and Technology (ICoMST), Copenhagen, Denmark.

Baruzzi, M., and Montanari, M. 1981. *Porci e porcari nel medioevo. Paesaggio Economia Alimentazione.* Bentivoglio: Museo della civiltà contadina di San Marino.

Benasaglio, G. 2017. I salumi: stili di servizio e forme di abbinamento. In *La grande salumeria italiana.* C. Cipolla. Milano: FrancoAngeli Srl.

Bittante, G., Sorato, O., Montobbio, P., and Gallo, L. 1990. Crossbreeding of Large White sows with Belgian Landrace, Duroc and Spotted Poland Boars: Effects on carcass traits and meat quality. *Zootecnica e Nutrizione Animale* 26:179–195.

Cantoni, C. 2003. Il salame Milano: tecnologia e sua origine probabile. *Premiata Salumeria Italiana* 1:45–49.

Casey, J., and Meyer, J. 1985. Presence of X-prolyldipeptidyl peptidase in lactic acid bacteria. *Journal of Dairy Science* 68:3212–3215.

Casiraghi, E., Pompei, C., Dellaglio, S., Parolari, G., and Virgili, R. 1996. Quality attributes of Milano salami, an Italian dry-cured sausage. *Journal of Agricultural and Food Chemistry* 44:1248–1252.

Cerutti, G. 1999. *Residui, additivi e contaminanti degli alimenti.* Milano: Edizioni Tecniche Nuove.

Cocolin, L., Manzano, M., Cantoni, C., and Comi, G. 2001. Denaturing gradient gel electrophoresis analysis of the 16S rRNA gene V1 region to monitor dynamic changes in the bacterial population during fermentation of Italian sausages. *Applied and Environmental Microbiology* 67:5113–5121.

Commission Regulation (EC) No 492/2003 of 18 March 2003 supplementing the Annex to Regulation (EC) No 2400/96 on the entry of certain names in the 'Register of protected designations of origin and protected geographical indications' provided for in Council Regulation (EEC) No 2081/92 on the protection of geographical indications and designations of origin for agricultural products and foodstuffs (Soprèssa Vicentina, Asparago verde di Altedo, Pêra Rocha do Oeste). *Official Journal of European Union* L 473, 19.03.2003, p. 3–4.

Conter, M., Zanardi, E., Ghidini, S., Pennisi, L., Vergara, A., Campanini, G., and Ianieri, A. 2007. Survey on typology, PRPs and HACCP plan in dry fermented sausage sector of Northern Italy. *Food Control* 18:650–655.

Coppola, R., Giagnacovo, B., Iorizzo, M., and Grazia, L. 1998. Characterization of lactobacilli involved in the ripening of soppressata molisana, a typical southern Italy fermented sausage. *Food Microbiology* 15:347–353.

Della Casa, G., Panciroli, A., Cavuto, S., Faeti, V., Poletti, E., Calderone, D., and Marchetto, G. 1999. Effects on pig fat quality of partially hydrogenated lard fed to growing-finishing heavy pigs. *Zootecnica e Nutrizione Animale* 25:51–62.

Enser, M. 1991. Animal carcass fats and fish oil. In *Analysis of oilseeds, fats and fatty foods,* ed.J. B. Rossel and J. L. R. Pritchard, 329. London: Elsevier Applied Science.

Federici, S., Ciarrocchi, F., Campana, R., Ciandrini, E., Blasi, G., and Baffone, W. 2014. Identification and functional traits of lactic acid bacteria isolated from Ciauscolo salami produced in Central Italy. *Meat Science* 98:575–584.

Feiner, G. 2016. Fermented salami: Non–heat treated In *Salami – Practical science and processing technology,* ed.G. Feiner, 111–176. Oxford: Elsevier Inc.

Franci, O., Pugliese, C., Acciaioli, A., Poli, B. M., and Geri, G. 1994. Comparison among progenies of Large White, Italian Landrace, Belgian Landrace, Duroc, Cinta Senese Boars and Large White sows at 130 and 160 kg live weight. 1. Performances in vita and at slaughter. *Zootecnica e Nutrizione Animale* 20:129–142.

Gardini, F., Suzzi, G., Lombardi, A., Galgano, F., Crudele, M. A., Andrighetto, C., Schirone, M., and Tofalo, R. 2001. A survey of yeasts in traditional sausages of southern Italy. *FEMS Yeast Research* 1:161–167.

Hocquet, J. C. 1999. *Denaro, Navi e Mercanti a Venezia 1200-1600.* Roma: Il Veltro Editrice.

Hocquet, J. C. 2003. *Le saline dei veneziani e la crisi del tramonto del Medioevo*. Roma: Il Veltro Editrice.
INSOR (Istituto Nazionale di Sociologia Rurale). 2002. *Atlante dei prodotti tipici – I salumi*. Roma: Agra Editrice.
Laszlo, P. 2004. *Storia del sale. Miti, cammini e saperi*. Roma: Donzelli Editore.
Meynier, A., Novelli, E., Chizzolini, R., Zanardi, E., and Gandemer, G. 1999. Volatile compounds of commercial Milano salami. *Meat Science* 51:175–183.
Micossi, E., Tubaro, F., Magnabosco, P., Ciani, F., and Urini, F. 1996. Analisi della insaturazione degli acidi grassi del tessuto adiposo del maiale. *Rivista di Suinicoltura* 37:97–101.
Ministero delle Politiche Agricole Alimentari e Forestali. Dipartimento delle Politiche Competitive della Qualità Agroalimentare e della Pesca. Direzione Generale per la Promozione della Qualità Agroalimentare. SAQ VII. Disciplinare di produzione della denominazione di origine protetta «Prosciutto di Parma». https://www.politicheagricole.it/flex/files/6/1/c/D.324fdfa38ad9aa3459c4/Disciplinare_Prosciutto_di_Parma_28.11.2013.pdf (accessed July22, 2020).
Miyakawa, H., Kobayashi, S., Shimamura, S., and Tomita, M. 1991. Purification and characterization of an X-prolyldipeptidyl aminopeptidase from Lactobacillus delbrueckii ssp. bulgaricus LBU-147. *Journal of Dairy Science* 74:2375–2381.
Montel, M. C., Reitz, J., Talon, R., Berdagué, J. L., and Rousset-Akrim, S. 1996. Biochemical activities of Micrococcaceae and their effects on the aromatic profiles and odours of a dry sausage model. *Food Microbiology* 13:489–499.
Mordenti, A., Rizzitelli, N., and Cevolani, D. 1992. *Manuale di alimentazione del suino*. Bologna: Edagricole.
Mottram, D. S. 1991. Meat. In *Volatile Compounds in Food and Beverages*, ed.H. Maarse, 107–177. New York: Marcel Dekker.
Novelli, E., Zanardi, E., Ghiretti, G. P., Campanini, G., Dazzi, G., Madarena, G., and Chizzolini, R. 1998. Lipid and cholesterol oxidation in frozen stored pork, salame Milano and mortadella. *Meat Science* 48:29–40.
Olesen P. T., and Stahnke, L. H. 2000. The influence of Debaryomyces hansenii and Candida utilis on the aroma formation in garlic spiced fermented sausages and model minces. *Meat Science* 56:357–368.
Osei Abunyewa, A. A., Laing, E., Hugo, A., and Viljoen, B. C. 2000. The population change of yeasts in commercial salami. *Food Microbiology* 17:429–438.
Papa, F., Zambonelli, C. and Grazia, L. 1995. Production of Milano style salami of good quality and safety. *Food Microbiology* 12:9–12.
Parente, E., Grieco, S., and Crudele, M. A. 2001. Phenotypic diversity of lactic acid bacteria isolated from fermented sausages produced in Basilicata (Southern Italy). *Journal of Applied Microbiology* 90:943–952.
Parente, E., Martuscelli, M., Gardini, F., Grieco, S., Crudele, M. A., and Suzzi, G. 2001. Evolution of microbial populations and biogenic amine production in dry sausages produced in Southern Italy. *Journal of Applied Microbiology* 90:882–891.
Piedrafita, J., Christian, L., and Lonergan, S. M. 2001. Fatty acid profiles in three stress genotypes of swine and relationships with performance, carcass and meat quality traits. *Meat Science* 57:71–77.
Rantsiou, K., and Cocolin, L. 2006. New developments in the study of the microbiota of naturally fermented sausages as determined by molecular methods: A review. *International Journal of Food Microbiology* 108:255–267.
Ranucci, D., Branciari, R., Acuti, G., Della Casa, G., Trabalza-Marinucci, M., and Miraglia, D. 2013. Quality traits of Ciauscolo salami from meat of pigs fed rosemary extract enriched diet. *Italian Journal of Food Safety* 2:49–52.

Rea, S., Pacifici, L., Stocchi, R., Loschi, A. R., Ceccarelli, M. 2003. Storia, caratteristiche e tecnologia di produzione del Ciauscolo. *Industrie Alimentari* 42:371–377.

Rebecchi, A., Crivori, S., Sarra, P. G., and Cocconcelli, P. S. 1998. Physiological and molecular techniques for the study of bacterial community development in sausage fermentation. *Journal of Applied Microbiology* 84:1043–1049.

Regulation (EU) No 1151/2012 of the European Parliament and of the Council of 21 November 2012 on quality schemes for agricultural products and foodstuffs. *Official Journal of European Union* L 343, 14.12.2012, p. 1–29.

Regulation (EU) No 1308/2013 of the European Parliament and of the Council of 17 December 2013 establishing a common organisation of the markets in agricultural products and repealing Council Regulations (EEC) No 922/72, (EEC) No 234/79, (EC) No 1037/2001 and (EC) No 1234/2007. *Official Journal of European Union* L 347, 20.12.2013, p. 671–854.

Russo, V., Lo Fiego, D. P., Bigi, D., Nanni Costa, L., and Pignatti, M. 1989. Relationship between carcass lean content and the technological and commercial characteristics of Parma ham. Book of Abstracts 40th Annual Meeting of European federation of Animal Science (EAAP), p. 273. Dublin, Ireland.

Russo, V., and Costa, N. 1995. Suitability of pig meat for salting and the production of quality processed products. *Pig News and Information* 16:17N–26N.

Silvestri, G., Santarelli, S., Aquilanti, L., Beccaceci, A., Osimani, A., Tonucci, F., and Clementi, F. 2007. Investigation of the microbial ecology of Ciauscolo, a traditional Italian salami, by culture-dependent techniques and PCR-DGGE. *Meat Science* 77:413–423.

Spisni, A. 2017. Lo stile Emiliano. In *La grande salumeria italiana*, ed.C. Cipolla. Milano: FrancoAngeli Srl.

Stahnke, L. H. 1994. Aroma components from dried sausages fermented with Staphylococcus xylosus. *Meat Science* 38:39–53.

Talon, R., and Montel, M. C. 1997. Hydrolysis of esters by staphylococci. *International Journal of Food Microbiology* 36:207–214.

Talon, R., Lebert, I., Lebert, A., Leroy, S., Garriga, M., Aymerich, T., Drosinos, E. H., Zanardi, E., Ianieri, A., Fraqueza, M. J., Patarata, L., and Laukova, A. 2007. Traditional dry fermented sausages produced in small-scale processing units in Mediterranean countries and Slovakia. 1: Microbial ecosystems of processing environments. *Meat Science* 77:570–579.

Trani, A., Gambacorta, G., Loizzo, P., Alviti, G., Schena, A., Faccia, M., Aquilanti, L., Santarelli, S., and Di Luccia, A. 2010. Biochemical traits of Ciauscolo, a spreadable typical Italian dry-cured sausage. *Journal of Food Science* 75:514–524.

UNI 10268:1996. Salame Milano. Definizione, composizione, caratteristiche. Milano: Ente Nazionale Italiano di Unificazione. Seconda edizione.

Viallon, C., Berdagué, J. L., Montel, M. C., Talon, R., Martin, J. F., Konjdoyan, N., and Denoyer, C. 1996. The effect of stage of ripening and packaging on volatile content and flavour of dry sausage. *Food Research International* 29:667–674.

Virgili, R., Degni, M., Schivazappa, C., Faeti, V., Poletti, E., Marchetto, G., Pacchioli, M. T., and Mordenti, A. 2003. Effect of age at slaughter on carcass traits and meat quality of Italian heavy pigs. *Journal of Animal Science* 81:2448–2456.

Zanardi, E., Novelli, E., Ghiretti, G. P., and Chizzolini, R. 2000. Oxidative stability of lipids and cholesterol in salame Milano, coppa and Parma ham: Dietary supplementation with vitamin E and oleic acid. *Meat Science* 55:169–175.

Zappa, A., and Pugliese, C. 1991. Determinazioni analitiche sul tessuto adiposo e loro importanza nella valutazione qualitativa della carne suina (Analytical measurements of the adipose tissue and their role in quality assessment of pork.). *Rivista di Suinicoltura* 11:37–41.

13 Salchichón (Spanish Dry-Cured Sausage)
An Integrated Point of View Through Culture, Technology and Innovation

José Angel Pérez-Alvarez, Manuel Viuda-Martos, and Juana Fernández-López
CYTED Healthy Meats network 119RT0568, IPOA Research Group, Agri-Food Technology Department, Orihuela Polytechnical High School, Miguel Hernández University, Carretera a Beniel Km. 3,2, C.P. 03312, Orihuela Alicante, Spain.

CONTENTS

13.1 INTRODUCTION

Food is a "nutrient vehicle" (proteins, lipids, vitamins, minerals…); its consumption provides the necessary substances to maintain human health. Recently, food has been seen from an integrated point of view since any food can also be something else (Pérez-Alvarez et al., 2020). This is how food is seen as art (gastronomy), culture (traditional festivals associated with slaughter, *fiesta de la matanza tradicional* with the participation of the slaughterer and the *mondonguera*), who oversaw the sausage formulation and elaboration process (Incoming Salamanca, 2019; Museu de l'embotit, 2020), tradition (anthropology Museu de l'embotit, 2020), even religious aspects (Halal, Kosher, Christian Lent), are included in its preparation and consumption. Meat products are a very important representation of these aspects of life and culture (Pérez-Alvarez et al., 2019). Especially the dry-cured sausages, since they are widely established in Mediterranean cultures, such as Spanish, Italian, Portuguese, and French, as well as Muslim cultures, such as Turkish or Syrian, where there are dry-cured meat products similar to those of the Western Mediterranean but whose meat (as raw material), is not pork. Cultures in which the consumption of pork is not allowed use other meat, such as that from sheep, goats, and cows. Fermented Mediterranean meat products (dry-cured sausages and dry-cured ham and loins) constitute some of the most exquisite delicacies. But all around the world, there are other dry-cured sausages with high acceptability but low commercial distribution and about which there have been few scientific studies (Aleu, 2017). Danilović and Savić (2017) mention that there are also some early records about fermented meat products in China.

13.1.1 GENERAL CONSIDERATIONS

Meat and meat products are rich in proteins of high biological value that are easily assimilated by the human body, and they provide all the essential amino acids. They are also an important source of vitamins (Salvá et al., 2009) and essential minerals such as iron. On the other hand, as there are different types of meats with different lipid profiles (saturated, unsaturated, polyunsaturated, essential fatty acids), meat products offer different options for consumption within a varied and balanced diet (Pérez-Alvarez et al., 2019).

Meat products in general, and particularly dry-cured sausages, evolved as an effort to economize and out of the necessity to preserve meat that could not be consumed fresh at slaughter (Marchello and Garden-Robinson, 2017), and through processing, increase the shelf life of raw material of high nutritional (Sirini et al., 2020) and economic value, as well as of great social and cultural relevance.

The region in which dry-cured sausages originated is uncertain but it seems that it was produced in the areas around the Mediterranean Sea, specifically in the Greek islands in the Ionian Sea. Production of these products was favoured by moderate winter temperatures and Mediterranean culinary practices. The English term *sausages* comes from the Latin word *salsus*, which means salty or, literally, meat preserved by salting. Thus, the production of sausages began with a simple process of salting and drying meat to preserve it when it could not be consumed immediately (Sayas-Barberá et al., 2007).

The first references to dry-cured sausages go back to ancient Greek and Roman times (Tauber, 1976). The great writers of imperial Rome wrote down the techniques to produce sausages (*farciminas*) (Jensen, 1953) and dry-cured loins and hams, and these are still applied today. These ancient sausages were flavoured with oriental spices, such as pepper, cinnamon, ginger and nutmeg. Dry-cured sausages also uniquely represent many different countries, such as Spain, where salchichón and chorizo are representative of Spanish culture and gastronomy.

There is a great diversity of dry-fermented sausages as a result of manufacturing with raw meat of different types and origins, and unique formulations, condiments, additives, meat grinding size, casing diameter, and smoking and drying periods (Toldrá et al., 2007). These meat products underwent development and innovation as spices that came from the New World, such as paprika, spread throughout Europe. Thus, in Spain, dry-cured sausages were classified into two types: red and white, depending on whether they had paprika in their formulation (Figure 13.1). Another very important milestone for dry-cured sausages occurred in the 1960s, in which the fusion of traditional sausage-making "art" together with advances in manufacturing man-made casing, started what could be called the modern Sausage Age. In this period, sausages were stuffed into attractive and functional man-made or natural casings that became popular all around the world.

Dry-cured sausages can be described as a processed meat products that are made with minced lean and fatty tissues, with added salt (marine or rock), to which curing salts (nitrites and/or nitrates, ascorbates, erythorbates) are added with other ingredients (carbohydrates: lactose, glucose, saccharose, potato starch, cooked rice,

FIGURE 13.1 Spanish traditional dry-cured meat products (dry-cured ham and dry-cured loin), white (Salchichón Ibérico Extra and Salchichón de Bellota Ibérico Extra, and salami) and red (Sobrasada, Chorizo, Chorizo Ibérico Extra, Chorizo de bellota Ibérico Extra and dry-cured sausages in a local supermarket in Spain.

dietary fibres; lemon juice, wine, vinegar, almonds, walnuts, pistachios, etc.); additives (phosphates) and spices (black and white pepper, anise, cinnamon, garlic, onion, and oregano, among others). After a process of mixing and resting, the mixture is pressed into natural or artificial casings and undergoes a process of ferementation (by means of autochthonous microbiota or starter cultures and, depending on the dry-cured sausages specialty, may be smoked) and then drying—maturation (Sayas-Barberá et al., 2007).

Today, this type of meat product is being reformulated due to studies related to red meat consumption and the effect of saturated fatty acids (SFA) on the health of consumers. Harrison et al. (2020), suggested that the intake of SFAs from low-nutritive-value foods, but not total SFA intake, can be associated with unhealthy eating that does not account for SFA intake. Also, Sheehy et al. (2020) reported that the consumption of processed and unprocessed red meat was associated with increased all-cause and cardiovascular mortality.

Different strategies are being developed to reduce the contribution of saturated fats to the traditional meat products of Mediterranean and Spanish culture and gastronomy.

One of the characteristics that consumers most appreciate is that dry-cured sausages, like salchichón, have a long shelf life. This is caused by different hurdles (Leistner, 1995), structural modifications, ingredient additions (salt, curing agents), microbiota metabolism (mainly from lactic acid bacteria—LAB), low redox potential, and intermediate water activity ($a_w \leq 0.90$) that take place during the manufacturing process.

The use of herbs and spices improves the sensory properties of these products, and the use of casing to facilitate their handling and dehydration allowed makers to take advantage of other dry-cured meat products (Parma, San Daniele, and Serrano hams, dry-cured loins, among others). Dry-cured sausages are influenced by local weather. Thus, only air-dehydration is used in Mediterranean dry-cured sausages, but in Central and Northern Europe smoking is required to facilitate dehydration and meat preservation. At the end of the 19th century, scientists began studying the curing processes of these types of meat products. In the last 5 decades great technological advances have improved safety, and now it is possible to produce these products all year, in all climates and all around the world, and standardize their quality (Sayas-Barberá et al., 2007). Most recently, wellness and the sustainability of manufacturing meat products (included the dry-cured sausages) are very important concerns for consumers (Pérez-Alvarez et al., 2020a).

Among dry-cured sausages, consumers can find a wide variety of unique offerings. In Spain, each region has its dry-cured sausage speciality: Salchichón, Longaniza de Pascua, Longaniza de Aragón, Salchichón Vela, Salchichón de Vic, among others, as do different countries (Sucuk from Turkey, Genoa salami from Italy, salame Colonia from Argentine, etc.). The variation in spices, meat particle size (2–22 mm), and the type of meat can be wide (pork, beef, lamb, goat, poultry, wild boar, foal deer, etc. are all used), as can anatomical origin (shoulder, loin, ham, breast, etc.). Casings can be from the small or large intestines or bladders of a variety of animals (beef, sheep, pork, goat, foal), and include whole or sewn casings of differing sizes (18–120 mm).

13.1.2 LEGISLATION

Salchichón is a traditional dry-cured sausage speciality well distributed in Spain. Traditional Spanish Salchichón normally contains lean pork, pork backfat, salt, and spices (Fonseca et al., 2015). The name *Salchichón* covers a wide variety of dry-cured sausages, and for this reason these sausages have a specific legal category in Spanish legislation and special appellations of these products can be found in Spain. Thus, Longaniza de Pascua, Longaniza de Aragón, Imperial de Lorca, Longaniza de Repisco (seca), Salchichón Ibérico, Salchichón de Vic, among others, can be found.

According to the Spanish Ministry of Agriculture in 2019, Spaniards eat 11.41 kg/year of processed meats, representing 25.2% of the consumed meat. Salchichón is the second-most-consumed dry-cured fermented sausage in Spain (0.42 kg/year/inhabitant); Andalusia is the only Spanish region with higher consumption of these types of meat products (12.09 kg/year/inhabitant) (Corsevilla, 2020). To avoid possible fraud to the consumer, Salchichón is regulated by Spanish law. For this reason, this dry-cured sausage must comply with the provisions of 1.1 of Annex I of Regulation 853/2004 of the European Parliament and of the Council of April 29, 2004. In Spain, Royal Decree 474/2014, (BOE 2014) sets meat products' quality standards. This document classifies meat products: their composition and quality characteristics, labelling, and safety standards (to control the manufacturing process, critical control points during processing, industrial auto-control rules, and traceability, are established). Table 13.1 shows the legal requirements for the different commercial salchichón types (BOE 2014).

13.1.3 TRADITIONAL SALCHICHÓN-TYPE DRY-CURED SAUSAGES

Spain is one of the countries with the longest tradition of pork-meat products in Europe, and is the most important in Europe in dry-cured meat products (dry-cured

TABLE 13.1

Commercial Categories, Bromatological Composition, and Legal Requirements for the Different Salchichón Types Found in the Spanish Market

Salchichón Type	Commercial Category	Fat (g/100 g D.M.)	Carbohydrates (g glucose/100 g D.M.)	Protein (g/100 g D.M.)	Collagen/ Protein Ratio (%)	Added Protein (g/100 g)
Commercial	Premium	≤57	≤9	≤30	≤16	≤1
Salami	Premium	≤68	≤9	≤22	≤25	≤1
Salchichón de Málaga	Premium	≤50	≤5	≤37	≤14	≤1
Iberian salchichón	Premium	≤65	≤5	≤22	≤25	≤1
Salchichón		≤70	≤10	≤22	≤30	≤3

D.M.: Dry matter

ham, loin, and fermented sausages). In each town or region of the country, there is a wide choice of this type of meat product. The different orographic and climatic conditions, agronomic resources, the use of local ingredients, etc., has made the compendium of these products enormous. At the same time, there are great differences between them, mainly due to the choice of spices, which varies from one area to another. For example, there are areas of Valencia where the Longaniza de Pascua contains anise (whole or ground), while in other areas it is not incorporated.

In Spain, there are several artisanal white fermented sausages, such as salchichón-type dry-cured sausages. Most of them are locally produced and do not have a wide commercial distribution. The origin of most of them is related to the ancient Aragon kingdom. This is one of the reasons that in Sicily, Corsica, Sardinia, Naples, and other regions of South Italy, similar salchichón-type dry-cured sausages are found.

Depending on their state of processing, mainly longanizas (a colloquial name given collectively to white and red small-diameter sausages) can be eaten, with a heat treatment (boiling water, oven, pan-fried, on barbecue, oven and roasted) or dehydrated. Thus, in Spain can be found, *Longaniza de Alicante*, *Longaniza blanca*, *Longaniza seca Murciana* (Murcia), *Longaniza Andaluza*, (Andalusia), *Longaniza de Salamanca*, *Longaniza extremeña* (Extremadura), *Longaniza de Salamanca* (Iberian pork meat), *Longaniza imperial de Lorca*, and *Longaniza de Pamplona*, among others.

As these dry-cured sausages are artisanal, there is not enough scientific and technological information about them. Some of the information collected regarding each of the most representative is detailed.

Fuet: This is a Catalonian traditional dry-cured sausage. It is industrially produced and is the second (first is Salchichón) most well-known sausage in Spain. Fuet is a type of fine sausage of a varying diameter that presents a characteristic whitish colour ("flor") on the sausage surface (coating). This is caused by mould growth during its manufacturing process (dry-ripening stage). The formula consists of lean pork, beef, mixtures of beef and pork meats, pork backfat as well as other meats such as wild boar and deer, and nonmeat ingredients including water, salt, curing agents, dextrose, ground black pepper and starter culture (*Penicilium candidum*) to obtain the flor, are included (Hospital et al., 2016; Latorre-Moratalla et al., 2010; Marcos et al., 2020).

Fuet is characterized by a higher pH, than other similar dry-cured sausages. This is caused by the control of fermentation temperature (usually below 15°C) and avoiding the addition of lactic acid starter cultures and sugars (in some commercial brands, dextrose is incorporated). Furthermore, Fuet usually has a smaller diameter than Salchichón. The meat batter is stuffed into pork natural casing, but collagen casings (40 mm) also are used. The sausages are ripened for 20 days: 6 days at 15°C and 85% relative humidity (RH), and 14 days at 12°C and 75–80% RH. The processing time usually takes 18 days. Most consumers considered Fuet a "perfect snack," ready to eat "wherever you want" (Ca curro 2020). In Figure 13.2, several types of commercial Fuet are shown.

Espetec: A traditional dry-cured sausage from Catalonia. Although there is no consensus about this dry-cured sausage, since its name seems to be the specific

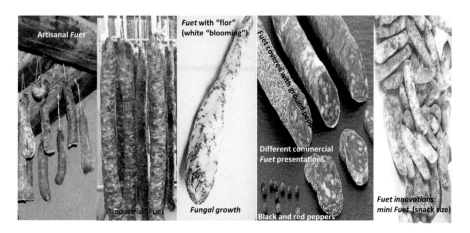

FIGURE 13.2   Visual characteristics of several commercial Fuets.

name of an adapted Fuet-type from Navarclés (Bages county). At this moment, Espetec is a well-known sausage in Spain and is industrially produced. This sausage is a low-acid, fermented dry-cured sausage, much appreciated in Mediterranean countries because of its moderate sour taste. The manufacture of these products is based on the use of low temperatures (<12°C) during ripening, thus avoiding intense fermentation and a strong acidic flavour (Sanz et al., 1998). During manufacture, low-acid fermented sausages undergo a moderate pH decrease, therefore allowing microbial growth that can affect both shelf-life and safety. Formulation of this sausage is quite different, thus a mixture of pork meats (pork shoulder, pork backfat, ham, lean meats, salt, lactose, spices (whole black pepper, ground white peppers and other species), starter cultures and curing agents. In some formulas only sodium ascorbate and nitrate, as curing agents, are used.

Secallona: Dry-cured sausages from the Aragonese and Catalonian Pyrenees (Jamonvillamon, 2020). This sausage is characterized by its small diameter. It is stuffed into lamb and pork casings (small intestine 18–33 mm diameter). The traditional shape of this sausage is called *vela* (candle shape, 30 cm length).

This sausage is distinguished by its flat shape, giving it the appearance of a flat sausage. It has a mild flavour and pronounced aroma. Although it is a traditional artisan sausage, it is considered suitable for commercialization when Secallona has lost 30% of its weight. Often, the consumer acquires it before its consumption, to control the best consumption point. Thus, the sausage finishes dehydrating in the consumer's home (this also occurs with other similar sausages such as Longaniza de pascua, and Longaniza seca murciana (Murcian dry-cured sausage). Due to its small size, dehydration is relatively quick (2–3 days after purchase). From a culinary and sensorial point of view, Secallona is ready when the sausage breaks when is folded. For its specific characteristics, most consumers considered Secallona a tasty (Casanoguera, 2020) snack.

Longaniza de Alicante: A dry-cured sausage from the Valencia Autonomous Community, it is made with pork meat (pork back fat: 70%. and shoulder and/or ham meats: 30%), sea salt, natural spices (mainly cinnamon and a mix of black and white peppers, among others), sugar without any additives (preservative and colourants). This sausage could be considered a clean label meat product. In the manufacturing process, meats (fatty and lean) are chopped by an 8 mm diameter plate, then the chopped meats are seasoned with salt and spices and mixed, then rested for 12 hours in refrigerated conditions. The meat batter is stuffed into natural pork casings (30–33 mm diameter) and dehydrated for 21 days until this sausage's weight has decreased by 30%. Normally it is sold in the *vela* shape with an approximate weight of 200 g.

Longaniza blanca: This dry-cured sausage is also traditionally from the Alicante province (Valencia Autonomous Community) and made using the following ingredients: pork back fat, salt, spices (mainly black and white peppers), antioxidants (sodium citrate and ascorbate), dextrose, and preservatives (sulphites). As with all of these sausages, Longaniza blanca can be eaten after heat treatment or dehydrated at a refrigerated temperature (5–10°C).

Salchichón de Malaga: Although its formulation is characterised by the presence of whole or ground black pepper (*Piper nigrum*), it could also include other spices such as nutmeg (*Myristica fragans*), white pepper, wild marjoram (*Origanum vulgare*), anise (*Pimpinella anisum*), and cinnamon (*Cinnamomum zeylanicum*), among others. (Ruíz Pérez-Cacho et al., 2005). Once Salchichón is mixed and stuffed, it is subjected to fermentation and ripening at 12–15°C for about 4–9 weeks. Its quality depends on the quality of the raw materials and the technology of production.

Longaniza imperial de Lorca: A typical Murcian dry-cured sausage made with pork that obtained the Quality Product of Murcia brand in 1990 (Bañón et al., 2009). It is considered as an Agri-Food Quality Guarantee Mark of the Region of Murcia (CARM, 2020). It is considered one of the youngest raw-cured meat products, since its production began in 1927 (Larrosa, 2009). Legally, this traditional sausage is considered a dry-cured sausage made by selecting, chopping, and mincing pork meat and fat, and it incorporates seasonings, spices, ferments and authorized additives that are kneaded and stuffed into a natural casing, and subjected to a controlled maturation and drying process that ensures good stability, as well as a characteristic colour, smell, flavour, and texture.

Longaniza de Pamplona: This meat product is representative of Navarre. Its composition is based on the mixture of meat and lean meat (ham and pork shoulder and/or ham) (67%) and pork back fat (33%), sea salt (2%), white pepper (0.2%) and fresh chopped garlic (0.5%). Normally it is consumed fresh, pan-fried, or dried. If this sausage will be sold dehydrated, curing agents must be incorporated. The dehydration can be like other dry-cured sausages.

Longaniza de Pascua: This dry-cured meat product was mainly produced at Easter in butcher stores. At this time, the weather conditions allowed the homemade elaboration of dry-cured sausages, without refrigeration. *Longaniza de Pascua* is a characteristic sausage from the Valencian Community (Mediterranean area) and in some Aragonese counties (Maestrazgo). For Valencian consumers, the traditional Longaniza de Pascua is made from lean meat and pork back fat, and is approximately 30 cm length and about the width of a finger when it is consumed dry (Pérez-Alvarez,

1996). Traditionally, this a sausage that was made in Alberique (Valencia). It is a sausage that can be made of different types of meat, according to economic or social circumstances, using goat, lamb, beef, foal meats, and more recently poultry (to give more options and be healthier). The recipe that is currently the most accepted uses shoulder pork meat and pork back fat, salt, black and white pepper, nutmeg, and in some areas, anise, stuffed into lamb casings (18–22 mm diameter).

When the industry needs to explore the suitability of potential functional ingredients, *Longaniza de Pascua* is the best option. During the COVID-19 pandemic, research and innovation using this sausage had a great increase. In the creation of healthy meat products, scientific–technological research is required. Thus, nutrients and bioactive compounds with functional properties must be assayed. In a short time, the effect that the addition of new ingredients or the change of formulation has on raw-cured sausages can be easily evaluated. The short manufacturing period makes it possible to estimate the effects that these compounds have on the chemical, physical, physico-chemical, microbiological, and sensorial characteristics of dry-cured sausages without having to wait for the effects upon other Salchichón with longer elaboration periods. That is why Longaniza de Pascua acts as a model system for this type of meat product (Pérez-Alvarez, 1996; Sánchez-Zapata et al., 2013; Sirini et al., 2020). The quality of this traditional dry-cured sausage is controlled and regulated by the Valencian Institute of Agrifood Quality, under Decree 91/1998, of the Valencian Government (DOGV, 1998).

Longaniza de Salamanca: These sausages are high-value products obtained from autochthonous Iberian pigs (Casquete et al., 2012). There are two types of traditional Salamanca dry-cured sausage. The first one is *Longaniza de Salamanca with a mix of meats* (Iberian pork lean meat, beef, and Iberian pork back fat: 50%, 25% and 25% respectively), and the other is *Pure Iberian pork meat Longaniza de Salamanca* in which sausages are made with 75% lean meat (Iberian pork shoulder and/or ham) and 25% pork back fat. The formulation also includes garlic (smashed or dehydrated), wine (normally white), salt (2.0%), and curing agents. To make the sausage, all the ingredients are mixed in a container (except garlic, wine and water), add them to the dough dissolved in water. All ingredients are mixed and rested for 24 hours in refrigerated conditions. Fermentation takes place at 17–18°C and 80–95% relative humidity for 24 hours. During the dry-curing stage, the drying chamber temperature is 12–14°C and 70–75% relative humidity. Depending on the diameter of these Longaniza de Salamanca, the dry-curing stage can be between 15 and 30 days.

The pigs used to make this type of Longaniza can be raised under three different feeding conditions. Montanera pigs are fed with acorns and grass, and generate products of the highest quality and commercial value (García et al., 1996; López-Bote, 1998). The other two feeding conditions for Iberian pigs are: (1) an additional mixed diet supplement, if the availability of acorns and grass is insufficient, called Recebo, and (2) a formulated diet only, in intensive conditions, the so-called Cebo. These feeding conditions are regulated by the Industry Quality Policy (BOE, 2007). Also, other Salchichón-related dry-cured sausages such as Salchichón de ajo (garlic Salchichón), Secallones, Somalles, Petadors, Salchichón de Málaga, Salami, Longaniza de Payés, Longaniza de Aragón, and Longaniza are regulated by Spanish Law.

13.1.4 Salchichón Manufacturing Process

Salchichón, like any other meat product, must comply with the hygiene and food safety standards for raw materials, production processes, storage, distribution, and commercialization, preferrably sustainable and eco-friendly.

However, some characteristics must be considered when this product is elaborated. For example, all ingredients must be carefully selected. Thus, meats must come from animals to which animal welfare standards have been applied and slaughtered in authorized facilities. Also, strict quality controls must be observed for other ingredients, additives, spices, casings, containers and packaging, and operating conditions such as processing times, temperatures, conservation, storage, and transport.

The main stages of Salchichón production can be summarized in the following steps:

i. **Raw material:** Salchichón, as a meat product, is prepared totally or partially with meat (normally, there is no offal in the Salchichón formulation) ,om the authorized sources (pork, beef, poultry, game, among others) mentioned in point 1.1 of Annex I of Regulation 853/2004 of the European Parliament and of the Council of April 29, 2004, which establishes specific hygiene rules for animal-based foods and subject to specific operations previously to consumption. Salchichón must be prepared as established in the European Community Food Regulations Laws.

ii. **Mincing–Mixing:** To produce Salchichón it is necessary to create several transformations in the meat structure (fatty and lean tissues). These changes can be due to mincing, in which all ingredients are incorporated to make Salchichón meat batter. In this, meat proteins, fats, salt, and other ingredients start all biochemical, chemical, physico-chemical, microbiological, physical, and sensorial changes necessary to reach the well-appreciated dry-cured Salchichón.

iii. **Stuffing:** Necessary to give the characteristic shape of Salchichón. In this stage, meat batter is inserted, by pressure, into a natural or artificial casing.

iv. **Casings:** A key element in the elaboration process of dry-cured meat products, technological (pH, water activity, among others) variables, and the correct microbiota development for which Salchichón depends on its characteristics and composition. Natural casings are obtained from the small and large intestines of bovine, ovine, caprine, porcine, and equine species, as well as game animals farmed for food purposes, and from the oesophagi and bladders of cattle and pork, which, after the necessary manipulations, serve as a technological container for food products. The artificial casing will be made of collagen, cellulose, or authorized polymeric materials.

v. **Resting:** A short-term stage (12–24 hours) this stage allows that the formulation water excess (the amount of water depends on the Salchichón type, but usually from 1% to 10%) can be eliminated through the casing pores. This water moves by osmosis to the sausage surface and thus can be eliminated by dehydration.

vi. **Fermentation:** In this stage, which is temperature sensitive, the microbiota growth (wild microbiota or a starter culture) is initiated. Microbiota metabolize natural or added sugars and produce lactic acid, which reduces Salchichón pH. Changes in pH produce several changes in the meat batter, thus biochemical (enzyme activities), chemical (protein and fat hydrolysis among others), physical (colour, texture), and physico-chemical (water holding capacity, oil holding capacity, water activity, etc.) changes take place. Microbiota metabolism also generates other chemicals, as well as physical, physico-chemical, and sensory changes that characterize this type of meat product. From an industrial point of view, this stage can be conducted in two ways: cold and warm fermentation. Cold fermentation (fermentation temperatures between 16°C and 18°C) and warm" fermentation (fermentation temperatures between 19°C and 28°C). The fermentation time will depend on the Salchichón's diameter (1–2 days in cold fermentations to 7 days in warm fermentations) and sensory characteristics of the sausages. In both fermentation types, the relative humidity in the fermentation chamber is high (90–95%). To avoid that the surface dehydrates and separation between meat batter and casing takes place. This aspect must be avoided during the changes from sol to gel in the stuffed sausage meat batter (fresh Salchichón).

vii. **Smoking:** This treatment is applied in few Salchichon types (mainly in rural areas from the north of Spain) as other dry-cured sausages (Chorizo gallego de carne, Chorizo de cebolla, Morcillas de calabaza, etc.). In some types of these dry-cured sausages, smoking is associated with the fermentation stage (takes place at the same time) since combustion generates the heat necessary to carry out fermentation stage (especially in warm fermentation). The raw materials for combustion are authorized aromatic herbs and woods (non-resinous woods). Combustion is both a heat source for fermentation and a supply of some types of industrial interest compounds (antimicrobial and flavour compounds). These types of compounds are chemically aldehydes, ketones, and organic acids, among others. The action of these compounds is to act as antimicrobials, especially for those that, in previous stages, have been growing on the sausage surface. The sausage casing also suffers modification with these compounds; several reactions (crosslinkings) take place between smoke compounds, casing, and meat batter. Casings also act as a supplier of these substances for later stages.

viii. **Dry-maturation:** The longest stage of the elaboration process. In this stage, several changes (chemical, physical, physico-chemical, microbiological and sensorial) take place. Under the industrial point of view, Salchichón must lose at least 30% of their original weight. Also, water activity (a_w) is a good indicator that the product is finished (a_w lower than 0.90). At this value, Salchichón can be considered as Intermediate Moisture Food, is self stable at room temperature, and guaranteed to be stable during commercialization.

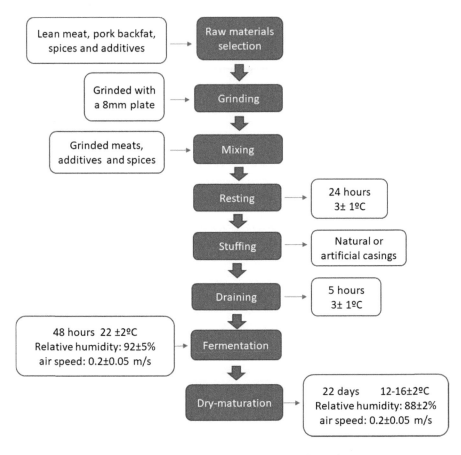

FIGURE 13.3   Commercial Salchichón elaboration process flow chart.

This stage is characterized by a slow and gradual reduction in humidity (90–80%), and an increase in pH caused by the microbiota evolution, the dynamic enzymatic activity (to produce aromatic compounds), and the most notorious physical change, sausage hardening, all of which are necessary to provide the desirable Salchichón characteristics.

In Figure 13.3, the commercial Salchichón elaboration process flow chart is shown.

13.1.5 INGREDIENTS

In addition to the lean and fatty meats in the Salchichón formula, some other ingredients and additives give this sausage its special characteristics. The chemical, physical, physico-chemical and techno-functional properties (water and oil holding capacity, emulsion capacity, emulsion stability, sol to gel transition of meat proteins and fats, salt, and curing agents (nitrite, nitrate, ascorbate/erythorbate and phosphates), among others, are excluded from this chapter. However, we will review several ingredients that are commonly used in the formulation of Salchichón (traditional and commercial). In Table 13.2, ingredients, additives and the energy value (kJ and kcal) of different commercial Salchichón types (and other dry-cured specialities associated with this sausage) found in the Spanish market.

13.1.5.1 Meat Extenders

From an economic point of view, the use of meat extenders (mainly animal and vegetable-based proteins) in this type of product is straightforward, for economic (lower prices than meat), technological (emulsion formation and stability, and holding capacity (water and oil, among others) reasons. From an economic point of view, the use of meat extenders reduces cost in a very competitive sector. But its use is bound by a set of constraints, which are imposed as a result of the great difference between the technological and techno-functional properties of the meat and the meat extenders, without forgetting the legal aspects. The main legal meat extenders used as ingredients in the processing of dry-cured sausages (Salchichón) are dairy and soy proteins. In Table 13.2, meats extenders used in Salchichón (and other related dry-cured sausages) formulations are shown.

Dairy ingredients: Dairy proteins are used extensively in meat technology as a meat extender, meat binder, and texturizing agent (Prabhu and Keeton, 2008). In white dry-cured sausages, such as Salchichón, the use of dairy products includes spray-dryer milk (usually skimmed milk), sodium caseinate, dairy proteins, and lactose. Andujar and colleagues (2013) reported that these ingredients are *meat extenders*. These authors reported that sodium caseinate is used for its water holding capacity (WHC), gelling, emulsifying, and stabilizing properties among its functional and nutritional properties. Its use has other technological advantages, such as its water solubility, and due to its excellent emulsifying capacity, it exerts a stabilizing action in meat batters. Dairy proteins are used in this type of meat product because they show excellent techno-functional properties, such as emulsifying properties and increased WHC, as well as being antifoams, gelling agents, and thickeners. Dairy proteins with low calcium content (0.5–0.8% Ca) are better emulsifiers for the meat industry and at the same time favour WHC. The use of these improves the texture of this type of dry-cured sausage.

Soy proteins: Soy, as a meat extender, has been widely used in meat products for a long time, although its initial use was to lower production costs, without the consumer opting for its addition. However, at this time, soy is having a very important boom as a partial substitute (desired by the consumer) for animal-based proteins, especially meat proteins. There are different food trends in which the reduction or elimination of proteins of animal origin (reducetarian, vegetarian, and vegan) in which the use of soy is widespread (Reducetarian Foundation, 2020).

Normally, soy protein isolates are used in meat products since, unlike concentrates and other presentations (flours, pasta and concentrates, etc.) the functionality and techno-functionality of the native proteins is preserved almost intact, which is of great interest for this type of meat product. In addition to the fact that with these soy preparations, oligosaccharides (stachyose and raffinose that can produce flatulence) and other soluble carbohydrates are not present. Andujar and colleagues (2000) found the soy "beany" flavour is less intense and doesn't affect Salchichón flavour.

TABLE 13.2

Consumer Information, According to Legal Requirements for Labelling of the Spanish Commercial Types of Salchichón (and Other Dry-cured Specialties Associated with this Sausage), Found in the Spanish Market

Salchichón	Meats*	Ingredients*	Energy kJ/100 g and/or kcal/100 g
Salchichón Montañes	Pork meats (lean and back fat)	Milk powder, salt (4.4%), corn maltodextrin, corn dextrose, soy protein, spices mix, sodium erythorbate, sodium nitrite, potassium nitrate, carminic acid, non-edible collagen casing	2092 kJ or 506 kcal
Longaniza de Pascua extra	Pork meats (lean and back fat)	Salt (3.9%), lactose, dextrin, spice mix, sodium citrate, alpha tocopherol, sodium ascorbate, dextrose, maltodextrin, flavouring agents, dehydrated vegetables, sodium glutamate, sodium nitrite, potassium nitrate, carminic acid, beef collagen casing	1897 kJ or 458 kcal
Longaniza de Payés extra	Pork meats (shoulder meat and back fat)	Skimmed milk powder, salt (3.6%), spices mix, curing agents (sodium nitrate, potassium nitrate and sodium ascorbate)	1461 kJ or 351 kcal
Salchichón extra with milk proteins	Pork meats (lean and back fat)	Salt (3.5%), dairy proteins, dextrose, spices mix, sodium diphosphate, sodium triphosphate, curing agents (sodium nitrite and ascorbate)	1569 kJ or 378 kcal)
Salami extra with milk proteins	Pork meats (lean and back fat)	Lactose, salt (3.5%), milk proteins, dextrose, spice mix, sodium diphosphate, sodium triphosphate, curing agents (sodium nitrite and ascorbate), carminic acid	1193 kJ or 462 kcal
Espetec extra	Pork meats (shoulder, back fat, ham and lean meat)	Salt (3.7%), lactose, spice mix, sodium ascorbate, potassium nitrate, edible casing	1794 kJ or 421 kcal
Salchichón ibérico extra	Iberian pork meats (lean and back fat)	Salt (3.5%), dextrose, soy protein, spice mix, pentasodium triphosphates, sodium nitrite, potassium nitrate, pork natural casing	2040kJ or 492.6 kcal
Salchichón cular de bellota ibérico	Acorn feed Iberian pork meats (lean and back fat)	Salt (2.7%), dextrose, soy protein, spice mix, flavouring agents, pentasodium triphosphates, sodium	1981.90 kJ or 478.40 kcal

(Continued)

TABLE 13.2 (Continued)

Salchichón	Meats*	Ingredients*	Energy kJ/100 g and/or kcal/100 g
		glutamate, sodium nitrite, potassium nitrate, pork natural casing	
Salchichón de pavo extra	Turkey (thighs lean and fat)	Lactose, salt (4.0%), skimmed milk powder, dextrose, spice mix, sodium diphosphate and triphosphates, milk protein, flavouring agents, curing agents (sodium nitrite and potassium nitrate and sodium ascorbate), carminic acid	1139 kJ or 273 cal
Longaniza de pavo extra	Turkey (thighs lean and fat)	Skimmed milk powder salt (4.0 %), lactose, flavouring agents, spice mix, sugars, sodium ascorbate, dextrose, sodium nitrite and potassium nitrate	1314 kJ or 315 kcal
Longaniza snack	Pork meats (lean and back fat)	Salt (3.7%), dextrin, dextrose, spice mix, flavouring agents, curing agents (sodium nitrite and ascorbate), beef edible collagen casing	2095 kJ or 505 kcal
Chicken Sticks	Chicken (lean and fat)	Salt (3.8%), maltodextrin, dextrin, dextrose, spice mix, flavouring agents, curing agents (sodium nitrite and ascorbate), beef edible collagen casing	1708 kJ or 410 kcal.

Notes

* Ingredient orders are according to the Spanish labelling legislation, from high to lower concentration (BOE, 2015)

Lactose: A milk disaccharide with similar technological applications to dextrose (Freixanet, 2020). For the dry-cured fermented meat products must be incorporated since act as specific "easy nutrient" for the heterofermentative LAB. This microbiota plays an important role in this type of meat product. But lactose malabsorption results from the inability to properly digest lactose, the disaccharide found in mammalian milk. Lactase is an enzyme in the brush border (microvilli) of the small intestine responsible for cleaving lactose into absorbable monosaccharides. Lactase enzyme deficiency leads to lactose malabsorption because the gut is unable to absorb the larger disaccharide.

Spices: The term *spices* means any aromatic vegetable substances in whole, broken, or ground form, whose significant function in food is seasoning rather than nutritional, and from which no portion of any volatile oil or other flavouring principle has been removed.

Several spices are used in the formulation of dry-cured meat products: black and white pepper, Hungarian and/or Spanish paprika (pimentón), oregano, clove, nutmeg, anise, and cinnamon, among others, are typical. Paprika is the spice that is specifically added to produce red dry-cured meat products, like chorizos and pepperoni, among others, and black pepper is usually used in white dry-cured meat products like Salchichón, Longaniza de Pascua, etc.

Black pepper *(Piper nigrum)*: This spice is one of the best known and used in many meat products. This spice gives a pleasant sensorial effect (pungency and characteristic flavour), and is a highly prized spice (Hu et al., 2018; Munekata et al., 2020). It has been used from ancient times to avoid spoilage of meat products. This is one of the main reasons that ground black pepper is used for dry-cured sausage surface impregnation (Figure 13.2). Also used as a natural insecticide, pepper is a natural repellent for the meat fly (*Sarcophaginae*). When the dry-cured sausage still "fresh," pepper prevents the meat fly from depositing eggs on dry-cured meat products, which results in deterioration and loss of the meat product.

Roncalés (1995) mentions that the spices used are what makes Spanish dry-cured sausages unique, and Martín-Bejarano (1992) and Roncalés (1995) observe that Salchichón and related dry-cured sausages are distinguished by the presence of black pepper. This spice can be used whole (Figure 13.1) or ground. In Salchichón, black pepper is also used as flavour enhancer (Alirezalu et al., 2020). The pungency of its extract, which contains alkaloid compounds, and the aroma of its essential oils give it value as food additive; the odour and flavour of meat can be protected by these components. Piperine shows lipid peroxidation inhibition and antioxidant effects (Vijayakumar et al., 2004; Andrade and Ferreira, 2013); the bioactive alkaloid present in black pepper is associated with many health beneficial physiological effects (Srinivasan, 2007). This compound also inhibits growth of both Gram-positive and Gram-negative bacteria such as *Bacillus substilis*, *B. sphaericus*, *Staphylococcus aureus*, *Klebsiella aerogenes*, and *Chromobacterium violaceum* (Zarringhalam et al., 2013). Zhang et al. (2016) reported that Gram-negative bacteria were more sensitive than Gram-positive bacteria to black pepper essential oils.

Although black pepper is the most popular spice, there are also less pungent white and green peppers with marginal commercial value (Abdulazeez et al., 2016). White pepper is also used in Salchichón related dry-cured sausages. The antioxidant and antimicrobial properties of the essential oil of black pepper could be used as a potential natural food preservative (Abdulazeez et al., 2016). Jeena et al. (2014) reported that antioxidant and antimicrobial activities are related to its terpenoids contents, highlighting the amounts of caryophyllene, limonene, α-terpinene, and α-pinene (Munekata et al., 2020).

Nutmeg *(Myristica fragrans Houtt)*: This spice is widely used as a culinary adjunct in meat and meat products and imparts a pleasing flavour to the product. In Spain, nutmeg is usually used in white dry-cured sausages. Nutmeg is used to prevent rancidity of meats (Šojić et al., 2015; Zakaria et al., 2015). This activity is related to its excellent activity in retarding lipid peroxides and malondialdehydes, protein oxidation, scavenging radicals, reducing metal ions, and inhibiting lipid oxidation (Gupta et al., 2013). Nutmeg is an important source of antioxidants such as α-pinene, β-pinene, p-cymene, β-caryophyllene and carvacrol, known to have anti-glycation and antimicrobial effects (Ashish et al., 2013; Kazeem et al., 2012; Gupta et al., 2013). Nutmeg retards the formation of adverse changes in colour and sensory qualitiesof dry-cured sausages. This spice is used ground.

Garlic (*Allium sativum*): This spice is widely used in dry-cured sausages, both in red and white fermented sausages. It is an ancient spice used in meat products. Technological (antioxidant and antimicrobial activities) and sensorial properties make it suitable for dry-fermented products. Several components of garlic and garlic extracts possess antioxidant properties and it is an effective hydroxyl radical scavenger. It is important to note that this activity is concentration-dependent (Yang et al., 1993). The French *saucisson a l'ail* (French garlic sausage) has the highest concentration of this spice. In its formulation, 8 g/kg of garlic is used (La Table d'Amelie, 2014). In Spanish Salchichón, the garlic concentration to be added depends on the dry-cured sausage type, thus 1–5 g/kg can be applied. Higher concentration gives garlic flavour to this sausage. Substances such as alliin, diallyl sulfide, allyl sulfide, and propyl sulfide account for the antioxidant effect. Likewise, garlic contains ascorbic acid and nitrates and nitrites (Aguirrezabal et al., 1998). These substances also showed antimicrobial activity against both food spoilage bacteria and food-borne pathogens (Unal et al., 2001; Leuschner and Ielsch, 2003) In Salchichón and related dry-cured fermented sausage formulations, garlic can be used fresh (crushed) and/or dehydrated. Industrially, garlic powder is the most common.

Carminic acid: A colourant ($C_{22}H_{20}O_{13}$) normally found on the labels of this type of product with the European Additives Nomenclature of E-120. Its specific characteristics related to its great stability to light and pH modifications during processing make it suitable for these types of products. In Figure 13.4, the different colour of Salchichón (same commercial category) with or without addition of E-120 can be observed.

13.1.5.2   Starter Cultures

Meat fermentation dates from the Roman Empire. Fraqueza (2015) mentioned that the manufacturing process for dry-fermented sausages involves fermentation driven by natural microbiota or intentionally added starter cultures and further drying. The most relevant fermentative microbiota are LAB such as *Lactobacillus, Pediococcus* and *Enterococcus, Leuconostoc, Lactococcus and Weissella* (Fontana et al., 2012) producing mainly lactate and contributing to product preservation. The great diversity of LAB in dry-fermented sausages is linked to manufacturing practices. Indigenous starters allow high sanitary and sensorial quality of sausage production and maintain the qualities of these well-appreciated meat products. LAB includes a diverse group of Gram-positive, non-spore-forming cocci, coccobacilli or rods, with common morphological, metabolic, and physiological characteristics (Batt, 2000). They are facultative anaerobic with variable oxygen tolerance in different species (Fraqueza, 2015). The same author mentioned that LAB growth depends on the presence of fermentable carbohydrates. They are classified as homofermentative or heterofermentative, based on the end products of glucose metabolism. While homofermentative LAB converts

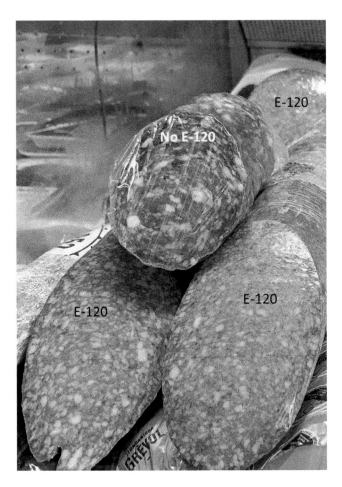

FIGURE 13.4 Spanish Salchichón commercial samples with the same legal category, with and without the addition of carminic acid (E-120).

glucose mainly to lactic acid using the glycolysis (Embden–Meyerhof–Parnas or Embden–Meyerhof) pathway, the heterofermentative LAB uses the phosphoketolase (6-phosphogluconate) pathway and converts glucose to lactic acid, carbon dioxide, and ethanol or acetic acid. Flores (2018) mentioned that the microbial fermentation of carbohydrates (dextrose, sucrose, lactose, and maltodextrin) in meat products produces the formation of organic acids such as lactate and acetate, contributors to meat product sourness. This carbohydrate fermentation is mainly achieved by LAB, resulting in a pH drop in fermented meats (Flores, 2017), while coagulase negative staphylococci are involved in the bacterial formation of aroma compounds (Ravyts et al., 2012). The catabolism of pyruvic and lactic acids by lactic acid bacteria (Liu, 2003) produces compounds such as acetic acid, formic acid, ethanol, acetaldehyde, 2,3-butanedione (diacetyl), 3-hydroxy-2-butanone (acetoin), and 2,3-butanediol that are responsible for the specific aroma.

Bacterial communities coexist in dry-fermented meat products, allowing for microbial diversity. Microbiota development will be influenced by technological particularities used in dry-cured sausage processing and will adapt to this special niche. The addition of ingredients such as sodium chloride, nitrate/nitrite, sugars, wine, and spices (garlic, black pepper), as well as the values of intermediate moisture food (0.85–0.92), temperature (12–18°C to 24–30°C), oxygen gradient during ripening and smoking application will select a microbiota able to develop in dry-cured sausages (Fraqueza, 2015). Also, another ecological determinant can influence the establishment of a specific microbiota consortium that will determine the rate of colonization. The selective influence of intrinsic (concentration and availability of nutrients, pH, redox potential, buffering capacity, a_w, meat structure) and extrinsic (temperature, relative humidity and oxygen availability) factors may determine differences in microbial ecosystem raw meat substrate (Fraqueza, 2015). The LAB predominant species in dry-fermented sausages are *Lactobacillus sakei*, *Lactobacillus curvatus*, *Lactobacillus plantarum*, *Leuconostoc mesenteroideus*, *Pediococcus* spp., and *Enterococcus* spp., which growth is modulated and adapted to the existing stringent conditions of processing (Federici et al., 2014, Wanangkarn et al., 2014). Other microbiota are also used in other dry-cured fermented sausages such as *Lactobacillus rhamnosus* (Erkkilä et al., 2001) and *Lactobacillus paracasei* (Salazar et al., 2009). Most commercial dry-fermented sausages use a combination of starter cultures to guarantee their unique flavours, and *Lactobacillus*, *Staphylococci* and/or *Micrococci* are generally used (Cheng et al., 2018). In Salchichón, commercial starter culture is added at 0.025 g/100 g (Bis-Souza et al., 2020). In dry-cured sausages, the carbohydrate concentration and microbiota fermentation activity must be controlled. The pH values lower than 5.0 must be avoided because they generate physico-chemical changes that create unpleasant sensorial characteristics.

13.1.5.3 Casings

Sausage is the oldest form of processed meat, and a variety of it is produced using the processed intestines of pigs and other animals. These so-called natural casings (the material that encloses the filling of a sausage) are made from the submucosa of the intestinal tract (Dibaba, 2019). Natural casings are still a very popular choice all over the world. However, they can be contaminated by enteric bacteria as well as exogenous microorganisms, mainly due to a lack of hygiene during slaughter and post-processing handling, and as the result of high storage temperatures. (Trigo and Fraqueza, 1998).

Today, many types of sausage casings are used, including natural, manufactured collagen, cellulose, and plastic as well as co-extruded casings made of collagen, alginate, or an alginate–collagen hybrid (Harper, 2013).

In the sausage-making process, several types of casings are used. In pork, the anatomical parts of the digestive system used as casings in Spanish sausages are the small and large intestines and, in a few regional sausages, the bladder. For Salchichón and related dry-cured sausages, the most important are the small intestines (sheep, pork and beef). In all-natural casings only the submucosa are used (Suurs and Barbut, 2020). In the dry-cured sausages elaboration process, the use of natural casings enhances quality, since the casings support, during the stuffing stage, stresses in all directions. This property allows casings to adapt to the extension and retraction of the sausage meat batter during the fermentation and dry-curing stages and its storage and commercialization.

Chen et al. (2019) reported that the sheep casing consists of a large number of thick collagen fibres oriented at ±45° from the longitudinal direction with a high-density interwoven network structure. This structural feature of sheep casing gives the special mouthfeel of *cracking bite* to sausages. These authors also reported that the layered structure is filled with fine collagen fibrils and large gaps in collagen film results in poor mechanical properties and higher swelling ratio in water. Suurs and Barbut (2020) mentioned that in pork casings, the submucosa contains a network of collagen and elastic fibres and blood vessels, and is less brittle than beef casing, which contains both the muscular and mucosa layers. There are two types of pork casings, A and B. The A-type (free of holes) is used for fin- emulsion sausages (Wien, Bratwurst, etc.) and the B-type is used for coarse-ground products like dry-cured sausages (Salchichón, Chorizo, Longaniza de pascua, Fuet, Salchicha de repisco, among others) and fresh sausages (Longaniza roja and Longaniza blanca, Chorizo fresco, Chorizo verde de Toluca, Chistorra, and Salchicha fresca among others).

While sheep and pork casings are typically eaten with the sausage, beef casings are usually removed before consumption as they are thicker and tougher/hard to chew (Wijnker et al., 2008).

Collagen casing: These types of casings are an alternative to the worldwide limited supply of animal casings and health-related animal diseases associated with them. Although the elaboration process started in the 1920s, it was not until 1960 that its use was popularized in the meat industry (Savic and Savic, 2016). They possess many similarities to traditional animal casings, including the fact that they are edible (Suurs and Barbut, 2020). Manufactured collagen casings are composed of both fibrous and solubilized material that is extracted from hides, bones, and connective tissue (Ratanavaraporn et al., 2008). Suurs and Barbut (2020) review the main properties of these types of casings in these 10 points: (i) better uniformity than natural casings, no size variations or weak spots, (ii) strength during filling, allowing stuffing at higher pressures and the use of high-speed stuffing machines, (iii) flexibility, (iv) cleaner and more hygienic, (v) easy to use, (vi) their uniformity makes portioning easy, (vii) slicing, (viii) less expensive, (ix) less labour, and (x) easy to work with.

Co-extruded casings: This type of sausage casing represents a great innovation in the dry-cured sausages sector. In this field, co-extrusion has a good implementation in this type of meat product. This technology enhances the research and development (R&D) in dry-cured sausages (formulation); thus, new dry-cured meat snacks appeared in the market, with diameters lower than traditional suasages (made with sheep or collagen casings, ≤ 18 mm).

A co-extrusion system is a fully automated system that simultaneously extrudes a continuous flow of meat batter and a thin layer of casing material. The most used co-extruded casing materials are made with calcium alginate (Marcos et al., 2020). The application of alginate co-extrusion for the manufacture of dry-fermented sausages has focused on the prevention of surface efflorescence (formation of white deposits on the surface of dry-fermented sausages) (Walz et al., 2018; Walz et al., 2019).

Casing properties (natural or artificial) are fundamental in the dry-cured sausages manufacturing process. Thus, casing thickness, tensile strength, elongation to break, and swelling ratio are analysed. These properties are measured in wet state and after boiling in water, using a tensile tester.

In today's sausage production, various types of casings are employed including natural, manufactured collagen, cellulose, and plastic, as well as the new type of co-extruded casings made of collagen, alginate, and alginate-collagen hybrids (Suurs and Barbut, 2020).

13.1.5.4   Sensorial Attributes of Salchichón and Related Dry-Cured Sausages

Ruiz Pérez-Cacho et al., (2005) determined the main sensory attributes of Salchichón and found 15 sensory attributes that describe this traditional dry-cured sausage. These authors mentioned the following sensory elements: 4 for appearance (luminance, presence of crust, fat/lean connection, and exudate); 4 attributes for odour (black pepper, lactic acid, mould and other spices); 2 for texture (hardness and initial juiciness), and 5 for flavour (black pepper aroma, mould aroma, other spices aroma, acid taste and salty taste). Recently, a similar study was made for Fuet (Marcos et al., 2020). These authors define Fuet sensory parameters as follows: (i) Appearance (made over the slice); (ii) colour intensity (red colour intensity on a transverse section, made on a fresh cut); (iii) Brightness (brightness intensity evaluated on a transverse section made on a fresh cut), (iv) Odour (made over the slice); (v) Intensity (overall odour intensity of fuet); (vi) Mold Intensity (mold odour); (vii) Ripened (pleasant odour characteristic of dry-fermented sausages); (viii) Taste and flavour (made over the slice); (ix) Flavour Intensity (overall flavour intensity of Fuet); (x) Sourness (basic flavour sensation elicited by lactic acid); (xi) Bitterness (basic taste sensation elicited by L-tryptophan); (xii) Ripened (pleasant flavour characteristic of dry-fermented sausages); (xiii) Texture (tactile to whole piece); (xiv) Hardness (amount of pressure required to completely compress the sample); (xv) Waxy (resembling wax when touched); (xvi) Texture (made over the slice); (xvii) Hardness (force required to bite through the sample); (xviii) Elasticity (degree of return to the original position of the sample when a compression force is applied between molars); (xix) Crumbliness (texture property characterized by the ease with which the product can be separated into smaller particles during chewing); and (xx) Gumminess (texture property of a sample which resembles the consistency of gum). In this case, there are five more parameters than are used to describe Salchichón.

13.1.6   Innovations in Dry-Cured Sausages

The serious health risks of obesity and related illnesses underscore the great importance of changing the lifestyle-related risk factors of these conditions, including sedentary behaviour and unhealthy diets (SweeGen, 2021). An excess of saturated fats in dry-cured sausages can both contribute to overall caloric density and impact the nutritional quality of the diet by providing an excess of energy, leading to weight gain and increased risk of chronic diseases.

Within the meat industries, there are several research and innovation strategies. One of them is the use of local raw materials and the rescue of autochthonous pork breeds with exceptional adaptation to the environment and that "know" how to optimize resources, producing meats with excellent technological and healthy properties (less content of saturated fats). That is the reason universities, research centres, consumer associations, chefs, and the meat industries (Research and Development multidisciplinary teams), are working intensively to develop healthy meat products, taking into account sustainability criteria and the conservation of genetic resources. This is the origin of the concept of *gourmet meat products*, and dry-cured sausages are no exception. Fortunately, in Spain, one of these trends has gained popularity in the market. This is related to meat products elaborated with Iberian pork meats, however other indigenous pig breeds are gaining market share, such as the Chato Murciano and Celtic Pig (Porco Celta) meats and meat products.

Gourmet dry-cured sausages: Among "industrial" meats (from white pig breeds, such as Belgium White, Landrace, and Pietrain, among others), there are several Mediterranean (Cerdo Ibérico, Preto Alentejano, and Chato Murciano) and Atlantic (Celtic breeds Porco Celta and Bisaro) autochthonous pig breeds (Teixeira and Rodrigues, 2013; Alvarez-Rodríguez and Teixeira, 2019). Their meats produce well-appreciated meat products (gourmet) known for their exceptional quality. They are fed with chestnuts, acorns, and grass, and are extensively reared. Its industrialization can be a good source of income and contributes to the Agenda of Sustainable Development Goals, favouring the circular economy and the rural environment's sustainability (United Nations, 2019).

The behaviour of these meats, used in the elaboration of high-value-added meat products, such as Salchichón among others (dry-cured ham, dry-cured loins, chorizo, etc.) is not fully understood. To improve the use of these autochthonous pork meats in traditional meat products (dry-cured ham, shoulder and loin, dry-fermented sausages and others) it is necessary to study in depth their meat properties, techno-functional properties (water holding capacity, oil holding capacity, emulsion capacity, and stability, among others), and the chemical, biochemical, and technological impact of processing since they have not yet been fully studied.

Meat characterization of native pig breeds is essential for processing. With the processing of these high commercial value, meats can be ensured and improved their qualities (technological and healthiness). Besides, it is widely held that the consumption of products derived from native pig breeds should not only satisfy consumer demand, but also promote the development of areas where these breeds are originated (Salazar et al., 2016). All of these Iberian pig breeds are used as a raw material in the making of Salchichón and all related dry-cured sausages. For example, meats from Chato Murciano or acorn fed (Montanera) Iberian breeds have higher values for unsaturated and polyunsaturated fatty acids and the n6/n3 index, lower values for saturated fat and the Saturated/Polyunsaturated fat ratio and a lower Atherogenesis Index (Auqui et al., 2019). These characteristics can be suitable for new consumers, who select meat and meat products not only according to the price and perceived eating quality, but also by their nutritional value and healthiness, issues of animal welfare, and the ethics underlying the production of meat as well as the level of environmental impact caused by the production system (Font-i-Furnols and Guerrero, 2014; Bergstra et al., 2017). Within the same consumer trend, another type of meat is used to elaborate Salchichón and related dry-cured sausages: poultry meats (chicken and turkey), and foal (Lorenzo et al., 2017), and also exotic meats such as wild boar, warthog (Mahachi et al., 2019), and venison (Utrilla et al., 2014), among others.

13.1.6.1 Dry-Cured Sausages as a Healthy Meat Product

Healthy meats are a big challenge and a goal for the meat industry and the scientific community. Dry-cured sausages, such as Salchichón, are not an exception. The international Healthy Meat network (CYTED 119RT0568) promotes the goal of producing healthier meat products. This objective can be reached through different strategies and interventions that have an impact on reducing the consumption of some additives, improving the nutritional and health benefits for consumers. That can be applied to these types of dry-cured sausages (Healthymeat, 2020).

In general, research, development, and innovation (R&D+i) have a lot of work to do, since there are, in general, a huge number of meat products of which dry-cured sausages (as Salchichón) are just a few. This work needs a multidisciplinary team in which several food-related professionals such as anthropologists, dieticians, food chemists, food technologists and engineers, gastronomists, microbiologists, lawyers, veterinarians, molecular biologists, and science journalists, among others, must be included.

This challenge requires: (i) studying the anthropological, religious, and social aspects of each meat to create a meat products inventory within each geographic region that presents any problems associated with the health of consumers and that therefore might be reformulated or innovated; (ii) collecting available technical and scientific information related to the physico-chemical, techno-functional, microbiological, sensorial, nutritional, and engineering (new technologies such as high pressure processing, pulsed-UV light, irradiation, among others) that can be utilized for processing and decontamination of meat products to improve process efficiency, product safety and product quality, and composition (Ojha et al., 2015), consumers' behaviour characteristics, of raw material, ingredients (traditional and new: chia, quinoa, animal and vegetable agroindustrial coproducts, etc.), additives, elaboration process (sustainable and ecoefficient technologies) and shelf life without forgetting mass and social media networks; (iii) looking for new ingredients/natural additives (antioxidants and antimicrobials) to reduce or eliminate sugars, salt, saturated fats, nitrates, and nitrites to maintain the best quality and consumer acceptance of these types of sausage.

13.1.6.2 Salchichón Innovations

As mentioned above, there are 44 main issues related to improving the healthiness of Salchichón: (i) fat reduction (dos Santos et al., 2021; Panea and Ripoll, 2021) and the addition of healthy fats (omega 3 fatty acids) (Rubio et al., 2007; Lorenzo et al., 2016); (ii) salt reduction (Guàrdia et al., 2008; Corral et al., 2014; Panea and Ripoll, 2021); (iii) additives reduction/substitution in the formulation, focus on nitrites

(Aquilani et al., 2018; Fernández-López, et al., 2021) and "artificial" antimicrobials and antioxidants (Martín-Sánchez et al., 2011; Munekata et al., 2017; Aquilani et al., 2018); (iv) adding nutrients such as dietary fibre (Fernández-López et al., 2008) and prebiotics (Neri-Numa et al., 2020), probiotics, the "healthy" microbiota (Sayas-Barberá et al., 2012; Sirini et al., 2020), and bioactive compounds (Neri-Numa, et al., 2020).

13.1.6.3 Healthy Salchichón

Consumers look for foods that provide nutrition and pleasure while safeguarding their health, the result of which is that they increasingly avoid foods containing cholesterol and saturated and trans fatty acids (Ospina-E et al., 2010). However, Teixeira and Rodrigues (2020) mentioned that consumer perceptions towards healthier meat products are now mainly associated with how meat is produced and processed; the physical and chemical composition; nutritional quality; sensory properties, and social, ethnic, or religious aspects.

In the future, 2020 will be remembered fundamentally as the year of the COVID-19 pandemic, where everyone was caught off guard in many respects and that has shaken many of the foundations of the Western lifestyle. Food has not been immune to the effect of COVID-19, causing some uncertainty in its production, processing, and consumption. This has been caused by the perception that food could affect the different actors associated with "farm to fork" programs. Thus, consumers, research and development teams (industrial and academic), food industries, transport, and groups of people affected by different pathologies (celiac disease, lactose intolerance, cancer, food-related allergies, etc.) all can be included (Pérez-Alvarez et al., 2020).

The pandemic is an opportunity that we, as a society, should not miss, since it is the perfect time for science and society to finally work together to improve consumers' wellness. The work to be carried out, and even more so in food industries, is arduous, but it is still interesting and exciting for everyone involved (consumers, researchers, industries, governments, etc.). If one specifically speaks about meat products, it is time to strengthen multidisciplinary teams between industries, universities, and research centres (Pérez-Alvarez et al., 2020a).

A new way of developing or innovating Salchichón and other related dry-cured sausages is to see them from a different point of view (integral vision concept). Thus, to be successful in their development and innovation, dry-cured sausages must be considered as part of art and tradition, way of life, religion, social activities related to these sausages, leisure, self-indulgence, and local or national identity, among others. All of these apply to the development of a new and/or healthy dry-cured sausages, like Salchichón. All these aspects are included in the Miguel Hernández 5Stars FOODS, (each star is related to Healthy, Safety, Tasty, Sustainable and Socially accepted) concept (Pérez-Alvarez et al., 2019).

The innovation of meat products is focused on reformulating meat products, and Salchichón is not an exception. There are several strategies to produce healthy dry-cured sausages. Nowadays, consumers, health authorities, and meat industries are in search of those compounds that are related to several chronic pathologies (blood cholesterol, diabetes, hypertension, cancer, etc.). At this moment, industry and health authorities are working together to reduce food compounds such as salt, fat, sugars, and additives. Dry-meat product reformulation is an important technological challenge, since many factors (microbiological, physical, and chemical) interact during the entire production process. Any modifications can alter the physical, chemical, and sensorial characteristics for which this meat product is well appreciated.

Reformulation is a good option to increase Research and Development or reformulated dry-cured sausages, like Salchichón, through innovation. It is possible to achieve dry-cured sausages with the same or better quality (technological, nutritional, etc.) as those that were already available for the consumer in the supermarkets. Teixeira and Rodrigues (2019) mentioned that the incorporation of healthier ingredients, natural antioxidants, modified fat profiles (lower saturated fats and cholesterol, and higher polyunsaturated fats), and reducing salt and nitrite will remain issues, and the meat industry must adapt and improve. To reach these goals, meat industries assume that is necessary to promote innovation among their associates to encourage them to offer healthier new meat products (for the consumer conscious of the relationship between food and health) on the market. Particularly with Salchichón (as other dry-cured meat products), the strategy focuses on reductions in the salt content, fats (content and type), and added sugars (in Salchichón; fortunately, this contribution is completely restricted by law, Table 13.1), without increasing calories. To reach these objectives, industries are seeking synergies with other sectors (academic, food machinery, etc.) and other companies (technological industries, food additives, among others) to optimize efforts to reach the agreements and fulfil all planned objectives. Using these agreements, it is possible to offer a wide variety of dry-cured meat products with an improved nutritional composition. Thus, Spanish consumers will have access to a healthier meat product in different establishments, settings, and services, and the healthier meat should also be a more desirable export.

At this moment, meat industries and Spanish health authorities (Ministry of Health, Consumption and Social Welfare) have reached, an agreement (BOE, 2019) and meat processors have incorporated some practices to make meat products healthier. These can be summarized in several points:

i. Low limits or absence of trans-fatty acids should be maintained. In dry-cured sausages, the addition of fat, rich in trans-fatty acids, as an ingredient is not usually used.

ii. According to the meat products, there are established several "reductions" for fat, salt and sugar content, and this reduction must be applied for 2020 (Table 13.3).

iii. If additional reductions are made in the same products and/or others, which would also provide health benefits, they will be framed within technological, food safety, acceptance, and legislation limits.

iv. If Salchichón-based new products will be developed, these should be aligned with the reductions in salt, fat, and sugar content.

At this moment, meat industries and Spanish health authorities (Ministry of Health, Consumption and Social Welfare) have reached an agreement (BOE, 2019), and meat processors have incorporated some practices to make meat products healthier.

TABLE 13.3

Agreements Between the Spanish Health Authorities and the Meat Industry to Reduce the Content of Salt, Fat, and Sugars in Salchichones Extra (Premium) Until 2020

Ingredient	Reduced Amount in Meat Products (%)	Average Concentration (g/100 g) 2016	Average Concentration (g/100 g) 2020
Salt (NaCl)	16	3.9	3.5[*]
Total fat	5	41.0	38.95
Total sugars	10	4.0	3.6

Notes

* To ensure Salchichón microbiological safety, a reduction of no more than 10% can be applied

The sensory and microbiological quality of meat products depends on the formulation, including salt, fat, and spices (Panea and Ripoll, 2021). In meat products, pork back fat reduction or substitution is quite complicated, thus Ospina-E et al. (2010) mention that back fat substitutes must have specific attributes with regard to colour, consistency, stability in the face of oxidation, taste, percentage of extractable fat, iodine index, C18:2 content and C18:0/C18:2 ratio, the double bond index, firmness, fatty acid composition, fusion behaviour, solid fat content (SFC), and crystallisation time (Hugo and Roodt, 2007; Pérez-Alvarez and Fernández-López, 2007; Wood et al., 2003). However, changing the fat composition of dry-fermented sausages, like Salchichón, is very difficult. Aside from its nutritional content, fat contributes to the quality and acceptability properties (flavour, texture, mouthfeel, etc.) of dry sausage (Ruiz-Capillas et al., 2012). On the other hand, the granulated (visible) fat also has a technological function in sausage production since it can facilitate the regular moisture release occurring during the fermentation process (Wirth, 1988). Several attempts were made to reduce saturated fat content. This goal can be achieved by substituting some of the animal fat with nuts, protein or carbohydrates (Sánchez-Zapata et al., 2011; Triki et al., 2013).

REFERENCES

Abdulazeez, M.A., Sani, I., James, B.D., Abdullahi, A.S. (2016). Black pepper (*Piper nigrum l.*) oils. In Essential Oils in Food Preservation, Flavor and Safety. Academic Press, Chapter 31, pp. 277–285.

Aguirrezábal, M., Mateo, J., Domínguez, M.C., Zumalacárregui J.M. (1998). Compounds of technological interest in the dry sausage manufacture. Sciences des Aliments, 18 (4): 409–414.

Aleu, G. (2017). Estudio de distintas matrices de formulación con propiedades funcionales de productos crudo-curados embutidos característicos de la Provincia de Córdoba. Ph.D. Thesis, Facultad de Ciencias Agropecuarias, Taholic University of Cordoba (Argentine), pp. 1–148.

Alirezalu, K., Pateiro, M., Yaghoubic, M., Alirezalu, A., Peighambardoust, S.H., Lorenzo, J.M. (2020). Phytochemical constituents, advanced extraction technologies and techno-functional properties of selected Mediterranean plants for use in meat products. A comprehensive review. Trends in Food Science & Technology, 100: 292–306.

Álvarez-Rodríguez, J., Teixeira, A. (2019). Slaughter weight rather than sex affects carcass cuts and tissue composition of Bisaro pigs. Meat Science, 154: 54–60.

Andrade, K.S., Ferreira R.S. (2013). Antioxidant activity of black pepper (*Piper nigrum L.*) oil obtained by supercritical CO_2. III Iberoamerican Conference on Supercritical Fluids, Cartagena de Indias (Colombia) (2013), pp. 1–5.

Andujar, G., Guerra, M.A., Santos, R. (2000). La utilización de extensores cárnicos. Experiencias de la industria cárnica cubana, Instituto de Investigaciones para la Industria Alimenticia Internal Report, La Habana, Cuba. http://www.fao.org/tempref/GI/Reserved/FTP_FaoRlc/old/prior/segalim/pdf/extensor.pdf. Accessed 01/15/2020.

Andujar, G., Guerra, M.A., Santos, R. (2013). Extensores en la industria cárnica. Énfasis Alimentario. 01/16/2013. http://www.alimentacion.enfasis.com/articulos/65984-extensores-la-industria-carnica. Accessed 01/15/2021.

Anónimo (2020). El Salchichón. Un embutido genuino del Bajo Guadalhorce. Sabor a Málaga. Diputación de Málaga. http://www.saboramalaga.es/producto/salchichon/. Accessed 12/23/2020.

Aquilania, C., Sirtori, F., Flores, M., Bozzi, R., Lebret, B., Pugliese, C. (2018). Effect of natural antioxidants from grape seed and chestnut in combination with hydroxytyrosol, as sodium nitrite substitutes in *Cinta Senese* dry-fermented sausages. Meat Science, 145: 389–398.

Ashish, D.G., Vipin, K.B., Vikash, B., Nishi, M. (2013). Chemistry, antioxidant and anti-microbial potential of nutmeg (*Myristica fragrans Houtt.*). Journal of Genetic Engineering & Biotechnology, 11 (1): 25–31.

Auqui, S.M., Egea, M., Peñaranda, I., Garrido, M.D., Linares, M.B. (2019). Rustic Chato Murciano pig breed: Effect of the weight on carcass and meat quality. Meat Science, 156: 105–110.

Bañón, S., Bedia, M., Martínez, J., López, C., Almela, E. (2009). Caracterización tecnológica de la Longaniza Imperial de Lorca. Eurocarne, 175: 1–8.

Batt, C.A. (2000). Lactococcus: Introduction. In Encyclopedia of Food Microbiology, Eds.R.K. Robinson, C.A. Batt, P.D. Patel. Academic Press, pp. 1164–1166.

Bergstra, T., Hogeveen, H., Kuiper, W.E., Lansink, A.G.J.M.O., Stassen, E.N. (2017). Attitudes of Dutch Citizens toward Sow Husbandry with regard to animals, humans, and the environment. Anthrozoös, 30: 195–211.

BOE. (2007). RD. 1469/2007, de 2 de noviembre. Norma de Calidad para la carne, el jamón, la paleta y la caña de lomo ibéricos. Spain. Boletín Oficial del Estado, 264:45087–45104.

Bis-Souza, C.V., Pateiro, M., Domínguez, R., Penna, A.L.B., Lorenzo, J.M., Silva Barretto, A.C. (2020). Impact of fructooligosaccharides and probiotic strains on the quality parameters of low-fat Spanish Salchichón. Meat Science, 159: 107936.

BOE. (2015). Real Decreto 126/2015, de 27 de febrero, por el que se aprueba la norma general relativa a la información alimentaria de los alimentos que se presenten sin envasar para la venta al consumidor final y a las colectividades, de los envasados en los lugares de venta a petición del comprador, y de los envasados por los titulares del comercio al por menor. Ministerio de la Presidencia, Gobierno de España. de 4 de marzo de 2015, 54: BOE-A-2015-2293.

BOE. (2019). Resolución de 8 de febrero de 2019, de la Secretaría General de Sanidad y Consumo, por la que se publica el Convenio entre la Agencia Española de Consumo, Seguridad Alimentaria y Nutrición y la Federación Empresarial de Carnes e Industrias Cárnicas, para el desarrollo de los acuerdos del plan de colaboración para la mejora de

la composición de los alimentos y bebidas y otras medidas 2017-2020. Ministerio de
Sanidad, Consumo y Bienestar Social. Boletín Oficial del Estado, miércoles 13 de
marzo de 2019. Gobierno de España. 62. Sec. III: 24134-24143.

Ca curro. (2020) Embutidos curados. http://www.cacurro.com/es/embutidos-curados-venta-
online/8-fuet.html. Accessed 01/05/2021.

CARM. (2020). Marca de Garantía de Calidad Agroalimentaria. Comunidad Autónoma de la
Región   de   Murcia.   http://www.carm.es/web/pagina?IDCONTENIDO=34703&
IDTIPO=100&RASTRO=c214$m1185. Spain. Accessed 12/19/2020.

Casanoguera   (2020).   Embutidos   tradicionales   catalanes   desde   1870.   https://
www.casanoguera.com/fuet/# Accessed 05/01/2020.

Casquete, R., Benito, M.J., Martín, A., Ruiz-Moyano, S., Aranda, E., Córdoba, M.G. (2012).
Microbiological quality of salchichón and chorizo, traditional Iberian dry-fermented
sausages from two different industries, inoculated with autochthonous starter cultures.
Food Control, 24 (1–2): 191–198.

Cheng, J.R., Liu, X.M., Zhang, Y.S. (2018). Characterization of Cantonese sausage fermented by a mixed starter culture. Journal of Food Processing and Preservation, 42 (6):
e13623.

Chen, X., Zhou, X.C.L., Xu, H., Yamamoto, M., Shinoda, M., Tada, I., Minami, S.,
Urayama, K., Yamane, H. (2019). The structure and properties of natural sheep casing
and artificial films prepared from natural collagen with various crosslinking treatments.
International Journal of Biological Macromolecules, 135: 959–968.

Clarke, M.W. (1994). Herbs and spices In Handbook of industrial seasonings, EdE.W.
Underriner. Springer Science & Business Media, pp. 43–61.

CorSevilla (2020). El consumo de jamón y paleta ibérica aumentó un 7% en 2019. https://
www.corsevilla.es/jamon-paleta-iberica-consumo-2019/. Accessed 12/28/2020.

Corral, S., Salvador, A., Belloch, C., Flores, M. (2014). Effect of fat and salt reduction on the
sensory quality of slow fermented sausages inoculated with Debaryomyces hansenii
yeast. Food Control, 45: 1–7.

Danilović, B., Savić, D. (2017). Microbial ecology of fermented sausages and dry-cured meats.
In Fermented Meat Products: Health Aspects, Ed.N. Zdolec. CRC Press, 127–166.

Dibaba, A.B. (2019). The risk of introduction of swine vesicular disease virus into Kenya via
natural sausage casings imported from Italy. Preventive Veterinary Medicine, 169:
104703.

DOGV. (1998). DECRETO 91/1998, de 16 de junio, del Gobierno Valenciano, por el que se
aprueba el Reglamento de la Marca de Calidad «CV» para Productos Agrarios y
Agroalimentarios. [1998/L5313]. Diario Oficial de la Comunitat Valenciana. DOGV
núm. 3273 de 26.06.1998) Ref. Base Datos 1265/1998 DOGV núm. 3273 de
26.06.1998.

DOGV. (2003). RESOLUCIÓN de 10 de noviembre de 2003, sobre la reglamentación de
calidad de la "longaniza de Pascua", para su distinción con la marca de calidad "CV".
[2003/X12358] Diario Oficial de la Generalitat Valenciana. DOGV núm. 4636 de
24.11.2003.

dos Santos, J.M., Ignácio, E.O., Bis-Souza, C.V., Andrea Carlada Silva Barretto, A.C.
(2021). Performance of reduced fat-reduced salt fermented sausage with added mi-
crocrystalline cellulose, resistant starch and oat fiber using the simplex design. Meat
Science, In Press (Online) 108433.

Erkkilä, S., Suihko, M.L., Eerola, S., Petäjä, E., Mattila-Sandholm, T. (2001). Dry sausage
fermented by Lactobacillus rhamnosus strains. International Journal of Food
Microbiology, 64 (1–2): 205–210.

Federici, S., Ciarrocchi, F., Campana, R., Ciandrini, E., Blasi, G., Baffone, W. (2014).
Identification and functional traits of lactic acid bacteria isolated from Ciauscolo
salami produced in Central Italy. Meat Science, 98: 575–584.

Fernández-López, J., Sendra, E., Sayas-Barberá, E., Navarro, C., Pérez-Alvarez, J.A. (2008). Physico-chemical and microbiological profiles of "salchichón" (Spanish dry-fermented sausage) enriched with orange fiber. Meat Science, 80 (2): 410–417.

Fernández-López, J., Viuda Martos, M., Pérez-Alvarez, J.A. (2021). Quinoa and chia products as ingredients for healthier processed meat products: technological strategies for their application and effects on the final product. Current Opinion in Food Science, 40: 26–32.

Flores, M. (2017). The eating quality of meat: III flavour. In Woodhead Publishing Series in Food Science, Technology and Nutrition, Ed. F. Toldrá, Elsevier Ltd., Lawrie's meat science, 8th edition, pp. 383–412.

Flores, M. (2018). Understanding the implications of current health trends on the aroma of wet and dry cured meat products. Meat Science, 144: 53–61.

Fonseca, S., Gómez, M., Domínguez, R., Lorenzo, J.M. (2015). Physicochemical and sensory properties of Celta dry-ripened "salchichón" as affected by fat content. Grasas y Aceites, 66 (1): 59.

Fontana, C., Fadda, S., Cocconcelli, P.S., Vignolo, G. (2012). Lactic acid bacteria in meat fermentations. In Lactic Acid Bacteria—Microbiological and Functional Aspects (4th ed.), Eds. S. Lahtinen, A.C. Ouwehand, S. Salminen, A. Von Wright. CRC Press, Taylor & Francis Group, pp. 247–264.

Font-i-Furnols, M., Guerrero, L. (2014). Consumer preference, behavior and perception about meat and meat products: an overview. Meat Science, 98: 361–371.

Fraqueza, M.J. (2015). Antibiotic resistance of lactic acid bacteria isolated from dry-fermented sausages. International Journal of Food Microbiology, 212 (6): 76–88.

Freixanet, L. (2020). Aditivos e ingredientes en la fabricación de productos cárnicos cocidos de músculo entero. Metalquimia. Internal report. http://es.metalquimia.com/upload/document/article-es-12.pdf. Accessed 01/15/2021.

García, C., Ventanas, J., Antequera, T., Ruiz, J., Cava, R., Alvarez, P. (1996). Measuring sensorial quality of Iberian ham by Rasch model. Journal of Food Quality, 19: 397–412.

Guàrdia, M.D., Guerrero, L., Gelabert, J., Gou, P., Arnau, J. (2008). Sensory characterisation and consumer acceptability of small calibre fermented sausages with 50% substitution of NaCl by mixtures of KCl and potassium lactate. Meat Science, 80 (4):1225–1230.

Gupta, A.D., Bansal, V.K., Babu, V., Maithil, N. (2013). Chemistry, antioxidant and anti-microbial potential of nutmeg (*Myristica fragrans Houtt*). Journal of Genetic Engineering and Biotechnology, 11: 25–31.

Harper, B.A. (2013). Understanding interactions in wet alginate film formation used for inline food processes. Ph.D. Thesis, The University of Guelph, Guelph, Ontario, Canada..

Harrison, S., Brassard, D., Lemieux, S., Lamarche, B. 2020. Dietary saturated fats from different food sources show variable associations with the 2015 healthy eating index in the canadian population. The Journal of Nutrition, 150(12): 3288–3295.

Healthymeats (2020). Productos cárnicos más saludables 119RT0568 (HEALTHYMEAT). CYTED network. http://cyted.org/content/119rt0568-objetivos. 12/30/2019. Accessed 01/14/2021.

Hospital, X.F., Hierro, E., Stringer, S., Fernández, M. (2016). A study on the toxigenesis by *Clostridium botulinum* in nitrate and nitrite-reduced dry fermented sausages. International Journal of Food Microbiology, 218 (2): 66–70.

Hu, L., Yin, C., Ma, S., Liu, Z. (2018). Assessing the authenticity of black pepper using diffuse reflectance mid-infrared Fourier transform spectroscopy coupled with chemo-metrics. Computer Electronic Agriculture, 154: 491–500.

Hugo, A., Roodt, E. (2007). Significance of porcine fat quality in meat technology: A review. Food Reviews International, 23:175–198.

Incoming Salamanca (2019). Fiesta de la matanza tradicional 2019-2020. Agencia de Turismo Gastronómico en España. Especialistas Gourmet Salamanca. http://www.incomingsalamanca.com/noticias/fiesta-de-la-matanza-tradicional-2019-2020#:~:text=Del%209%20de%20Noviembre%20al,la%20fiesta%20%2Ctradici%C3%B3n%20y%20turismo. Accessed 01/15/2021.

Jamónvillamon (2020). Secallonas. https://jamonvillamon.com/33-secallonas. Accessed 05/01/2020.

Jeena, K., Liju, V.B., Umadevi, N.P., Kuttan, R. (2014). Antioxidant, anti-inflammatory and antinociceptive properties of black pepper essential oil (*Piper nigrum Linn*). Journal of Essential Oil Bearing Plants, 17: 1–12.

Jensen, L.B. (1953). Man's Food. The Guarrad Press.

Kazeem, M.I., Akanji, M.A., Hafizur, R.M., Choudhary, M.I. (2012). Antiglycation, antioxidant and toxicological potential of polyphenol extracts of alligator pepper, ginger and nutmeg from Nigeria. Asian Pacific Journal of Tropical Biomedicine, 2 (9): 727–732.

Larrosa, P. (2009). Imperial de Lorca. Gastronomía. Diario La Verdad. 09/10/2009. https://www.laverdad.es/gastronomia/que-probar/la-imperial-lorca-20090909000000-nt.html. Accessed 12/19/2020.

Latorre-Moratalla, M.L., Bover-Cid, S., Vidal-Carou, M.C. (2010). Technological conditions influence aminogenesis during spontaneous sausage fermentation Meat Science, 85 (3): 537–541.

La Table d'Amelie, (2014). Saucisson a l'ail Maison (nature, fumé, aux pistaches..!). La table Lorraine´Amelie. Entre "Brimbelles et mirabelles". 02/22/2014. https://latabledamelie.blogspot.com/2014/02/saucisson-lail-maison-nature-fume-aux.html Accessed 01/10/2021.

Leistner, 1995. Stable and safe fermented sausages world-wide. In Fermented Meats, Eds. G. Campbell-Platt, P.E. Cook. Blackie Academic & Professional, pp. 160–175.

Leuschner, R.G.K., Ielsch, V. (2003). Antimicrobial effects of garlic, clove and red hot chilli on Listeria monocytogenes in broth model systems and soft cheese. International Journal of Food Sciences and Nutrition, 54:127–133.

Liu, S.Q. (2003). Practical implications of lactate and pyruvate metabolism by lactic acid bacteria in food and beverage fermentations. International Journal of Food Microbiology, 83:115–131.

López-Bote, C.J. (1998). Sustained utilisation of the Iberian pig breed. Meat Science, 49: S17–S27.

Lorenzo, J.M., Munekata, P.E.S., Pateiro, M., Campagnol, P.C.B., Domínguez, R. (2016). Healthy Spanish salchichón enriched with encapsulated n−3 long chain fatty acids in konjac glucomannan matrix. Food Research International, 89 (1): 289–295.

Lorenzo, J.M., Munekata, P.E.S., Campagnol, P.C.B., Zhu, Z., Alpas, H., Barba, F.J., Tomasevic, I. (2017). Technological aspects of horse meat products – A review. Food Research International, 102: 176–183.

Mahachi, L.N. Rudman, M., Arnaud, E., Muchenje, V., Hoffman, L.C. (2019). Development of semi dry sausages (cabanossi) with warthog (*Phacochoerus africanus*) meat: physicochemical and sensory attributes. LWT, 115: 108454.

Marchello, M., Garden-Robinson, J. (2017). The art and practice of sausage making. North Dakota State University, pp. 1–12.

Marcos, B., Gou, P., Arnau, J., Comaposada, J. (2020). Co-extruded alginate as an alternative to collagen casings in the production of dry-fermented sausages: Impact of coating composition. Meat Science, 169: 108184.

Martín-Sánchez, A.M., Chaves-López, C., Sendra, E., Sayas, E., Fenández-López, J., Pérez-Álvarez, J.A. (2011). Lipolysis, proteolysis and sensory characteristics of a Spanish fermented dry-cured meat product (salchichón) with oregano essential oil used as surface mold inhibitor. Meat Science, 89(1): 35–44.

Martín Bejarano, S. (1992). Hierbas, especias y eondimentos. In Manual práctico de la carne, Eds. Martín & Macias, España.

Ministerio de la Presidencia. (2014). Real Decreto 474/2014, de 13 de junio, por el que se aprueba la norma de calidad de derivados cárnicos. Ref.: BOE-A-2014-6435. Boletín Oficial del Estado. Gobierno de España. https://www.boe.es/buscar/pdf/2014/BOE-A-2 014-6435-consolidado.pdf Accessed 15/10/2020.

Munekata, P.E.S., Domínguez, Franco, D., Bermúdez, R., Trindade, M.A., Lorenzo, J.M. (2017). Effect of natural antioxidants in Spanish salchichón elaborated with encapsulated n-3 long chain fatty acids in konjac glucomannan matrix. Meat Science, 124: 54–60.

Munekata, P.E.S., Rocchetti, G., Pateiro, M., Lucini, L., Domínguez, R., Lorenzo, J.M. (2020). Addition of plant extracts to meat and meat products to extend shelf-life and health-promoting attributes: An overview. Current Opinion in Food Science, 31: 81–87.

Museu de l'embotit. (2020). Museu de l'Embotit. Catalunya Convention Bureau. Generalitat de Catalunya. https://www.catalunya.com/museu-de-lembotit-17-16001-210?language= es. Accessed 01/15/2021.

Neri-Numa, I.A., Silvano Arruda, H., Vilar Geraldi, M., Maróstica Júnior, M.R., Pastor, G.M. (2020). Natural prebiotic carbohydrates, carotenoids and flavonoids as ingredients in food systems. Current Opinion in Food Science, 33: 98–107.

Nishada, J., Koley, T.K., Varghese, E., Kaur, C. (2018). Synergistic effects of nutmeg and citrus peel extracts in imparting oxidative stability in meat balls. Food Research International, 106: 1026–1036.

Ojha, K.S., Kerry, J.P., Duffy, G., Beresford, T., Tiwaria, B.K. (2015). Technological advances for enhancing quality and safety of fermented meat products. Trends in Food Science & Technology, 44(1): 105–116.

Ospina-E, J.C., Cruz-S, A., Pérez-Álvarez, J.A., Fernández-López, J. (2010). Development of combinations of chemically modified vegetable oils as pork backfat substitutes in sausages formulation. Meat Science, 84 (3): 491–497.

Panea, B., Ripoll, G. (2021). Substituting fat with soy in low-salt dry fermented sausages. NFS Journal, 22: 1–5.

Pérez-Alvarez, J.A. (1996). Contribución al estudio objetivo del color en productos cárnicos crudo-curados. Ph.D Thesis, Polytechnical University of Valencia, Valencia, Spain.

Pérez-Alvarez, J.A., Fernández-López, J. (2007). Chemical and biochemical aspect of color in muscle foods. In Handbook of Meat, Poultry and Seafood Quality, Ed. L.M.L. Nollet. Blackwell Publishing, pp. 25–44.

Pérez-Alvarez, J.A., Viuda-Martos, M., Sayas-Barberá, E., Navarro-Rodríguez de Vera, C., Fernández-López, J. (2019). Alimentos 5S un nuevo concepto aplicado a la elaboración de productos cárnicos más saludables. I Congreso Iberoamericano de Maracas de Calidad de Carne y Productos Cárnicos. Eds.A. Teixeira and C. Sañudo. MARCARNE. Livro de Actas, pp. 231–235.

Pérez-Alvarez, J.A., Viuda-Martos, M., Fernández-López, J. (2020). La sostenibilidad uno de los pilares de los Alimentos 5S. Congress of Gastronomy, Sustainability and Development (ICAF2020). IV International Meeting of the UNESCO-UOC Chair on Food, Culture and Development & UNITWIN Network. Food, Technology and Sustainability section. Cáceres (Spain). 24/09/2020.

Pérez-Alvarez, J.A., Viuda-Martos, M., Fernández-López, J. (2020a). Retos y oportunidades para la industria alimentaria en tiempos del Covid-19. Tecnifood, 132: 80–84.

Ponnampalama, E.N., Jacob, J.L., Knight, M.I., Plozz, T.E., Butlere, K.L. (2020). Understanding the action of muscle iron concentration on dark cutting: An important aspect affecting consumer confidence of purchasing meat. Meat Science, 167: 108156.

Prabhu, G., Keeton, J. (2008). Aplicaciones de productos de suero y lactosa en carnes procesadas. Mundo Lácteo y Cárnico, 01/02: 18–25.

Ratanavaraporn, J., Kanokpanont, S., Tabata, Y., Damrongsakkul, S. (2008). Effects of acid type on physical and biological properties of collagen scaffolds. Journal of Biomaterials Science, Polymer Edition, 19 (7): 945–952.

Ravyts, F., Vuyst, L.D., Leroy, F. (2012). Bacterial diversity and functionalities in food fermentations. Engineering in Life Sciences, 2012 (12): 356–367.

Reducetarian Foundation. (2020). What we do. Improve human health, protect the environment, and spare farm animals from cruelty by reducing societal consumption of animal products. Reducetarian Foundation. https://www.reducetarian.org/what. Accessed 12/15/2020.

Roncalés, P. (1995). Tecnología, cambios bioquímicos y calidad sensorial de los embutidos curados. Alimentación, Equipos y Tecnología (Enero/Febrero), pp. 73–82..

Rubio, B., Martínez, B., Sánchez, M.J., García-Cachán, M.D., Rovira, J., Jaime, I. (2007). Study of the shelf life of a dry fermented sausage "salchichon" made from raw material enriched in monounsaturated and polyunsaturated fatty acids and stored under modified atmospheres. Meat Science, 76 (1): 128–137.

Ruiz-Capillas, C., Triki, M., Herrero, A.M., Rodriguez-Salas, L., F.Jiménez-Colmenero, F. (2012). Konjac gel as pork backfat replacer in dry fermented sausages: Processing and quality characteristics. Meat Science, 92 (2): 144–150.

Ruiz Pérez-Cacho, M.P., Galán-Soldevilla, H., León Crespo, F., Molina Recio, G. (2005). Determination of the sensory attributes of a Spanish dry-cured sausage. Meat Science, 71(4): 620–633.

Salazar, P., García, M.L., Selgas, M.D. (2009). Short-chain fructooligosaccharides as potential functional ingredient in dry fermented sausages with different fat levels. International Journal of Food Science and Technology, 44 (6): 1100–1107.

Salazar, E., Abellán, A., Cayuela, J.M., Poto, Á., Tejada, L. (2016). Dry-cured loin from the native pig breed Chato murciano with high unsaturated fatty acid content undergoes intense lipolysis of neutral and polar lipids during processing. European Journal of Lipid Science and Technology, 118(5): 744–752.

Salvá, B.K., Zumalacárregui, J.M., Figueira, A.C., Osorio, M.T., Mateo, J. (2009). Nutrient composition and technological quality of meat from alpacas reared in Peru. Meat Science, 82 (4): 450–455.

Sánchez-Zapata, E., Fernández-López, J., Peñaranda, M., Fuentes-Zaragoza, E., Sendra, E., Sayas, E., Pérez-Alvarez J.A. (2011). Technological properties of date paste obtained from date by-products and its effect on the quality of a cooked meat product. Food Research International, 44: 2401–2407.

Sánchez-Zapata, E., Díaz-Vela, J., Pérez-Chabela, M.L., Pérez-Alvarez, J.A., Fernández-López, J. (2013). Evaluation of the effect of tiger nut fibre as a carrier of unsaturated fatty acids rich oil on the quality of dry-cured sausages. Food and Bioprocess Technology, 6 (5):1181–1190.

Sanz, Y., Vila, R., Toldrá, F., Flores J. (1998). Effect of nitrate and nitrite curing salts on microbial changes and sensory quality of non-fermented sausages. International Journal of Food Microbiology, 42: 213–217.

Savic, Z., & Savic, I. (2016). Sausage casings (2nd ed.). Victus International GmbH, A-1130 Wien, pp. 47–87, 212-258..

Sayas-Barberá, E., Pérez-Alvarez, J.A., Fernández-Lopez, J., Oñate, M.D. (1998). Physical and physicochemical characterization of Longaniza imperial de Lorca. Alimentaria, 294: 27–33.

Sayas-Barberá, E., Fernández-López, J., Pérez-Alvarez, J.A. (2007). Elaboración de embutidos crudo-curados. In Industrialización de Productos de Origen Animal, Eds. E.

Sayas-Barberá, J. Fernández-López, J.A. Pérez-Alvarez. Universidad Miguel Hernández de Elche, pp. 77–101.

Sayas-Barberá, E., Viuda-Martos, M., Fernández-López, J., Pérez-Alvarez, J., Sendra, E. (2012). Combined use of a probiotic culture and citrus fiber in a traditional sausage 'Longaniza de Pascua'. Food Control, 27 (2): 343–350.

Sheehy, S., Palmer, J.R., Rosenberg, L. 2020. High consumption of red meat is associated with excess mortality among African-American women. The Journal of Nutrition, 150(12): 3249–3258.

Sirini, N., Roldán, A., Lucas-González, R., Fernández-López, J., Viuda-Martos, M., Pérez-Álvarez, J.A., Frizzo, L.S., Rosmini, M.R. (2020) Effect of chestnut flour and probiotic microorganism on the functionality of dry-cured meat sausages. LWT, 134: 110197.

Srinivasan, K. (2007). Black pepper and its pungent principle - piperine: A review of diverse physiological effects. Critical Reviews in Food Science and Nutrition, 47: 735–748.

Suursa, P., Barbut, S. (2020). Collagen use for co-extruded sausage casings – A review. Trends in Food Science & Technology, 102: 91–101.

SweeGen, (2021). Diets, health and sugar reduction in the spotlight during the era of COVID-19. White paper. SweeGen Co. http://Sweegen+WP+Sugar+Reduction+for +Wellness.pdf. Accessed 01/13/2021.

Šojić, B., Vladimir, T.K., Sunčica, T., Snežana, Š., Predrag, I., Natalija, D., ... , Snežan, K. (2015). Effect of nutmeg (Myristica fragrans) essential oil on the oxidative and microbial stability of cooked sausage during refrigerated storage. Food Control, 54: 282–286.

Tauber, W.F. (1976). The history of sausage, from Babylon to Baltimore: Sausage through the ages. 29th Annual Reciprocal Meat Conference of the American Meat Science Association, 55-60.

Teixeira, S. Rodrigues, S. (2013). Pork meat quality of Preto Alentejano and commercial large white landrace cross. Journal of Integrative Agriculture, 12 (11): 1961–1971.

Teixeira, A., Rodrigues, S. (2021). Consumer perceptions towards healthier meat products. Current Opinion in Food Science, 38: 147–154.

Toldrá, F., Nip, W.-K., Hiu, Y.H. (2007). Dry-fermented sausages: An overview. Handbook of Fermented Meat and Poultry, Ed. F. Toldra. Blackwell Publishing, 1st ed.,pp. 321–325.

Trigueros, N. (2016). ¿Cuál es el auténtico salchichón de Málaga? Sur. Empresas malagueñas. 02/13/2016 ¿Cuál es el auténtico salchichón de Málaga? Diario Sur Accessed 12/20/2020.

Triki, M., Herrero, A.M., Rodríguez-Salas, L., Jiménez-Colmenero, F., Ruiz-Capillas, C. (2013). Chilled storage characteristics of low-fat, n-3 pufa-enriched dry fermented sausage reformulated with a healthy oil combination stabilized in a konjac matrix. Food Control, 31: 158–165.

Trigo, M.J., Fraqueza, M.J. (1998). Effect of gamma radiation on microbial population of natural casings. Radiation Physics and Chemistry, 52:125–128.

Unal, R., Fleming, H.P., McFeeters, R.F., Thompson, R.L., Breidt Jr. F., Giesbrecht, F.G. (2001). Novel quantitative assays for estimating the antimicrobial activity of fresh garlic juice. Journal of Food Protection, 64:189–194.

United Nations (2019). Sustainable Development Goal. United Nation. https://www.un.org/ sustainabledevelopment/. Accessed 01/13/2021.

Utrilla, M.C., García Ruiz, A., Soriano, A. (2014). Effect of partial reduction of pork meat on the physicochemical and sensory quality of dry ripened sausages: Development of a healthy venison salchichon. Meat Science, 98 (4): 785–791.

Vijayakumar, R.S., Surya, D., Nalini, N. (2004). Antioxidant efficacy of black pepper (Piper nigrum L.) and piperine in rats with high fat diet induced oxidative stress. Redox Rep., 9: 105–110.

Walz, F.H., Gibis, M., Lein, M., Herrmann, K., Hinrichs, J., Weiss J. (2018). Influence of casing material on the formation of efflorescences on dry fermented sausages. LWT, 89: 434–440, 10.1016/j.lwt.2017.11.019.

Walz F.H., Gibis, M., Herrmann, K., Weiss J. (2019). Impact of smoking on efflorescence formation on dry-fermented sausages. Food Structure, 20 (4): 100111.

Wanangkarn, A., Liu, D.-C., Swetwiwathana, A., Jindaprasert, A., Phraephaisarn, C., Chumnqoen, W., Tan, F.-J. (2014). Lactic acid bacterial population dynamics during fermentation and storage of Thai fermented sausage according to restriction fragment length polymorphism analysis. International Journal Food Microbiology, 186: 61–67.

Wijnker, J.J., Tersteeg, M.H.G., Berends, B.R., Vernooij, J.C.M., Koolmees, P.A. (2008). Quantitative histological analysis of bovine small intestines before and after processing into natural sausage casings. Journal of Food Protection, 71(6): 1199–1204.

Wirth, F. (1988). Technologies for making fat-reduced meat products. What possibilities are there? Fleischwirtschaft, 68 (9):1153–1156.

Wood, J.D., Richardson, R.I., Nute, G.R., Fisher, A.V., Campo, M.M., Kasapidou, E., Sheard, P.R., Enser, M. (2003). Effects of fatty acids on meat quality: A review. Meat Science, 66(1): 21–32.

Yang, G.C., Yasaei, P.M., Page, S.W. (1993). Garlic as antioxidants and free radical scavengers. Journal of Food and Drug Analysis, 1 (4): 357–364.

Zakaria, M.P.M., Abas, F., Rukayadi, Y. (2015). Effects of *Myristica fragrans Houtt.* (nutmeg) extract on chemical characteristic of raw beef during frozen storage. International Food Research Journal, 22 (3): 902–909.

Zarringhalam, M., Zaringhalam, J., Shadnoush, M., Rezazadeh, S., Tekieh, E. (2013). Inhibitory effect of black and red pepper and thyme extracts and essential oils on Enterohemorrhagic *Escherichia coli* and DNase activity of *Staphylococcus aureus.* Iranian. Journal of Pharmacology Research, 12 (3): 1–7.

Zhang, J., Wang, Y., Pan, D-D., Cao, J-X., Shao, X-F., Chen, Y-J., Sun, Y-Y., Ou, C.R. (2016). Effect of black pepper essential oil on the quality of fresh pork during storage. Meat Science, 117: 130–136.

14 Manufacture of Whole Muscle Cook-In Ham

Vladimir Tomović,[1] Marija Jokanović,[1] Branislav Šojić,[1] and Igor Tomašević[2]
[1]University of Novi Sad, Faculty of Technology Novi Sad, Bulevar cara Lazara 1, 21000 Novi Sad, Serbia
[2]University of Belgrade, Faculty of Agriculture, Nemanjina 6, 11080 Belgrade, Serbia

CONTENTS

14.1 INTRODUCTION

Cooked cured products are high quality meat products consumed worldwide. According to Codex Alimentarius cooked cured ham shall be made of meat from the hind leg of a pig—divided transversely from the remainder of the side at a point not further anteriorly than the end of the hip bone. All bones and detached cartilage, tendons and ligaments shall be removed. Skin and fat may or may not be removed. The meat shall be cured and may be smoked, spiced and/or flavoured.

Although, the term *ham* specifically refers to meat from the leg, many cooked, cured products are made from other cuts of pork (loin, shoulder) or from other types of meat (beef, turkey, chicken). The final quality depends on both the raw materials and the processing technology. The most outstanding factors are the quality of meat, the type and amounts of ingredients and additives, the injected volume of brine, the rate and extent of mechanical treatment, and the cooking/cooling time and temperature. The processing steps for producing cooked ham should take place in sequence, each one following smoothly from the previous step. Brine promotes protein solubilisation and curing, increases sensory quality and gives a better yield and more uniform distribution

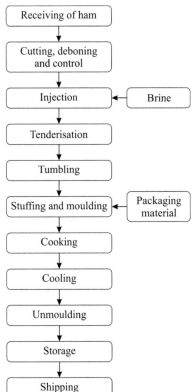

FIGURE 14.1 A process flow diagram illustrating the steps involved in the production of cooked hams.

of sodium chloride, polyphosphates, nitrites and other ingredients and additives. In general, the more water is injected into the ham, the poorer the quality. Mechanical treatment (tumbling) is important to distribute the brine into muscles. This mechanical process facilitates the extraction of proteins that cause binding of meat pieces. The main objectives of heat treatment are destruction of microorganisms and extension of shelf life, and development of sensory properties (Figure 14.1).

In Europe and most other parts of the world, cooked ham is produced as whole muscle product made from individual muscles free of visible fat and connective tissues. These products are usually packed in plastic films, and are also known as cook-in hams or uncanny hams. They are cooked in their final packaging (cook-in), which do not exuded juices during the cooking process.

Cook-in technology is a sophisticated technology that requires rigorous control and care in all phases of the process, with great technological know-how and proper selection of the machinery.

14.2   PREPARATION AND SELECTION OF RAW MATERIAL

In cooked, cured products the condition of the raw material has a decisive influence on optimal final quality. The selected fresh pork legs should be chilled to a temperature between 0°C and 4°C. In this product, "hot" meat or frozen and thawed meat may have negative effects on weight yield, colour development, cohesion of slice, tenderness, juiciness and/or shelf life. Where fresh chilled meat is used, all pieces of meat should be deboned in a way that avoids as much as possible making deep cuts into the muscle tissue. It is important to have good manufacturing practices during slaughtering and cutting to avoid high bacterial counts, which decrease shelf life and can produce sensory defects.

When cutting whole muscle that is to be processed by mechanical treatment (tumbled and/or massaged) after brine injection, it is important that the individual muscles are free of intermuscular fatty and connective tissues, as well as sinews, ligaments, tendons, glands, blood spots and bloody tissue, as they would be visible in the finished product and affect coherency or binding between individual muscles after cooking. This operation is done manually owing to the lack of appropriate machinery. Trimming of intermuscular fat and connective tissue facilitates extraction of the salt-soluble surface proteins, which increases the binding of muscles, improves the cohesiveness of the slices and decreases the risk of formation of brine pockets during injection and formation of gelatine between muscles during cooking. The pieces of meat are also more elastic and softer after tumbling and therefore easier to handle during subsequent stuffing than they woud be if the layer of intermuscular fat and connective tissues were still present. The amount of fat affects appearance of the slice and acceptability by consumers. In the case of fat, it is also important that the meat is marbled, which considerably improves the product's texture and flavour development. Slice coherency is at its best when there is activated surface protein on pieces of meat, and it is improved greatly by skinning off the membranes of individual muscles. Special machines are available for membrane skinning and the process is quick. This process removes the thin layer of connective tissue surrounding the muscle (perimysium) and also makes it easier to introduce brine into the muscle.

It is also important to select hams with a good technological quality. The main technological quality attributes of meat include: pH value, water-holding capacity (WHC) and colour. The pH of the raw pork is particularly important for the cooked ham manufacture, as it is evident that pH value is significantly related to both colour and WHC. The pH of muscle in a live animal is almost neutral, at 7.2. After carcass chilling, ultimate pH value from 5.4 to 5.8 is reached. In pork with normal glycolysis rates, most biochemical processes (*rigor mortis*) are completed within 6–12 hours *post mortem*. In PSE (pale, soft and exudative) pork, the pH decreases rapidly; one hour after slaughter it may be below 5.8, having poor WHC and pale colour. When making cooked cured products, a high WHC and thus a high pH are desirable. In the case of DFD (dry, firm, dark) pork, ultimate pH value is above 6.2. Because of high pH, WHC of DFD pork is very good (solubility of protein is high), but meat is usually too dark and is more prone to bacterial deterioration. Therefore, PSE and DFD pork are not suitable for the manufacture of cooked cured products. The optimal ultimate pH for cooked ham production is between 5.8 and 6.2. This represents a compromise between WHC, ability to take the cure, colour development, eating quality and shelf life. When processing pork for whole muscle ham products, the meat could be injected 36–76 hours after slaughter. Such matured meat exhibits a higher pH, which is beneficial for WHC and the solubility of protein. Meat with a pH of around 5.8–6.2 exhibits greater repulsion forces between the protein molecules and hence more added water can be immobilized within the protein structure, resulting in enhanced weight yields. Nowadays, many ham products are produced by injecting meat that has matured for only around 18–24 hours, resulting in a minimal rise in pH, but this is compensated for by additives such as phosphates and salt.

Different individual muscles have different redness, as can be seen in the finished products if a dark-coloured muscle is next to a light-coloured one. Best quality cooked hams are produced out of the same type of individual muscle, such as *m. semimembranosus* only, to achieve an even curing colour in the finished product. A two-tone effect can be seen occasionally, where light- and dark-coloured areas are present within the same muscle. These light- and dark-coloured areas are primarily associated with different levels of myoglobin and different pH levels within the same muscle.

14.3 BRINE COMPOSITION AND PREPARATION

The process of curing meat requires the addition of a number of ingredients and additives that are necessary to obtain the desired colour, texture and flavour. These substances are dissolved in water to form the brine. Ingredients and additives used in whole muscle products are primarily chosen based on the desired level of injection, which in most cases correlates with weight yield in the finished product.

Phosphates, salt and nitrites are generally always used; other ingredients and additives are selected according to the requirements of the manufacturer.

The main functions of salt are the solubilization of myofibrillar proteins and its contribution to typical salty taste. The addition of phosphates is primarily responsible for the solubilization of myofibrillar proteins. The processes of protein solubilization are extremely important for all meat products where water is introduced, because only solubilized proteins can immobilize large amounts of added water. The addition of phosphates and salt plays a vital role in the process of protein activation. Activated meat protein is still the best material for achieving good slice coherency, bite and texture in the finished product. The action of phosphates and salt in meat can be explained in several ways. Firstly, alkaline phosphates may affect the water holding capacity of *post rigor* muscle by increasing the pH of the muscle, which increases the net negative charge on muscle. These negative charges increase the electrostatic repulsion between fibres and ultimately increase the hydration of the muscle.

Secondly, phosphates act as electrolytes and increase ionic strength of the meat system. The phosphate anions can enhance meat hydration directly by increasing the number of binding sites for water. By increasing the ionic strength, phosphates increase the net negative charge on proteins. This causes a repulsion of adjacent molecules, which increases hydration. However, at the levels of phosphate normally used, the ionic strength may not be increased enough to be a big influence.

Thirdly, polyphosphate can dissociate the actomyosin complex into actin and myosin. Individually, actin and myosin are easier to solubilize at the phosphate levels used in most cured products than is the actomyosin complex. The addition of salt only enhances ionic strength and has the same effect on *post mortem* meat hydration as previously described for phosphates. The addition of citrate enhances the ionic strength but does not solubilize protein.

Summarizing, at the levels of salt and phosphate used in the production of cooked ham, the salt concentration increases ionic strength enough to spread the filaments but does not cleave the crossbridges between actin and myosin, while phosphates resolve the actomyosin structure but may not increase ionic strength enough to spread the filaments. It is the synergistic effect of salt combined with alkaline phosphates that improves yields and maximizes myofibrillar protein solubilization. For sensory and nutritional reasons, a mildly salted cooked ham is the one preferred today. A salt content of 18–25 g per kilogram in the whole-muscle products should be the aim if there is to be a positive effect on water binding and on the flavour development, but may be outside this range. Blends of phosphates contain predominately long-chained polyphosphates because of their excellent solubility in cold water, and occasionally a few shorter-chained phosphates, such as tetrasodium pyrophosphate, which acts quickly on muscular protein. Most legislation limits the use of added phosphates up to a level of 0.5%, or 5 g of phosphorus pentoxide (P_2O_5) per kilogram of product. A P_2O_5 content of 0.5% corresponds to around 8–9 g of added phosphate per kilogram of finished product, but such high levels of added phosphates do not result in any advantage from a technological point of view. No-added-phosphate hams do not make sense from a technological viewpoint because protein is the most valuable part of lean meat and by not using phosphates, the meat protein is not used effectively.

Curing colour is obtained by the addition of sodium nitrite at levels up to 150 mg per kilogram, calculated per kilogram of injected meat, as defined according to legislation. Nitrite contributes to formation of the desired typical light pink colour, as well as flavour development, antioxidant activity and preservation effect against pathogens (*Clostridium botulinum*). Pink colour is formed in the reaction between the myoglobin and NO (obtained from NO_2^- via HNO_2) which initially forms nitrosomyoglobin. Nitrosomyoglobin, upon heating, is converted to the stable cured meat colour pigment called nitrosohemochrome. Nitrite may be used only as a mixture with salt.

The most commonly applied colour enhancers are ascorbic acid, ascorbate and erythorbate (isoascorbate). Ascorbic acid must never be added to a ham brine containing nitrite because of their instant reaction. Sodium ascorbate (erythorbate) is a strong reducing agent, enabling the fast and direct formation of NO from residual nitrite, which results in enhanced levels of nitrosomyoglobin and thus stabilization of the cured colour obtained initially. Also, ascorbate helps to prevent the formation of cancer-promoting nitrosamines by blocking the formation of nitrosating agents (N_2O_3) being originated from the nitrous oxide. It is recommended to apply ascorbate up to 500 mg per kilogram.

Sugars (dextrose) are also commonly used for the manufacture of the best whole muscle meat products in order to round up the flavour. With well-balanced brines, dextrose can reach concentrations of more than 3% in the finished product without negatively affecting the product's taste.

TABLE 14.1

Commonly Used Combinations of Ingredients and Additives in Injected Ham

Level of Injection (%)	Functional Ingredients and Additives Utilized
10–30	Phosphates, salt
30–50	Phosphates, salt, carrageenan (or protein such as soy instead of carrageenan)
50–70	Phosphates, salt, carrageenan, protein
70–100	Phosphates, salt, carrageenan, protein, starch

In addition, highly injected cook ham products may also contain some non-meat ingredients like blood plasma, caseinate, soy protein isolates, pork rind powder, carrageenan (and other hydrocolloids), starch, flavour enhancers and other sugars (saccharose, maltodextrin) to improve water retention, texture and flavour.

To prepare the brine, it is necessary to decide the percentage to be injected. Table 14.1 shows the most commonly used combinations of ingredients and additives based on the level of injection.

The concentration of each ingredient in the brine can be estimated by means of the following formula (for cook-in ham):

$$B = [P * (100 + I)]/I$$

where: B = percentage of ingredient in the brine, P = percentage of ingredient in the final product, I = percentage of injection (g brine per 100 g uninjected muscle).

Water is added to ham products as a solvent for proteins in conjunction with phosphates, salt and other ingredients and additives. The water used to prepare brine must be of drinking quality. From a technological point of view, the water must be as soft as possible (free of Ca^{2+} and Mg^{2+} ions and heavy metals). The final temperature of the brine should be adjusted to 2–5°C. To ensure complete solution, polyphosphates should be dissolved into the water first; then the non-meat proteins, sugars, nitrite, salt, carrageenan and finally ascorbate. Generally, all additives in the brine must have a small enough particle size to ensure that they do not block the needles or filters during the injection process. A good measure of injectability is that the finished brine should pass easily through a filter with a grid size of 0.5–0.6 mm, which is fine enough for most injectors.

14.4 BRINE INJECTION

Correct brine addition into the meat that is being cured is of the greatest importance for final quality, especially for colour, texture and flavour. The percentage of brine to be added is determined by the desired quality of the finished product, according to quality categories that are commonly regulated by legislation.

Traditionally, bone-in hams were submerged in the brine and the curing salts diffused slowly into the meat. This process lasted several weeks and the cooked hams were not homogeneous. Submerged wet curing is also used for smaller pieces of deboned meat. Also, for bone-in hams, brine could be pumped into the meat through the arterial system, called the artery pumping method. For this method, it is essential that the relevant arteries are not damaged during slaughtering and cutting. The pumping pressure should not be more than 2.5 bar (0.25 MPa), to avoid tearing the blood vessels. The advantage of artery pumping is the even distribution of the brine right to the bone, but the process is labour-intensive and nowadays rarely undertaken.

Injection methods are generally used for large or whole pieces of boneless meat (ham). Brine is injected in such a way that the meat tissue is not torn and that the brine is distributed evenly into all parts of the muscles, reducing the process time to a couple of days. Injection rates differ widely regarding a range of varying final product quality. Therefore, it is important to achieve the desired injection rate, ranging from a very low level such as 5%, which presents difficulties in brine ingredients distribution, up to a rate of 100%, which presents capacity problems. The best quality products are generally produced with a low brine injection rate, up to 25–30%. When more brine is injected into the cooked ham, the quality drops, since water retention is facilitated by some binders like soy proteins, carrageenan and starch. The injection percentage is given in the following formula:

Injection(%) = (Fresh meat weight + Injected brine weight)/Fresh meat weight

As the quantity of the injecting brine to be used is previously defined, great attention should be paid to the amount of brine injected (as % of fresh meat weight) and to the homogeneity of injecting (accuracy).

A precise injection rate ensures a minimum standard deviation of the brine content in different meat pieces and thus increases quality. The less standard deviation there is, the greater is precision of injection, obtaining fewer under-injected pieces (which could cause problems with flavour, and water holding capacity due to a lower concentration of brine) and also fewer over-injected pieces (over legal analytical limits).

Uniform distribution of brine in the muscles is important to reduce to a minimum the time required for the brine to migrate to the uninjected areas (Figure 14.2). This improves the appearance (colour) and texture homogeneity, and above all the final product yield, since the brine ingredients responsible for the solubilisation of proteins are received by all the muscle fibres. A good brine distribution allows better brine retention and consequently reduces the drip loss. In contrast, an irregular distribution of brine results in a deficiency or excess of ingredients in differing zones, causing irregularities of colouring, binding and flavour.

The brine is injected direct into the muscles by a system of hollow multi-needles (Figure 14.3). Using the brine injector (also known as a pickle injector) meat on the conveyer belt is supplied with brine from injection needles ranged vertically in one or more rows. The distance between the needles should not be too great (up to 20 mm) if

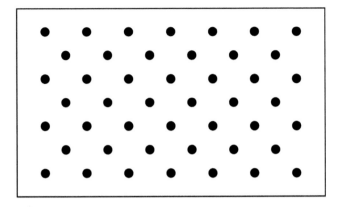

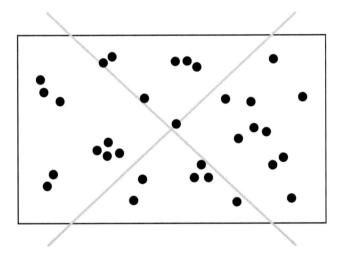

FIGURE 14.2   Correct and incorrect injection settings for optimum brine distribution.

brine distribution is to be good. The amount of injected brine and the capacity of the machine can be influenced by altering the brine pressure, the thrust of the conveyer belt and the lifting frequency (number of needle lifts per minute). So the choice of a brine injector is very important when deciding on a processing line, since this will directly condition the results of the production. Many different brine injector systems are available, but basically they fall into one of two categories: low pressure and high-pressure (spraying system) injection machines. The difference between the two types of machines lies principally in the method used to introduce brine into the meat. Low pressure (up to 2 bar; 0.2 MPa) injectors deposit the brine during the needle stroke through the meat, with needles usually having 2–4 holes of more than 1 mm in diameter, forming brine deposits which must then be distributed by mechanical

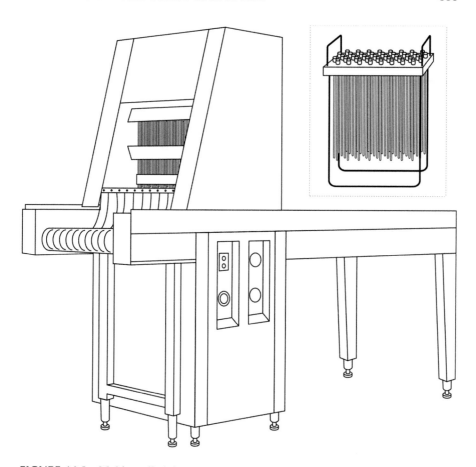

FIGURE 14.3 Multi-needle injector.

action. Modern injectors can easily inject 40–70% in one pass without destroying the muscle fibre structure.

If injecting at 2 bar does not result in the desired weight yield after one pass, injection can be repeated. Double handling can be avoided by having a second injector. Some injectors have two needle heads on the same machine, thus double injection can be achieved on one machine in one pass. High pressure (8–12 bar; 0.8–1.2 MPa) injectors introduce a dose of brine with a spraying effect. When the needles have completely penetrated the meat and are stopped at the end of their downstroke, the brine is injected into the meat by means of a spraying technique. The brine is distributed more homogeneously, since the needle holes (11–14) are distributed along the needles and the diameter of each hole is around 0.6 mm. High-pressure injection is more precise and gives a more uniform brine distribution with lower drip loss. Besides the characteristics of the injector, there are other factors that will influence the percentage reached. The most important of these are meat (size, temperature, fresh or frozen and thawed) and brine (viscosity, temperature) characteristics.

14.5 MECHANICAL TREATMENT

14.5.1 TENDERIZATION

Tenderization (also known as mechanical protein activation) is mechanical treatment applied immediately after brine injecting and before massaging and tumbling, but it is also possible to apply this treatment to the meat before injecting. In some very specific high-quality products, in which the fibrous structure of the meat should be preserved as much as possible, this technique is usually not applied. Instead, the product goes directly to the massaging and tumbling cycle.

The effect of tenderizing treatments is to greatly increase the surface extraction of muscle protein (myofibrillar proteins) of the injected meat, which positively affects the activation of protein during massaging and tumbling, and eliminates the necessity of thorough trimming, since the structure of the connective tissue is broken down, and consequently the effect of retraction of the connective tissue during cooking is reduced. This is a particularly important point for cook-in ham. The cuts also puncture the brine collection areas between muscles, facilitating brine distribution. Tenderization reduces drip loss, binding defects and appearance of holes, and makes the whole production process considerably shorter. In the final cooked product, tenderizing the injected meat improves yield (the brine can be better absorbed by the open structure of the meat), slice coherency and firmness.

The most common methods of tenderization use needles or rollers. With needle tenderizing, an extra set of needles penetrates the injected meat, introducing vertical light cuts into the muscle and enlarging the muscle surface, softening it but without tearing or separating the muscles. The tenderizer needle head is often a part of the injection machine itself, and the tenderizing needles are sharpened at an angle for easy penetration. Tenderizing needles do not have holes. They are used principally for whole pieces of high quality meat, but depending on the model of needle, are also very useful for high-yield products, which require a high degree of mechanical processing. There are generally several types of rollers used, depending on the product. A tenderizer with rollers has two sprung rollers between which the meat is passed. Some have blades or knives that make superficial cuts in the meat surface, obtaining a high degree of protein extraction; some have rollers with prongs that make deep cuts in the muscle. The system with prongs allows a strong tenderizing effect without breaking the muscular surface. Knife tenderizing is more effective than needle tenderizing.

Not all products require the same degree of tenderizing. The degree of tenderization can be adjusted by adjusting the depth (distance between the rollers) to which the blades or prongs penetrate the meat. This will depend on the desired yield and on the type of product being processed. As a general rule of thumb, the depth of the cut during tenderization correlates to the level of injection. Higher levels of injection commonly require deeper cuts to create more surface area and to allow greater amounts of protein to be activated. For extremely tenderized meat, applied when extension is high, the tenderizer blades almost cut the pieces of meat apart. For products where the primary objective is only to improve tenderness, the tenderization will be light. Where only a light degree of tenderization is required, the knives only cut into the meat to a depth of 2–4 mm. If lean minced meat (up to 5%, 3–4 mm blade) is to be added during tumbling, for instance to highly extended products, the injected meat should be well tenderized first so that the minced meat mixes well with tenderized larger pieces of meat.

14.5.2 MASSAGING AND/OR TUMBLING

These are mechanical treatments widely used in the manufacture of whole muscle cooked products, as well as for large products made out of an individual muscle. In this way, a uniform distribution of previously injected brine ingredients, like sodium chloride, polyphosphates, nitrite, sugars and spices, through the entire piece may be achieved. It improves yield of injected meat, colour development, tenderness, juiciness, firmness and slice coherency in the finished product.

Two of the most important characteristics regarding the quality of cooked products are the binding of muscles and water (brine) absorption. The muscular components responsible for these two characteristics are the myofibrillar proteins that, once extracted and solubilized, form what is called the exudate (a kind of mechanically activated protein film on the surface of the meat) with a glue-like effect between the muscles. For water retention to take place, it is necessary that proteins remain "open" so that water can penetrate them. The brine ingredients' purpose is fundamentally that of solubilizing and loosening the proteins. Salt and polyphosphates, both of which increase the ionic strength of the medium and the pH, make possible the swelling of chains and the extraction of proteins. The muscular structure is loosened by means of mechanical working, breaking up the cells and making the cellular membranes (sarcolemma) more permeable, facilitating the distribution and absorption of the brine. This process increases the mobilization of myofibrillar muscle proteins, and the solubilized protein increases its water content and is activated and released in the intercellular spaces and on the surface for the fixing in place of water and the binding of muscles.

There are basically two types of machines, massagers and tumblers. The massagers consist of a stationary, upright container in which there is a vertical stirring arm with paddles. The vertical stirring arm has a horizontal movement at several levels. The pieces of meat are mechanically worked by compression and tension produced by the rotation of the arm in the quadrangular container. The arm can be raised and lowered as it turns. This mechanical treatment is through friction between different muscles, with the walls and paddles of the massaging machine producing a gentle massage (gentler than the tumbling system). This form of massage by stirring is chiefly used for small pieces of meat that are being prepared for use in the manufacture of restructured hams. This particular type of machine usually works without vacuum and cooling.

The second group, the tumblers, consist of a drum with baffles that rotate around its longitudinal axis (Figure 14.4). The meat pieces are lifted by baffles to the upper part of the machine and then fall onto the meat mass below, producing an intense mechanical treatment (severer than the massaging system) suitable for high-yield products. During tumbling, massaging effects (friction) are also present. This type of mechanical treatment results in great cellular breakage and therefore optimum

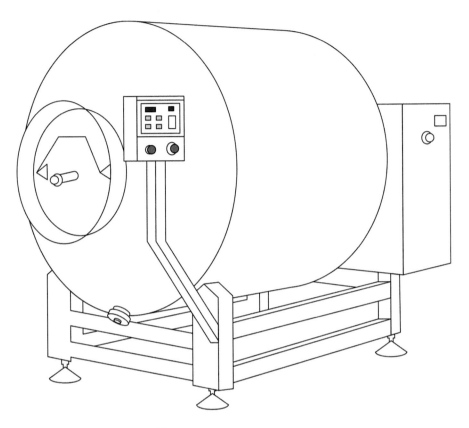

FIGURE 14.4 Vacuum tumbler.

solubilization and extraction of proteins. Tumbling of meat is usually performed
using a vacuum tumbler with cooling. Therefore, tumbling is more beneficial than
massaging for whole muscle ham products, because most massaging machines tend
more or less to tear the pieces of meat apart. These two models of mechanical
treatment correspond to very different types of products and there are many others
to which an intermediate process, or a combination of the two, must be applied. The
higher the level of injection, the more mechanical treatment is required.

Beside the previously mentioned parameters (quality of the meat, degree of
comminution of the pieces of meat, amount of brine injected, brine composition,
tenderization), the tumbling effect depends a great deal on the tumbler construction
and its capacity (size–diameter of the drum, types of baffle inside the drum, rotation
speed, optimum filling level of the drum). The diameter of the drum plays a crucial
role in calculating the amount of tumbling required. Larger tumblers require fewer
tumbling revolutions in order to obtain a comparable tumbling effect to a smaller
tumbler. Different tumblers have different types of baffle inside the drum. The more
gentle the baffles, the longer tumbling time is required. A higher rotation speed will
result in a greater extraction of proteins but also in greater damage to the muscle
structure. Practical experience also shows that the tumbling drum should not be

completely horizontal, and an angle between 5° and 15° is ideal. The tumbler should be between 60% and 80% full, based on total capacity, and the best tumbling effect occurs at a filling level of around 75–80% of total capacity. Therefore, the best working conditions must be determined specifically for each machine regarding the type of product and desired quality.

To achieve a good weight yield and water binding, a certain number of revolutions during tumbling, that is 4000–8000 revolutions, is required. The speed of rotation of the drum should be up to 20 r.p.m.; this corresponds to about 3–7 hours of pure tumbling. Rotation speed of more than 20 r.p.m. considerably increases the meat "dust." Two thousand revolutions are needed to significantly improve the slice cohesion of whole hams. Up to 8000 revolutions increase weight yield. Over-tumbling must be avoided, otherwise the meat structure is lost and is replaced by something resembling an emulsion. Brine added during tumbling to adjust the injection weight gain should be at a minimum (<5%) to prevent the hams swimming about in the brine, which hinders the desired tumbling effect.

There are two different methods for tumbling based on the total time available for the entire process. When tumbling continuously for a short time the meat is intensively treated in the tumbler without a pause. Long interval tumbling consists of tumbling and resting sessions. This stop-and-go process is typically continued for a total period of 10–16 hours, mostly overnight. The disadvantage of interval tumbling is that the tumbler is in use for a long time on the same batch of meat. Interval tumbling is based on a predetermined total number of tumbling revolutions and a period of time allocated for the entire tumbling process. For example, if a total of 5120 tumbling revolutions are required over a period of 16 hours (960 minutes) and the speed of the tumbler is 16 r.p.m., then the meat has to be tumbled 320 minutes (5120/16). This leaves 640 minutes (960–320) for resting periods within the 16 hours. Therefore, the tumbling programme could be 10 minutes of tumbling and 20 minutes of rest repeated 32 times, which would result in 960 minutes of tumbling. Combining a certain tumbling time with several hours of pause has given excellent results, allowing the solubilization of proteins that form the exudate to take place. Generally, the longer tumbling time applied, the greater the effect, because increased solubilization and extraction of myofibrillar proteins will be obtained; also interval tumbling is preferable to continuous tumbling for a short time. But excessive massaging time can produce results contrary to those desired, affecting the water holding capacity as well as the appearance of the slice.

The tumbling tends to increase the temperature of the meat and although efficiency of tumbling is greater at higher temperatures, there is also a risk of bacterial contamination. The ideal working temperature is between 4°C and 8°C, but that requires working with very cold brines and meat, which makes injection difficult, or having a cooling circuit in the machine itself. If tumbling is carried out without direct cooling, the optimal room temperature is between –2°C and 0°C, which maintains a temperature of around 0–4°C in the tumbled meat. Keeping the temperature low is also important to prevent bacteria from growing.

Tumbling produces foam from exuding protein, principally due to the emulsifying effect of the meat's proteins themselves. In the finished products this foam appears as undesirable air bubbles between and inside the muscles. Foam production can be limited by tumbling under vacuum. Removing atmospheric pressure by applying a vacuum allows the fibrous proteins to swell much faster and more effectively. The curing colour develops more quickly in the absence of oxygen, and therefore tumbling under vacuum results in a stronger curing colour in the finished product. Additional benefits of tumbling under a vacuum include the anaerobic conditions, which dramatically slow the growth of aerobic spoilage bacteria and help to keep the temperature low for longer because heat transfer is slower under vacuum than in air.

14.5.3 STUFFING AND MOULDING

Tumbled meat for whole muscle products is stuffed and usually placed into moulds to give the product the desired shape during cooking. The correct placing of the pieces of muscle (direction of fibre) is important when stuffing whole pieces of muscles into moulds. Stuffing can be done manually or automatically. When whole pieces are used, each piece is simply manually placed in a mould. If the products are stuffed automatically, with a stuffing machine, muscles will be distributed at random in the finished product.

Moulds are used to unite the different muscles during cooking and give the product an aesthetically pleasing shape. These receptacles allow for the moulding of pieces whose weight usually ranges from 3–10 kg. They can be of various materials (aluminium, stainless steel) and shapes. The most common shapes are square, rectangular or cylindrical due to design simplicity; sometimes they resemble the original pear shape of the natural ham.

Whatever the material of the mould is, whole muscle hams must be stuffed into special protective layers of polyethylene before they are placed in the mould. This prevents the meat from sticking to the mould, allows a better water retention (zero water loss) and hygienic storage and distribution, especially if the product is to be repackaged after cooking.

When a product is to be cooked in the same packaging in which it will be sold, the meat mass must be introduced into a flexible shrinkable or thermoformed plastic under vacuum. Thermoshrinkable plastic is the material most commonly used in the industry for stuffing cook-in meat products. Depending on the stuffing process, the plastic material can be in the format of individual bags or plastic rolls. It is a multilayer plastic material that retracts when exposed to thermal treatment. This retraction allows the plastic to mould itself perfectly around the product, exerting pressure on it that is essential for obtaining a product with zero cooking loss. If thermoshrinkable plastic is used, after the stuffing process the bag must be vacuum sealed, either by a clipping system or thermosealing. Thermoformable plastic consists of two multilayer plastic films, which will be thermosealed once the meat mass is contained inside. The lower thick film is first thermoformed and then filled with the meat mass. If moulding has been done in a thermoformer, after the stuffing process the next procedure is thermosealing of the upper film under vacuum. For thermosealing, the seal area must be very clean to avoid leaking seals.

14.5.4 COOKING/COOLING

Cooking is the term commonly used for meat products that are heat treated, but pasteurizing is the more precise description. Pasteurizing normally refers to the boiling point of water, which is 100°C at sea level. For whole meat products, pasteurisation takes place at a temperature range between 72°C and 85°C, until a core temperature of 69–72°C is reached. This process destroys all vegetative pathogens (but spores may survive this heat treatment), denatures proteins and improves the texture, stabilizes the curing colour and intensifies the flavour. Cooking and core temperature specifications are highly debated among experts because a compromise between product safety, economic factors (for example yield), stabilization of curing colour and functionality of additives has to be reached.

The heating effect is made up of the parameters temperature and time. For whole muscle cooked hams, the time required for microbiological aspects of thermal processing should be equivalent to heating the centre for 40–60 minutes at a constant temperature of 70°C (F70 values). The initial bacterial count of the meat material and the standard of hygiene during the whole cooked ham manufacturing process play an important part and may demand a different F-value. Cooked ham can be stored for a prolonged period of time (around 90 days) under refrigeration temperatures below 4°C.

The muscle proteins extracted and solubilized by the combined effect of certain ingredients and additives (phosphates and salt) and by the mechanical treatment undergo a process of denaturation due to the effect of heat, which reduces the intercellular spaces, compacts the denaturalized fibres and forms a three-dimensional network able to hold water, giving the finished product consistency, firmness and muscular binding.

Heat is the cause of the denaturation of the red pigment in cured meat transforming it into the finally stable pink pigment characteristic of these products; nitrosomyoglobin is decomposed into globin and nitrosomyochromogen. Stabilization of this pigment is produced basically in the final phase of cooking, and the minimum temperature for this to occur is 65°C.

The aromatic aspect of cooked cured meat is developed and stabilized in two consecutive stages of the manufacturing process: prior to cooking and during cooking. Endogenous muscle enzymes such as proteolytic and lipolytic enzymes contribute to the formation of the aromatic precursors (fatty acids, amino acids, triglycerides, phospholipids, peptides) prior to cooking. But muscle protease and lipase are sensitive to temperature and are inactivated during cooking. Formation of the aromatic precursors and their transformation, by means of heat, into aromatic compounds (alkanes, alkenes, aldehydes, ketones, alcohols, esters, terpenes, sulphur compounds, furans, etc.) are key aspects in flavour generation. Several enzymatic reactions, oxidations, Maillard reactions, etc., take place in the hams during cooking, and all of them contribute to the final development of flavour characteristics typical of cooked hams. Development of flavour is optimum at temperatures of around 60–65°C.

The packaged and moulded hams are usually placed in cooking trolleys or baskets and cooked in steam cookers or water baths. In steam cookers it is important to allow a good distribution of the steam; recirculation is advisable in water baths. There are two heat transfer mechanisms: convection (heat transfer from the heating medium to the ham surface) and conduction (heat transfer from the ham surface to the inner areas). Water baths are quicker than steam cookers.

The speed of temperature increase and its control during cooking is important. Three types of cooking can be used: cooking at constant temperature, step-by-step cooking and delta-T cooking.

Cooking at constant temperature is the most common method of thermal treatment for whole muscle ham products and large individual pieces of meat in plastic films, where there is no cooking loss and the desired core temperature is reached in the fastest possible time. With this method, the product is exposed to a constant cooking temperature right from the beginning of the cooking process. The cooking temperature used is generally between 74°C and 80°C. Thus, the constant cooking temperature is around 6–10°C above the desired final core temperature. If the difference in temperature between cooking and core temperature is too small, for example 2–4°C, the cooking process is significantly prolonged. The advantage of cooking at constant temperature is that the desired core temperature is reached in a shorter period of time than other methods of cooking. The disadvantage is a negative impact on sensory properties of the product's surface, creating problems of overcooking in this area of the piece.

In step-by-step cooking, the applied temperature is increased in stages. In the first stage, large products should be treated at 60–65°C for at least 1 hour before increasing the temperature. In the second stage, temperature is increased to 70°C for a specified time (another hour or so) and finally up to 74–80°C until the desired core temperature is reached (Figure 14.5). This type of cooking produces good results, above all in zero cooking loss products (cook-in ham), although the cooking time is longer than in the system using a constant temperature.

In delta-T cooking, the temperature is increased continuously in relation to the increase in temperature at the core of the product (e.g., delta-T = 25°C). For delta-T = 25°C, the cooking temperature remains constant at, for example, 60°C during the initial stage of the cooking process, until the temperature inside the product rises to 35°C.

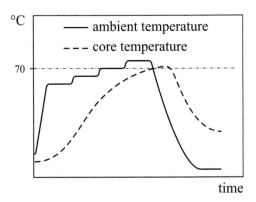

FIGURE 14.5 Step-by-step cooking.

At this point the difference between the cooking temperature and the temperature of the core of the product is the required 25°C. Once this point is reached, the cooking temperature starts to rise and the core temperature inside the product rises at the same rate. The predetermined delta-T, that is the difference between the cooking temperature and the temperature of the core of the product, of 25°C is always maintained. When the cooking temperature reaches a certain value, it is maintained constant, as in the constant temperature heating method, until the desired core temperature is reached. In this system, overcooking of the surface is minimized, but the cooking time is longer.

After cooking, the cook-in hams are cooled and should not be removed from mould until they are completely cool. This phase has a strong influence on the final characteristics and quality of the finished product. Cooling of the meat products, and the way in which this has been carried out, can affect the wholesomeness of the hams and the degree of pasteurization.

Cooling must bring the ham below 4°C, and this may be achieved by air blast, immersion in cold water, or with cold water showers. Combinations of these three chilling systems are used sometimes. At the end of the cooking process, the product is immersed in water or showered with cold water prior to transfer to a chilling room in order to minimize growth of surviving microorganisms. The final cooling from 40°C to 15°C is considered the most crucial period and should be restricted to less than 4 hours when possible. It is required that the product must remain in the chilling room for a minimum of 48 hours before dispatch, in order to make sure that the colour and the other sensory properties of the cooked meat product have stabilized.

Cook-in hams must be removed from their moulds and may be sold as either entire pieces for slicing at the retailer shop at consumer request, or as packaged slices ready to be consumed. A wide variety of vacuum and modified atmosphere packages containing different numbers of slices are typically found in supermarkets.

REFERENCES

Arnau Arboix, J. 2004. Cooked ham. In *Encyclopedia of meat sciences*, ed.W. K. Jensen, D. Carrick, and M. Dikeman, 562–567. Oxford: Elsevier Ltd.

Casiraghi, E., Alamprese, C., Pompei, C. 2007. Cooked ham classification on the basis of brine injection level and pork breeding country. LWT- Food Science and Technology, 40:164–169.

Delahunty, C. M., McCord, A., O'Neill, E. E., Morrissey, P. A. 1997. Sensory characterisation of cooked hams by untrained consumers using free-choice profiling. Food Quality and Preference, 8:381–388.

Desmond, E. M., Kenny, T. A., Ward, P., Sun, D. W. 2000. Effect of rapid and conventional cooling methods on the quality of cooked ham joints. Meat Science 56:271–277.

Desmond, E., Kenny, T., Ward, P. 2002. The effect of injection level and cooling method on the quality of cooked ham joints. Meat Science, 60:271–277.

Feiner, G. 2006. Meat products handbook: Practical science and technology. Cambridge: Woodhead Publishing Limited and CRC Press LLC.

Freixanet, L. Manufacturing process for whole muscle cooked meat products IV: Stuffing and moulding. Additives and ingredients in the manufacture of whole muscle cooked meat products. Spray injection of meat. Influence of the brine pressure in the quality of injected products. http://www.metalquimia.com (accessed April14, 2004).

Lagares, J. Manufacturing process for whole muscle cooked meat products V: Cooking. http://www.metalquimia.com (accessed April14, 2004).

Müller, W. D. 1989. The technology of cooked cured products. Fleischwirtsch, 69:1425–1428.

Nute, G. R., Jones, R. C. D., Dransfield, E., Whelehan, O. P. 1987. Sensory characteristics of ham and their relationships with composition, visco-elasticity and strength. International Journal of Food Science and Technology, 22:461–476.

Scheid, D. 1984. Manufacture of cook-in ham. Fleischwirtschaft, 64: 1077–1080.

Scheid, D. 1986. Cooked ham manufacture. Pumping, mechanical treatment and heat treatment. Fleischwirtschaft, 66:1022–1026.

Standard For Cooked Cured Ham. CODEX STAN 96-1981. Adopted 1981. Revision: 1991, 2014 and 2015. Codex Alimentarius International Food Standards, FAO, WHO.

Toldrá, F. 2010. Cooked ham. In *Handbook of meat processing*, ed. F. Toldrá, L. Mora, and M. Flores, 301–311. Ames: Wiley-Blackwell Publishing.

Tomović, V. 2009. Effect of chilling rate of carcasses, time of deboning *post mortem* and way of curing on quality and safety of cooked ham. PhD Thesis, Faculty of Technology, University of Novi Sad (in Serbian), Novi Sad.

Tomović, V., Jokanović, M., Petrović, Lj., Tomović, M., Tasić, T., Ikonić, P., Šumić, Z., Šojić, B., Škaljac, S., Šošo, M. 2013. Sensory, physical and chemical characteristics of cooked ham manufactured from rapidly chilled and earlier deboned *M. Semimembranosus*. Meat Science, 93:46–52.

Tomović, V., Žlender, B., Jokanović, M., Tomović, M., Šojić, B., Škaljac, S., Tasić, T., Ikonić, P., Šošo, M., Hromiš, N. 2014. Technological quality and composition of the *M. semimembranosus* and *M. longissimus dorsi* from Large White and Landrace Pigs. Agricultural and Food Science, 23: 9–18.

Válková, V., Saláková, A., Buchtová, H., Tremlová, B. 2007. Chemical, instrumental and sensory characteristics of cooked pork ham. Meat Science, 77:608–615.

Xargayó Teixidor, M. Manufacturing process for whole muscle cooked meat products II: Injection and tenderization. Manufacturing process for whole muscle cooked meat products III: Massage. http://www.metalquimia.com (accessed April14 2004).

15 Bacon

The Processing, Shelf Life, and Macronutrient Composition of Bacon Manufactured from Pork Bellies

Benjamin M. Bohrer
Department of Food Science, University of Guelph, Guelph, Ontario, N1G 2W1, Canada

CONTENTS

15.1 DEFINING BACON

Bacon, a challenging meat product to fully define, is generally described as a salt-cured pork product that is marketed and served in thin slices. Depending on how and where the product is being manufactured and distributed, bacon can originate from several different meat cuts within the pork carcass. For instance, in the United States of America, the USDA Food and Labeling Policy Book describes bacon as follows:

> *The term "bacon" is used to describe the cured belly of a swine carcass. If meat from other portions of the carcass is used, the product name must be qualified to identify the portions, for example, "Pork Shoulder Bacon."* (USDA, 2005)

In light of the success the pork industry has seen with bacon in the past several decades, other meat commodity groups have developed their own bacon-like products. The USDA Food and Labeling Policy Book goes on to describe "bacon-like products":

> *Bacon-like products, including poultry bacon, labeled with "bacon" in the name must follow the same requirements as those applied to pork bacon. These requirements include, but are not limited to, limits on restricted ingredients and the requirement that the bacon must return to green weight.*
> *Beef bacon is a cured and smoked beef product sliced to simulate regular bacon. It is prepared from various beef cuts and offered with a variety of coined names, including "Breakfast Beef," "Beef Bacon," etc. A common or usual name is required, for example, "Cured and Smoked Beef Plate," and should be shown contiguous to the coined name.*

> *Poultry bacon products are acceptable and may be designated as (Kind) Bacon. However, a true descriptive name must appear contiguous to (Kind) Bacon without intervening type or design, in letters at least one-half the size of the letters used in the (Kind) Bacon, and in the same style and color and on the same background. An example of an acceptable designation is "Turkey Bacon—Cured Turkey Breast Meat—Chopped and Formed."* (USDA, 2005).

In addition to the definitions of bacon provided by governing agencies in the United States of America, bacon is classified by different standards of identity throughout the world. These standards of identity include the separation of bacon terminology described by the Canadian Food Inspection Agency (CFIA, 2019a), which defines *bacon* as boneless pork belly, *back bacon* as boneless pork loin, and *Wiltshire bacon* as boneless pork loin with a portion of the belly attached. In European, Asian, and Australasian nations, bacon can be labeled under several different designations. In the United Kingdom and Ireland, bacon is commonly referred to as *rashers*, which is synonymous with North American back bacon (boneless pork loin), or *streaky bacon*, which is synonymous with North American bacon (boneless pork belly) (USDA, 2013). In Italy, bacon is commonly referred to as *pancetta* and sold/consumed more like a deli-meat or charcuterie-style product (USDA, 2013). In Germany, bacon is generally referred to as *smoked fatback* (*Rückenspeck*) and consumed exclusively as a deli-meat product. In Japan, bacon is manufactured from the pork belly, yet is sold to the consumer as a precooked product with deli-meat properties. In Australia and New Zealand, bacon is generally referred to as *middle bacon*, which is synonymous with North American Wiltshire bacon (boneless pork loin with a portion of the belly attached), *short cut bacon*, which is synonymous with North American back bacon (boneless pork loin), or *streaky bacon*, which is synonymous with North American bacon (boneless pork belly) (KR Castlemain, 2009).

For the purpose of this book chapter, the focus will be on bacon as defined and described as the cured and smoked belly of a pork carcass, which is then marketed and served in thin slices.

15.2 PROCESSING TECHNIQUES

A variety of techniques can be used when manufacturing bacon. Important processes where different facets should be considered include (1) quality, type, and quantity of the raw materials; (2) curing methodology; (3) smoking/thermal processing; and (4) slicing. Here, considerations for each of these processes will be addressed.

15.2.1 RAW MATERIALS

15.2.1.1 Meat and Fat Quality

Both meat quality and fat quality should be considered when manufacturing bacon (Sheard, 2010). Considerations for meat quality include hydration ability of the lean tissue, dimensions of the belly, and belly firmness. Considerations for fat quality include quantity of fat, level of unsaturation of the fat, and firmness/softness of the fat.

The hydration ability of the lean tissue in its own right (before processing or addition of non-meat ingredients) is primarily influenced by the rate of pH decline during the conversion of muscle to meat and postmortem aging protocol (Offer et al., 1989; den Hertog-Meischke et al., 1997; Pearce et al., 2011). If pork experiences abnormal pH decline, resulting in pale, soft, and exudative (PSE) conditions (as outlined by Barbut et al., 2008), then there will be challenges with proper brine uptake during bacon processing.

Compositionally, pork bellies can be as high as 55–60% lipid (Person et al., 2005; Kyle et al., 2014; Soladoye et al., 2015), which undoubtedly makes fat quality extremely important in the manufacture of bacon. Beyond the amount of subcutaneous and seam fat within a pork belly, considerations for fat quality is primarily defined by a single characteristic, which is the amount, or ratio, of unsaturated fatty acids to saturated fatty acids. This is oftentimes reported as a single figure in the pork industry and referred to as the calculation of the iodine value (Seman et al., 2013; Goehring, 2015; Soladoye, 2017). The iodine value terminology stems from laboratory titration methodology, yet this value is usually predicted with various fatty acid calculations (AOCS, 2009; Meadus et al., 2010). Iodine value is generally thought of as a very meaningful prediction of belly quality with a direct influence on bacon processing yields (Seman et al., 2013; Tavárez, 2014; Overholt et al., 2019). When discussing bacon processing yields, the terminology of bacon slicing yields is often used, which simply is the amount of saleable bacon slices originating from a belly before processing. There are conflicting reports on the influence of iodine value on bacon processing yield. Kyle et al. (2014) reported iodine value as well as fatty acid composition values were weakly correlated $(r < |0.19|)$ with bacon slicing yield. While Overholt et al. (2019), which studied pork with extreme range of iodine values, reported bacon slicing yields decreased linearly as iodine value increased from 61.7 to 98.6. Using step-wise regression in their study, Overholt et al. (2019), reported iodine value alone accounted for 60% of variation observed in bacon slicing yield.

In addition to the aforementioned considerations of lean tissue and fat tissue quality, dimensions of the belly are often used by pork processing facilities to sort and market fresh bellies. Parameters used by the USDA Daily National Carlot Meat Report to market pork bellies include whether or not the belly has had the skin/rind removed (de-rind versus skin-on) and categorization by weight (USDA, 2019). Other dimensions that bacon processors may request from pork processors include belly thickness, belly width, belly length, and belly softness. Depending on the size and scope of a commercial pork processing facility, these measurements may be conducted with advanced technology, like imaging software, for sorting purposes (Kapper, 2012; Čítek et al., 2015; Lee et al., 2018), or these measurements may be conducted by trained personnel with more subjective techniques (Trusell et al., 2011; Soladoye et al., 2015, 2017).

Whether using imaging technology or subjective scoring with trained personnel, it is important to ensure that evaluation criteria and location (within the carcass and within the belly) are consistent. Studies conducted by Trusell et al. (2011) and Knecht et al. (2018) reported that sampling location can affect most major belly quality parameters. It is generally accepted that these parameters, most notably belly thickness and belly softness, have significant influence on processing yields and consumer evaluation of bacon (Person et al., 2005; McLean et al., 2017). As Person et al. (2005) reported, thin bellies had reduced processing yields and thick bellies had reduced consumer appeal, thus bellies of intermediate thickness (approximately 2.5 cm thick) were most ideal for bacon manufacture. It is likely, as the industry has shifted to marketing pigs at heavier weights, that an intermediate thickness for bellies is closer to 3.5–4.0 cm, as documented by recent research (Tavárez et al., 2016; Soladoye et al., 2017b; Lowell et al., 2018). McLean et al. (2017) that reported bacon slice fat:lean ratio and slice shape were among the parameters that consumers were most critical of in terms of visual appraisal of uncooked bacon slices. Even so, Kyle et al. (2014) reported that belly weight (r = 0.17), belly thickness (r = 0.27), belly width (r = −0.01), belly length (r < 0.01) and belly flop (a prediction of belly softness; r = 0.18) were all weakly correlated with bacon slicing yields.

Regarding fresh belly softness, various tests have been used in the industry and by meat science researchers to determine this attribute. Previous belly softness testing includes subjective methods such as finger testing or fat manipulation by hand and the aforementioned belly flop test, which is usually described as an arrangement of bend angle or flexibility quantifications (Trusell et al., 2011; Seman et al., 2013; Soladoye et al., 2017; as seen in Figure 15.1). Generally subjective firmness and belly flop tests have been strongly correlated with chemical composition (moisture and lipid content), belly weight, and belly thickness, but weakly correlated with other belly dimensions (width and length) (Trusell et al., 2011; Kyle et al., 2014; Soladoye et al., 2017).

One final consideration for raw materials for bacon manufacturers is the cut specifications of belly that is being used. Bacon manufacturers can elect to follow guidelines from the Meat Buyers Guide (NAMI North American Meat Institute, 2014) and work with minimally trimmed natural fall bellies (similar to IMPS #409

(a)   (b)

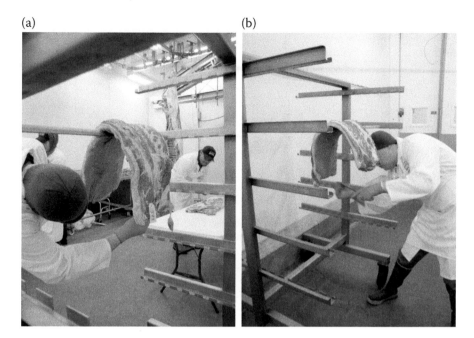

FIGURE 15.1 An example of belly flop testing; the value between the blade and flank edges when the belly is hung over a 2.5 cm smokehouse bar is recorded.
(Photo credit: B.M. Bohrer, M.A. Tavárez, R.T. Herrick, K.L. Little, A.C. Dilger, and D.D. Boler).

Pork Belly, Skinless) or choose to work with variants of trimmed and squared bellies (similar to IMPS #409B Pork Belly, Center-Cut, Skinless). This differentiation is typically identified on product packaging (i.e., center cut bacon).

With the conflicting reports that are currently available, there certainly should be more research efforts devoted to the theme of fresh belly quality, specifically identifying the influence that fresh belly parameters have on bacon processing yields and consumer acceptance.

15.2.1.2 Ingredients

As with most processed meat products, the known characteristics and attributes of a product are dependent on the ingredients and the processing techniques used to incorporate those ingredients. With bacon, these ingredients are usually incorporated with a curing solution, which is generally synonymous with the terms *brine* and *pickle*.

Past reviews that have discussed functional non-meat ingredients used by the meat processing industry have categorized non-meat ingredients into the following groups: water, salt (typically in the form of sodium chloride), phosphates, curing agents (nitrate or nitrite), cure accelerators (sodium ascorbate, sodium erythorbate, ascorbic acid, and erythorbic acid), and organic acids (for preservation) (Sebranek, 2009; Sebranek, 2015). In addition to these functional ingredients, which are important for quality and yield purposes when manufacturing bacon, a variety of spices and flavorings can be added to bacon during the curing process, many of which give specific

TABLE 15.1

An Example of a Commercial Bacon Curing Blend and Mixing Instructions (Assuming Bacon is Manufactured Using Pump/injection Processing Techniques)

Ingredients: Salt, sugar, sodium phosphate, sodium erythorbate, sodium nitrite
Components:

Salt	60.00%
Sugar	27.60%
Sodium phosphate	10.95%
Sodium erythorbate	1.00%
Sodium nitrite	0.45%

Mixing instruction: The following table contains the weight (g/1 kg water) of this product required at various pumped rates based on federal regulations. Regulations are dependent on the country, and figures may require recalculation according to federal regulations in each respective country.

Targeted pump uptake percentage:	g/1 kg of water
15% uptake	285 g
20% uptake	255 g
25% uptake	225 g
30% uptake	195 g

brands or types of bacon their characteristic organoleptic properties (Soladoye, 2017). For further simplification, Table 15.1 provides an example of a typical commercially available curing solution unit.

Water, particularly in whole-muscle processed meat products like bacon, serves as the carrier system for the other ingredients in the curing solution. Water, although simple in its structure, is unique in its properties as a functional ingredient. Hydrogen bonds allow water to easily bind to itself and other molecules, which makes water a highly effective solvent with direct roles as a carrier and dispersing agent for other non-meat ingredients (Ruan and Chen, 1998). The amount of water in a curing solution is dependent on targeted pump uptake. If greater pump uptake is desired, there would likely be a greater ratio of water to other ingredients in a curing solution, in an effort to dilute the amount of the other ingredients in the final product.

Salt (usually in the form of sodium chloride) has a multifunctional role in processed meat products, including intricate roles in preservation (by reducing water activity), texture (mainly attributed to heat-set gelation and reduced water activity), and flavor (attributed to both perceived saltiness and enhancement of other flavors) (Matthews and Strong, 2005; Ruusunen and Puolanne, 2005). Desmond (2006) discussed the challenges that sodium chloride reduction presents to the meat industry and highlighted that alteration in the aforementioned characteristics can create challenges in many processed meat products, bacon included. That is not to say that research and development has not been successful in reducing sodium chloride content in bacon. Research dating back to the 1970s and 1980s first reported the successes with reduction in sodium chloride content in bacon formulations (Mottram and Rhodes, 1973;Kimoto et al., 1976;

Mandigo et al., 1981; Gray and Pearson, 1984). The challenge with reducing sodium chloride content in bacon products is mainly described as poor flavor profile and reduced processing yields during pumping and cooking.

Several subsequent studies in that time period (1970s and 1980s) began investigating the use of other salt sources, namely potassium chloride, yet challenges persisted with flavor as potassium chloride presented a bitter flavor profile and reduced perceived saltiness (Murray, 1983; Jeremiah et al., 1996). In more recent work, Aaslyng et al. (2014) and Lowell et al. (2017) investigated sodium chloride reduction in bacon products. Aaslyng et al. (2014) reported that reducing sodium chloride levels in bacon (from 2.80% of the final product to 2.06% or 1.41% of the final product) did not affect processing yield or microbial stability, but significantly altered sensory properties (saltiness, juiciness, and texture). Lowell et al. (2017) reported that different sodium chloride levels in bacon (1.2%, 1.5%, or 1.8% of the final product) did not have meaningful effects on processing yields or most storage and sensory attributes, but lower sodium chloride levels did slightly affect bacon slice composition (lower moisture) and perceived saltiness (lower saltiness). Overall, it is plausible to imply that sodium chloride can be significantly reduced (not eliminated) in the curing solution formulation of most bacon products, yet negative impacts on perceived saltiness should be expected in those products.

The primary role of phosphates in processed meat products is improved water retention through multiple mechanisms including its impact on pH, chelating activity, and ionic bonding (Sofos, 1986; Lewis et al., 1986; Petracci et al., 2013). Phosphate ingredients can be classified by the number of phosphorus atoms in the molecule (for example monophosphates, diphosphates, tripolyphosphates, polyphosphates). Alkaline phosphates, such as sodium tripolyphosphate, are most commonly used in the meat industry due to their improved effectiveness compared with non-alkaline phosphates (Petracci et al., 2013). Most governing agencies regard phosphate ingredients as GRAS (generally recognized as safe) compounds; however, their inclusion levels are closely monitored. For example, the Canadian Food Inspection Agency regulations have a maximum permitted usage level of 0.5% as calculated in the final product using sodium phosphate dibasic in the calculation (CFIA, 2019b). Despite GRAS status and strict limits on the inclusion level of phosphates in meat products, some consumers demand processed meat products have labels that are free of synthetic compounds (that is "clean label"), and therefore the commercial meat processing industry is exploring alternatives to synthetic phosphates (Petracci et al., 2013; Asioli et al., 2017). While ingredients such as vegetable starches and fibers, citrus fiber, and dairy isolates have been tested as potential replacements or complements to phosphates (Weiss et al., 2010; Resconi et al., 2016; Powell et al., 2019), there remains work to be completed on this research theme. Specific to bacon, there is limited scientific literature available that has focused on evaluating the effects of replacing phosphate with alternative ingredients; however, there are many commercially available bacon products that are formulated without synthetic phosphates.

Curing agents, which typically are in the form of nitrate or nitrite salts such as sodium nitrate ($NaNO_3$) or sodium nitrite ($NaNO_2$), are multifunctional ingredients with specific roles in cured color formation, antioxidant/antimicrobial inhibition, and flavor development (Honikel, 2008; Pegg and Shahidi, 2008). The use of sodium nitrate and sodium nitrite are often under scrutiny by health-conscious consumers, as these compounds can result in the development of carcinogenic N-nitroso compounds such as N-nitrosamines during final product preparation/cooking (Pegg and Shahidi, 2008). For this reason, the application and inclusion level of nitrate and nitrite salts are closely monitored by governing agencies. For example, nitrate salts can only be used in slow cured products like dry cured hams (in order to allow ample time for the conversion of nitrate (NO_3-) to nitrite (NO_2-) to nitric oxide (NO)), whereas nitrate salts are not permitted for use in rapid-cured products (CFIA, 2019c). Further, the *Food and Drug Regulations Act* states that in products other than side bacon, the maximum inclusion level of nitrite salts is 20 g per 100 kg of meat product (that is 200 mg/kg or 200 ppm), whereas, in the curing of side bacon, the maximum inclusion level of sodium nitrite salts is 12 g per 100 kg of pork bellies (that is 120 mg/kg or 120 ppm) (CFIA, 2019c).

Regardless of the tight regulations on nitrite salts implemented by governing agencies, the meat industry has recently experienced increased use of "natural" alternatives to nitrite salts. Common alternatives to nitrite salts are vegetable extracts that contain high levels of naturally occurring nitrate/nitrite like celery, spinach, radish, and lettuce (Shahidi and Pegg, 1995; Alahakoon et al., 2015). Like many other meat products, research efforts focusing on the use of naturally occurring nitrate/nitrite sources as an alternative to nitrite salts have been conducted for bacon (Sindelar, 2006). Products that are formulated with naturally occurring nitrate/nitrite and without nitrite salts can be labeled as "uncured" in many countries such as the United States of America and Canada, which can create confusion among consumers. Sindelar et al. (2007) reported that there were only minimal differences in color, lipid stability, aroma, and sensory (flavor and texture) traits in commercially sourced bacon that was formulated with plant-based sources of nitrate/nitrite compared with bacon that was formulated with nitrite salt. Yet there appeared to be high variation among brands of commercially sourced bacon that were formulated with plant-based sources of nitrate/nitrite (Sindelar et al., 2007).

Curing accelerators react with curing agents (nitrate or nitrite) in order to reduce nitrate or nitrite compounds into nitric oxide at a more rapid rate, which enhances efficiency during production of cured meat products (Sebranek, 2015). Curing accelerators are categorized as reductants or acidulates based on their mode of action. Reductants (for example ascorbic acid/sodium ascorbate and their molecular stereoisomers erythoribic acid/sodium erythorbate) provide for accelerated formation of nitric oxide from nitrite and are used in processing facilities to maintain continuous processing throughput during the injection and cooking stages of processing (Sebranek, 2009). USDA–FSIS inspection regulations state that, in bacon, sodium ascorbate or sodium erythorbate is required at 55 g per 100 kg of pork bellies (that is 550 mg/kg or 550 ppm) (USDA, 2013). The reasoning for this stipulation is to accelerate the conversion of nitrite to nitric oxide in bacon. This conversion was outlined by Woolford and Cassens (1977) and the lack of this conversion can cause high levels of residual nitrate and/or nitrite in bacon products. The other type of curing accelerator used in meat processing is acidulants (for example, fumaric acid, sodium acid pyrophosphate, and glucono delta-lactone); however, these ingredients are rarely used in the formulation of bacon (Sebranek, 2009).

The primary role of organic acids in processed meats is to provide antimicrobial inhibition, but organic acids are also believed to have roles in flavor preservation (Theron and Lues, 2007; Mani-López et al., 2012). Examples of organic acids that are commonly used in meat processing include acetic acid, citric acid, lactic acid, propionic acid, malic acid, succinic acid, and tartaric acid (Theron and Lues, 2007; Mani-López et al., 2012). Organic acids are not often used in bacon curing solutions. The role of organic acid as a functional ingredient has not been fully investigated in bacon products; this is still an under-researched area.

Beyond the aforementioned functional ingredients that can be found in bacon curing solution (water, salt, phosphates, curing agents, curing accelerators, and organic acids), the other category of ingredients that are often used in the manufacture of bacon are flavors and spices. This area of bacon formulation can be viewed as more of an art than a science; common bacon flavors/spice ingredients include sugars (brown sugar or maple syrup) and spice blends (pepper or barbecue flavorings). Sugars can be added to the curing solution, while spice blends may be added to the curing solution or included after the product has been smoked and partially cooked.

The processing intervention of smoking can also affect bacon flavor and bacon product differentiation. The smoking process can be accomplished by generation of wood smoke or application of liquid smoke. Popular hardwoods such as apple, beech, cherry, hickory, maple, mesquite, oak, pecan, and walnut can be used to generate unique flavor and aroma profiles before, during, or following the cooking process. Liquid smoke is usually applied before, during, or following the cooking process and generates characteristic bacon flavor, aroma, and visual color. Several recent research efforts have been dedicated to exploring the volatile flavor and aroma compounds generated with different smoking techniques (Saldaña et al., 2019a; Saldaña et al., 2019b; Saldaña et al., 2019c).

In summary, there remains a significant need to merge industry ingredient formulation trends with new scientific findings that are then made available to the research community and general public. Valuable research themes include reduction in sodium chloride, replacement or partial replacement of synthetic phosphates, replacement of nitrite salts with plant-based ingredients that contain naturally occurring nitrate/nitrite sources, and an expansion in the flavoring offerings for bacon products.

15.2.2 CURING TECHNIQUES

The act of curing is typically defined and described by the addition of nitrite compounds to a meat product (Sebranek, 2009); however, for the purposes of bacon production, this terminology can be used to describe the methodology in which the non-meat ingredients are incorporated into or come into contact with the pork belly. This is accomplished with three primary methods: injection with a curing solution (curing solution can also be referred to as brine or pickle), dry curing, and immersion curing (Knipe and Beld, 2014). Traditional Wiltshire curing techniques (as described by Sheard (2010)) involve all three of these methods (injection, immersion, and maturation). Yet the typical methodology used today for bacon curing generally utilizes only a single curing method.

Injection with a curing solution is the most common method of curing pork bellies for bacon production in North America and in most places around the world due to the increased throughput and processing efficiency associated with this technique. It can also be speculated that this curing method would be advantageous in repeatability and consistency. As outlined by Knipe and Beld (2014), the order in which ingredients are added in the curing solution and proper agitation of the final curing solution can affect pump uptake of ingredients. Phosphate ingredients should be included first and followed by other ingredients to ensure ingredients are dissolved properly. In large bacon processing facilities, automated needle injection equipment is used to inject bellies to targeted pump uptake levels, therefore a proper mixture (that is, consistency and dissolution) of ingredients is crucial. The amount of curing solution permitted to be retained in cooked bellies varies from country to country. In the United States of America, regulations dictate that the weight of cured pork bellies immediately before slicing cannot exceed the weight of the fresh uncured pork bellies (US GOP 9 CFR 319.107 – bacon; USDA, 2010). However, other countries do not have a restriction on the amount of curing solution permitted to be retained in bellies following thermal processing (Knipe and Beld, 2014). For example, the standard of identity for bacon in Canada is grouped into a broad spectrum with little to no limitations on pump uptake retention levels. Recent research conducted by Sivendiran et al. (2018) reported that bacon slice composition and sensory attributes of bacon from bellies with pump retention levels exceeding 100% of green weight were largely unaffected when compared with bellies with pump retention levels at or below 100% of green weight.

Dry curing can be described as rubbing a mixture of the aforementioned curing ingredients on the surface of the belly (Knipe and Beld, 2014). Little et al. (2015) evaluated injected bacon and dry-cured bacon, and while direct comparisons were not statistically assessed in the study, it can be inferred that dry cured bacon was saltier and had greater flavor intensity, which was likely caused by reduced cook yields and lower moisture content (cooked yields of approximately 98% were reported for injection bacon and cooked yields of approximately 92% were reported for dry-cured bacon).

Immersion curing is described as submerging pork bellies into a curing solution for a set period of time (Knipe and Beld, 2014). The period of time used in immersion curing can vary. Knipe and Beld (2014) describe this period of time as 2–3 days, while traditional approaches to Wiltshire curing as described by Sheard (2010) stated the length of immersion can be several days. During the immersion process, the curing solution can develop a unique microflora of bacteria, which has garnered significant attention from microbiologists (Woods et al., 2019), and this topic could be further explored.

15.2.3 SMOKING AND THERMAL PROCESSING

The thermal processing of bacon can generally be described as smoking followed by partial cooking (an example of a common cooking cycle is described in Table 15.2). Wood or liquid smoking techniques elicit characteristic bacon flavor profiles. While the differences between wood smoking and liquid smoking have not been fully quantified, research efforts have been dedicated to several aspects of this topic. Research conducted by Zhao et al. (2013) evaluated volatile flavor compounds in bacon from wood smoking and liquid (fluid) smoking, and reported that

TABLE 15.2

An Example of a Bacon Smokehouse Cooking Cycle from the University of Guelph Meat Science Laboratory[1]

Stage Number	Stage Name	Time (minutes)	Temperature (°C)	Humidity (%)
Stage 1	Preheat	10	54	0
Stage 2	Drying	5	54	0
Stage 3	Smoke Ignition	4	54	0
Stage 4	Smoking	14	54	0
Stage 5	Smoke Discharge	5	54	0
Stage 6	Smoking	18	54	0
Stage 7	Smoke Discharge	8	54	0
Stage 8	Cooking	15	60	30
Stage 9	Cooking	15	65	40
Stage 10	Cooking	15	69	50
Stage 11	Cooking	15	70	60
Stage 12	Cooking	15	72	60

Notes

[1] The University of Guelph Meat Science Laboratory uses a Scott mini single cage vertical air flow smokehouse (ScottPec, Guelph, Ontario)

the variety and quantity of volatile flavor compounds were significantly greater in traditional wood smoked products when compared with modern liquid smoked products. Other research efforts have been dedicated to the antimicrobial and antioxidant advantages of liquid smoking (Coronado et al., 2016; Soares et al., 2016). While both applications are common, there appear to be consumer-driven initiatives to market bacon that has been wood smoked at a greater premium than bacon that has not been wood smoked (McLean et al., 2017).

Knipe and Beld (2014) stated that the common endpoint cooking temperature of bacon is approximately 52–53°C; however, this endpoint cooking temperature can certainly range in value. This temperature range aligned closely with several research studies that have manufactured bacon in a commercial setting. For instance, Kyle et al. (2014), Tavárez et al. (2014), and Lowell et al. (2017) stated that bellies were smoked and thermally processed using a step-up cooking cycle for approximately 4 hours until internal belly temperature was 53.3°C. Yet a study conducted by Scramlin et al. (2008) described that bellies were cooked to an internal temperature of 60°C. In a study that conducted two separate "extremes" in bacon cooking cycles in an effort to determine effects of pump uptake retention levels, bellies were cooked to internal cooking temperatures of 55°C and 62°C (Sivendiran et al., 2018). When thermally processing bellies to different endpoint cooking temperatures, processors should expect to see differences in processing yields. Sivendiran et al. (2018) reported bellies that were pumped to a normal target uptake level (target of 115% of green weight) and then removed from the smokehouse at an endpoint cooking temperature of 55°C had an average cooked yield of 101.52%; bellies that were pumped to a normal target uptake level (target of 115% of green weight) and then removed from the smokehouse at an endpoint cooking temperature of 62°C had an average cooked yield of 94.74%.

15.2.4 SLICING

Following smoking and thermal processing, bellies are usually sprayed with cold water (cold showered) and gradually chilled. As indicated by studies that researched bacon processed in a commercial setting, bellies are often chilled over a long period of time to a point near the freezing point of the cured/cooked belly. For instance, Kyle et al. (2014), Tavárez et al. (2014), and Lowell et al. (2018) reported thermally processed bellies were chilled for at least 24 hours to an internal temperature between −5.6°C and −4.4°C. It is important that this chilling process be conducted in a slow and gradual manner. Too-rapid chilling can cause breaks or shatters in the fat portion of bacon slices as indicated by Robles (2004) (illustrated in Figure 15.2). On the same note, the level of tempering for bellies immediately prior to slicing also must be carefully considered. Typically, bellies remain slightly below the freezing point of water (and close to the freezing point of the bellies) at around −5.5°C and −3.3°C for slicing (Knipe and Beld, 2014); however, the fatty acid composition of bellies (and inherent melting point of the predominant fats in the bellies) and slicing equipment

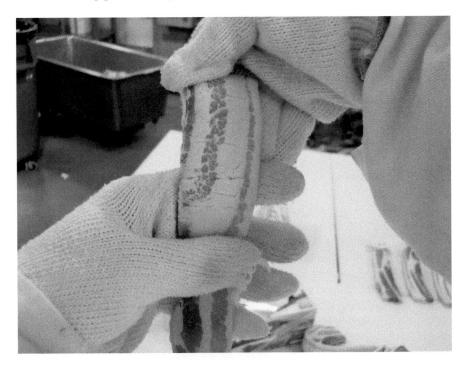

FIGURE 15.2   An example of bacon slice shatters caused by rapid tempering prior to slicing. (Photo credit: B.M. Bohrer, J.M. Kyle, and D.D. Boler).

being used must be carefully considered when selecting a slicing temperature, as fat smearing and orientation of the belly are key contributors to both bacon slice quality and bacon slicing yields. Commercial processing facilities will typically utilize advanced high-speed slicing equipment, so consistency in belly firmness and belly dimensions are very important. A processing intervention typically used in commercial processing facilities is pressing and forming of cooked and chilled bellies prior to slicing. Pressing and forming is usually conducted with a hydraulic belly press machine and should correspond with the width of the belly slicer being used (Knipe and Beld, 2014).

Finally, once a belly is tempered to the desired slicing temperature and pressed to the desired dimensions, the slicing process will be initiated. The major differentiating factors that must be chosen at this time are the desired slice thickness and type of packaging that will be used. Slice thickness can be designated as thick slice (10–16 slices per pound or 22–35 slices per kg), regular slice (16–22 slices per pound or 35–48 slices per kg), or thin slice (22–28 slices per pound or 48–62 slices per kg) (Knipe and Beld, 2014). McLean et al. (2017) reported in a consumer survey study that slice thickness affected price and visually motivated preferences, with thick-slice bacon being marketed at a premium and preferred by consumers when compared with regular-slice bacon. Thin-sliced bacon is rarely seen in the retail setting and is usually reserved for the food-service sector or the pre-cooked bacon sector (Knipe and Beld, 2014).

15.3 PACKAGING AND DISTRIBUTION

15.3.1 PACKAGING TYPES

Following slicing, bacon slices are arranged systematically, depending on the type of packaging that has been selected. The type of packaging can vary, depending on storage conditions and which sector of the industry the product is destined for (that is, retail or food service). While bacon can be further cooked following slicing and before packaging if the product is destined to become pre-cooked bacon (that is, microwave-ready bacon), the discussion herein will only include packaging of partially cooked bacon. That being said, three major types of packaging are common for partially cooked bacon. The first is vacuum-sealed shingle packaging (partial overlap between slices, shown in Figure 15.3), the second is vacuum-sealed stack packaging (slices arranged in a slab, shown in Figure 15.4), and the third is vacuum-sealed or open layout packaging (shown in Figure 15.5) (Lukas, 2016).

Shingle packaging, as first depicted in the 1952 U.S. Patent by Louis B. Uehlein (Uehlein, 1952), was later modified into the common L-board style packaging that is commonly seen in the retail setting (shown in Figure 15.6). This design was created by a team of food scientists at Oscar Mayer (now a subsidiary of The Kraft Heinz Company) in 1974 (Seiferth et al., 1974) and then later improved by a packaging company in 1992 (Nedblake and Johnson, 1992). Shingle packaging allows the consumer to visually inspect a portion of all the slices in the package, while still offering protection of the product from oxidation and physical contamination. This type of packaging typically contains approximately 500 g of bacon and the packages are usually stored and sold under refrigeration.

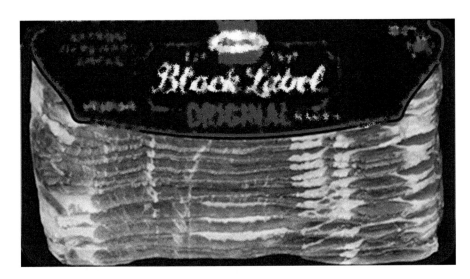

FIGURE 15.3   An example of vacuum-sealed shingle-style packaged bacon.

FIGURE 15.4   An example of vacuum-sealed stack-style packaged bacon.

Stack packaging displays the bacon slices arranged in a slab, with the entirety of the front and back slices exposed. This type of packaging requires less packaging waste (no L-boards), while still offering protection of the product from oxidation and physical contamination. One additional advantage of this style of packaging is the opportunity to include resealable options. This type of packaging contains a wide range of weights ranging from 500 g to 2 kg of bacon and the packages are usually stored and sold under refrigeration. A limitation of both of the aforementioned options (shingle packaging and stack packaging) is that often non-consecutive slices are not visually acceptable in these packages, so end slices from a belly cannot be used. While these two types of packaging dominate the retail sector, the food-service sector often prefers to work with layout packaging.

Layout packaging can be described as the placement of bacon slices edge to edge in a single layer on parchment paper and then stacking the layers of bacon and parchment paper on top of one another. This enables ease of loading into boxes and then an extra level of convenience during preparation at a restaurant or cafeteria.

FIGURE 15.5 An example of open layout–style packaged bacon.

Layout packaging can either be vacuum-sealed or simply covered with plastic wrap inside a box. Layout packaged bacon is usually sold by box weight or by total number of slices, and products are usually sold and stored under frozen conditions until use.

15.3.2 STORAGE PERIOD

The storage period of bacon is dependent on a number of factors, most of which are derived from the composition of the raw materials (especially level of unsaturated fats), how the bacon was processed (that is cured, cooked, sliced), and then how the bacon was packaged and stored. Successful storage is defined by adequate oxidative and microbial stability. Previous research studies (Lowe et al., 2014; Herrick et al., 2016; Matney et al., 2019; Overholt et al., 2019), where commercial bacon was manufactured and shelf-life testing was conducted, indicated that with proper storage, bacon can be successfully stored up to 120 days under vacuum-sealed, refrigerated conditions and up to 90 days in open-layout frozen conditions. At the same time, both lipid oxidation and microbial growth can become increasing issues when bacon is stored for long periods of time. Li et al. (2019) reported microorganism growth was positively related to storage time, and high levels of many different categories of microorganisms were observed in bacon that was vacuum-sealed and stored under refrigeration for up to 45 days. Li et al. (2019)

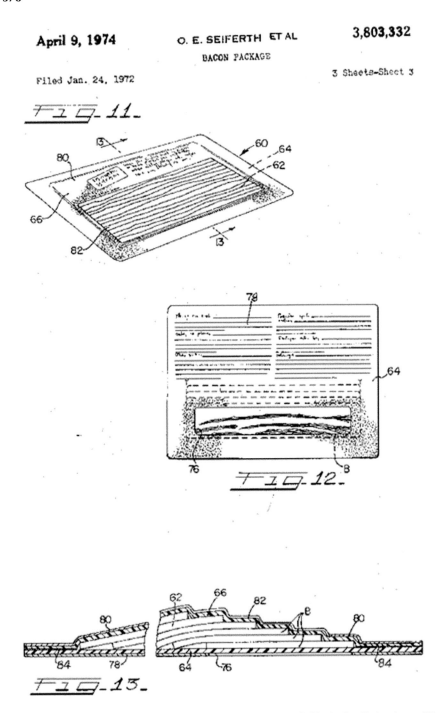

FIGURE 15.6 L-board packaging from the following patent: Seiferth, O., G. Austin, and D. Paul. 1974. U.S. Patent No. 3,803,332. Washington, DC: U.S. Patent and Trademark Office.

further reported that lactic acid bacteria, mainly *Leuconostoc* and *Lactobacillus*, domi-nated the microbial population after only 7 days of vacuum packaged storage at 0–4°C, and the specific species *Lactobacillus sakei* and *Lactobacillus curvatus* were most pre-valent at 45 days of storage, which could indicate that these are responsible for spoilage of bacon products.

Bacon that is produced from bellies that have high levels of unsaturated fats, as indicated from high iodine values, is thought to be more susceptible to lipid oxidation. However, there is conflicting research on this assumption. Matney et al. (2019) studied bellies with normal iodine values (IV range of 58–83) and reported that iodine value had no effect on lipid oxidation in either anaerobic or aerobic packaged con-ditions. While Overholt et al. (2019) studied bellies with very extreme iodine values (IV range of 61.7–98.6) and reported that bellies with iodine values greater than 80 were more prone to lipid oxidation in aerobic packaging. Beyond the level of un-saturated fats, prior storage of raw materials is an additional consideration. Even though bellies are not typically frozen and stored for an extended period of time before bacon is manufactured, this can happen. Lowe et al. (2014) reported freezing bellies for 2 months, 5 months, or 7 months before processing may exacerbate lipid oxidation as the storage time of bacon after processing is extended.

Processing considerations that affect storage length of bacon include many dif-ferent factors, but generally can be described by intricate details during the curing, cooking, and slicing steps during processing. While research is limited as it relates to the ingredient formulation of bacon and its effect on storage length, research related to lowering salt (sodium chloride) content and replacing nitrite salt with alternative sources of nitrate/nitrite in other processed meat products may have some carry-over to bacon. Aaslyng et al. (2014) reported that salt cannot simply be reduced in bacon without significant losses in sensory properties; however, this study did report that processing yields and microbial stability were maintained when reducing salt content. Research conducted by Pham-Mondala et al. (2019) evaluated the effectiveness of rosemary and green tea extracts as a replacement to synthetic nitrites in bacon, and reported these ingredients were successful in extending the shelf life of bacon through 90 days of open layout packaging in frozen conditions. Research conducted by Dempster (1972) reported that processing methods (fresh curing solution, recycled curing solution, and Wiltshire curing methods) and storage temperature (0°C and 10°) had great impacts on microbial populations and sensory properties of bacon.

Finally, packaging and storage conditions can have a significant effect on storage capabilities of bacon. A research study conducted by Lowe et al. (2014) investigated the differences in oxidation and sensory characteristics for bacon that was vacuum-sealed and stack packaged at refrigeration temperatures (2°C) for 60 or 120 days of storage when compared with bacon that was open layout packaged at frozen tem-peratures (−33°C) for 45 or 90 days of storage. This study (Lowe et al., 2014) concluded that as storage times were extended, vacuum-sealed stack packaged bacon stored at refrigeration temperatures had lower levels of lipid oxidation (as indicated by the thiobarbituric acid reactive substances assay) compared with open layout packaged bacon stored at frozen temperatures, yet off flavor and oxidized odor (evaluated by a trained sensory panel) were greater as storage times were extended in vacuum-sealed stack packaged bacon stored at refrigeration temperatures compared with open layout packaged bacon stored at frozen temperatures. Overall, there is still a need for greater research efforts in the area of bacon packaging, both from the industry and the academic community.

15.4 COMPOSITION

15.4.1 NUTRIENT COMPOSITION

The macronutrient composition of bacon is one of the most unique aspects of this processed meat product when compared with other types/categories of processed meat products. Typically characterized by high lipid content, the macronutrient composition of bacon can range quite drastically. An average of twenty-one bacon samples from the USDA nutrient composition database (2020) indicated that fully prepared pan-fried bacon was comprised of 510.05 ± 85.83 kcal of energy/100-g serving, 29.57 ± 8.43 g of protein/100-g serving, 41.53 ± 8.72 g of total lipid/100-g serving, 0.06 ± 0.29 g of carbohydrates/100-g serving, and 15.48 ± 3.07 g of saturated fat/100-g serving (Table 15.3). These data are supported to some degree by research studies that have evaluated proximate composition of bacon samples as well as research studies that have evaluated visual lean:fat. Previous studies reported unprepared/uncooked bacon was 12–15% protein and 31–42% lipid (Lowe et al., 2014; Kyle et al., 2014; Sivendiran et al., 2018). Additionally, unprepared/uncooked bacon has been reported to have the following fatty acid profile: SFA – 32–36%, MUFA – 44–50%, and PUFA – 14–21% (Kyle et al., 2014; Lowell et al., 2018).

The clear driver of the large range in macronutrient values is lipid content, and these values can range within individual bellies and even within the same belly

TABLE 15.3

Pan-fried Bacon Macronutrient Composition as Reported in Multiple Reports on the Usda Nutrient Composition Database[1,2]

Macronutrient, Expressed per 100 g	Average	Standard Deviation	Minimum	Maximum
Energy, kcal	510.05	85.83	205.00	615.00
Protein, g	29.57	8.43	7.14	40.00
Total lipid, g	41.53	8.72	15.23	53.85
Carbohydrates, g (by difference)	0.06	0.29	0.00	1.32
Saturated fat, g	15.48	3.07	6.62	20.83
Sodium, mg	1660.52	509.93	566.00	2333.00

Notes

[1] Source: USDA Food Composition Database. 2020. USDA food composition database. Accessed January 2020. https://ndb.nal.usda.gov/ndb/

[2] Food Data Central IDs included: 358449, 366388, 396500, 400999, 445240, 470662, 470662, 470663, 471871, 495990, 505595, 511244, 514023, 519696, 529220, 543628, 568347, 568713, 570563, 588432, and 588441

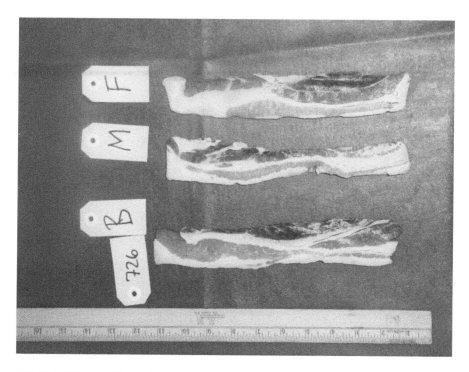

FIGURE 15.7 An illustration of bacon slices originating from the blade end of the belly (B), the center of the belly (M), and the flank end of the belly (F).
(Photo credit: B.M. Bohrer, J.M. Kyle, K.L. Little, and D.D. Boler).

depending on where the slice originated from and how the bacon was processed. Previous research has indicated that belly thickness can significantly influence the proximate composition of bacon (Brewer et al., 1995; Person et al., 2005). Previous research has indicated that whether bacon originates from the blade end of the belly, the center of the belly, or the flank end of the belly can significantly influence the proximate composition and visual lean:fat of bacon (Clark et al., 2014; as indicated in Figure 15.7). Previous research has indicated that pump uptake retention levels and endpoint cooking temperature in the smokehouse can influence proximate composition as well (Sivendiran et al., 2018). These compositional differences have also been established to influence sensory characteristics (Brewer et al., 1995; Person et al., 2005; Sivendiran et al., 2018) and storage capabilities of bacon.

A significant consideration for bacon manufacturers and consumers is the amount of sodium found in bacon products. An average of 21 bacon samples from the USDA nutrient composition database (2020) indicated that prepared pan-fried bacon had 1660.52 ± 509.93 mg of sodium/100-g serving. This is alarming, as the recommended daily allowance for sodium for adults is 2,300 mg/day (DeSalvo et al., 2016). To put this value into context, the USDA nutrient composition database reports that a typical serving size of bacon is two pan-fried bacon slices, which is roughly equivalent to 18 g. Using this value as the standard serving size, 1 serving of bacon would equate to 298.89 ± 91.79 mg of sodium, accounting for roughly 9.0–17.0% of the recommended daily allowance for sodium. This is a challenge stemming from product formulation and product identity that certainly warrants further investigation.

15.5 CONCLUSION

To summarize, many advancements have been achieved by meat processors and meat scientists concerning bacon, yet there is still much to be learned in the area of raw material, processing, and storage considerations and their impact on processing yields, bacon quality, bacon eating satisfaction, and nutrition. Greater research focusing on these considerations using multi-disciplinary and systematic approaches are warranted. Even so, bacon remains an exciting processed meat product and the future is bright for bacon enthusiasts.

REFERENCES

Aaslyng, M. D., C. Vestergaard, and A. G. Koch. 2014. The effect of salt reduction on sensory quality and microbial growth in hotdog sausages, bacon, ham and salami. *Meat Science* 96(1):47–55. https://doi.org/10.1016/j.meatsci.2013.06.004

Alahakoon, A. U., D. D. Jayasena, S. Ramachandra, and C. Jo. 2015. Alternatives to nitrite in processed meat: Up to date. *Trends in Tood Science & Technology* 45(1):37–49. https://doi.org/10.1016/j.tifs.2015.05.008

AOCS. 2009. Recommended Practice Cd 1c-85 Calculated Iodine Value. Official Methods and Recommended Practices of the AOCS.

Asioli, D., J. Aschemann-Witzel, V. Caputo, R. Vecchio, A. Annunziata, T. Naes, and P. Varela. 2017. Making sense of the "clean label" trends: A review of consumer food choice behavior and discussion of industry implications. *Food Research International* 99:58–71. https://doi.org/10.1016/j.foodres.2017.07.022

Barbut, S., A. A. Sosnicki, S. M. Lonergan, T. Knapp, Daniel C. Ciobanu, L. J. Gatcliffe, E. Huff-Lonergan, and E. W. Wilson. 2008. Progress in reducing the pale, soft and exudative (PSE) problem in pork and poultry meat. *Meat Science* 79(1):46–63. https://doi.org/10.1016/j.meatsci.2007.07.031

Brewer, M. S., C. R. Stites, F. K. McKeith, P. J. Bechtel, J. E. Novakofski, and K. A. Bruggen. 1995. Belly thickness effects on the proximate composition, processing, and sensory characteristics of bacon. *Journal of Muscle Foods* 6(3):283–296. https://doi.org/10.1111/j.1745-4573.1995.tb00573.x

CFIA. 2019a. Canadian standards of identity. Volume 7 – meat products. Canadian Food Inspection Agency. Available online: https://www.inspection.gc.ca/about-the-cfia/acts-and-regulations/list-of-acts-and-regulations/documents-incorporated-by-reference/canadian-standards-of-identity-volume-7/eng/1521204102134/1521204102836 (accessed on 11November2019).

CFIA. 2019b. Canadian meat processing controls and procedures. Chapter 4 – meat products. Available online: https://www.inspection.gc.ca/food/archived-food-guidance/meat-and-poultry-roducts/manual-of-procedures/chapter-4/eng/1367622697439/1367622787568 (accessed on 11 November 2019).

CFIA. 2019c. Preventive control recommendations on the use of nitrites in the curing of meat products. Canadian Food Inspection Agency. Available online: https://www.inspection.gc.ca/preventive-controls/meat/nitrites/eng/1522949763138/1522949763434 (accessed on 11November2019).

Clark, D. L., B. M. Bohrer, M. A. Tavárez, D. D. Boler, J. E. Beever, and A. C. Dilger. 2014. Effects of the porcine IGF2 intron 3-G3072A mutation on carcass cutability, meat quality, and bacon processing. *Journal of Animal Science* 92(12):5778–5788. https:// doi.org/10.2527/jas.2014-8283

Čítek, J., R. Stupka, M. Okrouhlá, K. Vehovský, L. Stádník, D. Němečková, and M. Šprysl. 2015. Prediction of pork belly composition using the computer vision method on transverse cross-sections. *Annals of Animal Science* 15(4):1009–1018. https://doi.org/ 10.1515/aoas-2015-0034

Coronado, S. A., G. R. Trout, F. R. Dunshea, and N. P. Shah. 2016. Effect of dietary vitamin E, fishmeal and wood and liquid smoke on the oxidative stability of bacon during 16 weeks' frozen storage. *Meat Science* 62(1):51–60. http://en.cnki.com.cn/Article_en/ CJFDTotal-SPKX201306041.htm

Dempster, J. F. 1972. Vacuum packaged bacon; the effects of processing and storage temperature on shelf life. *International Journal of Food Science & Technology* 7(3):271–279. https://doi.org/10.1111/j.1365-2621.1972.tb01662.x

DeSalvo, K. B., R. Olson, and K. O. Casavale. 2016. Dietary guidelines for Americans. *Jama* 315(5):457–458. https://doi.org /10.1001/jama.2015.18396

Desmond, E. 2006. Reducing salt: A challenge for the meat industry. *Meat Science* 74(1):188–196. https://doi.org/10.1016/j.meatsci.2006.04.014

den Hertog-Meischke, M. J. A., R. J. L. M. Van Laack, and F. J. M. Smulders. 1997. The water-holding capacity of fresh meat. *Veterinary Quarterly* 19(4):175–181. https://doi. org/10.1080/01652176.1997.9694767

Goehring, B. L. 2015. Investigation of factors that influence belly quality and of cooked bacon characteristics. Dissertation. Kansas State University, Manhattan, Kansas: U.S. https://krex.k-state.edu/dspace/handle/2097/20550

Gray, J. I., and A. M. Pearson. 1984. Cured meat flavor. In *Advances in food research* (Vol. 29, pp. 1–86).Academic Press. https://doi.org/10.1016/S0065-2628(08)60055-5

Herrick, R. T., M. A. Tavárez, B. N. Harsh, M. A. Mellencamp, D. D. Boler, and A. C. Dilger. 2016. Effect of immunological castration management strategy on lipid oxidation and sensory characteristics of bacon stored under simulated food service conditions. *Journal of Animal Science* 94(7):3084–3092. https://doi.org/10.2527/jas.2016-0366

Honikel, K. O. 2008. The use and control of nitrate and nitrite for the processing of meat products. *Meat Science* 78(1):68–76. https://doi.org/10.1016/j.meatsci.2007.05.030

Jeremiah, L. E., R. O. Ball, B. Uttaro, and L. L. Gibson. 1996. The relationship of chemical components to flavor attributes of bacon and ham. *Food Research International* 29(5-6):457–464. https://doi.org/10.1016/S0963-9969(96)00058-0

Kapper, C. 2012. Rapid prediction of pork quality: correlation of fresh meat measurements to pork water holding capacity and its technological quality. Dissertation. Wageningen University and Research Centre, Wageningen, The Netherlands. https://library.wur.nl/ WebQuery/wurpubs/431630

Kimoto, W. I., A. E. Wasserman, and F. B. Talley. 1976. Effect of sodium nitrite and sodium chloride on the flavor of processed pork bellies. *LWT Lebensmitt Wissensch Technol* 9(2): 99–101.

Knecht, D., K. Duziński, and A. Jankowska-Mąkosa. 2018. Variability of fresh pork belly quality evaluation results depends on measurement locations. *Food Analytical Methods* 11(8):2195–2205. https://doi.org/10.1007/s12161-018-1205-2

Knipe C. L. and J. Beld. 2014. Bacon production. In *Encyclopedia of meat sciences* (2nd ed., pp. 54–57), ed. M. Dikeman, and C. Devine. Academic Press, Elsevier.

KR Castlemain. 2009. Bacon. In food service products. https://web.archive.org/web/ 20091001024943/http://www.krcastlemaine.com.au/foodservice/product_info.php?category_id=1&category_name=Bacon (accessed on 11November2019).

Kyle, J. M., B. M. Bohrer, A. L. Schroeder, R. J. Matulis, and D. D. Boler. 2014. Effects of immunological castration (Improvest) on further processed belly characteristics and commercial bacon slicing yields of finishing pigs. *Journal of Animal Science* 92(9):4223–4233. https://doi.org/10.2527/jas.2014-7988

Lee, E. A., J. H. Kang, J. H. Cheong, K. C. Koh, W. M. Jeon, J. H. Choe, K. C. Hong, and J. M. Kim. 2018. Evaluation of whole pork belly qualitative and quantitative properties using selective belly muscle parameters. *Meat Science* 137:92–97. https://doi.org/10.1016/j.meatsci.2017.11.012

Lewis, D. F., K. H. Grooves, and J. H. Holgate. 1986. Action of polyphosphates in meat products. *Food Microstructure* 5(1):53–62. http://digitalcommons.usu.edu/foodmicrostructure/vol5/iss1/7

Li, X., C. Li, H. Ye, Z. Wang, X. Wu, Y. Han, and B. Xu. 2019. Changes in the microbial communities in vacuum-packaged smoked bacon during storage. *Food Microbiology* 77:26–37. https://doi.org/10.1016/j.fm.2018.08.007

Little, K. L., B. M. Bohrer, H. H. Stein, and D. D. Boler. 2015. Effects of feeding high protein or conventional canola meal on dry cured and conventionally cured bacon. *Meat Science* 103:28–38. https://doi.org/10.1016/j.meatsci.2014.12.007

Lowe, B. K. B. M. Bohrer, S. F. Holmer, D. D. Boler, and A. C. Dilger. 2014. Effects of retail style or food service style packaging type and storage time on sensory characteristics of bacon manufactured from commercially sourced pork bellies. *Journal of Food Science* 79(6):S1197–S1204. https://doi.org/10.1111/1750-3841.12480

Lowell, J. E., B. N. Harsh, K. B. Wilson, M. F. Overholt, R. J. Matulis, A. C. Dilger, and D. D. Boler. 2017. Effect of salt inclusion level on commercial bacon processing and slicing yields. *Meat and Muscle Biology* 1(1):8–17. https://doi.org/10.22175/mmb2016.11.0005

Lowell, J. E., B. M. Bohrer, K. B. Wilson, M. F. Overholt, B. N. Harsh, H. H. Stein, A. C. Dilger, and D. D. Boler. 2018. Growth performance, carcass quality, fresh belly characteristics, and commercial bacon slicing yields of growing-finishing pigs fed a subtherapeutic dose of an antibiotic, a natural antimicrobial, or not fed an antibiotic or antimicrobial. *Meat Science* 136: 93–103. https://doi.org/10.1016/j.meatsci.2017.10.011

Lukas, P. 2016. Why supermarket bacon hides its glorious fat. https://www.bloomberg.com/features/2016-bacon-package-product-design/ (accessed on 26January2020).

Mandigo, R. W., T. S. Janssen, M. L. Lesiak, and D. G. Olson. 1981. Will consumers accept low sodium pork?. *Nebraska Swine Report EC81–219*, 21–22.

Mani-López, E., H. S. García, and A. López-Malo. 2012. Organic acids as antimicrobials to control Salmonella in meat and poultry products. *Food Research International* 45(2):713–721. http://dx.doi.org/10.1016/j.foodres.2011.04.043

Matney, M., A. Tapian, B. Goehring, J. Unruh, B. Gerlach, G. McCoy, D. Mohrhauser, T. O'Quinn, J. Gonzalez, T. Houser, and R. Johnson. 2019. Influence of iodine value and packaging type on quality characteristics of food service packaged bacon slices. *Meat and Muscle Biology* 192:80. https://doi.org/10.221751/rmc2016.076

Matthews, K., and M. Strong. 2005. Salt–its role in meat products and the 'industry's action plan to reduce it. *Nutrition Bulletin* 30(1):55–61. https://doi.org/10.1111/j.1467-3010.2005.00469.x

McLean, K. G., D. J. Hanson, S. M. Jervis, and M. A. Drake. 2017. Consumer perception of retail pork bacon attributes using adaptive choice-based conjoint analysis and maximum differential scaling. *Journal of Food Science* 82(11):2659–2668. https://doi.org/10.1111/1750-3841.13934

Meadus, W. J., P. Duff, B. Uttaro, J. L. Aalhus, D. C. Rolland, L. L. Gibson, and M. E. R. Dugan. 2010. Production of docosahexaenoic acid (DHA) enriched bacon. *Journal of Agricultural and Food Chemistry* 58(1):465–472. https://doi.org/10.1021/jf9028078

Mottram, D. S., and D. N. Rhodes. 1973. Nitrite and the flavour of cured meat. In *Proceedings of the International Symposium on Nitrite in Meat Products*, Zeist, pp. 161–170. https://edepot.wur.nl/320439#page=148

Murray, D. G. 1983. Functional flavorings which reduce the bitterness of potassium chloride. Activities report of the R and D Associates, USA.

NAMI (North American Meat Institute). 2014. *NAMP meat 'buyer's guide* (8th rev. ed.). North American Meat Association.

Nedblake, G. W., and Johnson, L. E. (1992). U.S. Patent No. 5,087,498. Washington, DC: U.S. Patent and Trademark Office. https://patents.google.com/patent/US5087498A/en

Offer, G., P. Knight, R. Jeacocke, R. Almond, T. Cousins, J. Elsey, N. Parsons, A. Sharp, R. Starr, and P. Purslow. 1989. The structural basis of the water-holding, appearance and toughness of meat and meat products. *Food Structure* 8(1):17. https://digitalcommons. usu.edu/foodmicrostructure/vol8/iss1/17

Overholt, M. F., J. E. Lowell, G. D. Kim, A. C. Dilger, and D. D. Boler. 2019. Use of iodine value to predict commercial slicing yield and shelf-life of bacon. *Meat and Muscle Biology* 1(3):114-114. https://doi.org/10.221751/rmc2017.109

Pearce, K. L., K. Rosenvold, H. J. Andersen, and D. L. Hopkins. 2011. Water distribution and mobility in meat during the conversion of muscle to meat and ageing and the impacts on fresh meat quality attributes – A review. *Meat Science* 89(2):111–124. https://doi.org/10.1016/j.meatsci.2011.04.007

Pegg, R. B., and F. Shahidi. 2008. *Nitrite curing of meat: The N-nitrosamine problem and nitrite alternatives.* John Wiley & Sons. https://www.wiley.com/en-us/Nitrite+Curing+of+Meat %3A+The+N+Nitrosamine+Problem+and+Nitrite+Alternatives-p-9780470384862

Person, R. C., D. R. McKenna, D. B. Griffin, F. K. McKeith, J. A. Scanga, K. E. Belk, G. C. Smith, and J. W. Savell. 2005. Benchmarking value in the pork supply chain: Processing characteristics and consumer evaluations of pork bellies of different thicknesses when manufactured into bacon. *Meat Science* 70(1):121–131. https://doi. org/10.1016/j.meatsci.2004.12.012

Petracci, M., M. Bianchi, S. Mudalal, and C. Cavani. 2013. Functional ingredients for poultry meat products. *Trends in Food Science & Technology* 33(1):27–39. https://doi.org/10. 1016/j.foodres.2017.07.022

Pham-Mondala, A., R. Boyle, L. Bond, A. Vanek, and P. Joseph. 2019. Utilization of ro- semary and green tea extracts as clean label antioxidant solutions in bacon formula- tions. *Meat and Muscle Biology* 2(2):178. https://doi.org/10.221751/rmc2018.156

Powell, M. J., J. G. Sebranek, K. J. Prusa, and R. Tarté. 2019. Evaluation of citrus fiber as a natural replacer of sodium phosphate in alternatively-cured all-pork Bologna sausage. *Meat Science* 157:107883. https://doi.org/10.1016/j.meatsci.2019.107883

Resconi, V. C., D. F. Keenan, M. Barahona, L. Guerrero, J. P. Kerry, and R. M. Hamill. 2016. Rice starch and fructo-oligosaccharides as substitutes for phosphate and dextrose in whole muscle cooked hams: Sensory analysis and consumer preferences. *LWT - Food Science and Technology* 66:284–292. https://doi.org/10.1016/j.lwt.2015.10.048

Robles, C. C. 2004. The effect of fresh and frozen bellies on bacon processing characteristics and bacon quality. Dissertation. University of Nebraska-Lincoln, Lincoln, Nebraska, USA. https://digitalcommons.unl.edu/cgi/viewcontent.cgi?article=1018&context= animalscidiss

Ruan, R. R., and P. L. Chen. 1998. Water in foods and biological materials: a nuclear magnetic resonance approach. Technomic Publishing. Lancaster, Pennsylvania, USA. https://trove.nla.gov.au/version/46518133

Ruusunen, M., and E. Puolanne. 2005. Reducing sodium intake from meat products. *Meat Science* 70:531–541. https://doi.org/10.1016/j.meatsci.2004.07.016

Saldaña, E., L. Saldarriaga, J. Cabrera, J.H. Behrens, M. M. Selani, J. Rios-Mera, and C. J. Contreras-Castillo. 2019a. Descriptive and hedonic sensory perception of Brazilian consumers for smoked bacon. *Meat Science* 147:60–69. https://doi.org/10.1016/j. meatsci.2018.08.023

Saldaña, E., L. Saldarriaga, J. Cabrera, R. Siche, J. H. Behrens, M. M. Selani, M. A. de Almeida, L. D. Silva, J. S. Silva Pinto, and C. J. Contreras-Castillo. 2019b. Relationship between volatile compounds and consumer-based sensory characteristics of bacon smoked with different Brazilian woods. *Food Research International* 119:839–849. https://doi.org/10.1016/j.foodres.2018.10.067

Saldaña, E., M. M. Martins, B. S. Menegali, M. M. Selani, and C. J. Contreras-Castillo. 2019c. Obtaining the ideal smoked bacon: What is the influence of the product space and multivariate procedure to construct the external preference mapping?. *Scientia Agropecuaria* 10(1):29–37. http://dx.doi.org/10.17268/sci.agropecu.2019.01.03

Scramlin, S. M., S. N. Carr, C. W. Parks, D. M. Fernandez-Dueñas, C. M. Leick, F. K. McKeith, and J. Killefer. 2008. Effect of ractopamine level, gender, and duration of ractopamine on belly and bacon quality traits. *Meat Science* 80(4):1218–1221. https://doi.org/10.1016/j.meatsci.2008.05.034

Sebranek, J. G. 2009. Basic curing ingredients. In *Ingredients in meat products* (pp. 1–23). Springer. https://doi.org/10.1007/978-0-387-71327-4_1

Sebranek, J. G. 2015. An overview of functional non-meat ingredients in meat processing: the current toolbox. In *Reciprocal Meat Conference 2015 Proceedings*, 42–46. https://lib.dr.iastate.edu/ans_conf/4/

Seiferth, O., G. Austin, and D. Paul. 1974. U.S. Patent No. 3,803,332. Washington, DC: U.S. Patent and Trademark Office.

Seman, D. L., W. N. G. Barron, and M. Matzinger. 2013. Evaluating the ability to measure pork fat quality for the production of commercial bacon. *Meat Science* 94(2):262–266. https://doi.org/10.1016/j.meatsci.2013.01.009

Shahidi, F., and R. B. Pegg. 1995. Nitrite alternatives for processed meats. In *Developments in food science* (Vol. 37, pp. 1223–1241). Elsevier. https://doi.org/10.1016/S0167-4501(06)80231-1

Sheard, P. R. 2010. Chapter 18. Bacon. In *Handbook of meat processing,* ed. F. Toldrá., (pp. 327–336) John Wiley & Sons.

Sindelar, J. J. 2006. Investigating uncured no nitrate or nitrite added processed meat products. Dissertation. Iowa State University, Ames, Iowa: U.S. https://lib.dr.iastate.edu/cgi/viewcontent.cgi?article=4087&context=rtd

Sindelar, J. J., J. C. Cordray, D. G. Olson, J. G. Sebranek, and J. A. Love. 2007. Investigating quality attributes and consumer acceptance of uncured, No-nitrate/nitrite-added commercial hams, bacons, and frankfurters. *Journal of Food Science* 72(8):S551–S559. https://doi.org/10.1111/j.1750-3841.2007.00486.x

Sivendiran, T., L. M. Wang, S. Huang, and B. M. Bohrer. 2018. The effect of bacon pump retention levels following thermal processing on bacon slice composition and sensory characteristics. *Meat Science* 140:128–133. https://doi.org/10.1016/j.meatsci.2018.03.007

Soares, J. M., P. F. da Silva, B. M. S. Puton, A. P. Brustolin, R. L. Cansian, R. M. Dallago, and E. Valduga. 2016. Antimicrobial and antioxidant activity of liquid smoke and its potential application to bacon. *Innovative Food Science & Emerging Technologies* 38:189–197. https://doi.org/10.1016/j.ifset.2016.10.007

Sofos, J. N. 1986. Use of phosphates in low-sodium meat products. Food Technology, USA. http://agris.fao.org/agris-search/search.do?recordID=US8726886

Soladoye, P. O., P. J. Shand, J. L. Aalhus, C. Gariépy, and M. Juárez. 2015. Pork belly quality, bacon properties and recent consumer trends. *Canadian Journal of Animal Science* 95(3):325–340. https://doi.org/10.4141/cjas-2014-121

Soladoye, O. P. 2017. Predicting the physicochemical properties of pork belly and the effect of cooking and storage conditions on bacon sensory and chemical characteristics. Dissertation. University of Saskatchewan, Saskatoon, Saskatchewan, Canada. https://harvest.usask.ca/handle/10388/7981

Soladoye, O. P., B. Uttaro, S. Zawadski, M. E. R. Dugan, C. Gariépy, J. L. Aalhus, P. Shand, and M. Juárez. 2017. Compositional and dimensional factors influencing pork belly firmness. *Meat Science* 129:54–61. https://doi.org/10.1016/j.meatsci.2017.02.006

Tavárez, M. A. 2014. Evaluating fat quality and bacon slicing yields in immunoloigcally castrated barrows. Dissertation. University of Illinois, Urbana-Champaign, Illinois, USA. https://www.ideals.illinois.edu/handle/2142/49562

Tavárez, M. A., B. M. Bohrer, M. D. Asmus, A. L. Schroeder, R. J. Matulis, D. D. Boler, and A. C. Dilger. 2014. Effects of immunological castration and distiller's dried grains with solubles on carcass cutability and commercial bacon slicing yields of barrows slaughtered at two time points. *Journal of Animal Science* 92(7):3149–3160. https://doi.org/10.2527/jas.2013-7522

Tavárez, M. A., B. M. Bohrer, R. T. Herrick, M. A. Mellencamp, R. J. Matulis, M. Ellis, D. D. Boler, and A. C. Dilger. 2016. Effects of time after a second dose of immunization against GnRF (Improvest) independent of age at slaughter on commercial bacon slicing characteristics of immunologically castrated barrows. *Meat Science* 111:147–153. https://doi.org/10.1016/j.meatsci.2015.09.005

Theron, M. M., and J. F. Lues. 2007. Organic acids and meat preservation: a review. *Food Reviews International* 23(2):141–158. https://doi.org/10.1080/87559120701224964

Trusell, K. A., J. K. Apple, J. W. Yancey, T. M. Johnson, D. L. Galloway, and R. J. Stackhouse. 2011. Compositional and instrumental firmness variations within fresh pork bellies. *Meat Science* 88(3):472–480. https://doi.org/10.1016/j.meatsci.2011.01.029

Uehlein, L. B. 1952. U.S. Patent No. 2,596,514. Washington, DC: U.S. Patent and Trademark Office. https://patents.google.com/patent/US2596514A/en

USDA. 2005. Food standards and labeling policy book. United States Department of Agriculture Food Safety and Inspection Service, (May 2003), 1–202. http://www.fsis.usda.gov/OPPDE/larc/Policies/Labeling_Policy_Book_082005.pdf (accessed on 11 November 2019).

USDA. 2010. Regulatory information, code of federal regulations. 9 CFR 319.107 – Bacon. https://www.gpo.gov/fdsys/granule/CFR-2010-title9-vol2/CFR-2010-title9-vol2-sec319-107 (accessed on 21 December2019).

USDA. 2013. Bacon and food safety. United States Department of Agriculture Food Safety and Inspection Service. https://www.fsis.usda.gov/wps/portal/fsis/topics/food-safety-education/get-answers/food-safety-fact-sheets/meat-preparation/bacon-and-food-safety/ct_index (accessed on 11 November2019).

USDA. 2019. Daily National Carlot Meat Report. United States Department of Agriculture Agricultural Marketing Services, Des Moines, Iowa, USA. https://www.ams.usda.gov/mnreports/lsddb.pdf (accessed on 11 November2019).

USDA Food Composition Database. 2020. USDA food composition database. Accessed January2020. https://ndb.nal.usda.gov/ndb/

Weiss, J., M. Gibis, V. Schuh, and H. Salminen. 2010. Advances in ingredient and processing systems for meat and meat products. *Meat Science* 86(1):196–213. https://doi.org/10.1016/j.meatsci.2010.05.008

Woods, D. F., I. M. Kozak, S. Flynn, and F. O'Gara. 2019. The microbiome of an active meat curing brine. *Frontiers in Microbiology* 9:3346. https://doi.org/10.3389/fmicb.2018.03346

Woolford, G., and R. G. Cassens. 1977. The fate of sodium nitrite in bacon. *Journal of Food Science* 42(3):586–589. https://doi.org/10.1111/j.1365-2621.1977.tb12555.x

Zhao, B., L. Ren, W. Chen, X. Qiao, J. Li, and Y. Zhao. 2013. Effects of different smoking methods on volatile flavor compounds in bacon. *CNKI Food Science* 6. http://en.cnki.com.cn/Article_en/CJFDTotal-SPKX201306041.htm

APPENDIX FIGURES

Figures 15.8, 15.9, 15.10

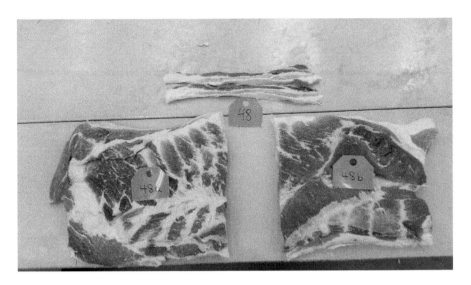

FIGURE 15.8   An illustration of an unprocessed pork belly depicting the blade end (left, identified as 48a) and the flank end (right, identified as 48b). (Photo credit: B.M. Bohrer).

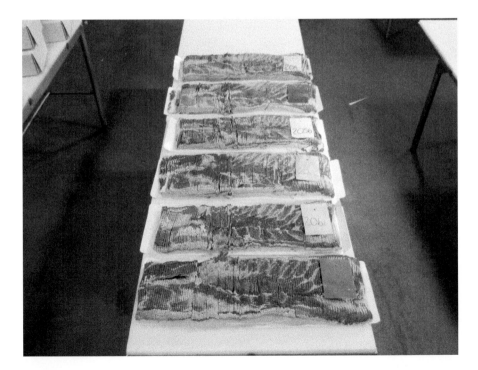

FIGURE 15.9   Processed bacon from a research study, notice anatomical orientation of the entire pork belly was maintained during processing. (Photo credit: B.M. Bohrer, J.M. Kyle, and D.D. Boler).

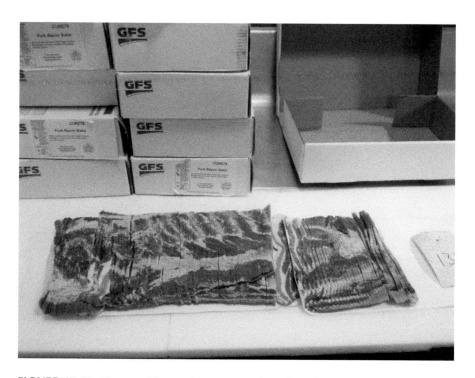

FIGURE 15.10 Processed bacon from a research study, prior to selection of individual bacon slices for further evaluation.
(Photo credit: B.M. Bohrer, J.M. Kyle, and D.D. Boler).

16 Bologna Sausages

Manufacturing Process, Physico-chemical Composition, and Shelf Life

Paulo Cezar Bastianello Campagnol,[1]
José M. Lorenzo,[2,3] Jordana Lima da Rosa,[1]
Bibiana Alves Dos Santos,[1] and
Alexandre José Cichoski[1]

[1]Universidade Federal de Santa Maria, CEP 97105-900, Santa Maria, Rio Grande do Sul, Brazil
[2]Centro Tecnológico de la Carne de Galicia, Parque Tecnológico de Galicia, San Cibrán das Viñas, Rúa Galicia N 4, Ourense, Spain
[3]Área de Tecnología de los Alimentos, Facultad de Ciencias de Ourense, Universidad de Vigo, 32004Ourense, Spain

CONTENTS

16.1   INTRODUCTION

Bologna sausages are consumed and enjoyed worldwide, and various factors contribute to the success of this meat product, including practicality, convenience, sensory quality, and affordable price. This product represents an important segment of the meat industry due to the great added value to the raw material.

The basic formulation of Bologna sausages consists of lean meat, fat, NaCl, and water. However, several other non-meat ingredients are commonly used. Its manufacturing process involves the elaboration of an emulsion containing water as a continuous phase, fat as a discontinuous phase, and meat myofibrillar proteins as emulsifying agents. The emulsion is stabilized through cooking, leading to the denaturation of myofibrillar proteins and the formation of a three-dimensional gelled protein network wherein fat and water are physically entrapped.

Pork meat is one of the most used raw materials for the manufacture of Bologna sausages, which makes it a good source of proteins of high biological value, B vitamins, and minerals such as iron and zinc. However, Bologna sausages should be eaten in moderation due to its high contents of animal fat and sodium.

Bologna sausages has a water activity (a_w) between 0.96 and 0.99 and a pH close to 6.0, which allows this product to be an excellent substrate for microbial growth. In addition, it is also very susceptible to lipid oxidation due to its high content of unsaturated fatty acids. The correct choice of raw materials, ingredients, and additives, as well as adequate manufacture and storage, can minimize these intrinsic factors and increase the shelf life and product safety.

In this context, this chapter will address the manufacturing process of Bologna sausages, with emphasis on the main ingredients and additives, the selection of the raw material, and the adequate procedures to obtain high quality and safe products. The physicochemical composition of Bologna sausages produced around the world and the shelf life-limiting factors will also be discussed.

16.2   MANUFACTURING PROCESS

16.2.1   MAIN INGREDIENTS AND ADDITIVES

16.2.1.1   Water

Water is a very important ingredient for the quality of Bologna sausages. It solubilizes dry ingredients incorporated into the sausage formulations, such as NaCl and phosphate, improving the extraction of myofibrillar proteins, which increases the production yield and improves the texture of the product. In addition, water in the form of ice allows reducing the batter temperature during the emulsification process, preventing the emulsion from breaking down due to protein denaturation. Water is also important to reduce the manufacturing costs and the energy value of the product, besides contributing to an improvement in tenderness and juiciness (Beggs, Bowers, and Brown 1997).

16.2.1.2   Sodium Chloride (NaCl)

NaCl is a key ingredient in the manufacture of Bologna sausages, commonly used at levels ranging from 2% to 3%. One of the main functions of NaCl is the solubilization of myofibrillar proteins of meat, increasing hydration and water holding capacity, thus reducing cooking losses, and increasing tenderness and juiciness. The role of NaCl in water retention is due to the tendency of chloride ions to penetrate into the intra-fluids around myofilaments, causing swelling of myofibrils (Hamm, 1986). Simultaneously, the expansion of the negative net charge within the myosin filaments leads to the disintegration of filaments at an ionic strength of 0.8 M (without the addition of phosphate salts) or 0.4 M (in the presence of phosphate salts). For sausage formulations without the addition of phosphate, the water holding capacity increases almost linearly with the NaCl concentration up to ~4% (Hamm, 1986). The addition of phosphate leads to a marked increase in water retention in the range of 1.0–1.5% NaCl, due to the breakdown of actomyosin bonds, facilitating the swelling of filaments (Offer and Knight, 1988). The fat retention also increases with the addition of NaCl, due to the fat emulsification by the solubilized proteins. Thus, NaCl is responsible for three important interactions with the myofibrillar proteins: water–protein (water retention), protein–protein (bond between the meat particles), and protein–fat (fat retention) (Hutton, 2002). The reduction of NaCl concentration leads to a decrease in the emulsion stability, as reported by Pinton et al. (2020), who studied the salt reduction from 2.5–1.25% in meat emulsions.

In addition to the technological functions, NaCl plays an important preservative effect by reducing the a_w, which is a dimensionless parameter that indicates the amount of free, unbound water in the system, that is available for use by the microorganisms. NaCl completely ionizes in water into sodium and chloride ions, which interact with water molecules, making them unavailable for microbial growth. The reduction of a_w increases the latent period, which is the interval during which the microorganisms adapt to the environmental conditions before growth. It is known that each microorganism has a critical a_w, below which growth cannot occur. At these a_w values, the water moves out of the cells to a higher solute concentration, leading to plasmolysis, and the plasmolyzed cells lose their rigid structure, which impairs the microbial growth (Jay, 2005).

In addition to providing the characteristic salty taste of meat products, NaCl can also improve the flavor and aroma of other components and reduce the perception of other stimuli, such as the bitter taste of some compounds (Coultate, 2002). This behavior is probably due to the effect of NaCl on a_w, which allows an increase in the concentration of other components, improving its volatility and sensory perception (Hutton, 2002). NaCl can also affect the texture of meat products, conferring a softer texture in reduced-salt products, as reported by Leães et al. (2020) in meat emulsions.

16.2.1.3   Phosphate

Phosphates are salts of phosphoric acid or orthophosphoric acid, widely used in meat processing industries. These salts are composed of positively charged metal ions and negatively charged phosphate ions from the corresponding acid by the loss of H+ (Feiner, 2006). Phosphates can be divided into mono- or orthophosphates, according to the number of phosphorus atoms. The dimers (two P atoms) are pyrophosphates, followed by triphosphates or tripolyphosphates (three P atoms) and tetraphosphates (four P atoms). Molecules with 5–15 P atoms are classified as oligophosphates. In general, phosphates with three or more P atoms are considered polyphosphates (Molins, 1991). According to the European legislation, phosphates are not permitted in fresh meat and may be added to meat products in a maximum amount of 0.5% P_2O_5 (Long and Buňka, 2011; Ritz et al., 2012).

At least ten specific phosphates can be used in meat products, which vary according to their functional properties and can have an impact on pH and solubility (Molins, 1991). Different phosphate blends are commonly used, as they work significantly better than a single type of phosphate, and the solubility of the different types must be taken into account. Mono and diphosphates increase the water holding capacity; however, its use is limited due to its low and slow solubility, especially in cold brine (Dusek et al., 2003). Although monophosphates have no effect on muscle protein, they exhibit good buffering capacity, which contributes to stabilizing the pH of the final product (Long and Buňka, 2011). Diphosphates have the ability to dissociate the actomyosin complex from meat. Tripolyphosphates and polyphosphates contribute to the activation of meat proteins by partially chelating Mg^{+2} and Ca^{+2} bound to the protein, leading to solubilization of myosin and actin, and polymerization into thick and thin filaments (Puolanne and Halonen, 2010).

Short-chain phosphates are effective for the manufacture of Bologna sausages due to their ability to assist NaCl in the solubilization of myofibrillar proteins. Phosphates alone cannot solubilize proteins since their function is to dissociate the actomyosin complex. When the actomyosin complex is dissociated into actin and myosin, the addition of salt and water leads to a swelling of the protein fibers and consequent protein solubilization (Ritz et al., 2012). Long-chain phosphates are occasionally added to phosphate blends for the production of Bologna sausages, providing more elastic and softer properties. Longer chain phosphates contribute to an increase in fat emulsification and are used in phosphate blends in formulations containing low meat and high fat and water contents (Ritz et al., 2012). Phosphates also delay the onset of lipid oxidation by binding metals such as iron and copper, stabilizing them in an inactive or insoluble form (Allen and Cornforth, 2010). In addition, phosphates also preserve the color of the products, preventing iron oxidation by lipolytic radicals (Vissa and Cornforth, 2006).

16.2.1.4 Nitrite
Nitrite is widely used in meat products to inhibit the growth of foodborne pathogens. However, its mechanism of action against bacteria is not completely known. One widely held hypothesis is that nitrite contributes to bacterial stress by being the precursor of peroxynitrite, which is a strong oxidant. Certain species are more resistant than others to nitrite. The most resistant are gram-negative aerobic/facultative anaerobic bacteria (*Escherichia coli, Salmonella*), while the most fragile are gram-positive anaerobic bacteria (*Clostridium botulinum*) (Majou and Cristieans, 2018).

This additive also plays an important role in color development and stabilization through the reaction between nitric oxide (NO) from the degradation of nitrite and myoglobin to form NO-myoglobin. The NO-myoglobin (red pigment) is unstable and the application of heat converts it into a stable pink pigment known as nitrosyl-hemochrome (Jo et al., 2020). In addition, nitrite has an important role in reducing lipid oxidation due to its ability to stabilize the heme iron of meat pigments during curing, acting as a metal chelator (Eskandari et al., 2013). It is also very important for improving the flavor of meat products, although this mechanism remains unclear. One hypothesis is that nitrite reduces the formation of aldehydes, mainly hexanal, which masks the sulfur-containing compounds responsible for the cured meat flavor (Thomas et al., 2013). Another hypothesis is that nitrite can induce the formation of Strecker aldehydes, which are related to the development of flavor in foods (Villaverde et al., 2014).

It is worth emphasizing that, despite improving the quality of meat products, nitrite can be a precursor of carcinogenic nitrosamines. Thus, to reduce the risk of nitrosamine formation, the maximum limit of nitrite in meat products, established by the legislation of each country, must be respected, and the use of salts of ascorbic acid is recommended as discussed below.

16.2.1.5   Salts of Ascorbic Acid

The salts of ascorbic acid, such as ascorbate and isoascorbate (erythorbate), are widely used to accelerate the color reactions in meat products. These compounds facilitate the transformation of nitrite into nitric oxide (NO) by accelerating the formation of NO-myoglobin (Sebranek and Bacus, 2007). Consequently, the time required for curing is lower and the product can be cooked right after filling.

In addition to speeding up the color reactions, the salts of ascorbic acid can also increase the toxicological safety of products, as their addition reduces the risk of nitrosamine formation in nitrite-containing products. Although the mechanism responsible for this reduction is not completely elucidated, it is probably due to the reduction of residual nitrite by the reaction with the salts of ascorbic acid. The recommended concentration ranges from 0.2 to 0.4 g/kg to reduce the risk of nitrosamine formation.

16.2.1.6   Other Ingredients and Additives

Non-meat proteins obtained mainly from soybean and milk are widely used in the manufacture of Bologna sausages to replace lean meat, which can reduce the manufacturing costs without affecting the technological quality. Polysaccharides, such as xanthan gum, carrageenan, carboxymethylcellulose, beta-glucan, guar gum, gellan gum, locust bean gum, and starch are also widely used to reduce costs and improve product quality. Due to their gelling properties, polysaccharides increase water and fat retention during and after cooking, improving juiciness, texture, and yield. Lactate (sodium or potassium) is another compound widely used to extend the shelf life of Bologna sausages due to its ability to reduce the a_w of these products.

Regarding the sensory quality, the main condiments used in the manufacture of Bologna sausages include garlic, white pepper, black pepper, nutmeg, and coriander. In addition to improving the aroma and taste, these condiments extend the shelf life of the product due to the presence of antimicrobial and antioxidant compounds. Flavor enhancers, such as monosodium glutamate, are also widely used to improve the sensory acceptance of Bologna sausages. Natural dyes, such as cochineal carmine and beetroot red are used to standardize the color of Bologna sausages.

16.2.2 SELECTION OF RAW MATERIALS: PORK MEAT AND PORK BACK FAT

The use of quality raw materials is fundamental to obtaining a product with extended shelf life and greater consumer acceptance. Pork meat is a rich source of nutrients, including water, proteins, lipids, and minerals (Pellissery et al., 2020). These nutrients are excellent substrates for the growth of microorganisms. The metabolic activities of microorganisms in meat can cause changes in color, odor, and texture when the bacterial counts reach high levels (> 10^7 CFU/g). In addition to leading to food deterioration, high bacterial counts can cause public health problems (Jay, 2005), thus the bacterial load of the raw material should be as low as possible (ideal < 10^3 CFU/g).

The pork quality is closely related to the pre-slaughter and slaughter management steps; inefficiency in the implementation of these steps can lead to defects such as the incidence of PSE (pale, soft, and exudative) and DFD (dry, firm, and dark) meat. One of the main economic losses in the swine industry is related to PSE pork. This defect is due to a rapid drop in the muscle pH at the beginning of the postmortem period (pH < 5.8 after 45 minutes). The low pH value in combination with a high carcass temperature (37–38°C) leads to protein denaturation, reducing water retention and emulsification capacities (Barbut et al., 2008). Thus, the use of PSE meat in Bologna sausages is not recommended, as the production yield, texture, and color of the products can be drastically impaired (Haddad et al., 2018). DFD meat is characterized by the insignificant amount of glycogen left in the muscle at the point of slaughter, when very little or no lactic acid is produced after the animal's death, resulting in an insufficient decline in meat pH. This event has a negative impact on meat tenderness since important proteases that act in the breakdown of the actomyosin complex during rigor mortis are not released or activated. On the other hand, a pH close to neutral gives the meat great water holding capacity, thus the DFD meat can provide technological benefits to Bologna sausages. However, levels greater than 30% of DFD meat can considerably reduce the shelf life of the product because the pH is favorable to microbial growth.

Another raw material of paramount importance in the manufacture of Bologna sausages is pork back fat. Bologna sausages contain a high content of pork back fat (20–35%), which plays an important role in reducing costs and improving technological and sensory quality. The pork back fat stabilizes the emulsion, reducing cooking losses, in addition to improving the tenderness, juiciness, flavor, and aroma of emulsified products. Monitoring the oxidation level is very important to selecting the pork back fat. Although a white color and the absence of rancid aroma are good indicators of low oxidation, a laboratory test is recommended to assess the oxidation degree. One of the most used analyses for this purpose is the TBAR test, which quantifies the content of malonaldehyde (secondary oxidation product) in the sample. Pork back fat with TBARS values greater than 0.5 mg malonaldehyde/kg of the sample should not be used in the sausage formulations, as the sensory quality and the shelf life of the product can be severely compromised.

16.2.3 Cleaning and Grinding of Raw Materials

Before the emulsification process, fragmentation of muscle and fat tissues under the effect of cutting forces is recommended, through the use of mechanical energy to disorganize tissue structures. This step can be performed with the aid of grinding machines or the cutter–grinder. The temperature of the raw material must be close to 0°C, once the heat generated through the mechanical energy increases the process temperature. Before grinding, the meat must undergo a cleaning step to remove tendons, connective tissues, lymph nodes, fats, cartilage, bones, skin debris, and blood vessels, among others. After cleaning, the meat is cut into pieces or strips for grinding, according to the diameter of the grinder plates. After grinding, the meat is weighed according to the Bologna sausage formulation.

16.2.4 Emulsification Process

The cutter is the commonly used equipment for the production of meat emulsions. It has extremely sharp knives that rotate at high speeds (5000–6000 per minute), cutting the raw material into very small pieces, while homogenizing and emulsifying the batter. The use of a vacuum cutter is preferred to prevent incorporating air into the batter, thus reducing the lipid oxidation rate.

The sequence of addition of the raw materials, additives, and ingredients is very important during the emulsification process. First, lean meat, salt, phosphate, and ice are mixed in the cutter for effective solubilization of the myofibrillar proteins. Then the other additives and ingredients are added, and fat is included only at the end of the process. This procedure is necessary to prevent the formation of small fat globules, because it requires a greater amount of myofibrillar proteins to cover these particles, and this can lead to emulsion breakdown, especially in high-fat low-protein products. It is fundamental that the cutter time is as short as possible to obtain an adequate extraction of myofibrillar proteins and correct emulsification of fat and water. At the end of the emulsification process, the batter should be homogeneous, with no visible meat or fat particles.

The batter temperature is another very important parameter to be controlled in the emulsification process, because temperature rapidly increases due to the heat generated by the high rotation of the cutter knives. It is recommended that the batter temperature does not exceed 12°C to prevent the denaturation of myofibrillar proteins, which leads to a lower emulsification capacity with consequent emulsion breakdown before or after cooking. In addition, increasing the temperature can decrease the batter viscosity, which also impairs the emulsion stability due to fat migration to the surface. To control the batter temperature, the raw material should be close to 0°C at the beginning of the process. The use of ice instead of water is also recommended to control the batter temperature and to prevent a decrease in viscosity.

16.2.5 Filling

After the emulsification process, the batter is stuffed into casings and cooked as quickly as possible to stabilize the emulsion. Plastic casings are the most used in the manufacture of Bologna sausages for several reasons, including the absence of microbiological risks, high thermal resistance, and expandability, which prevents the casing from bursting during heat exposure. In addition, the waterproof casing allows the cooking of the sausages in water. The major disadvantage is the artificial appearance, which can lead to consumer rejection.

To obtain a product of high quality and extended shelf life, the filling should be performed using vacuum fillers. These devices prevent air incorporation during filling, providing a firmer consistency, as well as contributing to color formation and conservation, and reducing the oxidative reactions that impair the sensory quality of the product during its shelf life.

16.2.6 COOKING, COOLING, AND PACKAGING

After filling, the Bologna sausages are cooked in an oven or a water bath. The most modern ovens can use either dry heat or a combination of heat and controlled relative humidity, and this flexibility brings significant technological advantages. The temperatures and times used in the cooking stage can vary according to the equipment used, as well as the diameter and the weight of the piece. The cooking temperature should not exceed 80°C to prevent emulsion breakdown (Terra et al., 2004). A cooking schedule is recommended, which should start at lower temperatures (50°C–60°C) with a gradual increase to 80°C. This approach avoids the formation of crusts in large-scale products because these affect the heat-transfer efficiency. The internal (center) temperature of the piece must reach 72°C, which corresponds to the pasteurization temperature, contributing to the increase in shelf life and product safety.

In addition to reducing the microbial load, the coagulation of myofibrillar proteins represents a major transformation during cooking. It is very important for product stability, as it leads to the formation of a three-dimensional network capable of trapping fat and water, which confers an adequate texture. During cooking, the formation and stabilization of the nitrosyl-hemochrome pigment gives the pink color characteristic of Bologna sausages.

After cooking, the Bologna sausages are cooled down with cold water. The temperature of the products cannot remain in the range between 20°C and 40°C for a long time, to prevent the development of microorganisms that may have survived the cooking stage and which can compromise shelf life and product safety. One of the great risks is the germination of spoilage microorganisms. For the cooling step, showers or water tanks can be used, and the water temperature should be close to 0°C to provide quick cooling. After reaching a temperature below 10°C, the product should be stored in a cold chamber between 0°C and 4°C. In addition, vacuum packaging and refrigerated storage (0–4°C) are recommended to increase shelf life.

16.3 PHYSICO-CHEMICAL COMPOSITION

Table 16.1 shows the formulation (percentages of meat, fat, water, and NaCl) and the physico-chemical composition (moisture, protein, and fat) of Bologna sausages produced in Brazil, Spain, Germany, Denmark, and the USA. These data were

TABLE 16.1
Formulation And Chemical Composition of Bologna Sausages Produced in Brazil, Spain, Germany, Denmark and USA

Country	Formulation (%)					Chemical composition (%)			Reference
	Pork	Beef	Pork back fat	Water	Salt	Moisture	Protein	Fat	
Brazil	42	18	20	17.03	1.25	62.8	15.9	17.1	Câmara et al. (2020)
	–	55	30	7.4	1.2	55.1	13.0	25.6	Pires et al. (2020)
	30	45	15	6.6	2	70	17.5	14.9	da Silva et al. (2020)
	63	–	20	14.3	2	64.3	13.1	18.2	Paglarini et al. (2019)
	45.8	16	13	12	2	NR	NR	NR	Júnior et al. (2019)
	–	60	17	15.4	1.2	65.8	13.6	12.5	Pires et al. (2019)
	65	–	20	11.9	2.5	65.5	15.4	22.8	Da Silva et al. (2019)
	–	55	30	7.38	1.375	54.6	13.8	24.6	Pires et al. (2017)
	65	–	20	11.0	2.5	59.4	13.5	17.0	Dos Santos Alves et al. (2016)
Spain	41.3	–	41.3	12.4	2.0	61.2	13.9	20.1	Fernández-López et al. (2020a)
	41.4	–	41.4	12.1	2.1	NR	NR	NR	Fernández-López et al. (2020b)
	45.5	–	33.4	9.5	2.5	NR	NR	26.5	Berasategi et al. (2014)
	52.5	–	33.4	9.5	2.5	NR	NR	NR	Berasategi et al. (2011)
	41.3	–	41.3	12.4	2.0	NR	NR	NR	Viuda-Martos et al. (2010a)
	41.1	–	41.1	12.3	2.1	65.6	13.1	21.6	Viuda-Martos et al. (2010b)
	55	–	30	10	NR	57.4	12.4	23.1	Cáceres et al. (2008)

(*Continued*)

TABLE 16.1 (Continued)

Country	Formulation (%)					Chemical composition (%)			Reference
	Pork	Beef	Pork back fat	Water	Salt	Moisture	Protein	Fat	
Germany	29	17	33	21	NR	61.5	12.1	23.6	Nowak et al. (2007)
	29	17	33	NR	1.8	NR	NR	NR	Riel et al. (2017)
Denmark	49.6	–	29.8	18.7	1.8	NR	NR	NR	Jongberg et al. (2013)
USA	46.08	–	30.3	19.11	1.53	56.5	12.6	26.11	Powell et al. (2019)

NR: not reported

obtained from scientific papers indexed in the Scopus Base using the terms *Bologna sausage*, *Bologna-type sausage*, and *mortadella* as search criteria. Data refer to the control treatments, which are representative of commercial samples from each country. Studies about Bologna sausages made from poultry and exotic meats were not included.

As shown in Table 16.1, in general, pork meat is the raw material primarily used for the manufacture of Bologna sausages. In some countries, like Brazil and Germany, beef meat is also used. However, the Bologna sausage produced in Spain is made exclusively with pork as a source of lean meat. These differences may be due to cultural factors, and the availability and cost of the raw materials in each country.

Pork back fat is the main fat source used in Bologna sausage formulations (Table 16.1). The concentrations of pork back fat used by the Brazilian meat industries vary from 13% to 30% and are close to 30% in Germany, Denmark, and the USA. The products made in Spain have slightly higher pork back fat content, reaching values close to 40%.

The amount of water used in Bologna sausage formulations varies according to the pork back fat content. In general, products with higher pork back fat have lower water content and vice versa. As shown in Table 16.1, most formulations have a water content between 10% and 15%, with minimum and maximum values close to 6% and 20%, respectively.

Concerning the salt concentrations, the NaCl content of most formulations is close to 2%, with values ranging from 1.2% to 2.5% (Table 16.1). Therefore, the consumption of 50 g of Bologna sausage made with 2% NaCl represents about 20% of the daily intake recommended by the World Health Organization (WHO, 2012). Thus, this product should be consumed in moderation, since studies have shown that high salt intake may be responsible for about 3 million deaths per year and 70 million disability-adjusted life-years worldwide (Afshin and Murray, 2019).

Another nutritional challenge of Bologna sausages is its high fat content. As can be seen in Table 16.1, the fat contents of Bologna sausages vary according to the countries where it is produced, but in general, it is close to 20%. Pork back fat is the main fat source used in this product, as previously discussed. The improvement in the sensory and technological quality and the reduction of production costs are the main reasons for the wide use of pork back fat. However, it has a high content (20–28%) of palmitic acid (16:0) and a high n-6/n-3 ratio (10–17). Therefore, excessive consumption of this product can increase the risk factors related to the appearance of cardiovascular diseases (Heck et al., 2021).

Despite the negative nutritional aspects of Bologna sausages, this product stands out for its high protein content. As shown in Table 16.1, Bologna sausages contained above 12% protein. In addition, many Bologna sausages can be claimed as "high in protein" according to European legislation, as more than 20% of their energy value is provided by protein (European Parliament, 2006). It is worth emphasizing that these proteins come from pork and/or beef, therefore they are of high biological value since they have all essential amino acids in proportions suitable for maintaining the life and growth of new tissues (Bohrer, 2017; Pereira and Vicente, 2013). The high concentrations of pork and/or beef used in the Bologna sausages (Table 16.1) contribute to high levels of micronutrients such as iron, selenium, zinc, and B vitamins (Biesalski 2005).

There are few studies on the physico-chemical composition of the Bologna sausages marketed in each country. Barbieri et al. (2013) evaluated the physico-chemical composition of 39 samples of Bologna sausage marketed in the north and central regions of Italy and found on average 14.6% protein, 25.79% fat, and 2.26% NaCl. These values represent a reduction of 5% fat and 10% NaCl when compared to Bologna sausages marketed in Italy 20 years ago. According to the authors, this reduction was performed to meet the needs of consumers and recommendations of nutritional and health experts.

16.4   SHELF LIFE

The shelf life of Bologna sausages is directly related to the quality of the raw materials, the manufacturing processes, and the storage conditions. In this meat product category, the lipid oxidation reactions and the microbial growth are the primary factors in determining the shelf life of the product.

Lipid oxidation is considered one of the main reactions responsible for the loss of nutritional and sensory quality, and the formation of toxic compounds. The composition and processing characteristics of Bologna sausages can accelerate these reactions, leading to a reduction of shelf life. Oxidation initiates during the manufacturing process, with the incorporation of oxygen and rupture of muscle membranes during grinding, emulsification, heat treatment, and storage. These steps cause the release of unsaturated fatty acids from phospholipids, as well as metal ions (Fe^{+2}) from myoglobin, which act as pro-oxidants of lipid oxidation (Addis, 2015).

Several factors can accelerate the lipid oxidation, including the meat raw material rich in water, proteins, and lipids used in the formulations, and the addition of pro-oxidizing ingredients such as NaCl. Other catalysts for lipid oxidation reactions such as light, moisture, pH, and lipolytic enzymes also play an important role in the oxidation process (Addis, 2015). In general, the lipid oxidation can be initiated by enzymatic reactions involving the enzymes naturally present in the product (such as lipases, esterases, and phospholipases) and/or enzymes from a microbial source, as well as the self-oxidation of unsaturated fatty acids (Alparslan et al., 2016).

All these reactions and the favorable conditions for the occurrence of lipid oxidation can lead to the formation of peroxide radicals, hydroperoxides, and volatile and non-volatile secondary compounds such as alcohols, hydrocarbons, aldehydes, and ketones, which contribute to rancid odor, changes in color, and nutrient loss. The TBARs test is the most commonly used method to monitor the development of lipid oxidation in Bologna sausages, and TBARs values of 0.5 mg MDA/kg sample are considered the detection limit for the rancid taste perceived by consumers (Gray and Pearson, 1987).

As can be seen in Table 16.2, the sensory quality and the lipid oxidation of Bologna sausages are correlated with the storage time, temperature, and type of packaging. Studies have shown that the storage time of up to 35 days at 4°C ± 1°C did not affect the shelf life of Bologna sausages, with TBARs values lower than 0.5 mg MDA/g sample, with no changes in the sensory quality of the products (Da Silva et al., 2019; De Almeida et al., 2015; Pinton et al., 2020; Jin, Choi, and Kim, 2021; Verma et al., 2018).

In contrast, other authors reported that an increase in the storage time to 42 days at 10°C ± 1°C reduced the sensory acceptance of sliced, vacuum-packed Bologna sausages (Khorsandi et al., 2019). The attributes of color, aroma, and flavor were the most affected by the increase in lipid oxidation, due to the formation of toxic compounds such as malonaldehyde and cholesterol oxides. Similarly, Viuda-Martos et al. (2010c) found low sensory acceptance of sliced Bologna packed in three types of packaging (vacuum, modified atmosphere, and airbag packaging) stored for 24 days at 4°C. The authors reported that the sliced Bologna became more susceptible to lipid oxidation, which was not affected by the types of packaging studied, with high TBARs values (>6.30 mg MDA/kg sample) observed for all types of packaging at the end of 24-day storage at 4°C.

Microbial growth can also reduce the shelf life of Bologna sausages, and the microbiological quality of the sausages is directly related to the quality of the raw materials and adequate hygiene during processing and storage. Bologna sausages are rich in carbon, nitrogen, and minerals, have water activity between 0.96 and 0.99, and pH close to 6.0, thus they are excellent substrates for the growth of spoilage and pathogenic microorganisms. Lactic acid bacteria (LAB) are the main microorganisms responsible for the deterioration of Bologna sausages. LABs affect the quality of Bologna sausages through the production of acids, undesirable development of odors, the formation of sticky exudates, and bursting (Horita et al., 2016). Some genera of gram-negative bacteria, such as *Pseudomonas*, *Acinetobacter*, and *Enterobacter* (Khorsandi et al., 2019) are also considered spoilage microorganisms in this product, as they lead to the formation of surface slime, strange aroma, and changes in color. A total viable count of 7 log CFU/g is the limit to determine the end of the shelf life of these products (ICMSF International Commission on Microbiological Specifications for Foods, 1986).

Microbial growth during the shelf life of Bologna sausages is well documented in the literature (Table 16.2). De Almeida et al. (2015) found low mesophiles and

TABLE 16.2

TBARs Values, Sensory Properties, Color Stability, and Microbiological Counts of Bologna Sausages at The End of Storage

Reference	Raw material (%)	Storage	Packing	TBARs (ug MDA/g sample)	Sensory properties	Color stability	Microbiological counts (log CFU/g)
Júnior et al. (2019)	45.83% pork, 16% loin, 13% bacon	90 days at 4°C	NR	3.1[*]	NR	Without changing L^*, a^* and b^*.	NR
Jin et al. (2021)	61.5% lean pork, 20.1% fat	28 days at 4°C	Vacuum	<0.3	No sensory change	Without changing $L^* < a^*$ and $> b^*$.	Mesophilic aerobes – <3.5 Lactic acid bacteria – <0.3
Pinton et al. (2020)	75% beef, 15% pork back fat	21 days at 4°C	NR	<0.6	No sensory change	NR	NR
Da Silva et al. (2020)	45% beef, 30% pork, 15% pork back fat	60 days at 5°C ± 1°	NR	0.181	NR	Without changing red color index (a / b^*) and color difference (ΔE)	Mesophilic aerobes – 2.78, Psychotrophic aerobes – 2.55, Lactic acid bacteria – 2.39
Da Silva et al. (2019)	65% lean pork meat, 20% pork back fat	35 days at 4°C	NR	0.49	Soft taste and aroma, ideal color and pleasant aroma.	Without changing L^*, $< a^*$ and $> b^*$	NR
Khorsandi et al. (2019)	60% lean catte meat	42 days at 10°C ±1°C	200g sliced vacuum	NR	Low acceptance for color, odor and global acceptance	Without changing L^*, a^* and b^*	Mesophilic aerobes – 8.92, Lactic acid bacteria – 8.39
Verma et al. (2018)	83.5% pork meat, 10% fat	6 days at 4°C ± 2°C	NR	1.29	NR	Without changing de L^*, a^* and b^*	Mesophilic aerobes – > 6.4, Coliforms – < 2.5Y, easts – <1.8

(Continued)

TABLE 16.2 (Continued)

Reference	Raw material (%)	Storage	Packing	TBARs (ug MDA/g sample)	Sensory properties	Color stability	Microbiological counts (log CFU/g)
De Almeida et al. (2015)	71.06% pork, 15% pork back fat	35 days at 4°C	vacuum	0.501	No sensory change	Without changing L^*, $< a^*$ and b^*	Mesophilic aerobes – 5.60, Lactic acid bacteria – 3.50
Horita et al. (2011)	50.50% lean beef, 32.35% pork	60 days at 5°C	Vacuum	0.21	NR	Without changing L^*, a^* and b^*	NR
Viuda-Martos et al. (2010b)	50% pork, 50% pork back fat	24 days at 4°C	Vacuum	6.30	Low acceptance for general quality and color intensity attributes	Without changing L^*, $< a^*$ and b^* in all types of packaging	Mesophilic aerobes and lactic acid bacteria <6 log UFC/g for vacuum and modified atmosphere packaging. Higher count of aerobic mesophiles and lactic acid bacteria in the air pouches packaging.
			Modified atmosphere (80% N_2 and 20% CO_2)	6.36	Low acceptance for general quality and color intensity attributes		
			Air pouches	7.38	Low acceptance for the attributes color intensity and homogeneity, general quality, shine and global appearance		

Notes
* Values referring to the end of the storage period for the control treatments.
NR: not reported

LAB counts in vacuum-packed Bologna sausages at the end of 35-day storage at 4°C. Similarly, Da Silva et al. (2020) evaluated the shelf life of Bologna sausages for 60 days at 5°C and reported mesophilic, psychrotrophic, and LAB counts around 2 log CFU/g at the end of storage. Some authors evaluated the microbiological quality of vacuum-packed sliced Bologna stored at 10°C for 42 days and reported that the storage temperature affected the shelf life of the products. The authors also reported the depreciation of sensory quality and aerobic mesophiles, and LAB counts higher than 8 log CFU/g, which suggests the end of the shelf life of Bologna sausages (Khorsandi et al., 2019).

REFERENCES

Addis, M. 2015. Major causes of meat spoilage and preservation techniques: A review. *Food Science and Quality Management* 41:101–114.

Afshin A., and Murray, C. J. L. 2019. Uncertainties in the GBD 2017 estimates on diet and health – Authors' reply. *The Lancet* 394:10211:1802–1803.

Allen, K., and Cornforth, D. 2010. Comparison of spice-derived antioxidants and metal chelators on fresh beef color stability. *Meat Science* 85:613–619.

Alparslan, Y., Yapıcı, H. H., Metin, C., Baygar, T., Günlü, A., and Baygar, T. 2016. Quality assessment of shrimps preserved with orange leaf essential oil incorporated gelatin. *LWT – Food Science and Technology* 72:457–466.

Alves, L. A. A. S., Lorenzo, J. M., Gonçalves, C. A. A., Dos Santos, B. A., Heck, R. T., Cichoski, A. J., and Campagnol, P. C. B. 2016. Production of healthier Bologna type sausages using pork skin and green banana flour as a fat replacers. *Meat Science* 121:73–78.

Barbieri, G., Bergamaschi M. Ge. B., and Franceschini, M. 2013. Survey of the chemical, physical, and sensory characteristics of currently produced mortadella Bologna. *Meat Science* 94:3:336–340.

Barbut, S., Sosnicki, A. A., Lonergan, S. M., Knapp, T., Ciobanu, D. C., Gatcliffe, L. J., and Wilson, E. W. 2008. Progress in reducing the pale, soft and exudative (PSE) problem in pork and poultry meat. *Meat Science* 79: 46–63.

Beggs, K. L. H., Bowers, J. A., and Brown, D. 1997. Sensory and physical characteristics of reduced-fat turkey Frankfurters with modified corn starch and water. *Journal of Food Science* 62:1240–1244.

Berasategi, I., De Ciriano, M. G.-Í., Navarro-Blasco, Í., Calvo, M. I., Cavero, R. Y., Astiasarán, I., and Ansorena, D. 2013. Reduced-fat Bologna sausages with improved lipid fraction. *Journal of the Science of Food and Agriculture* 94:4.

Berasategi, I., Legarra, S., De Ciriano, M. G.-Í., Rechecho, S., Calvo, M. I., Cavero, R. Y., Navarro-Blasco, Í., Ansorena, D., and Astiasarán, I. 2011. "High in omega-3 fatty acids" Bologna-type sausages stabilized with an aqueous-ethanol extract of *Melissa officinalis*. *Meat Science* 88(4):705–711.

Berasategi, I., Navarro-Blasco, I., Calvo, M. I., Cavero, R. Y., Astiasarán, I., and Ansorena, D. 2014. Healthy reduced-fat Bologna sausages enriched in ALA and DHA and stabilized with Melissa officinalis extract. *Meat Science* 96(3): 1185–1190.

Biesalski, H. K. 2005. Meat as a component of a healthy diet – Are there any risks or benefits if meat is avoided in the diet? *Meat Science* 70:3: 509–524.

Bohrer, B. M. 2017. Nutrient density and nutritional value of meat products and non-meat foods high in protein. *Trends in Food Science & Technology* 65:103–112.

Cáceres, E., García, M. L., and Selgas, M. D. 2008. Effect of pre-emulsified fish oil – as source of PUFA *n*-3 – On microstructure and sensory properties of *mortadella*, a Spanish Bologna-type sausage. *Meat Science* 80:2:183–193.

Câmara, A. K. F. I., Geraldi, M. V., Okuro, K. O., Júnior, M. R. M., Da Cunha, R. L., and Pollonio, M. A. R. 2020. Satiety and *in vitro* digestibility of low saturated fat Bologna sausages added of chia mucilage powder and chia mucilage-based emulsion gel. *Journal of Functional Foods* 65:103753.

Coultate, T. P. 2002. *Food: The chemistry of its component.* Cambridge:The Royal Society of Chemistry.

Da Silva, J. S., Voss, M., De Menezes, C. R., Barin, J. S., Wagner, R., Campagnol, P. C. B., and Cichoski, A. J. 2020. Is it possible to reduce the cooking time of mortadellas using ultrasound without affecting their oxidative and microbiological quality? *Meat Science* 159:107947.

Da Silva, S. L., Amaral, J. T., Ribeiro, M., Sebastião, E. E., Vargas, C., Franzen, F. L., Schneider, G., Lorenzo, J. M., Fries, L. L. M., Cichoski, A. J., and Campagnol, P. C. B. 2019. Fat replacement by oleogel rich in oleic acid and its impact on the technological, nutritional, oxidative, and sensory properties of Bologna-type sausages. *Meat Science* 149:141–148.

De Almeida, P. L., De Lima, S. N., Costa, L. L., De Oliveira, C. C., Damasceno, K. A., Dos Santos, B. A., and Campagnol, P. C. B. 2015. Effect of jabuticaba peel extract on lipid oxidation, microbial stability and sensory properties of Bologna-type sausages during refrigerated storage. *Meat Science* 110: 9–14.

Dusek, M., Kvasnicka, F., Lukásková, L., and Krátká, J. 2003. Isotachophoretic determination of added phosphate in meat products. *Meat Science* 65:765–769.

dos Santos Alves, L. A., Lorenzo, J. M.,Gonçalves, C.A.A., Dos SantosB.A., Heck, R. T., Cichoski, A.J., Campagnol, P.C.B. 2016. Production of healthier bologna type sausages using pork skin and green banana flour as a fat replacers. *Meat Science* 121: 73–78.

Eskandari, M. H., Hosseinpour, S., Mesbahi, G., and Shekarforoush, S. 2013. New composite nitrite-free and low-nitrite meat-curing systems using natural colorants. *Food Science & Nutrition* 1:5:392–401.

European Parliament. (2006). Regulation (EC) no 1924/2006 of the European Parliament and of the council of 20 December 2006 on Nutrition and Health claims made on Foods. *Official Journal of the European Union* L 404/9–L 404/25.

Feiner, G. 2006. *Meat products handbook, practical science and technology.* Cambridge, England:Woodhead Publishing Series.

Fernández-López, J., Lucas-González, R., Viuda-Martos, M., Sayas-Barberá, E., Ballester-Sánchez, J., Haros, C. M., Martínez-Mayoral, A., and Pérez-Álvarez, J. A. 2020a. Chemical and technological properties of Bologna-type sausages with added black quinoa wet-milling coproducts as binder replacer. *Food Chemistry* 310:125936.

Fernández-López, J., Lucas-González, R., Roldán-Verdú, A., Viuda-Martos, M., Sayas-Barberá, E., Ballester-Sánchez, J., Haros, C. M., and Pérez-Álvarez, J. A. 2020b. Effects of black quinoa wet-milling coproducts on the quality properties of Bologna-type sausages during cold storage. *Foods* 9:3:274.

Gray, J. I., and Pearson, A. M. 1987. Rancidity and warmed-over flavor. In *Advances in meat research, Restructured meat and poultry products*, ed.A. M. Pearson, and T. R. Dutson, 221–269. New York: Van Nostrand Reinhold Co.

Haddad, G. de B. S., Moura, A. P. R., Fontes, P. R., Cunha, S. de F. V. da, Ramos, A. de L. S., and Ramos, E. M. 2018. The effects of sodium chloride and PSE meat on re-structured cured-smoked pork loin quality: A response surface methodology study. *Meat Science* 137:191–200.

Hamm, R. 1986. Functional properties of the myofibrillar system. In *Muscle as food*, ed. P. J. Bechtel, 135–200. New York: Academic Press.

Heck, R. T., Lorenzo, J. M., Dos Santos, B. A., Cichoski, A. J., Menezes, C. R., and Campagnol, P. C. B. 2021. Microencapsulation of healthier oils: an efficient strategy to improve the lipid profile of meat products. *Current Opinion in Food Science* 40:6–12.

Horita, C., Farias-Campomanes, A., Barbosa, T., Esmerino, E., da Cruz, A. G., Bolini, H., Meireles, M. A. A., and Pollonio, M. A. R. 2016. The antimicrobial, antioxidant and sensory properties of garlic and its derivatives in Brazilian low-sodium frankfurters along shelf-life. *Food Research International* 84:1–8.

Horita, C. N., Morgano, M. A., Celeghini, R. M. S., and Pollonio, M. A. R. 2011. Physico-chemical and sensory properties of reduced-fat mortadella prepared with blends of calcium, magnesium and potassium chloride as partial substitutes for sodium chloride. *Meat Science* 89:426–433.

Hutton, T. 2002. Sodium: Technological functions of salt in the manufacturing of food and drink products. *British Food Journal* 104: 126–152.

ICMSF (International Commission on Microbiological Specifications for Foods). 1986. Micro-organisms in foods. Volume 2 "Sampling for microbiological analysis: Principles and specific applications". Canada: University of Toronto Press. ISBN: 0-802-05693-8.

Jay, J. M. I. 2005. *Microbiologia de Alimentos*. Artmed: Porto Alegre.

Jin, S.-K., Choi, J.-S., and Kim, G.-D. 2021. Effect of porcine plasma hydrolysate on phy-sicochemical, antioxidant, and antimicrobial properties of emulsion-type pork sausage during cold storage. *Meat Science* 171:108293.

Jo, K., Lee, S., Yong, H. I., Choi, Y.-S., and Jung, S. 2020. Nitrite sources for cured meat products. *LWT – Food Science and Technology* 129:109583.

Jongberg, S., Tørngren, M. A., Gunvig, A., Skibsted, L. H., and Lund, M. N. 2013. Effect of green tea or rosemary extract on protein oxidation in Bologna type sausages prepared from oxidatively stressed pork. *Meat Science* 93:3:538–546.

Júnior, M. M., Oliveira, T. P., Gonçalves, O. H., Leimann, F. V., Marques, L. L. M., Fuchs, R. H. B., Cardoso, F. A. R., and Droval, A. A. 2019. Substitution of synthetic anti-oxidant by curcumin microcrystals in mortadella formulations. *Food Chemistry* 300:125231.

Khorsandi, A., Eskandari, M. H., Aminlari, M., Shekarforoush, S. S., and Golmakani, M. T. 2019. Shelf-life extension of vacuum packed emulsion-type sausage using combination of natural antimicrobials. *Food Control* 104:139–146.

Leães, Y. S. V., Pinton, M. B., Rosa, C. T. A., Robalo, S. S., Wagner, R., De Menezes, C. R., Barin, J. S., Campagnol, P. C. B., and Cichoski, A. J. 2020. Ultrasound and basic electrolyzed water: A green approach to reduce the technological defects caused by NaCl reduction in meat emulsions. *Ultrasonics Sonochemistry* 61:104830.

Long, N. H. B. S., and Buňka, F. 2011. Use of phosphates in meat products. *African Journal of Biotechnology* 10:86:19874–19882.

Majou, D., and Cristieans, S. 2018. Mechanisms of the bactericidal effects of nitrate and nitrite in cured meats. *Meat Science* 145:273–284.

Molins, R. A. 1991. *Phosphates in food*. USA: CRC Press.

Nowak, B., Mueffling, T. V., Klein, J. G. G., and Watkinson, B.-M. 2007. Energy content, sensory properties, and microbiological shelf life of german bologna-type sausages produced with citrate or phosphate and with inulin as fat replacer. *Journal of Food Science* 72:9.

Offer, G., and Knight, P. 1988. The structural basis of water holding in meat. In *Developments in meat science*, ed. R. Lawrie, 163–171. London: Elsevier.

Paglarini, C. S., Furtado, G. F., Honório, A. R., Mokarzel, L., Vidal, V. A. S., Ribeiro, A. P. B., Cunha, L. R., and Pollonio, M. A. R. 2019. Functional emulsion gels as pork back fat replacers in Bologna sausage. *Food Structure* 20:100105.

Pellissery, A. J., Vinayamohan, P. G., Amalaradjou, M. A. R., and Venkitanarayanan, K. 2020. Spoilage bacteria and meat quality. In *Meat Quality Analysis*. 307–334. Academic Press.

Pereira, P. M. de C. C., and Vicente, A. F. dos R. B. 2013. Meat nutritional composition and nutritive role in the human diet. *Meat Science* 93:3:586–592.

Pinton, M. B., Dos Santos, B. A., Correa, L. P., Leães, Y. S. V., Cichosky, A. J., Lorenzo, J. M., Dos Santos, M., Pollonio, M. A. R., and Campagnol, P.C.B. 2020. Ultrasound and low-levels of NaCl replacers: A successful combination combination to produce low-phosphate and low-sodium meat emulsions. *Meat Science* 170:108244.

Pires, M. A., Dos Santos, I. R., Barros, J. C., and Trindade, M. A. 2019. Effect of replacing pork backfat with *Echium* oil on technological and sensory characteristics of Bologna sausages with reduced sodium content. *LWT — Food Science and Technology* 109:47–54.

Pires, M. A., Munekata, P. E. S., Baldin, J. C., Rocha, Y. J. P., Carvalho, L. T., Dos Santos, I. R., Barros, J. C., and Trindade, M. A. 2017. The effect of sodium reduction on the micro-structure, texture and sensory acceptance of Bologna sausage. *Food Structure* 14:1–7.

Pires, M. A., Rodrigues, I., Barros, J. C., Carnauba, G., De Carvalho, F. A. L., and Trindade, M. A. 2020. Partial replacement of pork fat by *Echium* oil in reduced sodium Bologna sausages: technological, nutritional and stability implications. *Journal of the Science of Food and Agriculture* 100:1:410–420.

Powell, M. J., Sebranek, J. G., Prusa, K. J., and Tarté, R. 2019. Evaluation of citrus fiber as a natural replacer of sodium phosphate in alternatively-cured all-pork Bologna sausage. *Meat Science* 157:107883.

Puolanne, E., and Halonen, M. 2010. Theoretical aspects of water-holding in meat. *Meat Science* 86:151–165.

Riel, G., Boulaaba, A., Popp, J., and Klein, G. 2017. Effects of parsley extract powder as an alternative for the direct addition of sodium nitrite in the production of mortadella-type sausages – Impact on microbiological, physicochemical and sensory aspects. *Meat Science* 131:166–175.

Ritz, E., Hahn, K., Ketteler, M., Kuhlmann, M. K., and Mann, J. 2012. Phosphate additives in food – A health risk. *Deutsches Ärzteblatt International* 109:4:49–55.

Sebranek, J. G., and Bacus, J. N. 2007. Cured meat products without direct addition of nitrate or nitrite: What are the issues? Meat Science 77:1:136–147.

Terra, N. N., Terra, A. B. M., and Terra, L.M. 2004. *Defeitos nos produtos cárneos: origens e soluções*. Portugal: Varela.

Thomas, C., Mercier, F., Tournayre, P., Martin, J. L., and Berdagué, J. L. 2013. Effect of nitrite on the odourant volatile fraction of cooked ham. *Food Chemistry* 139:432–438.

Verma, A. K., Chatli, M. K., Mehta, N., and Kumar, P. 2018. Assessment of physico-chemical, antioxidant and antimicrobial activity of porcine blood protein hydrolysate in pork emulsion stored under aerobic packaging condition at 4 ± 1°C. *LWT – Food Science and Technology* 88:71–79.

Villaverde, A., Ventanas, J., and Estévez, M. 2014. Nitrite promotes protein carbonylation and Strecker aldehyde formation in experimental fermented sausages: Are both events connected? *Meat Science* 98:4:665–672.

Vissa, A., and Cornforth, D. 2006. Comparison of milk mineral, sodium tripolyphosphate, and vitamin E as antioxidants in ground beef in 80% oxygen modified atmosphere packaging. *Food Chemistry and Toxicology* 71:2:66–68.

Viuda-Martos, M., Ruiz-Navajas, Y., Fernández-López, J., and Pérez-Álvarez, J. A. 2010a. Effect of orange dietary fibre, oregano essential oil and packaging conditions on shelf-life of Bologna sausages. *Food Control* 21:4:436–443.

Viuda-Martos, M., Ruiz-Navajas, Y., Fernández-López, J., and Pérez-Álvarez, J. A. 2010b. Effect of adding citrus fibre washing water and rosemary essential oil on the quality characteristics of a Bologna sausage. *LWT – Food Science and Technology* 43(6):958–963.

Viuda-Martos, M., Ruiz-Navajas, Y., Fernández-López, J., and Pérez-Álvarez, J. A. 2010c. Effect of added citrus fibre and spice essential oils on quality characteristics and shelf-life of mortadella. *Meat Science* 85:568–576.

WHO. 2012. Guideline: Sodium intake for adults and children. Geneva, Switzerland: Department of Nutrition for Health and Development, World Health Organization.

17 Morcilla and Butifarra Sausage
Manufacturing Process, Chemical Composition and Shelf Life

Eva María Santos López,[1] Ana María Díez Maté,[2] Isabel Jaime Moreno,[1] Magdalena Isabel Cerón Guevara,[1] Javier Castro Rosas,[1] Beatriz Melero Gil,[2] and Jordi Rovira Carballido[2]

[1]Universidad Autónoma del Estado De Hidalgo, Área Académica de Química, Crta. Pachuca-Tulancingo Km 4.5 s/n, Col. Carboneras, Mineral de la Reforma, 42183, Hidalgo, Mexico.
[2]Department of Biotechnology and Food Science, University of Burgos, Pza Misael Bañuelos s/n, 09001 Burgos, Spain

CONTENTS

17.1   INTRODUCTION

Morcilla and butifarra sausages are very well-known sausages produced throughout Spain, although they are not as popular internationally as other Spanish meat products, like Serrano ham or chorizo, which are protected under several food quality labels like protected geographical indication (PGI) or protected designation of origin (PDO). Morcilla and black butifarra are cooked sausages where blood, mainly from swine or cattle, is the characteristic ingredient. Morcillas are generally consumed fried or roasted, as a tapa or an ingredient in some typical stews like *fabada asturiana*, while cooked butifarra is more frequently consumed in thin slices, although it can be also used as an ingredient in other more elaborate dishes.

Traditionally, production has been intimately linked to the slaughter of home-reared pigs in rural areas during the winter months, although nowadays, it is produced during the entire year. These kinds of products are mainly produced by artisans and small producers using traditional recipes for local distribution and consumption. However, the European Union considers the promotion of local high-quality products, especially in the agrifood sector, as a successful policy to fight poverty and to support the inhabitants of rural areas (European Commission, 2008). In this sense, traditional and ethnic foods have gained relevance and a greater visibility in the last decades, becoming gourmet products with a strong identity. So the production of these blood sausages has moved from household and small craft manufacturers to industrial plants. This trend has generated the adaptation of the recipes to industrial production, including the careful selection of the basic raw materials and additives. In addition, these products use meat by-products, such as blood, trimmings and offal, which has the advantage of making them sustainable.

Morcilla is the Spanish name for blood sausages. However, different blood sausages can be found around the world, such as the English black pudding (Adesiyun and Balbirsingh, 1996), *cavournas* in Greece (Arvanitoyannis et al., 2000), *blutwurst* in Germany, *kaszanka* in Poland, *biroldo* in the San Francisco Bay area (USA), *krvavica* in Slovenia (Gasperlin et al., 2012), *morcella de Assar* and *morcella de arroz* in Portugal (Roseiro et al., 1998; Pereira et al., 2015), and *sanganel* in the northeastern region of Italy (Iacumin et al., 2017). Also, blood sausages are commonly found in Latin America, resembling the Spanish and Portuguese homologues, since they are local adaptations from traditional recipes from the Iberian Peninsula. So morcilla can be found in Argentina (Oteiza et al., 2003) and Ecuador, and in Mexico where it is called *moronga* or *rellena* (Domínguez and Lorenzo, 2020).

Due to the use of blood, color of this product varies from light or dark brown to black, depending also on the type of natural casing used (small intestine from cattle or large intestine from swine) (Figure 17.1).

In the case of butifarra sausage, which is mainly produced in the Catalonia region, Valencia and the Balearic Islands, two main varieties are considered, fresh and cooked butifarra with different variants according to the ingredients used. The fresh butifarra is ideal fried, roasted, or cooked in a stew, and no blood is used in its formulation. It is made of ground pork meat with pork back fat or jowl fat mixed with few spices like white pepper and stuffed in natural casing. In general, fresh butifarra resembles *saucisse* from Toulouse in composition. There are also the

FIGURE 17.1 Some examples of morcillas.

cooked butifarras, where different varieties can be distinguished according to the ingredients and region of elaboration. Mainly, two classes of cooked butifarra can be established: white butifarra and black butifarra. This last one receives this name because of the dark color, since blood is included in its elaboration (Martín Bejarano, 2001). Several varieties are well known, like butifarra from Figueras, llengua from campo de Tarragona or danegal from la Segarra. In the Balearic Islands, *camaiots* and *xarn i xua* are widely recognized (MAPA, 2020).

As with morcilla, this product was also regionalized in some Latin American countries, especially in the northern part of South America, like Colombia, and even in some regions of Mexico (Gonzalez-Schnake and Nova, 2014).

17.2 INGREDIENTS

The enormous richness and variety of these products relies not only on the great diversity of ingredients included, but also in the proportions used. So even when there are common ingredients in the product belonging to a specific region (morcilla from Burgos, morcilla from Aragon, etc.) the recipes could vary from one producer to another, giving each a unique flavor.

The main ingredients in the elaboration of these products are listed below.

Blood: The main characteristic of morcillas and black butifarra is the use of pork or beef blood with lard or even tallow. The use of whole animal blood in the elaboration of blood sausages, considered a by-product of the slaughtering process, reduces the need to dispose of blood, which has serious environmental implications (Ofori and Hsieh, 2014). The blood also enriches the protein, and especially the iron content, of the blood sausages, considering that heme iron from blood is better absorbed than non-heme iron (Ofori and Hsieh, 2014).

Fat: Apart from blood, another important ingredient both in morcilla and butifarra is the fat, mainly from lard, although tallow is used in some morcillas, and also pork back fat. Even pork head or jowl fat are used in some specialties.

Starch and vegetable sources: Some cereals or vegetables are added to get the typical consistency of these products, frequently rice and onion. Usually, two classes of morcillas are considered based on the main ingredient used, rice morcillas and onion morcillas, even though its production is so extended throughout Spain, that morcillas generally receive the name of the region or place where they are elaborated. In the elaboration of morcilla from Burgos (PGI) the use of the Horcal onion is mandatory. Horcal is a regional variety of onion, grown on the riversides near Burgos. This onion has an elliptical shape and bigger size compared to others kinds of onions. Its less firm texture and mild pungent flavor gives morcilla a remarkable sensory quality. However, the main limitation of this crop is its seasonality, necessitating careful storage. In the case of morcilla from Burgos, the high amount of Horcal onion used in the product affects the pH, total sugar content and dietary fiber parameters to a greater extent (Santos et al., 2003), influencing also the composition and intensity of some volatile compounds. Sulfur compounds from onions give morcilla from Burgos its peculiar smell and piquant flavor, totally different from morcillas elaborated with rice (specifications proposal from PGI morcilla de Burgos) (ITACYL, 2020). So, apart from morcilla from Burgos and León, morcilla from Asturias, Galicia, the Canary Islands, Aragón, and the Basque region are also recognized. Depending on the availability in each zone, leek, pumpkin, figs, sweet potato, apples, raisins, and even nuts, pine seeds and almonds are also included in some varieties. Sometimes breadcrumbs substitute or accompany the rice.

Spices: It is in the use of spices that the unique sensory profiles of each type of sausage are triggered, the most common being black pepper, paprika, oregano, cumin, clover, anise, and garlic. Some morcillas are more aromatic than others because of the use of different spices. Even if salt is usually added in morcilla, some morcillas have a sweet note because of the addition of sugar and cinnamon.

Different ingredients used in morcillas, according to the origin or production region, are listed in Table 17.1. Similar ingredients to morcilla de Burgos are used in the morcilla from Valladolid, although less onion and more paprika are used (Martín Bejarano, 2001). Varieties with pine seeds are typically made in areas with extensive pinewoods, and sometimes pumpkin or raisins are also added to the mix (Andrés, 2018). However, morcilla from Leon is basically onion and blood with breadcrumbs and some spices at lower concentration. The morcilla from Granada, *Gallega* and *dulce Canaria* stand out for their high blood content. *Morcilla gallega* and *Canaria* from Aragon also attract attention because of the sweet taste due to sugar in the formulation.

TABLE 17.1

Ingredients Used in Morcillas According to the Origin or Production Region

	Ingredients	Minority Ingredients and Spices Commonly Used
Morcilla de Burgos (ITACYL, 2020)	Horcal onion > 35%, rice (raw or precooked) 15–30%, lard or tallow 10–22%, blood > 12%, salt	Black pepper, paprika, cumin, oregano
Morcilla de Valladolid (Martín Bejarano, 2001)	Onion 20%, rice (raw or precooked) 41%, lard 29%, blood 10%, salt	Paprika, white pepper, oregano, cloves, (pine seeds in a typical variety)
Morcilla from León (Cabeza et al. (2004)	Onion 65–75%, lard and/or tallow 10–20%, blood 10–20%, rice/bread 2–5%, salt	Garlic, paprika
Morcilla asturiana (Martín Bejarano, 2001)	Onion (raw or precooked) 40%, pork back fat 50%, blood 10%, salt	Sweet and spicy paprika, oregano
Morcilla from Basque Country (Diaz Yubero, 2016; Martín Bejarano, 2001)	Onion (precooked) 25%, pork head and lard 15%, blood 25%, rice, leeks, pumpkin 35%, salt	Spicy paprika, black pepper, cinnamon, pine seeds and almonds (some recipes)
Morcilla de cebolla from Valencia (Martín Bejarano, 2001)	Cooked onion 60%, lard or pork back fat 30%, blood 10%, salt	Black or white pepper, oregano, cloves, anise
Morcilla de Picaro, Murcia (Martín Bejarano, 2001)	Cooked onion 45%, lard 35%, blood 20%, salt	Sweet paprika, oregano, cloves, pine seeds,
Morcilla andaluza" (Martín Bejarano, 2001)	Lard/back fat/pork jowl 60%, blood 40%, salt	Sweet paprika, white pepper, cloves, cumin, oregano, garlic
Morcilla gallega (Martín Bejarano, 2001)	Onion 2%, lard 16%, blood 30%, bread crumbs 15%, figs, nuts or raisins 20%, sugar 15%	Depending on the region, potatoes or apples can be added
Morcilla de Aragón	Onion, rice (raw or precooked) lard, blood, salt/sugar	Black pepper, anise, cinnamon, cloves, bread crumbs, pine seeds
Morcilla dulce canaria (Martín Bejarano, 2001)	Lean pork 20%, lard 12%, blood 55%, sweet potato 13%, sugar	Cinnamon, thyme, oregano, bread crumbs, raisins or almonds

The high variability in the ingredients, as well as the proportions, have made it difficult for these products to apply for legal protection labels like PDO or PGI. Morcilla from Burgos could be considered the referent of blood sausages since it was the first Spanish blood sausage to obtain a PGI, and more scientific information about it has been published (Santos et al., 2003, 2005a, 2005b; Diez et al., 2008a,

FIGURE 17.2   PGI logo for morcilla from Burgos.

2008b, 2009a, 2009b, 2009c). Morcilla from Burgos, produced in region of Burgos (in the north of Spain), is the only Spanish morcilla that has received a PGI, on August 29, 2018, in class 1.2, Meat products (European Union, 2018) (Figure 17.2).

Morcilla from Aragon is also under the quality label Aragon Calidad Alimentaria, which forces the producers to keep certain standards concerning the ingredients and the final physico-chemical characteristics (BOA, 1985).

In the case of butifarra, composition of white cooked butifarras is similar to fresh butifarra with local differences according to the traditional customs and procedures. Formulation comprises the use of meat, pork back fat and/or jowl, pork head, sometimes pork rind previously cooked, mixed with salt, some additives like ascorbate, caseinate or potato starch, dextrose, and ground white pepper (Martín Bejarano, 2001; Magrinyà et al., 2012). Other spices usually used are clove, nutmeg, occasionally cinnamon, pine seeds, and white wine. There is a typical butifarra in Lerida called butifarra trufada because it includes black truffles between 5 and 10 g/kg (Martín Bejarano, 2001). In the butifarra from Perol (Gerona), offal like heart, lungs, stomach and tongue from pork are finely comminuted with pork head and rinds accounting of 75% of the formulation with 25% pork jowl. Some specialties include from 2 to 5 beaten raw eggs per kg of formulation, like *butifarra blanquet* from Alicante or *butifarra de huevos* in Girona (Martín Bejarano, 2001). The black butifarra is the one that includes blood in its composition and the main ingredients are similar to the white butifarra (pork loin, back fat, in some regions lungs, pork rinds, pork jowl or head) but in general ground white pepper is substituted for ground black pepper combined with other spices like cinnamon, clove or paprika (Martín Bejarano, 2001).

17.3   MANUFACTURING PROCESS

The elaboration processes of morcilla and butifarra could be considered simple and similar. The main production steps include: weighting; mincing or grinding of some ingredients like onion, vegetables, fat (mainly lard, sometimes mixed with tallow), meat, trimmings, and offal; mixing with rice (sometime precooked), blood, additives and spices; stuffing in natural casings of beef or pork (around 35–45 mm in diameter); and a cooking process in hot water to give the desired characteristics (Figure 17.3).

Grinding: The operations that come before the mixing step can be very specific for different products. In general, onions have to be peeled and chopped before

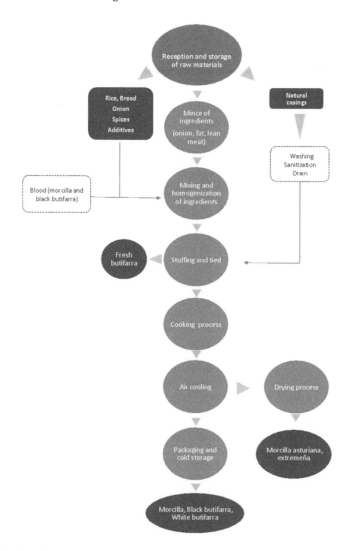

FIGURE 17.3   General scheme for elaboration process of morcilla and butifarra.

being added in the formulation, and in some cases, like morcilla from the Basque region, morcilla de cebolla from Valencia or morcilla from Pícaro, the onion is previously cooked. Meat, trimmings, offal and pork skin are also minced or ground before being added to morcilla or butifarra.

Mixing: The rice can be added raw to the mix but some recipes use precooked rice. Fat can be used in some formulations to stir-fry some vegetable components. The mixing process can be done at once or in several steps. In general, spices are previously mixed to be added to the mix, and blood is usually incorporated at the end to obtain a homogeneous mixture.

Stuffing: The mixture is generally stuffed into natural casings from swine and beef, although artificial collagen casings can be also employed (Gómez-Rojo et al., 2011). Then sausages are tied with string or thread, or directly stapled as single units, or joined by a string to facilitate the cooking process. The casings should be filled to around 70% of capacity in the case of morcillas, especially when raw materials are included, to avoid bursting of the sausages during the cooking process because of the swelling of rice, for example, or hydration of some components (Ramos et al., 2013).

Cooking: The thermal treatment is always done by soaking the product in boiling water and keeping the temperature between 75°C and 95°C, during enough time to properly cook the rice and other ingredients, and to assure the safety of the product. In general, 65–75°C is the temperature reached in the inner part (Ramos et al., 2013). When no rice is included in the formulation, the cooking process is usually shorter, around 10–20 minutes in boiling water. Some producers have developed special metal baskets to handle the complete batch instead of using strings to join all the products.

Air cooling: After cooking, morcillas and butifarras are air cooled horizontally to reach 8–10°C and stored chilled at 4°C. In some regions (morcilla asturiana, morcilla extremeña) the product is dried for 1 or 2 weeks.

Packaging: The product used to be sold unpackaged, although nowadays packaging is becoming more frequent. In the case of morcilla asturiana, many of the producers are grouped under a quality label to assure the traditional smoking procedure with precious woods (oak or chestnut), a differentiating element from other morcillas, which confers a characteristic flavor (Figure 17.4).

17.4   CHEMICAL COMPOSITION

When so many ingredients are used to elaborate these sausages, it is likely to find differences in the chemical composition. In general, morcilla and butifarra share a high humidity content, which ranges over 60% (Santos et al., 2003; Magrinyà et al., 2012; Ramos et al., 2013). Also, the pH of these products is usually around neutral, since no fermentation process is carried out (Santos et al., 2003; Ruiz-Capillas and Jiménez-Colmenero, 2004). These values are in accordance with those reported by Gašperlin et al. (2014) and Iacumin et al. (2017) for the typical blood sausages in Slovenia or Sanganel in Italia, respectively. On the contrary, Cachaldora et al. (2013) reported a pH lower than 5 and a water activity below 0.87 for blood sausage, but this product did not included rice in the formulation and a ripening process of around 21 days was applied. In Table 17.2, the chemical composition of two morcillas and a butifarra are shown.

In general, these cooked products are also characterized by a high content of fat, due to the lard, or back fat used. The use of rice, onion or other vegetables in these products improves the carbohydrate intake. The onion, an important ingredient in most of morcillas contributes to increase sugar and fiber contents (Santos et al., 2003), as well as other vegetable ingredients like pumpkin, leeks, etc. For example, the high percentage (65–75%) of onion used in the formulation of morcilla from Leon is reflected in a higher fiber content compared with morcilla de Burgos (Table 17.2). As

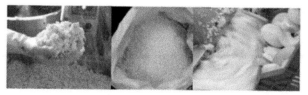

Conditioning of raw materials   Mixing step

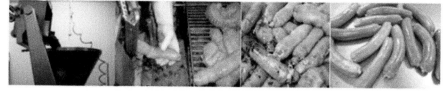

Stuffing in natural casings and tied

Cooking process   Air cooling   Packaging

FIGURE 17.4 Elaboration process of morcilla from Burgos.

TABLE 17.2

Physico-chemical Parameters and Composition (Expressed on a Dry Matter Basis) of Morcilla from Burgos, Morcilla from Leon and Butifarra

	Morcilla from Burgos (Santos et al., 2003)	Morcilla from Leon (Cabeza et al., 2004)	Butifarra (Magrinyà et al., 2012)
pH	6.39 (0.35)[a]	6.1 (0.4)	5.88 (0.05)
a_w	0.984 (0.004)	0.972 (0.003)	
Moisture	62.21 (4.05)	66.2 (4.8)	68.61 (0.08)
Fat	28.65 (4.92)	42.1 (6.6)	32.17
Protein	13.09 (2.30)	17. 1 (2.8)	
Starch	46.51 (5.89)	27.0 (6.1)	
Total sugar	4.61 (1.60)		
Ash	4.26 (0.49)	6.6 (1.4)	
Fiber	1.71 (0.68)	7.2	

Notes
[a] Mean (Standard Deviation)

happens with other meat products, the presence of meat and blood contributes to high biological protein, but the most important component supplied by blood is the iron content. Ramos et al. (2013) reported 33.24 ± 10.96 mg/100 g of Fe in dry basis in morcilla from Leon while in morcilla from Burgos 18.36 ± 5.78 mg/100 g was reported (Santos et al., 2003). Regardless, of the amount of Fe provided by morcillas or butifarra, the most interesting aspect is that Fe is provided as heme-iron, highly assimilable. Therefore, morcilla and black butifarra consumption could be considered a good strategy in treating iron deficiency anemia.

17.5   PRESERVATION METHODS APPLIED TO MORCILLA AND BUTIFARRA, AND SHELF LIFE

Food preservation by means of more "natural" methods has become an important opportunity for food industries to satisfy needs of and demand from consumers. Following this tendency, current research trends on microbiology and food technology are focused on soft physical preservation techniques and the use of natural antimicrobial compounds. It is clear that natural alternatives are not always as powerful as specific chemicals. So the intelligent use of combined processes may be the solution for obtaining optimal results.

As with many other cooked meat products, the shelf life of blood sausages is limited, affecting its commercial or market distribution. In general, the high water content, high water activity ($a_w > 0.970$), pH over 6.0 and low salt content (around 2%) mean these products are highly perishable and susceptible to microbial growth, making cold storage essential (Pereira et al., 2015).

Although these products used to be traditionally commercialized as fresh products, vacuum packaging (VP), modified atmosphere packaging (MAP) and post-packaging pasteurization were introduced in the 90s to extend shelf life and to increase the potential market and satisfy consumer demands for more natural products (Santos et al., 2005a). Recently, natural antimicrobial preservatives or high hydrostatic pressure (HHP) have been applied to increase shelf life by reducing microbial spoilage.

17.5.1   UNPACKAGED PRODUCT

Traditionally these products have been locally commercialized unpackaged. Although it is the most environmentally sustainable option, it is also the least safe microbiologically, since cross contamination is likely to occur after the cooking process. After the thermal processing of these cooked products, microbial vegetative cells have been eliminated and only thermo-resistant microorganisms survive, such as spore-forming microorganisms (*Bacillus* spp. and *Clostridium* spp.). However, the air-cooling process as well as the handling prior to packaging leads to post-cooking microbial contamination, where a mixed microflora is developed. Lactic acid bacteria (LAB), pseudomonads, enterobacteria, spore-forming bacteria, molds and yeasts are frequently found in these products (Santos et al., 2005a; Pereira et al., 2015). When morcilla is commercialized as a fresh product, the estimated shelf life is around two weeks until molds or yeasts develop on the surface

(Santos et al., 2005a). A significant growth of different microbial groups (pseudomonads, LAB, enterobacteria, molds and yeasts) was observed, with *Pseudomonas* spp. the predominant group, reaching counts over 7 Log CFU/g after 21 days of storage and a shelf life of 17 days. Moreover, the external desiccation of the product provokes an undesirable appearance, unless the morcillas are dried (like morcilla asturiana and occasionally fresh butifarra), where the water content is lower and the shelf life can be longer.

17.5.2 VACUUM AND MODIFIED ATMOSPHERE PACKAGING

In general, the use of packaging technology can extend the shelf life of meat products by improving safety and stability along the shelf life, being cost-effective and easy to apply, especially when only vacuum packaging (VP) is applied. When the atmosphere within the package is modified by lowering the oxygen concentration and increasing the content of carbon dioxide and/or nitrogen, the shelf life of perishable food products at chill temperatures can be significantly improved (Parry, 1993). This is the reason why VP and modified atmosphere packaging (MAP), along with refrigeration, have become increasingly popular preservation techniques to extend the shelf life of meat and meat products, allowing major changes in storage, distribution, and marketing of raw and processed products (Özogul et al., 2004).

In the case of VP or MAP morcillas and butifarras, the shelf life will depend mainly on the microbial load on the surface of the product, especially the content and composition of lactic acid bacteria. The absence of oxygen and cold temperatures promote the growth of lactic acid bacteria, causing slime and blown packaging, apart from a sour aroma and flavor, which determines the end of the shelf life. Also, the VP could provoke product deformation (Santos et al., 2005a). This phenomenon is not exclusive to morcilla or butifarra but is also seen in many cooked meat products and blood sausages (Korkeala and Björkroth, 1997; Pereira et al., 2015; Iacumin et al., 2017). Blown packaging has been related to the development of some species of *Clostridium* (Wambui and Stephan, 2019; Zhang et al., 2020). When the effect of VP polyamide/polyethylene and MAP (30%, 50%, and 80% of CO_2 complemented with nitrogen) in morcilla from Burgos were studied, LAB were the predominant microbial group decreasing the pH of the product below 5.0, while other microorganisms (enterobacteria, psedomonads, molds and yeast), remained during the cold storage (Santos et al., 2005a). Higher concentrations of CO_2 were more effective against pseudomonads and enterobacteria, increasing the shelf life of this product to 32 days compared with the 22 days reached by vacuum and 30% CO_2 packaged morcillas (Santos et al., 2005a).

Pereira et al. (2015, 2019) reported that MAP (80% CO_2/20% N_2) was the best atmosphere to increase shelf life in morcella de arroz at 4±1°C, reaching 44 days of storage. However, unpackaged samples presented a shelf life of 11.6 days and VP of 27.8 days. According to the results, LAB were the dominant spoilage microbiota in VP samples, just as happens in other meat products stored under similar conditions (Korkeala and Björkroth, 1997). *Pseudomonas* spp. were present in samples without packaging. However, MAP (80% CO_2/20% N_2) had a positive influence on the reduction of LAB, enterobacteria and *Pseudomonas* spp. The authors also attributed the microbial counts observed immediately after processing to post-cooking contamination.

The works of Ruiz-Capillas and Jiménez-Colmenero (2004) and Cepero et al. (2006) showed counts of LAB in the range 3.5–5 Log CFU/g in butifarra VP. When VP was applied to butifarra studied by Cepero et al. (2006), shelf life was around 21 days, almost double compared to fresh product and similar to the results obtained in morcilla from Burgos. However, when Cachaldora et al. (2013) studied the effect of different MAP ($15{:}35{:}50/O_2{:}N_2{:}CO_2$; $60{:}40/N_2{:}CO_2$ and $40{:}60/N_2{:}CO_2$) and VP on the shelf-life of morcilla from the north of Spain without rice, no significant differences were observed among packaging conditions. Moreover, the shelf life of this product was greater than 8 weeks under all packaging conditions, even though initial counts of pseudomonads, lactic acid bacteria, and mold and yeast were higher than those reported by Santos et al. (2005a). In this work, a significant decrease was observed for LAB and pseudomonads, while mold and yeast counts remained constant. However, it is necessary to point out that the blood sausage reported on by Cachaldora et al. (2013) had been subjected to a ripening process, which could have promoted a natural fermentation process contributing to the preservation of the sausages. In order to increase the shelf life of these refrigerated vacuum packed products, the combination of the following techniques has been tested and explained.

17.5.3 POST-PACKAGING PASTEURIZATION

Pasteurization is an economical, mild heat treatment that involves the destruction of organisms in a vegetative state, responsible for diseases or alterations in foods. In the case of morcilla de Burgos, reaching that extended shelf life, over 8 weeks, was only possible after a post-packaging thermal treatment at 95°C for 5 minutes. When a thermal treatment of 75°C for 10 minutes was applied, the shelf life was 40 days, almost double that of vacuum packaged items (Santos et al., 2007). Results in butifarra after a post-packaging thermal treatment at 80–85°C for 15 minutes were even better, increasing the shelf life from 21 days to 138 days (Santos et al., 2014). Post-packaging pasteurization has been successfully applied in cooked meat products like ham or frankfurters. However, the specific properties of the natural casing can provide a kind of thermal protection and surviving LAB could develop with time.

17.5.4 HIGH HYDROSTATIC PRESSURE

According to the current trend of using mild technologies to obtain foods with high nutritional and sensory qualities, microbiological safety and a long shelf life, high dydrostatic pressure (HHP) has received special attention due to its versatility in application.

In vacuum packaged morcilla from Burgos treated by HHP (300, 500 and 600 MPa for 10 minutes), the populations of enterobacteria and *Pseudomonas* spp. were reduced below the detection limits (1 and 2 Log CFU/g, respectively) with all treatments applied (Diez et al., 2008a). However, gram-positive bacteria such LAB, were only slightly reduced (between 1 and 2 Log CFU/g), being the treatment 600 MPa-10 minutes, the most effective in extending the shelf life up to 35 days. HHP also affected texture and taste, although the changes were not considered negative in the sensory evaluation (Diez et al., 2008a).

Respect to the butifarra treated with HHP, this technology had no effect either on the total count, or on lactic acid bacteria counts (Ruiz-Capillas and Jiménez-Colmenero, 2004). Despite the fact that HHP causes microorganism inactivation, the antimicrobial effect depends on factors associated with the type of microorganism, the nature of the product (composition, pH, etc.) and the processing parameters (pressure level, treatment time, temperature) (Cheftel, 1995; Lopéz-Caballero, Carballo, and Jiménez-Colmenero 1999; Hugas, Garriga, and Monfort 2002).

17.5.5 Natural Organic Acids and Salts

Organic acids and their salts (OAS) are also considered natural preservatives, and they are preferred to satisfy consumer demand for more natural foods. Sodium lactate (E-325) has been used for several years in the meat industry because of its ability to increase flavor, shelf life and the microbiological safety of these products. The antimicrobial effect of lactates is based on their ability to lower water activity and to the inhibitory effect of a dissociated molecule, the lactate ion (Koos, 1992; Houtsma et al., 1993). However, one of the problems of the use of sodium lactate is the salty taste, when used in concentrations higher than 3%. This problem may be overcome by using other OAS such as potassium lactate (E-326) and adjusting the salt content in the product (Wilmink, 2000). Others studies have shown the synergistic effect of the combined use of lactate and diacetate (E-262) against several foodborne pathogens (Glass et al., 2002; Mbandi and Shelef, 2002; Barmpalia et al., 2005). The addition of 3% sodium lactate/potassium lactate to vacuum packaged morcilla from Burgos was slightly more effective against lactic acid bacteria, giving a shelf life of 28 days, while the combination of 2.50% potassium lactate/sodium diacetate showed a greater antimicrobial effect against pseudomonads (Diez et al., 2008a). When acid organic salts were combined with the application of HHP, no improvement in shelf life was observed (Diez et al., 2009a).

17.6 SPOILAGE LACTIC ACID BACTERIA ASSOCIATED WITH MORCILLA

In general, considering that plants and vegetables are natural habitats for many LAB, it is likely that spices and dried vegetables could be a source of LAB contamination in blood sausage production (Säde et al., 2016). When spoilage of morcilla from Burgos has been studied, *Weissella* and *Leuconostoc* genera have been present since the onset of spoilage (Diez et al., 2008b). *Weissella viridescens* is frequently found at the beginning, while *Leuconostoc mesenteroides* is the species isolated during the whole storage. Despite these two species being intimately and synergistically associated in the spoilage of morcilla from Burgos, both play different roles. When the morcilla has been inoculated with these two species, *L. mesenteroides* has been discovered to grow faster than *W. viridescens*, causing faster spoilage by the decrease in pH, production of milky exudate and acid odor, while *W. viridescens* participated more in the vacuum loss (Diez et al., 2009b).

Related to the volatile compounds produced during the spoilage, *L. mesenteroides* is more connected to the presence of acetic and hexanal acids, while *W. viridescens* to ethanol, acetoin and diacetyl production in morcillas from Burgos inoculated with these species (Diez et al., 2009b). When the behavior of both bacteria was evaluated under the application of OAS, HHP and post packaging pasteurization, spoilage characteristics remained the same, but the intensity was lower and the date of spoilage appearance was delayed. When a HHP treatment was applied, *L. mesenteroides* needed more time to recover than *W. viridescens* (Diez, Björkroth, Jaime, and Rovira, 2009b, 2009c). It seemed that the stress produced by high pressure was reversible; some cells were injured by the treatment but not inactivated, and during the subsequent preservation, the damaged cells recovered and the survivors multiplied (Cheftel, 1995).

Leuconostoc mesenteroides was also the dominant species in vacuum packaged morcilla without rice studied by Chenoll et al. (2007), although species of *L. sakei* and *Enterococcus* spp. were also identified during the storage.

17.7 CONCLUSION

Morcilla and butifarra fall into the current trend that demands more sustainable foods, since their production is based on an economical use of by-products. Despite that, there is no doubt about the importance of production of morcilla and butifarra in the maintenance of the customs and food traditions of various Spanish regions, as well as the economic support of certain local areas, although these products still remain little known to the scientific community. Producers need products with long shelf lives to extend markets and increase production. However, the shelf life will heavily depend on the initial microflora deposited on the product after the cooking treatment and the preservation technologies available. In this sense, more scientific studies about the characteristics of the products, as well as the behavior of the physico-chemical, microbiological and sensory properties during storage, are necessary to boost the development and international trade of these products.

REFERENCES

Adesiyun, A. A., and Balbirsingh, V. 1996. Microbiological analysis of 'black pudding', a Trinidadian delicacy and health risk to consumers. *International Journal of Food Microbiology* 31(1–3): 283–299.

Andrés, D. 2018. La sangre de la tierra. Diario de Valladolid 15/10/2018. http://www.diariodevalladolid.es/noticias/laposada/sangre-tierra_131710.html consulted 23rd January 2020.

Arvanitoyannis, I. S., Bloukas, J. G., Pappa, I., and Psomiadou, E. 2000. Multivariate data analysis of Cavourmas – A Greek cooked meat product. *Meat Science* 54: 71–75.

B.O.A. 1995. Orden de 5 de Julio de 1995, del Departamento de Agricultura, Ganadería y Montes, por la que se aprueba el reglamento de utilización de la marca -Aragón Calidad Alimentaria-para el embutido – "Morcilla" de Aragón. *Boletín oficial de Aragón*, 14 de Julio de 1995.

Barmpalia, I. M., Koutsoumanis, K. P., Geornaras, I., Belk, K. E., Scanga, J. A., Kendall, P. A., Smith, G. C., and Sofos, J. N. 2005. Effect of antimicrobials as ingredients of pork bologna for *Listeria monocytogenes* control during storage at 4 or 10°C. *Food Microbiology* 22: 205–211.

Cabeza, E., Zumalacárregui, J. M., Antiduelo, A., and Mateo, J. 2004. pH, water activity, and proximate composition of "morcilla" de León, a traditional European blood sausage. In *Proceedings 50th International Congress of Meat Science and Technology*: 71–73, Helsinki, Finland.

Cachaldora, A., García, G., Lorenzo, J. M., and García-Fontán, M. C. 2013. Effect of modified atmosphere and vacuum packaging on some quality characteristics and the shelf-life of "Morcilla", a typical cooked blood sausage. *Meat Science* 93(2): 220–225.

Cepero, Y., Beldarraín, T., and Campos A. 2006. *Uso de la tecnología de barreras en la conservación de una butifarra fresca* 16(2): 7–13. Instituto de Investigaciones para la Industria Alimentaria. ISSN: 0864-4497.

Cheftel, J. C. 1995. High-pressure, microbial inactivation and food preservation. *Food Science and Technology International* 1: 75–90.

Chenoll, E., Macián, M. C., Elizaquível, P., and Aznar, R. 2007. Lactic acid bacteria associated with vacuum-packed cooked meat product spoilage: population analysis by rDNA-based methods delicacy and health risk to consumers. *International Journal of Food Microbiology* 31(1–3): 283–299.

Diaz Yubero, I. 2016. España a través de sus chacinas: Pais Vasco. Carnimad. https://carnimad.es/noticias/espana-a-traves-de-sus-chacinas-pais-vasco/ consulted 29 January 2020.

Diez, A. M., Santos, E. M., Jaime, I., and Rovira, J. 2008a. Application of organic acid salts and high-pressure treatments to improve the preservation of blood sausage. *Food Microbiology* 25(1): 154–161.

Diez, A. M., Santos, E. M., Jaime, I., and Rovira, J. 2009a. Effectiveness of combined preservation methods to extend the shelf life of "morcilla" from Burgos. *Meat Science* 81(1): 171–177.

Diez, A. M., Björkroth, J., Jaime, I., and Rovira, J. 2009b. Microbial, sensory and volatile changes during the anaerobic cold storage of "morcilla" from Burgos previously inoculated with *Weissella viridescens* and *Leuconostoc mesenteroides*. *International Journal of Food Microbiology* 131(2–3): 168–177.

Diez, A. M., Jaime, I., and Rovira, J. 2009c. The influence of different preservation methods on spoilage bacteria populations inoculated in "morcilla" from Burgos during anaerobic cold storage. *International Journal of Food Microbiology* 132(2–3): 91–99.

Diez, A. M., Urso, R., Rantsiou, K., Jaime, I., Rovira, J., and Cocolin, L. 2008b. Spoilage of blood sausages "morcilla" from Burgos treated with high hydrostatic pressure. *International Journal of Food Microbiology* 123(3): 246–253.

Domínguez R., and Lorenzo J. M.. 2020. Catálogo de productos cárnicos iberoamericanos. 1a Edicion. Red Healthy Meat (CYTED119RT0568). ISBN: 978-989-54435-1-2.

European Commission. 2008. Poverty and social exclusion in rural areas. Final Study Report. Directorate-General for Employment, Social Affairs and Equal Opportunities. Unit E2.

European Union. 2018. Commission Implementing Regulation (EU) 2018/1214 of 29 August 2018 entering a name in the register of protected designations of origin and protected geographical indications ("morcilla de Burgos" (PGI)). Official Journal of the European Community, 5 September 2018, num 224.

Gašperlin, L., Skvarča, M., Žlender, B., Lušnic, M., and Polak., T. 2014. Quality assessment of slovenian krvavica, a traditional blood sausage: Sensory evaluation. *Journal of Food Processing and Preservation* 38(1): 97–105.

Glass, K. A., Granberg, A., Smith, A., McNamara, A. M., Hardin, M., Mattias, J., Ladwig, K., and Johnson, E. A. 2002. Inhibition of *Listeria monocytogenes* by sodium diacetate and sodium lactate on wieners and cooked bratwurst. *Journal of Food Protection* 65: 116–123.

Gómez-Rojo, E. M., Wilches-Pérez, D., Melero, B., Diez, A.M., Jaime, I., and Rovira, J. 2011. Comparison between natural and collagen synthetic casings in "morcilla de Burgos" blood sausage. In *Proceedings 57th International Congress of Meat Science and Technology*: 126–129, 7–12 August, Ghent, Belgium.

Gonzalez-Schnake, F., and Nova, R. 2014. Ethnic meat products: Brazil and South America. In Michael Dikeman (Ed.), *Encyclopedia of meat science* (2nd ed.). USA: Carrick Devine.

Houtsma, P. C., de Wit, J. C., and Rombouts, F. M. 1993. Minimum inhibitory concentration (MIC) of sodium lactate for pathogens and spoilage organisms occurring in meat products. *International Journal of Food Microbiology* 20: 247–257.

Hugas, M., Garriga, M., and Monfort, J. M. 2002. New mild technologies in meat processing: high pressure as a model technology. *Meat Science* 62: 359–371.

Iacumin, L., Manzano, M., Stella, S., and Comi, G. 2017. Fate of the microbial population and the physico-chemical parameters of "sanganel" a typical blood sausages of the friuli, a North-East Region of Italy. *Food Microbiology* 63: 84–91.

ITACYL. 2020. Instituto Tecnológico Agrario de Castilla y León, "morcilla" from Burgos IGP. Pliego de condiciones vigente. http://www.itacyl.es/-/marcas-igp-morcilla-de-burg-1, consulted 14 January 2020.

Koos, I. J. T. 1992. Lactic acid and lactates. Preservation of food products with natural ingredients. *Food Market Technology* 6: 5–11.

Korkeala, H. J., and Björkroth, J. 1997. Microbiological spoilage and contamination of vacuum-packaged cooked sausages. *Journal of Food Protección* 60(6): 724–731.

Lopéz-Caballero, E., Carballo, J., and Jiménez-Colmenero, F. 1999. Microbiological changes in pressurized, prepackaged sliced cooked ham. *Journal of Food Protection* 62: 1411–1415.

Magrinyà, N., Bou, R., Rius, N., Codony, R., and Guardiola, F. 2012. Effect of fermentation time and vegetable concentrate addition on quality parameters of organic botifarra catalana, a Cured-Cooked Sausage. *Journal of Agricultural and Food Chemistry* 60(27): 6882–6890.

MAPA. 2020. Ministerio de agricultura, pesca y alimentación. Ficha técnica de butifarra. https://www.mapa.gob.es/es/ministerio/servicios/informacion/butifarra_tcm30-102 679.pdf consulted 23rd January 2020.

Martín Bejarano, S. 2001. *Enciclopedia de la carne y los productos cárnicos*, vol II. Spain:Ediciones Martín & Macías.

Mbandi, E., and Shelef, L. A. 2002. Enhanced antimicrobial effects of combination of lactate and diacetate on *Listeria monocytogenes* and *Salmonella* spp. in beef bologna. *International Journal of Food Microbiology* 76: 191–198.

Ofori, J. A., and Hsieh, Y.-H. 2014. Issues related to the use of blood in food and animal feed. *Critical Reviews in Food Science and Nutrition* 54: 687–697.

Oteiza, J. M., Giannuzzi, L., and Califano, A. N. 2003. Thermal inactivation of *Escherichia coli* O157:H7 and *Escherichia coli* isolated from "morcilla" as affected by composition of the product. *Food Research International* 36: 703–712.

Özogul, F., Polat, A., and Özogul, Y. 2004. The effects of modified atmosphere packaging on chemical, sensory and microbial changes of Sardines (*Sardina pilchardus*). *Food Chemistry* 85: 49–57.

Parry, R. T. 1993. Introduction. In R. T. Parry (Ed.), *Principles and application of modified atmosphere packaging of food* (pp. 1–17). Glasgow: Blackie Academic and Professional.

Pereira, J. A., Dionísio, L., Patarata, L., and Matos, T. J. S. 2019. Multivariate nature of a cooked blood sausage spoilage along aerobic and vacuum package storage. *Food Packaging and Shelf Life* 20: 100304.

Pereira, J. A., Dionisio, L., Patarata, L., and Matos, T. J. S. 2015. Effect of packaging technology on microbiological and sensory quality of a cooked blood sausage, "morcela" de Arroz, from Monchique region of Portugal. *Meat Science* 101: 33–41.

Ramos, D. D., Villalobos-Delgado, L. H., Cabeza, E. A., Caro, I., Fernández-Diez, A., and Mateo, J. 2013. Mineral composition of blood sausages – A two-case study. Food Industry, 93–107. IntechOpen.

Roseiro, L. C., Santos, C., Almeida, J., and Vieira, J. A. 1998. Influence of packaging and storage temperature on cured pork blood sausages shelf-life. In Proceedings *44th International Congress of Meat Science and Technology*: 430–431, 30 August–4 September, Barcelona, Spain.

Ruiz-Capillas, C., and Jiménez-Colmenero, F. 2004. Biogenic amine content in Spanish retail market meat products treated with protective atmosphere and high pressure. *European Food Research Technology* 218: 237–241.

Säde, E., Lassila, E., and Björkroth, J. 2016. Lactic acid bacteria in dried vegetables and spices. *Food Microbiology* 53: 110–114.

Santos, E. M., Diez, A. M., González-Fernández, C., Jaime, I., and Rovira, J. 2005a. Microbiological and sensory changes in "morcilla" from burgos preserved in air, vacuum and modified atmosphere packaging. *Meat Science* 71(2): 249–255.

Santos, E. M., Gonzalez-Fernandez, C., Jaime, I., and Rovira, J. 2003. Physicochemical and sensory characterisation of "morcilla" from burgos, a traditional Spanish blood sausage. *Meat Science* 65(2): 893–898.

Santos, E. M., Diez, A. M., González-Arnaiz, L., Jaime, I., and Rovira, J., 2007. Microbiological changes in "morcilla" from Burgos pasteurized by hot water immersion. In *Proceedings 53rd International Congress of Meat Science and Technology*: 39, Beijing, China.

Santos, E. M., Jaime, I., Rovira, J., Lyhs, U., Korkeala, H., and Björkroth, J. 2005b. Characterization and identification of lactic acid bacteria in "morcilla" from Burgos. *International Journal of Food Microbiology* 97(3): 285–296.

Santos, R., Ramos, M., Beldarraín, T., Rodríguez, F., Vergara, N., Carrillo, C., and Casañas, C. 2014. Repasteurización: método para extender la durabilidad de embutidos finos empacados al vacío. *Ciencia y Tecnología de Alim*entos 24(3): 18–24.

Wambui, J., and Stephan R. 2019. Relevant aspects of *Clostridium estertheticum* as a specific spoilage organism of vacuum-packed meat. *Microorganisms* 7(5): 142.

Wilmink, M. 2000. Solving a meat problem. *International Food Ingredients*, 6: 52–53.

Zhang, P., Ward, P., McMullen, L. M., and Yang, X. 2020. A case of "blown pack" spoilage of vacuum-packaged pork likely associated with *Clostridium estertheticum* in Canada. *Letters in Applied Microbiology* 70(1): 13–20.

18 Morcela de Arroz Sausage
Manufacturing Process, Chemical Composition and Shelf Life

Jorge A. Pereira,[1,2] Lídia Dionísio,[2,3]
Luis Patarata,[4] and Teresa J.S. Matos[5,6]
[1]Universidade do Algarve, Instituto Superior de Engenharia, Campus da Penha, Estrada da Penha, Faro 8005-139, Portugal
[2]MED – Mediterranean Institute for Agriculture, Environment and Development, Universidade do Algarve, Campus de Gambelas, Faro 8005-139, Portugal
[3]Universidade do Algarve, Faculdade de Ciências e Tecnologia, Campus de Gambelas, Faro 8005-139, Portugal
[4]CECAV, Centro de Ciência Animal e Veterinária, Universidade de Trás-os-Montes e Alto Douro, Quinta de Prados, Vila Real 5000-801, Portugal
[5]LEAF, Tapada da Ajuda, Lisbon 1349-017, Portugal
[6]Instituto Superior de Agronomia, University of Lisbon, DCEB, Tapada da Ajuda, Lisbon 1349-017, Portugal

CONTENTS

18.1   INTRODUCTION

Traditional foods, produced on a small scale, which rely on manufacturing processes with a limited degree of mechanization are strongly identified with their
place or region of origin (Talon et al., 2007). This gastronomic heritage (Toldrá and
Navarro, 2000) is strongly associated with tourism, driven by renewed interest by
consumers in typical regional foods.

Blood sausages are traditional/ethnic meat products of popular consumption in
several countries around the world. Most representative are Morcelas in Portugal
(Roseiro et al., 1998); Morcillas in Spain (Cachaldora, García, Lorenzo and García-
Fontán, 2013; Santos et al., 2003) and Argentina (Oteiza et al., 2006); Relleno,
Prieta, Moronga and Mocillón in México, Colombia, and Peru (Ramos et al., 2013);
Sanganel in Italy (Iacumin et al., 2017); black pudding in the United Kingdom and
Ireland (Fellendorf, O'Sullivan and Kerry, 2017); Blutwurst and Thüringian in
Germany (Feiner, 2006); Boudin Noir in France (Baracco et al., 1990); and
Krvavica in Slovenia (Gašperlin et al., 2014).

Morcela de Arroz is a Portuguese blood sausage that is unique because, beyond pork
and blood, it has rice in its composition. It is usually cooked, not smoked, and ready to
eat (Pereira et al., 2017). This blood sausage is part of a historic and cultural heritage
that is important to preserve. The European Union has promoted the protection of
quality traditional foods from specific regions, with the aim of improving rural areas
and supporting local livestock production (Commission Implementing Regulation (EU)
No 668/2014; Council Regulation (EEC) No 2081/1992; Regulation (EU) No 1151/
2012) and Morcela de Arroz is mainly produced by families of rural areas, from north
to south in Portugal, each region with its own ingredients and recipes.

18.2   MANUFACTURING PROCESS

Processing technology of blood sausages differs as a result of type, region and
manufacturer; therefore, among blood sausages, some are ready to eat, which includes smoked and dry-cured, like Chouriço Preto, Paio Preto (Laranjo et al., 2017);
and Cacholeira (Santos, Gomes and Roseiro, 2011); others are cooked and dry-
cured, like Morcela from Algarve, Alentejo and Trás-os-Montes; Morcilla
(Cachaldora, García, Lorenzo and García-Fontán, 2013) and Chorizo de Cebolla, a
variant of black pudding (Castaño et al., 2002) from Spain; Beloura from Alentejo
(Todorov et al., 2010); and Moura and Chouriço Doce from Trás-os-Montes
(Roseiro et al., 2012), which are normally submitted to a short smoking period after
the cooking process. Blood sausages also can be fully cooked, like Morcela de
Arroz from Monchique region of Algarve, which is cooked in *molho de moura*, a
broth prepared with water, whole onions, whole garlic, salt and laurel leaves
(Pereira et al., 2017) and Morcilla de Burgos (Santos et al., 2003); or blanched in
hot water, like Chouriço de Sangue (consisting of blood and pig fat with spices)

Meats selection and preliminary preparation of raw materials (e.g. premixing, scalding, curing)

Grinding

Seasonings → Mixing

Stuffing

Boiling

Cooling

Chilling

Packaging

Expedition

FIGURE 18.1   Processing technology of Morcela de Arroz.

from Portugal (Marcos et al., 2016) and Morcillas from Argentina, until the blood has coagulated, but further thermal treatment is necessary before consumption (Oteiza et al., 2006). Finally, in some blood sausages, meats and starch components are boiled before the sausages are stuffed, as in Krvavica from Slovenia (Gašperlin et al., 2014), onion and rice with the lard/tallow in Morcilla de León (Ramos et al., 2013), rice in Morcilla de Burgos (Santos et al., 2003) from Spain, and blood in Sanganel from Italy (Iacumin et al., 2017).

The most common method of manufacturing Morcela de Arroz is presented in Figure 18.1.

18.2.1   INGREDIENTS

The ingredients and formulations/recipes (Table 18.1) used in the preparation of blood sausages vary according to the country and specificity/tradition of the producing region and manufacturer (Gašperlin et al., 2014; Marcos et al., 2016; Pereira et al., 2015; Ramos et al., 2013; Santos et al., 2003). Blood, which is the common ingredient in all types of blood sausages and, usually, the main ingredient in black pudding, for example, often a meatless blood sausage (Anjos et al., 2019; Feiner, 2006), although the typical black pudding consumed in Ireland and the United Kingdom contains lean pork meat and fat (Fellendorf, O'Sullivan and Kerry, 2017). In Austria, black pudding contains meat materials such as cooked ham, and cured and cooked jowls, as well as cooked pork skin (Feiner, 2006). Blood added is normally from pigs but, in some types of Morcilla, beef blood is used (Santos, Zaritzky and Califano, 2008). Besides blood, blood sausages may include in their formulations lean pork (e.g., minced), pork fatty trimmings, lard, in some typical products tallow (Santos et al., 2003), pork skin and/or offal (liver and entrails), and an enormous variety of ingredients of vegetable origin, including spices, onions, leeks, and other less frequent ingredients such as rice, barley, cabbage, millet, flour, bread, nuts, pine

TABLE 18.1

Representative Formulations of Blood Sausages

Product Ingredients (%)	MA[a]	SM[b]	MB[c]	BP[d]	SA[e]	KR[f]
Pork meat	54.0	64.0	–	35.6	Meat and fat 57.9	*
Lard/fat	4.0	–	17.0	15.38	Value included with meat	*
Entrails	–	–	–	–	–	*
Pork skin	–	6.4	–	–	–	*
Pork blood	6.0	–	8.0	–	38.6	*
Blood powder	–	0.1	–	3.0	–	–
Rice	13.0	–	17.0	–	–	*
Onion	3.0	12.7	54.0	2.5	–	–
Salt	1.0	1.3	1.8	1.02	3.0	*
Water	17.0	12.5	–	27.0	–	–
Spices	2.0	1.7	2.2	1.1	0.5	*
Supplements (sugars, skimmed milk powder, sodium nitrite E250)	–	1.3	–	–	–	–
Cereals (oats, barley, wheat, among others)	–	–	–	14.4	–	*
Fruits	–	–	–	–	–	*
References	1	2	3, 4	5	6	7

Notes
[a] MA—Morcela de Arroz; 1- Adapted from Pereira, Ferreira-Dias, Dionísio, Patarata and Matos, 2017.
[b] SM—Spanish Morcilla; 2- Adapted from Cachaldora, García, Lorenzo, and García-Fontán. 2013.
[c] MB—Morcilla de Burgos; 3- Adapted from Diez, Urso, Rantsiou, Jaime, Rovira and Cocolin, 2008; 4- Diez, Santos, Jaime and Rovira, 2008.
[d] BP—Black Pudding; 5- Adapted from Fellendorf, O'Sullivan and Kerry, 2016.
[e] SA—Sanganel; 6- Adapted from Iacumin, Manzano, Stella and Comi, 2017.
[f] KR—Krvavica; 7- Adapted from Gašperlin, Skvarča, Žlender, Lušnic and Polak, 2014.
* Contains (values were not available).

nuts, dried raisins, wine, honey and fruits (Gašperlin et al., 2014; Iacumin et al., 2017; Marcos et al., 2016; Pereira et al., 2015; Ramos et al., 2013), that are usually stuffed into natural casings, which give the product the identity/specificity associated with the producing region. Casings that are used can be natural, usually from pork (small intestine, large intestine, stomach and bladder), or cattle (small intestine), fresh, salted and dried, although artificial casings are sometimes substituted (Marcos et al., 2016). In general, the use of natural casings is preferred in order to ensure tradition and originality of the products (Patarata, Saraiva and Martins, 1998).

18.2.2 THERMAL PROCESS

The thermal porcessing of blood sausages also varies according to the specificity/ tradition of the producing region and manufacturer (Cachaldora, García, Lorenzo and García-Fontán, 2013; Diez, Santos, Jaime and Rovira, 2008; Diez, Urso, Rantsiou, Jaime, Rovira and Cocolin, 2008; Fellendorf, O'Sullivan and Kerry, 2016; Gašperlin et al., 2014; Iacumin et al., 2017; Pereira et al., 2017). Nevertheless, cooked blood sausages, at some point of the manufacture process, are normally boiled above 85°C for 30 minutes to 4 hours (depending on the diameter of the product) (Table 18.2).

Thermal processing of Morcela de Arroz from the Monchique region of Portugal was extensively studied by Pereira et al., (2017). These authors observed, on average, a core product temperature higher than 90.0°C for 50.3 minutes, which largely exceeds the most conservative safe temperature of 90.0°C/10 minutes, which extends the refrigerated shelf life of this product (1 week at 0–5°C). The traditional thermal process, although following empirical procedures, involving temperatures above 95°C for about 38 minutes in the core of Morcela de Arroz (Table 18.3), is capable of destroying several spores, namely *B. cereus* and *C. perfringens* and nonproteolytic *C. botulinum*. However, this thermal process cannot theoretically destroy the most heatr-esistant proteolytic strains of *C. botulinum*, which might be a problem if the sausages are not stored at the correct temperature. Good hygiene and manufacturing practices must continue to be included in the control of the process, with particular attention to the thermal process and storage temperature. Nevertheless, Morcela de Arroz's thermal process complies with the recommendations for refrigerated, processed foods of extended durability.

Additionally, Pereira et al., (2017) obtained a very good fit of unsteady-state heat transfer equations (USHTE) to the experimental time/temperature data of the thermal process of Morcela de Arroz, allowing us to predict the adequacy of the product thermal process in terms of food safety.

18.3 CHEMICAL COMPOSITION

The chemical composition of Morcela de Arroz (Table 18.4) shows that water is the major component (53.9%), but lower than the amounts found by Santos et al. (2003) in Morcilla de Burgos, a Spanish cooked blood sausage (62.2%) or in Krvavica (61.5%), a Slovenian cooked blood sausage (Gašperlin et al., 2014). This difference can be explained by the use in both products of precooked rice and other starch components, with a high water content, before mixing with the rest of the ingredients, while in Morcela de Arroz formulation, the added rice (more than 10%) is not precooked (Pereira, Ferreira-Dias, Dionísio, Patarata and Matos, 2017).

The chemical composition of Morcela de Arroz shows that it is a nutritious meat product (Table 18.4). Protein content of Morcela de Arroz is higher than that found in similar cooked blood sausages, namely in Sanganel (14.6%) (Iacumin et al., 2017); Morcilla de Burgos (13.1%) (Santos, González-Fernández, Jaime and Rovira, 2003); Relleno de Tumbes (11.9%), a typical blood sausage from Northern Peru (Ramos et al., 2013); Krvavica (10.4%) (Gašperlin et al., 2014); Morcilla (cooked blood sausage with onion) (7.7%,) (Jiménez-Colmenero et al., 2010) and Morcilla de León (5.2%) (Herrera, 2006), which may be due to the high content of meat (more than 50% meat from the Alentejano pig breed, namely loin, shoulder, neck and bacon) in the Morcela de Arroz formulation. Sanganel is made with

TABLE 18.2

Representative Thermal Processes Applied in Cooked Blood Sausages

Product Phase of the Thermal Process	MA[a]	SM[b]	MB[c]	BP[d]	SA[e]	KR[f]
Ripening	–	21 days, 10–12°C, 75–85% of RH	–	–	–	–
Smoking	–	First 5 days of ripening with oak wood	–	–	–	–
Boiling/ cooking	91 ± 5°C for 1.5 hours	95–96°C for 0.5 hours	95–96°C for 1 hour	Using 100% steam at 85°C until sausage reaches 75°C at the core and held for 15 minutes	Boiled in water for 45 minutes	85–95°C for 1 hour to 4 hours depending on the diameter of the product
Cooling	2 hours at room temperature and stored at 4 ± 1°C	20°C for 20 minutes	Air cooled to 8–10°C	Cooled and stored at 4°C	Stored at 4 ± 2°C	Air cooled to 8–10°C and stored at 4°C
Type and diameter of casings	Natural beef casings, 60–65 mm of ∅	Natural pork casings, 32–35 mm of ∅	Natural Beef casings, 35–45 mm of ∅	Polyamide casings	Pork bowels	Natural Casings (hog or beef round casings)
References	1	2	3, 4	5	6	7

Notes

[a] MA—Morcela de Arroz; 1- Adapted from Pereira, Ferreira-Dias, Dionísio, Patarata and Matos, 2017.

[b] SM—Spanish Morcilla; 2- Adapted from Cachaldora, García, Lorenzo, and García-Fontán. 2013.

[c] MB—Morcilla de Burgos; 3- Adapted from Diez, Urso, Rantsiou, Jaime, Rovira and Cocolin, 2008; 4- Diez, Santos, Jaime and Rovira, 2008.

[d] BP—Black Pudding; 5- Adapted from Fellendorf, O'Sullivan and Kerry, 2016.

[e] SA—Sanganel; 6- Adapted from Iacumin, Manzano, Stella and Comi, 2017.

[f] KR—Krvavica; 7- Adapted from Gašperlin, Skvarča, Žlender, Lušnic and Polak, 2014.

bloody pork (bacon rinds and pig's head tender muscles, lungs and kidneys), lard and boiled blood (Iacumin et al., 2017). Morcilla de Burgos includes a high content of onion and rice, and it does not include meat in its formulation, only blood, up to 20% (Santos et al., 2003). Relleno de Tumbes includes blood (around 30%), pork

TABLE 18.3
Thermal Processing of Morcela de Arroz (Mean ± Standard Deviation of Four Batches) (Adapted from Pereira, Ferreira-Dias, Dionísio, Patarata and Matos, 2017)

Phase	Parameter	Mean ± SD
Heating	Heating medium initial T (°C)	85.9 ± 1.6
	Heating medium average processing T (°C)	93.7 ± 1.7
	Heating medium T_{max} (°C)	98.6 ± 0.1
	Product average processing T (°C)	77.2 ± 4.1
	Product T_{max} (°C)	97.9 ± 0.3
Cooking	Time to reach product T_{max} (minutes)	87.2 ± 5.4
	Time at T product >90°C (minutes)	50.3 ± 11.8
	Time at T product >95°C (minutes)	37.5 ± 12.2
	Time of cooking (minutes)	96.4 ± 6.8
Cooling	Air room medium average T (°C)	22.3 ± 1.6
	Time of product cooling to 35.0°C (minutes)	72.2 ± 7.9
	Total time of product cooling (minutes)	85.8 ± 14.8
	Cooled product T (°C)	31.6 ± 3.7
Chilling	Air chilling medium average T (°C)	4.6 ± 0.6
	Time of product chilling to 5.0°C (minutes)	266.5 ± 30.5
	Total duration of thermal process (minutes)	362.9 ± 28.2

lard fat (10%) and chopped cabbage (40%) (Ramos et al., 2013). Krvavica includes blood (up to 20%) and other offal, pork skin, entrails, meat from the pig head, trimmings and cracklings (10%), and a major level of starch components (Gašperlin et al., 2014), and Morcilla de León includes a mixture of chopped onion (65–75%), fat (10–20% of lard and/or tallow), blood (up to 20%), and rice or breadcrumbs (up to 10%) (Ramos et al., 2013).

Morcela de Arroz presents approximately half of the fat content (14%) of Morcilla de Burgos (29%) (Santos et al., 2003), Argentinean morcilla (32%) (Oteiza, Giannuzzi and Califano, 2003) and Spanish morcilla (35%) (Jiménez-Colmenero et al., 2010), and of Sanganel (34.5%) (Iacumin et al., 2017), but similar fat and higher carbohydrate contents compared to Krvavica (Gašperlin et al., 2014) and Morcilla de León (Herrera, 2006; Ramos et al., 2013).

Therefore, the chemical composition of blood sausages can vary in a wide range according to the ingredients and proportions used in their preparation (Marcos et al., 2016) according to the specificity/tradition of the producing county/region and manufacturer.

Due to its intrinsic parameters, namely rich nutrient composition, high water activity (a_w) that reaches values higher than 0.97, and pH (about 6.0), cooked blood sausages are susceptible to fast spoilage, representing a limited shelf life. The increase of consumer demand for traditional/ethnic specialties has renewed the

TABLE 18.4
Physico-Chemical Properties and Proximate Composition (Mean ± Standard Deviation) of Cooked Blood Sausages

Product Parameter	MA[a]	SM[b]	SA[c]	AM[d]	KR[e]	MB[f]	ML[g]	BP[h]	RT[i]
pH	6.4 ± 0.06	–	6.4 ± 0.2	6.2	6.7 ± 0.19	6.4 ± 0.35	6.0 ± 0.20	–	–
a_w	0.97 ± 0.01	–	0.97 ± 0.01	–	–	0.98 ± 0.00	0.98 ± 0.00	–	–
Moisture (g/100 g)	53.9 ± 1.15	42.9 ± 0.05	49.4 ± 1.5	52.0	61.5 ± 4.51	62.2 ± 4.05	67.1 ± 5.77	54.2 ± 0.3	71.8 ± 6.9
Protein (g/100 g)	15.8 ± 0.64	7,7 ± 0.06	14.6 ± 1.3	15.0	10.4 ± 1.72	13.1 ± 2.30	5.2 ± 0.92	11.3 ± 0.1	11.9 ± 2.8
Fat (g/100 g)	14.1 ± 1.82	35.1 ± 0.25	34.5 ± 1.5	32.0	14.0 ± 2.22	28.7 ± 4.92	14.2 ± 3.94	17.9 ± 0.1	9.4 ± 4.0
Carboh[j] (g/100 g)	14.3 ± 0.92	12.1 ± 0.23	–	1.0	11.0 ± 2.38	–	8.2 ± 3.40	–	3.6 ± 1.6
Ash (g/100 g)	1.9 ± 0.11	2.2 ± 0.02	3.0 ± 1.2	–	1.9 ± 0.36	4.3 ± 0.49	1.9 ± 0.13	2.7 ± 0.0	2.1 ± 0.9
SFA (% tfacid[k])	–	37.4 ± 0.21	–	–	–	–	–	–	–
MUFA (% tfacid[k])	–	45.4 ± 0.22	–	–	–	–	–	–	–
PUFA (% tfacid[k])	–	16.9 ± 0.22	–	–	–	–	–	–	–
Reference	1	2	3	4	5	6	7; 9	8	9

Notes

[a] MA—Morcela de Arroz; 1- Adapted from Pereira, Ferreira-Dias, Dionísio, Patarata and Matos, 2017.

[b] SM—Spanish Morcilla (cooked blood sausage with onion); 2- Adapted from Jiménez-Colmenero, Pintado, Cofrades, Ruiz-Capillas and Bastida, 2010.

[c] SA—Sanganel; 3- Adapted from Iacumin, Manzano, Stella and Comi, 2017.

[d] AM—Argentinean Morcilla; 4- Adapted from Oteiza, Giannuzzi and Califano, 2003.

[e] KR—Krvavica; 5- Adapted from Gašperlin, Skvarča, Žlender, Lušnic and Polak, 2014.

[f] MB—Morcilla de Burgos; 6- Adapted from Santos, González-Fernández, Jaime and Rovira, 2003.

[g] ML—Morcilla de León; 7- Adapted from Herrera, 2006; 9- Adapted from Ramos, Villalobos-Delgado, Cabeza, Caro, Fernández-Diez and Mateo, 2013.

[h] BP—Black Pudding (containing 15% fat and 0.8% salt); 8- Adapted from Fellendorf, O'Sullivan and Kerry, 2016.

[i] RT—Relleno de Tumbes; 9- Adapted from Ramos, Villalobos-Delgado, Cabeza, Caro, Fernández-Diez and Mateo, 2013.

[j] Carboh—Carbohydrate.

[k] % tfacid—% total fatty acids.

 - Data not available.

interest in these meat products, leading to a need to assure product safety and longer shelf life in an expanding market. In this context, the application of packaging technologies has been performed in Morcela de Arroz (Pereira, Dionísio, Patarata and Matos, 2015, 2019; Pereira, Silva, Matos and Patarata, 2015), using Quantitative Descriptive Analysis (QDA), consumer tests and spoilage indicators (microbiological and chemical), in whole or sliced product.

18.4   SPOILAGE AND SHELF LIFE

18.4.1   LOSS OF BIOLOGICAL SAFETY

In blood sausages there are four pathogens of major concern due to their psychrotrophic behavior, which allows them to grow in these meat products, usually with a high a_w and pH, and when they are stored in refrigeration. They are *Listeria monocytogenes*, *Yersinia* spp., *Bacillus cereus* and *Clostridium botulinum* (Coorey et al., 2018; Pereira et al., 2017). If any of these bacteria are likely to be present in the blood sausage at the end of processing, it is of utmost importance to consider whether storage conditions and time will result in multiplication to levels considered hazardous to the consumer over the entire shelf life (Laranjo et al., 2017). This concern is particularly important if the blood sausage keeps it sensory acceptability while supporting the growth of pathogens (Ricci et al., 2018). When the pathogen involved is proteolytic, as with some strains of *C. botulinum*, the accumulation of compounds derived from the catabolism of the amino acids will impart a putrid odor to the blood sausage that will be detected by the consumer as a warning not to consume it (Juneja and Marks, 1999).

Nowadays the main concern with blood sausages and other cooked meat products is *L. monocytogenes*. Due to its ability to grow under anaerobic conditions and at temperatures near 0°C, the presence of this pathogen must be absent at the end of the processing (Ricci et al., 2018). If the food safety system used by the producer does not allow it to guarante that absence, even if there are very low odds of contamination, the shelf life must be limited to the time the pathogens remain in the lag phase, and be limited by the moment the growth reaches 1 Log CFU/g (Pereira, Silva, Matos and Patarata, 2015). A microbiological challenge test with *L. monocytogenes* was conducted by the authors with sliced Morcela de Arroz. They found that at 3 ± 1°C the pathogen remains in the lag phase until the 20 days after the test, but when the temperature was 7 ± 1°C the growth of *L. monocytogenes* became of concern 4 days after the slices were packaged, and after only one day if no package was used. Considering that domestic refrigerators are frequently above 3°C (Roccato, Uyttendaele and Membré, 2017), the contamination of blood sausages with this pathogen should be avoided, particularly in those, like Morcela de Arroz, that are ready to eat.

18.4.2   MICROBIAL SPOILAGE

When the safety of blood sausages is guaranteed by the correct application of Hazard Analysis Critical Control Point (HACCP) methodology, the shelf life is limited by the growth of spoilage microorganisms (Lee et al., 2019). The vast majority of microorganisms that multiply in food are not pathogenic, so their presence in large numbers, besides spoiling the food and making it unpleasant for consumption, does not represent any risk to the consumer's health (Kolbeck et al., 2019). Spoilage becomes obvious when changes, mostly in appearance and odor, are detected by the consumer. Appearance changes might include color changes, moldy aspect or visible bacterial biofilm (Diez et al., 2009). Modifications to smell can be very diverse, depending on the food components modified: proteins, lipids or carbohydrates (Bolívar-Monsalve et al., 2019). All of these changes occur in food during the storage, and can take days, weeks or even months to be noticeable (Hough et al., 2003).

The multiplication of a particular microorganism or group of microorganisms during the storage of blood sausages depends on several factors, intrinsic to the food and extrinsic. We must take into account the initial contamination, the physico-chemical and nutritional properties, the moisture content, the a_w, the pH, the presence of chemical preservatives and the manufacturing technology, as well as the environment surrounding the product (Lauritsen et al., 2019). In blood sausages, most of the microflora is eliminated during thermal processing. Non-spore-forming pathogens are eliminated and spoilage microbiota experiences a considerable reduction (Pereira et al., 2017). Among the more or less varied total microflora, specific spoilage microorganisms (SSM) will be selected by the restringing conditions of the food, and during the shelf life they will dominate the microbial ecosystem of the product (Piotrowska-Cyplik et al., 2017). The microorganisms associated with blood sausages spoilage are (1) gram-negative bacteria (*Pseudomonas* spp., *Enterobacteriaceae*, (2) gram-positive, non-spore lactic acid bacteria (LAB) and coagulase-negative cocci, (3) spore-forming gram-positive bacteria and (4) fungi (Diez, Björkroth, Jaime and Rovira, 2009; Kreyenschmidt et al., 2010; Pereira, Dionísio, Patarata and Matos, 2015). The association of these microorganisms is very specific to certain products, due to their natural and environment microflora, processing and storage conditions. Consequently, their importance in each type of blood sausage, when made with different recipes and/or technologies, might be considerably different. Since these microorganisms have different physiological needs and use diverse metabolic pathways to obtain energy and other nutrients, the sensory consequences of their growth are also different (Borch, Kant-Muermans and Blixt, 1996; Pothakos et al., 2015). Usually, sensorily detectable metabolite accumulation coincides with the time when the total microflora of the food is composed almost exclusively of the SSM, which also coincides with the beginning of the stationary phase of its multiplication (Huis in't Veld, 1996).

The thermal processing of Morcela de Arroz assures that most of the non-spore-forming bacteria are eliminated. Despite that immense microbial reduction, LAB might be the major cause of spoilage of blood sausages due to the favorable conditions for the multiplication of psychrotrophic strains, and the ability to grow in the absence of oxygen and lack of competition make them the dominant microflora during the shelf life. In Morcela de Arroz, but also with other blood sausages, LAB is the most abundant group of spoilage microorganisms found (Cachaldora, García, Lorenzo and García-Fontán, 2013; Diez, Björkroth, Jaime and Rovira, 2009; Diez, Santos, Jaime and Rovira, 2009; Pereira et al., 2019). *Weisella viridescens*,

Leuconostoc spp., and *Weissella confusa* were the main LAB associated with the spoilage of Morcilla de Burgos (Santos et al., 2005), and *Leuconostoc mesenteroides* and *Lactobacillus sakei* were isolated from unspecified morcilla (Aznar and Chenoll, 2006).

The dominance of *Weissela* species has probably a strong influence on the sensory characteristics of the spoiled product, once it is obligatory heterofermentative, producing CO_2 and D- or D- and L-lactic acid as major end products of fermentation. (Fusco et al., 2015). This dominance of LAB might be favorable for the industry once these microorganisms have a more limited impact on sensory characteristics than other spoilage microorganisms with a stronger proteolytic and lipolytic activity (Demirok-Soncu et al., 2018; Kreyenschmidt et al., 2010). However, the accumulation of lactic acid and other organic acids results in a fermented odor of the product and in an acidic taste, and the eventual release of diacetyl results in a buttery odor that can be perceived as unpleasant by some consumers (Diez, Björkroth, Jaime and Rovira, 2009; Pothakos et al., 2014). In Morcela de Arroz it was observed that the accumulation of the D-isomer of lactic acid was responsible for the perception of freshness loss due to the fermented odor and acid taste. This was associated with the growth of LAB and, even in lower numbers, to *Enterobacteriaceae* (Pereira et al., 2019).

Pseudomonas spp. are associated with the spoilage of fresh meat, and die during the manufacturing process of Morcela de Arroz (Pereira, Dionísio, Patarata and Matos, 2015). However, if they are present, even in low counts, they can contribute to the spoilage due to the strong catabolic activity on the nitrogen fraction, resulting in ammonia and other volatile compounds usually recognized as putrid (Nychas et al., 2008). That association between *Pseudomonas* spp. count and non-protein nitrogen (NPN) catabolism, evaluated by the Total Basic Volatile Nitrogen (TBVN), was found in blood sausages, but it was not always detected by the sensory analysis (Iacumin et al., 2017; Pereiraet al., 2019). The slime on the surface could be due to exopolysaccharides produced by microorganisms, namely *Pseudomonas* spp. (Piotrowska-Cyplik et al., 2017), and was one important attribute found to justify the rejection of Morcilla de Burgos after longer periods of storage (Santos et al., 2005).

18.4.3 PHYSICO-CHEMICAL SPOILAGE

Physical spoilage of blood sausages is due mainly to the excessive loss of moisture during the storage of unpackaged products. The relative humidity in refrigerators is usually low (Vorst et al., 2018), favoring the loss of moisture from the blood sausages. In Morcela de Arroz, weight losses of nearly 20% were observed during storage at 4°C for 44 days. The use of packaging reduces or eliminates this problem. However, products manufactured in small units for local distribution, frequently in butcher shops and small neighborhood grocery stores, might have that problem (Pereira, Dionísio, Patarata and Matos, 2015). Another physical problem that may arise is the loss of volatile compounds, resulting in products with a less intense odor. That loss happens in unpackaged blood sausages, but might also occur in packaged products, since some packaging material is permeable to some molecules or can even adsorb some of these molecules (Sajilata et al., 2007).

The chemical changes that blood sausages experience during the spoilage process result from reactions that occur between food components, or between these and external conditions. The major chemical changes that occur in blood sausages are lipid oxidation (Cachaldora et al., 2013; Diez, Björkroth, Jaime and Rovira, 2009; Silva et al., 2014). Lipids are an important fraction of blood sausages and can usually be modified through hydrolytic or oxidative phenomena, both happening in sequence. Fiirst, the hydrolysis promotes the liberation of fatty acids from the triglycerides and other complex lipids, and second, the free fatty acids are oxidized (Honikel, 2009). While the oxidation of fatty acids might have a positive role in the characteristic flavor development of cured meat products, in blood sausages it will be mainly associated with a rancid odor, eventual color changes and the formation of potentially toxic compounds (Flores, 2018). In Morcela de Arroz, the rancid odor was one of the main spoilage traits detected in products stored without packaging, reaching 3.75 on a five-point scale at the end of 44 days of cold storage (Pereira, Dionísio, Patarata and Matos, 2015). The substrates of oxidation reactions are mostly unsaturated fatty acids; the higher the degree of unsaturation, the greater the susceptibility to oxidation (Bernardi et al., 2016). This might be a concern if Morcela de Arroz is made from pigs raised with high proportions of unsaturated fatty acids in the feed, as are those fed acorns or oilseeds (Serrano et al., 2019). Each unsaturated fatty acid produces specific hydroperoxides, which in turn break down into specific by-products with important aromatic impact, namely aldehydes, ketones, alcohols, acids and esters. The profile and quantities in which they are present determine the odor of the blood sausage, and hence the degree of product acceptability (Amaral, Solva and Lannes, 2018).

Both the myoglobin of the meat and hemoglobin of the blood used in the preparation of blood sausages have high amounts of iron. That iron interacts with lipids, as a catalyst for oxidation or because it is oxidized by the radicals formed during the auto propagation of fatty acids oxidation. The high amount of iron in the blood used in Morcela de Arroz, and the denaturation of the protein due to the heating, exposing the heme ring, make blood sausages particularly prone to lipid oxidation (Grunwald and Richards, 2006; Zareian et al., 2019).

18.4.4 SENSORY DETECTION OF SPOILAGE AND CONSUMER ACCEPTANCE

The modifications that have sensory implications are the main reasons to establish the end of the shelf life (Manzocco, 2016). Several methodologies are used to study the adequate length of time for storage, both indirect, such as microbial counts or chemical indicator assessments, and direct, using sensory judges or consumers' perceptions of freshness and acceptability. Despite their pertinence, indirect methods have some weaknesses, namely, the costs of some approaches and their indirect nature, so that they need to be correlated with the perception of freshness by the final user of the product—the consumer (Hough and Garitta, 2012).

The use of consumers can be an interesting approach to defining shelf life. Consumers are asked if they are willing to consume, or willing to purchase, a given product with a given storage time, taking into consideration its freshness. The proportion of consumers accepting the product is used to limit the shelf life, and the shelf life is usually set at the point when large-consumption foods reach a level of acceptability of 50%. At the end of the shelf life, only a small part of the batch will still be on retail shelves, and the producer assumes the risk of half of the consumers buying these residual packages without considering them completely fresh (Giménez, Gagliardi and Ares, 2017; Pereira et al., 2019). Once the consumer tests are made at defined times, usually at predefined intervals that could be a few days to weeks, it is possible to estimate the exact cut-off point of the end of shelf life using the mathematical approach of survival analysis (Giménez, Ares and Ares, 2012; Hough, 2010). The use of that methodology allows us to know when consumers still accept the Morcella de Arroz, while with descriptive sensory analysis, the characteristics of spoilage are studied, but there is always some doubt on where to cut in a freshness evaluation scale (Pereira et al., 2019). Moreover, as shown by the previous authors, Morcela de Arroz has a multifactorial pattern of spoilage, involving both microbiological and chemical modifications, that are recognized sensorially and impact the decision of the consumer who is judging its freshness.

REFERENCES

Amaral, A., M. Solva, and S. Lannes. 2018. Lipid oxidation in meat: Mechanisms and protective factors – A review. *Food Science and Technology* 38:1–15.

Anjos, O., R. Fernandes, S. M. Cardoso, et al. 2019. Bee pollen as a natural antioxidant source to prevent lipid oxidation in black pudding. *LWT - Food Science and Technology* 111:869–875.

Aznar, R., and E. Chenoll. 2006. Intraspecific diversity of *Lactobacillus curvatus, Lactobacillus plantarum, Lactobacillus sakei,* and *Leuconostoc mesenteroides* associated with vacuum-packed meat product spoilage analyzed by randomly amplified polymorphic DNA PCR. *Journal of Food Protection* 69:2403–2410.

Baracco, P., Y. Berger, D. Chansac, et al. 1990. L'encyclopédie de la charcuterie. Thiais: Soussana.

Bernardi, D., T. Bertol, S. Pflanzer, V. Sgarbieri, and M.Pollonio. 2016. ω-3 in meat products: Benefits and effects on lipid oxidative stability. *Journal of the Science of Food and Agriculture* 96:2620–2634.

Bolívar-Monsalve, J., C. Ramírez-Toro, G. Bolívar, and C. Ceballos-González. 2019. Mechanisms of action of novel ingredients used in edible films to preserve microbial quality and oxidative stability in sausages – A review. *Trends in Food Science and Technology* 89:100–109.

Borch, E., M. Kant-Muermans, and Y. Blixt. 1996. Bacterial spoilage of meat and cured meat products. *International Journal of Food Microbiology* 33:103–120.

Cachaldora, A., G. García, J. M. Lorenzo, and M. C. García-Fontán. 2013. Effect of modified atmosphere and vacuum packaging on some quality characteristics and the shelf life of morcilla, a typical cooked blood sausage. *Meat Science* 93:220–225.

Castaño, A., M. C. García-Fontán, J. M. Fresno, M. E. Tornadijo, and J. Carballo. 2002. Survival of Enterobacteriaceae during processing of Chorizo de cebolla, a Spanish fermented sausage. *Food Control* 13:107–115.

Commission Implementing Regulation (EU) No 668. 2014. Laying down rules for the application of Regulation (EU) No 1151/2012 of the European Parliament and of the Council on quality schemes for agricultural products and foodstuffs. *Official Journal of the European Union* 179:36–61.

Coorey, R., V. Jayamanne, E. Buys, et al. 2018. The impact of cooling rate on the safety of food products as affected by food containers. *Comprehensive Reviews in Food Science and Food Safety* 17:827–840.

Council Regulation (EEC) No 2081. 1992. On the protection of geographical indications and designations of origin for agricultural products and foodstuffs. *Official Journal of the European Communities* 93:12–25.

Demirok-Soncu, E., B. Arslan, D. Ertürk, S. Küçükkaya, N. Özdemir, and A. Soyer. 2018. Microbiological, physicochemical and sensory characteristics of Turkish fermented sausages (sucuk) coated with chitosan-essential oils. *LWT - Food Science and Technology* 97:198–204.

Diez, A. M., E. M. Santos, I. Jaime, and J. Rovira. 2008. Application of organic acid salts and high-pressure treatments to improve the preservation of blood sausage. *Food Microbiology* 25:154–161.

Diez, A. M., E. M. Santos, I. Jaime, and J. Rovira. 2009. Effectiveness of combined preservation methods to extend the shelf life of Morcilla de Burgos. *Meat Science* 81: 171–177.

Diez, A. M., I. Jaime, and J. Rovira, 2009. The influence of different preservation methods on spoilage bacteria populations inoculated in morcilla de Burgos during anaerobic cold storage. *International Journal of Food Microbiology* 132: 91–99.

Diez, A. M., J. Björkroth, I. Jaime, and J. Rovira. 2009. Microbial, sensory and volatile changes during the anaerobic cold storage of morcilla de Burgos previously inoculated with *Weissella viridescens* and *Leuconostoc mesenteroides*. *International Journal of Food Microbiology* 131:168–177.

Diez, A. M., R. Urso, K. Rantsiou, I. Jaime, J. Rovira, and L. Cocolin. 2008. Spoilage of blood sausages *morcilla de Burgos* treated with high hydrostatic pressure. *International Journal of Food Microbiology* 123:246–253.

Feiner, G. 2006. Meat products handbook: Practical science and technology (pp. 516–518). Cambridge, UK: Woodhead Publishing.

Fellendorf, S., M. G. O'Sullivan, and J. P. Kerry. 2016. Impact of ingredient replacers on the physicochemical properties and sensory quality of reduced salt and fat black puddings. *Meat Science* 113:17–25.

Fellendorf, S., M. G. O'Sullivan, and J. P. Kerry. 2017. Effect of different salt and fat levels on the physicochemical properties and sensory quality of black pudding. *Food Science & Nutrition* 5:273–284.

Flores, M. 2018. Understanding the implications of current health trends on the aroma of wet and dry cured meat products. *Meat Science* 144:53–61.

Fusco, V., G. Quero, G. Cho, et al. 2015. The genus *Weissella*: Taxonomy, ecology and biotechnological potential. *Frontiers in Microbiology* 6:article 155.

Gašperlin, L., M. Skvarča, B. Žlender, M. Lušnic, and T. Polak, 2014. Quality assessment of Slovenian krvavica, a traditional blood sausage: Sensory evaluation. *Journal of Food Processing and Preservation* 38:97–105.

Giménez, A., A. Gagliardi, and G. Ares. 2017. Estimation of failure criteria in multivariate sensory shelf life testing using survival analysis. *Food Research International* 99:542–549.

Giménez, A., F. Ares, and G. Ares. 2012. Sensory shelf-life estimation: A review of current methodological approaches. *Food Research International* 49:311–325.

Grunwald, E., and M. Richards. 2006. Mechanisms of heme protein-mediated lipid oxidation using hemoglobin and myoglobin variants in raw and heated washed muscle. *Journal of Agricultural and Food Chemistry* 54:8271–8280.

Herrera, E. A. C. 2006. Aportaciones a la caracterización de la Morcilla de León y evolución de determinados parámetros físicos, químicos y microbiológicos durante su conservación a refrigeración. 383 f. Thesis (Doctorado en Higiene y Tecnologia de los Alimentos). León: Universidad de León.

Honikel, K.-O. 2009. Oxidative changes and their control in meat and meat products. In *Safety of meat and processed eat*, edited by F. Toldra, 313–340. New York: Springer.

Hough, G. 2010. Sensory shelf life estimation of food products. Boca Raton, FL: CRC Press.

Hough, G., K. Langohr, G. Gomez, and A. Curia. 2003. Survival analysis applied to sensory shelf life of foods. *Journal of Food Science* 68:359–362.

Hough, G., and L. Garitta. 2012. Methodology for sensory shelf-life estimation: A review. *Journal of Sensory Studies* 27:137–147.

Huis in't Veld, J. 1996. Microbial and biochemical spoilage of foods: an overview. *International Journal of Food Microbiology* 33:1–18.

Iacumin, L., M. Manzano, S. Stella, and G. Comi. 2017. Fate of the microbial population and the physico-chemical parameters of "Sanganel" a typical blood sausage of the Friuli, a north-east region of Italy. *Food Microbiology* 63:84–91.

Jiménez-Colmenero, F., T. Pintado, S. Cofrades, C. Ruiz-Capillas, and S. Bastida. 2010. Production variations of nutritional composition of commercial meat products. *Food Research International* 43:2378–2384.

Juneja, V., and H. Marks. 1999. Proteolytic *Clostridium botulinum* growth at 12–48°C simulating the cooling of cooked meat: Development of a predictive model. *Food Microbiology* 16:583–592.

Kolbeck, S., L. Reetz, M. Hilgarth, and R. Vogel. 2019. Quantitative oxygen consumption and respiratory activity of meat spoiling bacteria upon high oxygen modified atmosphere. *Frontiers in Microbiology* 10:article 2398.

Kreyenschmidt, J., A. Hübner, E. Beierle, L. Chonsch, A. Scherer, and B. Petersen. 2010. Determination of the shelf life of sliced cooked ham based on the growth of lactic acid bacteria in different steps of the chain. *Journal of Applied Microbiology* 108:510–520.

Laranjo, M., A. Gomes, A. C. Agulheiro-Santos, et al. 2017. Impact of salt reduction on biogenic amines, fatty acids, microbiota, texture and sensory profile in traditional blood dry-cured sausages. *Food Chemistry* 218:129–136.

Laranjo, M., R. Talon, A. Lauková, M. J. Fraqueza, and M. Elias. 2017. Traditional meat products: Improvement of quality and safety. *Journal of Food Quality* 2017:article 2873793.

Lauritsen, C., J. Kjeldgaard, H. Ingmer, M. Bisgaard, and H. Christensen. 2019. Microbiota encompassing putative spoilage bacteria in retail packaged broiler meat and commercial broiler abattoir. *International Journal of Food Microbiology* 300:14–21.

Lee, K., S. Baek, D. Kim, and J. Seo. 2019. A freshness indicator for monitoring chicken-breast spoilage using a Tyvek® sheet and RGB color analysis. *Food Packaging and Shelf Life* 19:40–46.

Manzocco, L. 2016. The acceptability limit in food shelf life studies. *Critical Reviews in Food Science and Nutrition* 56:1640–1646.

Marcos, C., C. Viegas, A. M. Almeida, and M. M. Guerra. 2016. Portuguese traditional sausages: different types, nutritional composition, and novel trends. *Journal of Ethnic Foods* 3:51–60.

Nychas, G., P. Skandamis, C. Tassou, and K. Koutsoumanis. 2008. Meat spoilage during distribution. *Meat Science* 78:77–89.

Oteiza, J. M., I. Chinen, E. Miliwebsky, and M. Rivas. 2006. Isolation and characterization of shiga toxin-producing Escherichia coli from precooked sausages (morcillas). *Food Microbiology* 23:283–288.

Oteiza, J. M., L. Giannuzzi, and A. Califano. 2003. Thermal inactivation of *Escherichia coli* O157:H7 and *Escherichia coli* isolated from morcilla as affected by composition of the product. *Food Research International* 36:703–712.

Patarata, L., G. Saraiva, and C. Martins. 1998. Processo de fabrico de produtos de salsicharia tradicional. In: 1ª Jornadas Geográficas de Queijos e Enchidos. Exponor p. 83–86 [In Portuguese].

Pereira J. A., L. Dionísio, L. Patarata, and T. J. S. Matos. 2015. Effect of packaging technology on microbiological and sensory quality of a cooked blood sausage, Morcela de Arroz, from Monchique region of Portugal. *Meat Science* 101:33–41.

Pereira J. A., L. Dionísio, L. Patarata, and T. J. S. Matos. 2019. Multivariate nature of a cooked blood sausage spoilage along aerobic and vacuum package storage. *Food Packaging and Shelf Life* 20:100304.

Pereira, J. A., L. Dionísio, T. J. S. Matos, and L. Patarata. 2015. Sensory lexicon development for a Portuguese cooked blood sausage – Morcela de Arroz de Monchique – To predict its usefulness for a geographical certification. *Journal of Sensory Studies* 30:56–67.

Pereira, J. A., P. Silva, T. J. S. Matos, and L. Patarata. 2015. Shelf life determination of sliced Portuguese traditional blood sausage – Morcela de Arroz de Monchique through microbiological challenge and consumer test. *Journal of Food Science* 80:M642–M648.

Pereira, J. A., S. Ferreira-Dias, L. Dionísio, L. Patarata, and T. J. S. Matos. 2017. Application of unsteady-state heat transfer equations to thermal process of Morcela de Arroz from Monchique region, a Portuguese traditional blood sausage. *Journal of Food Processing and Preservation* 41:e12870.

Piotrowska-Cyplik, A., K. Myszka, J. Czarny, et al. 2017. Characterization of specific spoilage organisms (SSOs) in vacuum-packed ham by culture-plating techniques and MiSeq next-generation sequencing technologies. *Journal of the Science of Food and Agriculture* 97:659–668.

Pothakos, V., C. Nyambi, B. Zhang, A. Papastergiadis, B. De Meulenaer, and F. Devlieghere. 2014. Spoilage potential of psychrotrophic lactic acid bacteria (LAB) species: *Leuconostoc gelidum* subsp. *gasicomitatum* and *Lactococcus piscium*, on sweet bell pepper (SBP) simulation medium under different gas compositions. *International Journal of Food Microbiology* 178:120–129.

Pothakos, V., F. Devlieghere, F. Villani, J. Björkroth, and D. Ercolini. 2015. Lactic acid bacteria and their controversial role in fresh meat spoilage. *Meat Science* 109:66–74.

Ramos, D. D., L. H. Villalobos-Delgado, E. A. Cabeza, I. Caro, A. Fernández-Diez, and J. Mateo. 2013. Mineral composition of blood sausages – A two-case study. January. http://doi.org/10.5772/53591.

Regulation (EU) No 1151. 2012. On quality schemes for agricultural products and foodstuffs. *Official Journal of the European Union* 343:1–29.

Ricci, A., A. Allende, D. Bolton, et al. 2018. *Listeria monocytogenes* contamination of ready-to-eat foods and the risk for human health in the EU. *EFSA Journal* 16:article 5134.

Roccato, A., M. Uyttendaele, and J. Membré. 2017. Analysis of domestic refrigerator temperatures and home storage time distributions for shelf-life studies and food safety risk assessment. *Food Research International* 96:171–181.

Roseiro, L. C., A. Gomes, L. Patarata, and C. Santos. 2012. Comparative survey of PAHs incidence in Portuguese traditional meat and blood sausages. *Food and Chemical Toxicology* 50:1891–1896.

Roseiro, L. C., C. Santos, J. Almeida, and J. A. Vieira. 1998. Influence of packaging and storage temperature on cured pork blood sausages shelf life. In Proceedings of the 44th International Congress of Meat Science and Technology, 30th August–4th September, Barcelona, Spain.

Sajilata, M., K. Savitha, R. Singhal, and V. Kanetkar. 2007. Scalping of flavors in packaged foods. *Comprehensive Reviews in Food Science and Food Safety* 6: 17–35.

Santos, C., A. Gomes, and L. C. Roseiro. 2011. Polycyclic aromatic hydrocarbons incidence in Portuguese traditional smoked meat products. *Food and Chemical Toxicology* 49:2343–2347.

Santos, E. M., A. M. Diez, C. González-Fernández, I. Jaime, and J. Rovira. 2005. Microbiological and sensory changes in "morcilla de Burgos" preserved in air, vacuum and modified atmosphere packaging. *Meat Science* 71:249–255.

Santos, E. M., C. González-Fernández, I. Jaime, and J. Rovira. 2003. Physichochemical and sensory characterization of "Morcilla de Burgos", a traditional Spanish blood sausage. *Meat Science* 65:893–898.

Santos, E. M., I. Jaime, J. Rovira, U. Lyhs, H. Korkeala, and J. Björkroth. 2005. Characterization and identification of lactic acid bacteria in "morcilla de Burgos." *International Journal of Food Microbiology* 97:285–296.

Santos, M.V., N. Zaritzky, and A. Califano. 2008. Modeling heat transfer and inactivation of *Escherichia coli O157:H7* in precooked meat products in Argentina using the finite element method. *Meat Science* 79:595–602.

Serrano, S., F. Perán, E. Gutiérrez de Ravé, A. Cumplido, and F. Jiménez-Hornero. 2019. Multifractal analysis application to the study of fat and its infiltration in Iberian ham: Influence of racial and feeding factors and type of slicing. *Meat Science* 148:55–63.

Silva, F., D. Amaral, I. Guerra, et al. 2014. Shelf life of cooked goat blood sausage prepared with the addition of heart and kidney. *Meat Science* 97:529–533.

Talon, R., I. Lebert, A. Lebert, et al. 2007. Traditional dry fermented sausages produced in small-scale processing units in Mediterranean countries and Slovakia.1: Microbial ecosystems of processing environments. *Meat Science* 77:570–579.

Todorov, S. D., P. Ho, M. Vaz-Velho, and L. M. T. Dicks. 2010. Characterization of bacteriocins produced by two strains of *Lactobacillus plantarum* isolated from *Beloura* and *Chouriço*, traditional pork products from Portugal. *Meat Science* 84:334–343.

Toldrá, F., and J. L. Navarro. 2000. Improved traditional foods for the next century. *Food Research International*, 33, 145.

Vorst, K., N. Shivalingaiah, A. Monge Brenes, et al. 2018. Effect of display case cooling technologies on shelf-life of beef and chicken. *Food Control* 94:56–64.

Zareian, M., T. Tybussek, P. Silcock, P. Bremer, J. Beauchamp, and N. Böhner. 2019. Interrelationship among myoglobin forms, lipid oxidation and protein carbonyls in minced pork packaged under modified atmosphere. *Food Packaging and Shelf Life* 20: 100311.

Index

Note: *Italicized* page numbers refer to figures, **bold** page numbers refer to tables